MOLECULAR METALS

NATO CONFERENCE SERIES

VI MATERIALS SCIENCE

MOLECULAR METALS

Edited by
William E. Hatfield

University of North Carolina
Chapel Hill, North Carolina

Published in coordination with NATO Scientific Affairs Division by

PLENUM PRESS · NEW YORK AND LONDON

Library of Congress Cataloging in Publication Data

Nato Conference on Molecular Metals, Les Arcs, France, 1978.
   Molecular metals.

   (NATO conference series: VI, Materials science; v. 1)
   Includes index.
   1. Organic conductors—Congresses. I. Hatfield, William E., 1937-   II. Nato
Special Program Panel on Materials Science. III. Title. IV. Series.
QD382 C66N37   1978   547'.84   79-4284

   ISBN-13:978-1-4684-3482-8   e-ISBN-13:978-1-4684-3480-4
   DOI:10.1007/978-1-4684-3480-4

Proceedings of the NATO Conference on Molecular Metals held at Les Arcs,
France, September 10—16, 1978, and sponsored by the NATO Special Program
Panel on Materials Science

© 1979 Plenum Press, New York
Softcover reprint of the hardcover 1st edition 1979

A Division of Plenum Publishing Corporation
227 West 17th Street, New York, N.Y. 10011

PREFACE

    During the past few years there has been intense research
activity in the design, synthesis, and characterization of materials
which are formed from molecular precursors, and which have high or
metal-like electrical conductivities, i.e. $d\sigma/dT < 0$. It has been
widely supposed that these new materials, which are commonly called
molecular metals, would be pressed into service, for example as
devices. Up to now, widespread, practical applications of these substances have not developed. The NATO Advanced Research Institute on
Molecular Metals at Les Arcs, France, September 10-16, 1978 was
organized to discuss the scientific and technological potential of
research and development in this field. The proceedings of the
Institute constitute this book.

    Several lectures were devoted to the assessment of the present
status of research on systems which serve to define major components
of the field. The systems which were discussed included TTF-TCNQ,
platinum chain compounds, $(SN)_x$, polyacetylene, polydiacetylene,
graphite intercalation compounds, and mercury chain compounds. One
lecture dealt with the status and potential of technological applications of molecular, semiconducting and metallic materials. Texts
accompanying these lectures are included in the book. In addition
to the lectures, poster presentations of new results were held;
these papers are also included here.

    The assigned work of the Institute was accomplished by inten-
sive discussions in small study groups. Each group had a chairman,
secretary, and five or more additional members. Summaries of their
discussions were compiled by the secretaries and presented to the
assembled Institute on the final day of the meeting. These study
group reports, which form an important part of this book, summarize
the conclusions that the unusual electrical, optical, magnetic, and
in some instances mechanical properties of molecular metals will
inevitably lead to technological applications of them. The ultimate potential of these substances was seen to lie not just in their
electrical conductivities, but in their unique combination of
several novel properties such as anisotropic optical and electrical

properties, low densities, ease of fabrication, sensitivity to their
environment, and the sensitivity of their properties to chemical
modification, thereby facilitating molecular engineering. The
study group reports call attention to numerous opportunities for
new and demanding research in this field and should provide helpful
directions to those who are already involved with this research,
to those who wish to enter the field, and to those who are responsible for making administrative decisions.

I wish to thank the members of the Institute for the cooperative, cheerful, and positive manner in which they approached the
assigned work, especially the difficult task of preparing the study-
group reports; to call special attention the input of the Organizing
Committee (Klaus Bechgaard, R. Comès, Peter Day, Heimo Keller, and
Alan MacDiarmid) and thank them again for their help, thoughtful
comments, and timely suggestions; to acknowledge the Special Program
Panel on Materials Science of the Scientific Affairs Division of the
North Atlantic Treaty Organization for the organizational and
financial support which made the Institute possible; and to thank
Ms. Sherry Handfinger for her very valuable secretarial assistance.

Chapel Hill, North Carolina   William E. Hatfield
December 1978

CONTENTS

STUDY GROUP REPORTS

THE CHEMISTRY OF TTF-TCNQ

Klaus Bechgaard

The H.C. Ørsted Institute, University of Copenhagen

Universitetsparken 5, DK-2100 Copenhagen, Denmark

The present paper is intended to briefly review some of the chemical properties of TTF, TCNQ and TTF-TCNQ. However, given the very large number of TTF and TCNQ derivatives which have appeared since the discovery of TTF-TCNQ[1] it is reasonable not to treat these in any detail, but present general results illustrated by a few examples.

TTF-TCNQ is a simple 1:1 compound crystallizing in segregated stacks of TTF and TCNQ molecules[2] (ions). The charge-transfer from donor to acceptor stack is incomplete[3] as demonstrated by diffuse X-ray scattering experiments.[3]

Room temperature single crystal conductivities of 500 (ohm·cm)$^{-1}$ are frequently found rising to more than 10^4 (ohm·cm)$^{-1}$ at 59 K (T_{max}). At 53, 47 and 38 K three consecutive transitions occur ultimately leading to 3-D order and an insulating state.

The electronic properties of TTF-TCNQ are highly anisotropic, since the solids are built of stacks of large flat molecules with extensive π interaction along the stacks and 1-D or quasi 1-D phenomena are to be expected. Finally the negative thermopower in the "metallic" regime indicates that the acceptor stack to some extent dominates the transport properties.

A most important feature of TTF-TCNQ is, however, that it is relatively easy to grow crystals of reasonable size and quality. This is probably why TTF-TCNQ has been the subject of a large number of studies during the past five years and it should be mentioned that most derivatives of TTF-TCNQ are more difficult to crystallize than the parent system itself.

1

TTF

A review of synthetic methods for preparing TTF and derivatives has recently been published.[28] Hurtley and Smiles[4] in 1926
prepared dibenzo TTF, but Wudl's preparation of TTF and TTF-chlor-
ide,[5] which exhibits a room temperature conductivity of 300 (ohm·
cm)$^{-1}$ started extensive work in the area of TTF and derivatives.
Notable successes are the preparation of tetraselenafulvalene[6]
(TSeF), HMTTF[7] and HMTSF.[8] TSeF-TCNQ is isomorphous with TTF-TCNQ,
whereas HMTTF-TCNQ and HMTSeF-TCNQ, due to a unique crystal structure at lower temperature, probably enter a semimetallic state[9]
rather than the usual insulating state. The TTF molecule is a
planar highly symmetric (D_{2h}) electron rich structure. It is easily oxidized to the cation radical and the dication, which are both
relatively stable. Most organic cation radicals react easily with
nucleophiles, and the chemical stability of TTF^{+}· is indeed important since in the crystallization of TTF-TCNQ a fraction of the TTF's
is present as TTF^{+}·.

TTF Derivatives

Symmetrical. The parent molecule has been substituted in a
symmetrical way with a variety of electron withdrawing groups such
as CF_3, CN or aryl. Increasing the IP of TTF, however, in most cases
leads, in the TCNQ complexes, to DA type structures rather than the
intended segregated stack structure. More success has been achieved
using weakly donating alkyl substituents (TMTTF, HMTTF etc.) Still
missing among the TTF (and TSeF) derivatives are the classical
strongly electron donating methoxy and dialkylamino substituted species.

Unsymmetrical. Coupling of unsymmetrical dithioles (or di-
selenoles) usually gives unseparable mixtures of cis and trans TTF's.
The main effect of the disorder introduced is smearing of the phase
tranistion(s) in the corresponding TCNQ salts.

Cross-coupling of different dithioles in contrast often leads
to mixtures of TTF's which can be separated by fractional crystallization or chromatography. In a recent example[10] it appears that
although the structure is ordered in the TCNQ salt, the molecular
"disorder" again results in smearing of the phase transitions.

Recently D. Green has developed a very useful synthetic route
to a variety of monosubstituted TTF's by lithiating TTF at low temperature and reacting with appropriate electrophiles.[22]

Three examples may serve as guidelines for changes occurring
when TTF derivatives are empolyed: a. Substituting S with Se.
TSeF-TCNQ shows 1 anomaly rather than the three seen in most sulphur
coumpounds and the thermopower goes hole-like. b. Disorder.

DEDMTSF-TCNQ[11] exhibits complete smearing of the metal to insulator transition. c. Size-matching. HMTTF-TCNQ and HMTSF-TCNQ have a chessboard-like structure, are less anisotropic than usual and probably undergo a metal to semimetal transition at lower temperature.

In conclusion the synthesis of TTF derivatives has reached a rather sophisticated level and the main effects of substitution on the physical properties are to some extent evaluated. More exotic TTF derivatives such as polymeric TTF[12] and mixed metal-bis-dithiolene-TTF polymers[13] have been made but few details are known since single crystals are not yet available.

TTF reacts with oxygen and forms TTF-oxide[14,15] which has recently been shown to interfere in the crystallization process of TTF-TCNQ.[15] Also TTF-TCNQ is surface oxidized as demonstrated by ESCA.[15] This complication has long been overlooked and the necessity of avoiding oxygen during crystallization and handling of TTF-TCNQ has to some extent been neglected. However, oxygen does not seem to enter the bulk of the crystals and meaningful physical measurement can still be made using surface oxidized TTF-TCNQ crystals.

TCNQ

Only three main routes to TCNQ and derivatives have appeared in the literature. In the original synthesis of TCNQ cyclohexane-1,4-dione is condensed with malonitrile followed by oxidation.[16] Alternatively p-halomethyl-benzenes can be converted in a series of reactions to p-bis-dicyanomethyl-benzenes which can then be oxidized to the corresponding TCNQ's. Finally direct substitution of halogen with the anion of tert-butyl malonitrile followed by pyrolysis and oxidation can be employed in special cases. In an excellent paper Wheland and Martin have reported synthetic methods for 21 TCNQ derivatives.[17]

Unsymmetrical TCNQ's are represented by methyl-TCNQ and "symmetrical" examples are 2,5-dialkyl- or 2,5-dihalo-TCNQ's. Also tetrasubstituted (tetrafluor) TCNQ has been made.

The derivatives of TCNQ have to some extent been neglected, probably because most synthetical procedures are unpleasant (except for TCNQ itself) and the fact that most chemists in the field are sulfur (and selenium) chemists.

TCNQ has a planar symmetric "electron poor" structure and is easily reduced to the anion radical and the dianion. The anion radical is remarkably stable in most solvents. TCNQ, however, has a tendency to "collect" cations when water is present and must be

sublimed preferentially onto teflon or quartz to ensure high purity.
The TCNQ's already synthesized represent a spectrum of acceptors
with rather different electron affinities, which to a first approximation can be represented by their half wave potentials in acetonitrile. Table I gives a few examples:

Table 1

Compound	$E_{\frac{1}{2}}$ V*
2,5-Dicyano TCNQ	+0.65
Tetrafluor TCNQ	+0.53
2,5-Dibromo TCNQ	+0.41
TCNQ	+0.17
2-Methyl TCNQ	+0.15
2,5 Diethyl TCNQ	+0.12
2,5-Dimethoxy TCNQ	-0.02

* Volts versus SCE in CH_3CN.

Much attention has been given to isomorphous series where the
donor has been changed (TFF-TCNQ, TSeF-TCNQ, DTDSeF-TCNQ). To the
present author isomorphous "acceptor-series" (TTF-TCNQBr$_2$; TTF-
TCNQCl$_2$) ought to be investigated as well, especially because the
effects of changing Br to Cl in TTF-TCNQBr$_2$ as judged from powder
conductivities[19] are orders of magnitude rather than "small" effect
seen on changing S to Se in the donor.

Bandfilling. The variety of TCNQ's available has recently
provided important experimental evidence as to which effects the
degree of bandfilling (i.e. charge transfer) can have on the transport properties of TTF-TCNQ salts. HMTSeF-TCNQ and HMTSeF-TCNQF$_4$
are isomorphous, and exhibit 0.74 e/mole[20] and presumably 1.0 e/mole
charge-transfer, respectively. HMTSeF-TCNQ has a room temperature
conductivity of 1500 or more (ohm·cm)$^{-1}$ whereas HMTSeF-TCNQF$_4$ is
a semiconductor; $\sigma_{RT} \sim 10^{-3}$ (ohm·cm). This result has been inter-
pretated in terms of HMTSeF-TCNQF$_4$ as a Mott-insulator.[21] Similar
results are obtained for HMTTF-TCNQF$_4$[22] and it is hoped that alloying HMTSeF-TCNQ and HMTTF-TCNQ with TCNQF$_4$ may provide some important
answers.

Guidelines. After preparing and investigating the new TTF-TCNQ
derivatives which have appeared during the last five years it is
felt appropriate to extract a few general results of interest for
future work. From a chemical point of view several important reactions in sulphur and selenium chemistry have been developed and
are synthetically useful. In principle any desired TTF or TCNQ

structure can be made in the search for new types of TTF-TCNQ
organic conductors and "special effects" can be built into the
molecules.

A few results of interest for the solid state properties which
are directly related to molecular properties are worth mentioning.

1. Substitution of sulphur with selenium raises the RT conductivity, lowers the metal to insulator transition and in most cases
make the donor stack dominate the transport properties. TTF and
TSeF salts with the same acceptor are normally isomorphous and
homogenous alloys can be made.

2. Bulky substituents on the constituent molecules pushes the
stacks apart (rises the anisotrophy).

3. Disorder (permanent dipole moments, or cis-trans mixtures)
smears the phase transition(s).

4. The IP-EA value of the DA pair is not very meaningful in
determining the final charge transfer in the solid.

5. It is not possible to predict crystal structure (i.e. the
effect of substituents) or when a TTF-TCNQ pair is more stable as
a mixed-stack structure. Often it is, however, possible to favor
the segregated stack structure by changing the crystal growth
conditions.

6. It seems to be possible to prevent the metal to insulator
transition.

Since it may be an ultimate goal for work in this field to
avoid the metal to insulator transition some features of the few
known systems which retains high conductivity to low temperature
are of interest. HMTSeF-TCNQ and HMTTF-TCNQ (under pressure) retain
conductivities in excess of 10 (ohm cm)$^{-1}$ at 1 K. Their structure
is unique and characterized by 4 short (Se-N or S-N) contacts for
each DA pair. Also the structure is slightly disordered[24] and relatively low anisotrophy is observed. The high conductivity at low
temperature appears to result from a crossover from a 1D metallic
state at high temperature to a 3D semimetallic state at lower temperature. Recently, however, in crystals which appear less disordered
from X-ray studies an anomaly has been detected at 24 K indicating
that also this system undergo a Peirls transition.[25]

A few other systems which are highly conducting at low temperature are known. HMTSeF-TNAP[26] and TScT$_2$Cl [27] have both recently
been investigated and more examples will hopefully appear as a
result of future work.

REFERENCES

1. L.B. Coleman, M.J. Cohen, D.J. Sandman, F.G. Yamagishi, A.F. Garito and A.J. Heeger, Sol. State Commun., 12, 1125 (1973). J.P. Ferraris, D.O. Cowan, V. Valatka and J.H. Perlstein, J. Amer. Chem. Soc., 95, 498 (1973).
2. T.J. Kistenmacher, T.E. Phillips and D.O. Cowan, Acta. Cryst., B30, 763 (1974).
3. F. Denoyer, R. Comes, A.F. Garito and A.J. Heeger, Phys. Rev. Letters, 35, 445 (1975).
4. W.R.H. Hurtley and S. Smiles, J. Chem. Soc., 1926, 2263 (1926).
5. F. Wudl, G.M. Smith and E.J. Hufnagel, J.C.S. Chem. Commun., 1970, 1453 (1970).
6. E.M. Engler and V.V. Patel, J. Amer. Chem. Soc., 96, 7376 (1974).
7. R.L. Greene, J.J. Mayrle, R.R. Schumaker, G. Castro, P.M. Chaikin, S. Etemad and S.J. Laplaca, Sol. State Commun., 20, 943 (1976).
8. A.N. Bloch, D.O. Cowan, K. Bechgaard, R.E. Pyle, R.H. Banks and T.O. Poehler, Phys. Rev. Letters, 34, 1561 (1975).
9. M. Weger, Sol. State Commun., 19, 1149 (1976).
10. D. Chasseau, J. Gaultier and C. Hauw, This report. 000.
11. C.S. Jacobsen, K. Mortensen, J.R. Andersen and K. Bechgaard, Phys. Rev. B, 18, 905 (1978).
12. Y. Ueno, Y. Masugama and M. Okawara, Chem. Letters, 1975, 603 (1975).
13. E.M. Engler, R.R. Schumaker and F.B. Kaufmann, This report, 000.
14. M.V. Laksmikantham, A.F. Garito and M.P. Cava, J. Org. Chem., in press.
15. L. Carlsen, K. Bechgaard, C.S. Jacobsen and I. Johansen, J.C.S. Perk. Trans II, in press.
16. D.S. Acker, R.J. Harder, W.R. Hertler, W. Mahler, L.R. Melby, R.E. Benson, and W.E. Mockel, J. Amer. Chem. Soc., 82, 6408 (1960).
17. R.C. Wheland and E.L. Martin, J. Org. Chem., 28, 3101 (1975).
19. R.C. Wheland and J.L. Gillson, J. Amer. Chem. Soc., 98, 3916 (1976).
20. C. Weyll, E.M. Engler, S. Etemad, K. Bechgaard and G. Jehanno, Sol. State Commun., 19, 925 (1976).
21. A.N. Bloch and D.O. Cowan, private communication.
22. J.B. Torrance and K. Bechgaard, unpublished results.
23. D.C. Green, J.C.S. Chem. Commun., 1977, 161 (1977).
24. T.E. Phillips, T.J. Kistenmacher, A.N. Bloch and D.O. Cowan, J.C.S. Chem. Commun., 1976, 334 (1976).
25. D. Jerome and J.P. Pouget, private communication.
26. K. Bechgaard, C.S. Jacobsen and N.H. Andersen, Sol. State Commun., 25, 825 (1978).
27. I.F. Schegolev and R.B. Lubjovskii, results presented at the "Int. Conf. on Quasi One Dimensional Conductors" Dubrovnik, 1978.
28. M. Narita and C.U. Pittman, Synthesis, 1976, 489 (1976).

PREPARATION OF ORGANIC METALS AND INSULATORS:

SOME IDEAS AND GUIDELINES

Jerry B. Torrance

IBM Research Laboratory

San Jose, California 95193, U.S.A.

I.   INTRODUCTION

Organic π—molecular donors and acceptors have been found to form a wide variety of charge transfer salts, with a wide variety of physical properties. Historically, interest has been sharply focused on those few which have a high d.c. electrical conductivity at room temperature and on how to design new organic metals. In the future, however, we anticipate that interest will broaden to include the understanding of how to make the various types of nonconducting salts. Considering the difficulty of making predictions for any material, any discussion of how to prepare an organic material with a particular property must be regarded as speculative, particularly in such a new and complex field as organic charge transfer salts. Thus, the ideas presented here should be regarded as a crude basis or starting point which must be tested, modified and improved.

In attempting to bridge the gap between the molecular properties of organic donors and acceptors, on the one hand, with the desired physical properties of the resultant compound, on the other, it is helpful to establish some "materials properties" as key intermediate goals. The most useful ones (in our opinion) are the following three:

(1)   Stacking

(2)   Stoichiometry

(3)   Strength of Ionicity (degree of charge transfer).

These "materials properties" have been used to classify charge transfer salts[1] and a number of attempts have been made[2-5] to understand the relationship between them and the physical properties. For example, <u>given</u> a charge transfer salt with 1:1 stoichiometry and uniform, segregated stacks and incomplete charge transfer (intermediate ionicity), it is expected to be highly conducting. But, this assumes that these particular properties have been <u>given</u>. What molecular factors determine stoichiometry, stacking, and strength of ionicity? The answers to these questions provide the connections between the molecular properties and these material properties and are the subject of this paper. Stack distortions are used as the basis of some classification schemes,[1,6] but should rather be viewed[7] as a consequence of the other three properties and not a fourth, independent property. Some discussion has been given previously[5,7-10] about the relationship between the degree of charge transfer and the molecular and solid state properties.

2. STACKING

The physical properties of a particular charge transfer salt depend strongly on which of the following two modes of stacking is adopted by the molecules in the crystal[1]:

(a) <u>Mixed Stacks</u>, in which donors (D) and acceptors (A) stack alternately face-to-face [D-A-D-A-D-A...]; or

(b) <u>Segregated Stacks</u>, in which the donors and acceptors form separate donor stacks [D-D-D-D...] and acceptor stacks [A-A-A-A...].

This distinction is important for the conductivity, for example, since it is believed that a salt with mixed stacks cannot be highly conducting. Indeed, all descriptions[2-5] of "how to design an organic metal" list "segregated stacks" as an essential requirement, but none of them give a hint or suggestion as to how to insure that the molecules will not form mixed stacks.

It has been noted[1] that the vast majority of charge transfer compounds do form with mixed stacks. This observation can be somewhat misleading, however, since the vast majority of these compounds are not ionic and have no unpaired electrons. For such neutral molecules, the attractive Mulliken charge transfer interaction between donors in a donor stack or between acceptors in an acceptor stack is very weak. The donor-acceptor interaction in mixed stacks is much stronger and they are therefore very highly favored. In fact, we are unaware of any segregated stacking neutral charge transfer compound. For the ionic salts, we have not gathered statistics, but it is clear that there are many examples of each type.

Recently two ideas concerning segregated vs. mixed stacking have been put forth.[11,12] Unfortunately, we have enough space here only to briefly discuss the view of Mayerle et al.,[12] which is simply that segregated stacks may be formed by default in cases where the D-A overlap is particularly weak. Qualitatively, the D-A overlap may be made "particularly weak" by, for example:

(a) Making the donor and acceptor very different in shape, e.g., alkali-metal salts with planar π-molecular acceptors; halide salts with planar organic donors; strongly bent donor with planar acceptor; donor with 2-fold symmetry combined with acceptor with 3-fold symmetry, etc.

(b) planar donors and acceptors the frontier orbitals of which have the opposite symmetry upon inversion; e.g., TTF-TCNQ and NMP-TCNQ. (Note TMPD has the same symmetry as TCNQ and chloranil and both salts have mixed stacks.)

At this stage, this idea is a crude, qualitative guide which needs to be substantiated by further experimental evidence and more quantitative calculations of specific cases in order to determine what is meant by "particularly weak."

3. STRENGTH OF IONICITY (Degree of Charge Transfer)

A number of review papers[2-5] have discussed "how to design an organic metal." These papers are each based on a preconceived physical model for TTF-TCNQ, NMP-TCNQ, and other TCNQ salts, with an emphasis on the importance of polarizability, or disorder, or Coulomb interactions, or distortions, etc. Aside from all this rhetoric, there is only one quantitative comparison[5] of a wide variety of salts, which is shown in Fig. 1. In this figure, the conductivity is plotted as a function of the electrochemical reduction potential of the cation of the known 1:1 TCNQ salts. Clearly, the cation of every one of the (few) highly conducting salts lies within a specific, narrow range of reduction potential.[5,13] Since this result (Fig. 1) cannot be a coincidence, we are forced to recognize the reduction potential of the cation as an essential molecular variable. There is probably only one reasonable interpretation[5,7,9,10] of Fig. 1: the charge transfer is the key variable and the organic metals are highly conducting because they are of mixed valence, with incomplete transfer of charge between donor and TCNQ. Thus, the metals have cations whose reduction potentials are between those favoring complete charge transfer and those favoring no charge transfer. Similarly, we can consider different anions and thus can account for the dramatic decrease of 10^6 in room temperature conductivity of HMTSF-TCNQF$_4$ compared with isostructural HMTSF-TCNQ[14] as being due to the increased electron affinity of TCNQF4 compared with TCNQ.

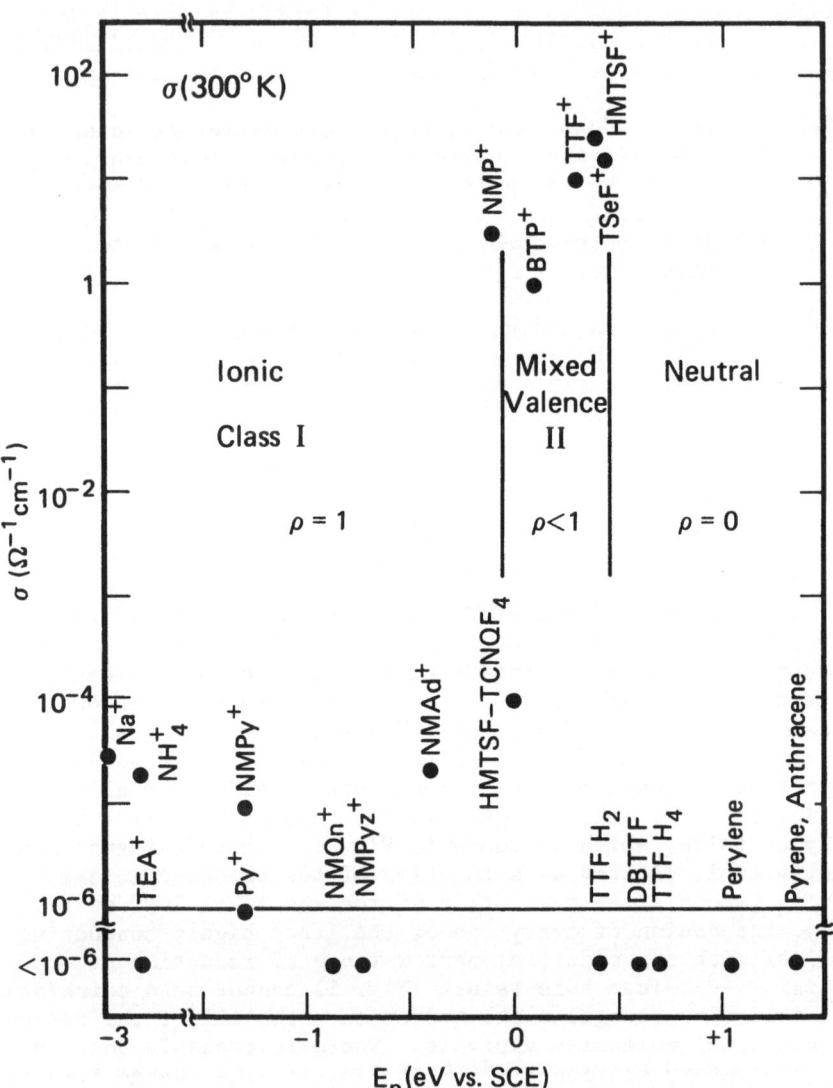

Fig. 1. The conductivity of the known 1:1 TCNQ salts plotted vs. the reduction potential of the cation. (See Ref. 5 for details).

It should be emphasized that such a simple comparison as
Fig. 1 is crude and should not be expected to work perfectly.
Firstly, the electrochemical redox potentials are dominated by
solvation effects. Secondly, the Madelung, polarization, and
other relevant energies have not been included. Apparently,
differences in these three energies (solvation, Madelung and
polarization) between different salts are not enough to
substantially smear out the boundaries between the ionic, mixed
valence and neutral phases.

The phase diagram of 1:1 charge transfer salts is more
complicated when we include the interrelationships between
stacking and strength of ionicity, as shown in Fig. 2. Note that
neutral segregated stacks are excluded, as discussed earlier,
since none have been found and none are expected. Similarly,
note that there is no mixed valence phase indicated for mixed
stacks between the neutral and ionic states. Early work[8] (which

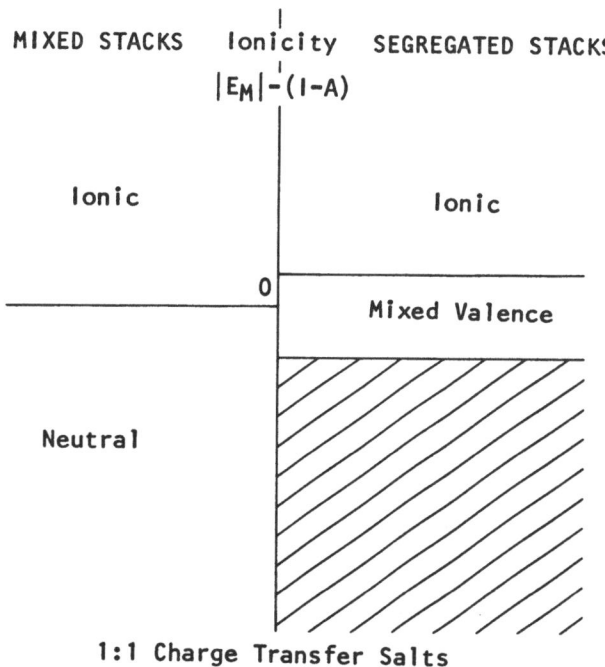

Fig. 2. The phase diagram showing the four different phases
found for 1:1 charge transfer systems and showing how these phases
are determined by the stacking and the strength of ionicity (the
degree of charge transfer).

neglected the effects of D–A overlap) showed that the ground
state in mixed stacks is either neutral or ionic, with a
discontinuous transition when the Madelung energy, $E_M = (I-A)$.
Including the effects of molecular overlap will clearly hybridize
the two states, as recent work has shown.[15] It is our view that
this hybridization or covalent interaction does not constitute
a mixed valence state and it should not be described as such.
In this sense, the term "incomplete charge transfer" is somewhat
misleading when applied to mixed stacks.

The position shown in Fig. 2 for the ionic-neutral phase
boundary for mixed stacks with respect to the mixed valence phase
is shown for TCNQ salts and is based on the reduction potentials
of the following cations which form mixed stacks with TCNQ:
dimethylphenazene and TMPD both form ionic salts, while TMTSF
forms a neutral charge transfer complex.

4.   STOICHIOMETRY

The stoichiometry is a particularly important variable for
determining the physical properties of a given compound. For
example, in the earliest work on TCNQ salts, several examples
were found[16] where 1:2 salts are much better conductors than the
1:1 salts of the same donor and acceptor. This is because if
the stoichiometry is not 1:1, one of the stacks will be of mixed
valence--even if the charge transfer is complete--due to the
stoichiometry. Another significant feature is that complex salts
(i.e., those where the stoichiometry is not 1:1) generally form
with segregated stacks. Thus, complex salts are all similar to
the mixed valence phase in Fig. 2. For a 1:2 TCNQ salt, for
example, there are two types of segregated stacks observed:
(1) where there are an equal number of donor and acceptor stacks,
but the lattice constant of the donor stack is twice as large as
that of the acceptor stack; or (2) where the two lattice constants
are the same, but there are twice as many acceptor stacks. It
appears that the latter type of structure tend to empirically
have higher conductivity,[7] but it is not known why.

Stoichiometry is by far the least understood of the three
properties discussed here. In the laboratory, one can try to
make a 1:2 salt, for example, by reacting the donor with a strong
excess of acceptor. If the 1:2 salt is stable, it may form. But
why do some donors form complex salts with TCNQ and others do
not? In cases where complex salts are formed, what factors govern
which of the two stacking types are adopted? In general, there
are no answers, and even no clues, to these questions.

A notable exception is the case of the defect halide salts
of TTF, e.g., $TTF-Br_{0.79}$, in which case a calculation has shown[10]
that the observed stoichiometry can be accounted for by I, A,

and the Madelung energy. The exact agreement is partially
fortuitious, but the significant point is that the Madelung energy
of this structure gave rise to a minimum. Furthermore, the
predicted stoichiometry varies as a function of the electron
affinity of the halide in quantitative agreement with experiment
for the Cl, Br, and I salts of TTF, indicating that most of the
essential factors determining the Br concentration were correctly
described. This calculation may have more relevance to the
stability of the charge transfer in 1:1 salts (when the charge
transfer can be continuously varied) than to the case of complex
salts, where steric factors generally restrict the stoichiometry
to 1:2 or 2:3. In any case, the role of the Madelung energy in
complex salts needs to be examined further.

5. TMTTF-BROMANIL

Before 1978, there were three general classes of solids which
exhibited segregated stacks:

(1) Alkali-metal salts (e.g., K-TCNQ, K-chloranil).

(2) Halide salts (TMPD-I, TTF-$Br_{0.79}$).

(3) Salts with TCNQ or one of its close derivatives,

and only two general classes of highly conducting salts:

(1) Halide salts (TTF-$Br_{0.79}$, TTT-$I_{1.5}$), and

(2) Salts of TCNQ or close derivatives (simple and complex).

From these statistics it appeared that neither segregated stacks
nor high conductivity were general phenomena; rather their
occurrence was restricted to a small subset of compounds. In
addition, it appeared that TCNQ was somehow special and unique.

It is in this context that the discovery[17] of TMTTF-bromanil
and the other TTF-tetrahalo-p-benzoquinones is significant.
TMTTF-bromanil has segregated stacks and a room temperature d.c.
conductivity of $5\Omega^{-1}cm^{-1}$ along the stacking axis. Thus, it is
the first highly conducting non-TCNQ 1:1 charge transfer salt.
This discovery substantiates the ideas of stacking and strength
of ionicity discussed above, since they were used in choosing
this system. Perhaps more importantly, it suggests that a much
wider variety of donors and acceptors may be used in forming
highly conducting solids.

6.   CONCLUSION

As can be seen from the phase diagram of Fig. 2, there are
four distinct phases of 1:1 charge transfer salts. The chances
of obtaining a compound from any desired phase can then be
considerably enhanced by choosing the factors which favor that
phase: the appropriate ionization potential (I) of the donor
with respect to the electron affinity (A) of the acceptor, the
factors (such as molecular size) which affect the Madelung energy,
and the factors (discussed above) which tend to favor the desired
stacking mode. At present, the factors which determine the
stoichiometry are largely unknown; new experimental (structural)
data and particularly new ideas are needed.

REFERENCES

1. Z. G. Soos, Ann. Rev. Phys. Chem. 25, 121 (1974).
2. A. F. Garito and A. J. Heeger, Accts. Chem. Res. 7, 232 (1974).
3. A. N. Bloch, in Energy and Charge Transfer in Organic
    Semiconductors, ed., K. Masuda and M. Silver (Plenum Press,
    New York, 1974) p. 159.
4. D. O Cowan, P. Shu, C. Hu, W. Krug, T. Carruthers, T. Poehler
    and A. Bloch, in "Chemistry and Physics of One-Dimensional
    Metals," NATO Advanced Study Institute Series, 25 (B-Physics),
    ed. H. J. Keller (Plenum Press, N.Y., 1977) p. 25.
5. J. B. Torrance, Accts. Chem. Res. (in press).
6. D. J. Dahm, P. Horn, G. R. Johnson, M. G. Miles and J. D.
    Wilson, J. Cryst. Mol. Struct. 5, 27 (1975).
7. J. B. Torrance, in Proceedings of Conference on Synthesis
    and Properties of Low-Dimensional Materials, ed. J. S. Miller
    and A. J. Epstein, Annals of N.Y. Acad. Sci. 313, 210 (1978).
8. H. M. McConnell, B. M. Hoffman and R. M. Metzger, Proc. Nat'l.
    Acad. Sci. USA 53, 46 (1965).
9. J. B. Torrance, B. A. Scott and F. B. Kaufman, Solid State
    Commun. 17, 1369 (1975).
10. J. B. Torrance and B. D. Silverman, Rev. B15, 788 (1977).
11. D. J. Sandman, J. Am. Chem. Soc. 100, 5230 (1978).
12. J. J. Mayerle, J. B. Torrance, V. Y. Lee and J. I. Crowley
    (to be submitted); and R. M. Metzger, J. B. Torrance, J. J.
    Mayerle and J. Crowley, Bull. Am. Phys. Soc. 23, 356 (1978).
13. R. C. Wheland and J. L. Gillson, J. Am. Chem. Soc. 98, 3916
    (1976); R. C. Wheland, J. Am. Chem. Soc. 98, 3926 (1976).
14. M. E. Hawley, T. O. Poehler, T. F. Carruthers, A. N. Bloch,
    D. O. Cowan and T. J. Kistenmacher, Bull. Am. Phys. Soc. 23,
    424 (1978); this volume, and to be submitted.
15. Z. G. Soos and S. Mazumdar, Chem. Phys. Lett. 56, 515 (1978).
16. W. J. Siemons, P. E. Bierstedt and R. G. Kepler, J. Chem.
    Phys. 39, 3523 (1963).
17. J. B. Torrance, J. J. Mayerle, V. Y. Lee, J. I. Crowley and
    K. Bechgaard, to be submitted.

THE DESIGN OF NEW ORGANIC ELECTRON ACCEPTORS

A.W. Addison, J.P. Barnier, V. Gujral, Y. Hoyano,
S. Huizinga, and L. Weiler

Department of Chemistry
University of British Columbia
Vancouver, B.C., Canada V6T 1W5

In the five years since the discovery of the first "organic metal", TTF-TCNQ (1) there has been widespread interest in the experimental properties and theoretical models for these materials. There is a wealth of carefully determined experimental data on these organic conductors and a number of theories have been proposed to rationalize and explain the properties of these materials. On the other hand chemists have had only limited success in designing new organic conductors. All of these organic conductors are charge transfer salts of electron donors and acceptors which structurally are very similar to TTF and TCNQ:

TTF   HMTTF

TSF   HMTSF

TCNQ

TNAP

In these materials the donors and acceptors, which are large flat molecules, stack face-to-face in segregrated stacks. In this arrangement the electrons are to a great extent constrained to move along the stacks. Early electrical conductivity measurements demonstrated the highly one-dimensional behaviour of TTF-TCNQ. X-ray crystallographic studies pointed to the importance of the overlap of the π orbitals within the separate donor and acceptor stacks. A number of experimental results including x-ray bond lengths, electron density plots, ESCA data, optical spectra, esr susceptibilities, and diffuse x-ray scattering results show that there is significant electron transfer in these solids. ESR studies on TTF-TCNQ (2), nmr results (3) and band structure calculations (4) all indicated that the room temperature conductivity of TTF-TCNQ is dominated by the TCNQ stacks. The aim of one aspect of our research is to design organic conductors with room temperature conductivities greater than that of TTF-TCNQ. Thus we asked why does TCNQ dominate the room temperature conductivity of TTF-TCNQ and what improvements can be made to the acceptor or the donor to increase the conductivity?

At present it appears that a one-electron band model and interelectron Coulombic interactions make important contributions to the electronic properties of these organic metals. Thus both of these effects must be considered in the design of new materials. First we consider how the conductivity could be affected by changes in the band structure. It is well known that the conductivity of a metal is related to the band-width. Thus we would expect that changes which lead to wider bands on either the donor or acceptor stacks would produce materials with higher conductivities than TTF-TCNQ. Our band structure calculation (4) indicated that in TTF-TCNQ the TCNQ band was significantly wider than the TTF band. In the one-electron band theory which we used, the band width is directly related to the overlap of highest occupied molecular orbitals (HOMO's) of adjacent TTF cations or TCNQ anions within a stack. Calculations (5) on dimers of TCNQ clearly showed that maximum overlap of the highest occupied π molecular orbitals occurs at exactly the observed geometry as found in TTF-TCNQ. Therefore, in this band picture, the TCNQ is making its maximum possible contribution to the conductivity.

On the other hand, our calculations (5) of TTF dimers showed that the overlap of HOMO's is poor for the geometry found in TTF-TCNQ. Interestingly, we found that there is excellent overlap if the TTF molecules are stacked exactly on top of each other. This is precisely the geometry of the conducting halides (6). Thus, we expect that by changing the size and shape of the <u>acceptor</u> we should be able to prepare new charge transfer complexes with TTF in which the TTF molecules are stacked directly over each other. We would predict that these complexes will have room temperatures conductivities greater than that of TTF-TCNQ.

A second chemical method to increase the MO overlap in the stacks, and hence increase the conductivity, is to replace the elements in donor or acceptor with elements lower in the periodic table. The atomic orbitals of these elements are usually more diffuse and larger than those of the second or third period and hence can provide better intermolecular overlap. A successful application of this occurs when the sulfur atoms of TTF are replaced by selenium atoms to give TSF (7). The room temperature conductivity of TSF-TCNQ, which is isostructural with TTF-TCNQ, is slightly greater than that of TTF-TCNQ (8). A band structure calculation (9) of TSF-TCNQ indicates that the donor TSF band width is comparable to the TCNQ band width. Since the orbitals on tellurium are even larger than those of selenium, we would suggest that the tellurium analog of TTF would yield a complex with TCNQ of even higher conductivity.

Now let us return to a discussion of the importance of electron repulsions in organic conductors. A number of recent experiments (10) and theoretical calculations (11) have pointed to the importance of Coulombic interactions in organic conductors. The most recent measurements suggest that in TTF-TCNQ the band width is approximately equal to the Coulombic repulsion, U, (10). In simple terms U is a measure of how successfully the acceptor molecule can accommodate two electrons or the donor molecule two holes. This energy has been calculated to be <u>ca.</u> 3 eV <u>in the gas phase</u>, but it must be substantially less in the solid state. Another way to look at U, is that it is a measurement of the free energy difference for the following equilibria:

$$2D^+ \longrightarrow D + D^{2+} \qquad\qquad 1$$

$$2A^- \longrightarrow A + A^{2-} \qquad\qquad 2$$

We would suggest that solution electrochemical data might be useful as a guide to the magnitude of U for various donors and acceptors; the better donors and acceptors having the lower values of U. In fact, it has been suggested that negative values of U may yield superconducting materials (12).

For processes 1 and 2 <u>in solution</u>:

$$\Delta G° = -96,500 \ \Delta\epsilon°$$

$$\ln K = \frac{\Delta\epsilon°}{.059} \qquad \text{(at 25°C)}$$

where $\Delta G°$ is the free energy change in joules for the process 1 or 2, and $\Delta\epsilon°$ is the difference between the first two oxidation potentials of the donor in 1 or the first two reduction potentials of the acceptor in 2. These values of $\Delta\epsilon°$ are readily available from electrochemical redox potentials and to a first approximation provide a guide to the value of U in charge transfer solids.

The electrochemical data for the common acceptors and donors are given in Table I. The magnitude of $\Delta\epsilon°$ for these donors and acceptors is very similar to the experimental values for U of 0.4 volts in TTF-TCNQ (10) and 0.35 volts in HMTSF-TCNQ (10). Also it is clear from this table that all of the common donors and acceptors, except TCNDQ (<u>vide infra</u>), are expected to have essentially the same value of U in charge transfer salts. Hence, it has not been possible from existing experimental data to determine what the effect of reducing U will have in organic charge transfer salts. will have in organic charge transfer salts.

There is an obvious need for the preparation of new donors and acceptors which can probe the effect of Coulombic effects in organic conductors and hence may offer additional insight into the

Table I.   Electrochemical Data for Donors and Acceptors[a]

	$\Delta\epsilon°$ (volt)
TCNQ	.42
TNAP	.38
TCNDQ	.17
TTF	.37
TSF	.27
HMTTF	.40
HMTSF	.31

[a] $\Delta\epsilon°$ is the difference between the first two reduction potentials of the acceptor or oxidation potentials of the donor in acetonitrile solutions.

nature of these unusual materials. Note that all of the known donors which produce organic conductors are all simple (although not in laboratory synthesis) substitutional analogues of TTF. There is even less structural variation in the known useful acceptors. Almost all organic conductors have TCNQ as the acceptor moiety. However, the complexes of TNAP with TTF (13) and with HMTSF (14) have both been prepared recently and they appear to be metallic at room temperature. There is great difficulty in obtaining crystals with TNAP as acceptor and other donors. This may be due to the lower symmetry of TNAP (D_{2h}) versus TCNQ (D_{2d}).

We were interested in synthesis of TCNDQ and its charge transfer salts. Unfortunately a number of attempts to prepare

T CNDQ

the neutral compound have failed (15). However, we have been able to prepare some alkali metal salts of TCNDQ$^-$ and TCNDQ^{2-} (16). The radical ion, TCNDQ$^-$, can be readily identified in solution by its complex ESR spectrum. In the solid state the alkali metal salts, M$^+$TCNDQ$^-$, have a single strong ESR signal reminiscent of the alkali metal TCNQ salts (17). We have also studied the solution electrochemistry of TCNDQ by measuring the oxidation potentials of TCNDQ^{2-}. This data is included in Table I. The key point is the very low value of $\Delta\varepsilon°$ for TCNDQ – this is one of the lowest for any known organic compound. This low value of $\Delta\varepsilon°$ indicates very small Coulombic repulsions in the anion TCNDQ^{2-} and is presumably a result of the increased delocalisation of the electrons in this system.

All of our attempts to prepare neutral TCNDQ from either of these anions of TCNDQ have been unsuccessful (15c) and indeed preliminary attempts to prepare charge transfer salts of TCNDQ by the metathetical reactions below have likewise been unsuccessful.

$$Na^+ \ TCNDQ^- + TTF^+Cl^- \ \xrightarrow{\quad} \ NaCl + TCNDQ\text{-}TTF$$

$$(Na^+)_2 TCNDQ^{2-} + TTF^{2+}(Br^-)_2 \ \xrightarrow{\quad} \ 2NaBr + TCNDQ\text{-}TTF$$

In all cases to date we obtain polymeric TCNDQ (15). Thus it would appear that although the Coulombic repulsions are much lower in the TCNDQ system polymerization prevents use of this

molecule in organic conductors. It will be necessary to put
substitutents on the basic TCNDQ skeleton to retain the favourable
electronic properties and supress its propensity for polymerization.

Acknowledgements: We are grateful to NATO and the Petroleum
Research Fund administered by the American Chemical Society for
support of this work.

1.   J. Ferraris, D. O. Cowan, V. Walatka, Jr., and J. H.
      Perlstein, J. Amer. Chem. Soc., 95, 948 (1973).

2.   Y. Tomkiewicz and A. Taranko, Phys. Rev. Lett., 36, 751
      (1976).

3.   G. Soda, D. Jérome, M. Weger, J. Alizon, J. Gallice, H. Robert,
      J. M. Fabre, and L. Giral, J. Physique, 38, 931 (1977).

4.   A. J. Berlinsky, J. F. Carolan, and L. Weiler, Solid State
      Commun., 15, 795 (1974).

5.   A. J. Berlinsky, J. F. Carolan, and L. Weiler, Solid State
      Commun., 19, 1165 (1976).

6.(a) F. Wudl, D. E. Schafer, W. M. Walsh, Jr., L. W. Rupp,
      F. J. DiSalvo, J. V. Waszczak, M. L. Kaplan, and G. A.
      Thomas, J. Chem. Phys., 66, 377 (1977);
  (b) R. B. Somoano, A. Gupta, V. Hadek, M. Novotny, M. Jones,
      T. Datta, R. Deck, and A. M. Herman, Phys. Rev. B, 15,
      595 (1977);
  (c) S. J. LaPlaca, P. W. R. Corfield, R. Thomas, and B. A.
      Scott, Solid State Commun., 17, 635 (1975).

7.   E. M. Engler and V. V. Patel, J. Amer. Chem. Soc., 96, 7376
      (1974).

8.   E. M. Engler, B. A. Scott, S. Etemad, T. Penney, and V. V.
      Patel, J. Amer. Chem. Soc., 99, 5909 (1977).

9.   F. Herman, Physica Scripta, to be published.

10.(a)D.Jérome, J. Physique Lett., 38, 489 (1977);
   (b)D.Jérome and M. Weger in "Chemistry and Physics of One-
      Dimensional Metals" edited by H. J. Keller, Plenum Press,
      1977.

11.   J. B. Torrance in "Chemistry and Physics of One-Dimensional
      Metals" edited by H. J. Keller, Plenum Press, 1977.

12.   D. Davis, H. Guefreund, and W. A. Little, Phys. Rev. B, <u>13</u>, 4766 (1976).

13.   P. A. Berger, D. J. Dahm, G. R. Johnson, M. G. Miles, and
      J. D. Wilson, Phys. Rev. B, <u>12</u>, 4085 (1975).

14.   K. Bechgaard, C.S. Jacobsen, and N.H. Andersen, Solid State Commun., <u>25</u>, 875 (1978).

15.(a) W. R. Hertler, U. S. Pat, 3,153,658 (1964);
   (b) D. J. Sandman and A. J. Garito, J. Org. Chem., <u>39</u>, 1165 (1974).;
   (c) Y. Hoyano, S. Huizinga, and L. Weiler, unpublished results.

16.   A. W. Addison, N. S. Dalal, Y. Hoyano, S. Huizinga, and L. Weiler, Can. J. Chem., <u>55</u>, 4191 (1977).

17.   T. Hibma and J. Kommandeur, Phys. Rev. B, <u>12</u>, 2608 (1975).

ORGANIC METALS: RECENT SYNTHETIC STUDIES

A.F. Garito[*]

M.P. Cava, and M.V. Lakshmikantham

Department of Physics* and Department of Chemistry
L.R.S.M., University of Pennsylvania, Phila, Pa. 19104

On the basis of a set of initial guidelines for organic and polymeric metals,[1] a series of π-donors have been synthesized for use with acceptors such as tetracyanoquinodimethan (TCNQ) (1).

The principal chemical and structural criteria for metallic behavior followed can be summarized by the basic conditions:

the existence of unpaired electrons within the molecular units;

molecular units exhibiting intramolecular electron correlation to allow charge density to reside on diametrically distant parts of the molecule;

highly polarizable molecular units having minimal size to diminish costly electron-electron repulsive interactions;

molecular donor units whose first ionization potential (electron affinity in the case of acceptors) does not assume large values on an established relative scale so as to avoid the Madelung energy contribution dominating the charge transfer process, thus favoring fractional charge transfer;

a uniform crystal structure consisting of linear, parallel stacked columns of flat planar molecules with signi-

ficant intermolecular overlap leading to bandwidths such
that, with minimized electron-electron repulsions, the
associated electronic band structure would be metallic.

Our snythetic studies have concentrated on structurally modified π-donors related to tetrathiafulvalene (TTF) (2)[2]. In brief,
the following organosulfur and in some cases organoselenium modifications of TTF have been successfully carried out: (i) symmetrical derivatives[3]; (ii) unsymmetrical derivatives[4]; (iii) vinylo-
gous derivatives[5]; and (iv) dithiinodithin isomers of TTF[6]. In this
report, three of our most recent synthetic studies are summarized.

(A) TETRATHIOALKYL SUBSTITUTED TETRATHIAFULVALENES

Two desirable features in the design of related TTF π-donors
for use in conducting charge-transfer salts are an increased polarizability and decreased band occupancy, both of which are favorable
toward diminishing energetically costly electron-electron repulsions
which inhibit electrical transport. In the rigid band approximation,
the polarizability (α) should decrease the on-site repulsion ener-
gy (U_{eff}) between charge carriers by approximately $\alpha e^2/\langle R_0^4 \rangle$, where
$\langle R_0 \rangle$ is the average distance between the carriers and polarizable
sites.

We have attempted to incorporate these features into TTF by
alkylthio substitution in the three tetrakis - (alkylthio) - tetra-
thiafulvalenes (3), (4a), and (4b)[3b].

(3)   (4)

a; $n = 2$
b; $n = 3$

(5)   (6)

a; $n = 2$
b; $n = 3$

Triethylphosphite coupling of the thiones (5), (6a), and (6b),
directly affords (3), (4a), and (4b) in high yield. All three TTF
derivatives yielded 1:1 black crystalline salts when heated in
solution with TCNQ.

The π-donor abilities of (3), (4a), and (4b) are smaller than
that of TTF as seen from comparing the $E_{\frac{1}{2}}$ values in Table 1. This
leads to smaller charge-transfer with TCNQ and hence should decrease
the band occupancy in the crystalline state. Moreover the difference

TABLE 1

	$E^1_{\frac{1}{2}}$a	$E^2_{\frac{1}{2}}$a	$\sigma(RT)/(\Omega cm)^{-1}$
(3)	+0.496	+0.668	10^{-5}c
(4a)	+0.532	+0.773	50b
(4b)	+0.589	+0.861	80b
(TTF)	+0.342	+0.721	20b

aReversible oxidations in MeCN with added Et₄NClO₄ (0.05M) vs. Ag/Ag⁺ (0.1N in MeCN) with a glassy carbon electrode as the working pellets (1:1 TCNQ salt) using silver-paint electrodes. cSame as footnote b except for single crystals (1:1 TCNQ salt).

between $E_{\frac{1}{2}}^1$ and $E_{\frac{1}{2}}^{\bar{2}}$ for these π-donors remains low which is important to allow desirable double-occupancy fluctuations in the conducting state.

The room temperature electrical conductivities measured on compressed pellets of the TCNQ salts of (3), (4a), and (4b) (see Table 1) when compared to the corresponding value of $20\Omega^{-1}cm^{-1}$ for TTF-TCNQ show a somewhat higher conductivity for the salts of (4a) and (4b) which within band theory may be associated with more favourable polarizability and band occupancy conditions caused by the π-donors (4a) and (4b). Clearly further detailed studies of the electronic and structural properties of the latter two TCNQ salts are required to confirm these conditions.

In contrast, single-crystal measurements on the TCNQ salt of (3) indicate relatively low conductivity. Since the first half-wave potential of (3) is lower than that of either (4a) or (4b), it is likely that the large difference in conductivity values results from special steric requirements of the rotating methylthio groups.

(B)   VINYLOGS OF TETRATHIAFULVALENE

Synthesis of vinylogs of TTF was initiated with the objective of localizing charge densities (hole carriers) resulting from charge transfer on diametrically distant parts of the molecular site to decrease the on-site repulsion (U_{eff}) between charge carriers. U_{eff} depends inversely on the distance across the molecule between the separated regions.

One system incorporating these features is the unknown p-quinodimethane analog of TTF, p-quinobis-(1,3-dithiole) (7).

(7)

The dibenzo derivative of (7), namely p-quinobis-(benzo-1, 3-dithiole) (8) was synthesized in pure crystalline form by a novel route[5]

Attempts to obtain a pure crystalline TCNQ complex of (8) have not yet succeeded, due to a combination of the great insolubility of (8) in organic solvents and by its ready decomposition in dilute solution in all solvents except carbon disulfide. Differential pulse polarography measurements of the stable bis(dithiolium) fluoroborate in acetonitrile showed two reductions at $E_{\frac{1}{2}}^1 = 0.330$ and $E_{\frac{1}{2}}^2 = +0.057$ V, corresponding to the monothiolium radical cation and the neutral compound (8). As evidenced by the irreversibility for polarographic reduction, both latter species appear highly unstable. Polarographic measurements of (8) were severely hampered by the presence of oxygen even though common precautionary procedures were followed, and stoichiometric TCNQ salt formation proved difficult due to the extremely low solubility of (8).

(C) S-OXIDES OF TETRATHIAFULVALENE

The one-, or two-electron, oxidation of TTF to give the radical cation, or dication, is indeed the only known reaction of the basic TTF system with the exception of the recently reported lithuim hydrogen interchange in TTF[7]. We have recently completed the first synthesis of a new type of TTF oxidation product namely TTF-S-oxide (9)

(9) R = H
(10) R = COOMe

(11)

Reaction of TTF with m-chloroperbenzoic acid in a cooled two phase system (CH_2Cl_2/aqueous Na_2HPO_4) gave the yellow TTF-S-oxide (9). In a similar manner, dibenzotetrathiafulvalene and tetrakis-carbomethoxy tetrathiafulvalene were oxidized to the lemon yellow (10) and orange crystalline S-oxide (11), respectively.

The first ($E_{\frac{1}{2}}^1$) and second ($E_{\frac{1}{2}}^2$) polarographic half-wave potentials and their difference ($\Delta E_{\frac{1}{2}}$) for the S-oxides are given in Table 2.

The $E_{\frac{1}{2}}^1$ values show that (9), (10), and (11) undergo oxidation to their respective monocations less readily relative to the corresponding unoxidized parent donors while the oxidation sequence due to substituent effects remains the same: (9)>(10)>(11). Further, a given sulfoxide monocation oxidizes to the dication more easily than the corresponding parent monocation. These systematic differences in oxidation properties of the parent donor and their S-oxides are related to the fact that the total free energy (ΔF) for oxidation in solution is a sum of electronic (ΔF_e), solvation (ΔF_s), and intramolecular distortion (ΔF_D) terms, $\Delta F = \Delta F_e + \Delta F_s + \Delta F_D$. The presence of the SO group would then change the molecular contributions to each of the three terms. For example in addition to overall changes in the molecular electronic states (ΔF_e), the pyramidal bonding around S at each S-O site would markedly distort the TTF ring structure (ΔF_D) and introduce larger dipole moments within each ring (ΔF_s).

TABLE 2

	$E_{\frac{1}{2}}^1$ a	$E_{\frac{1}{2}}^2$ a	$\Delta E_{\frac{1}{2}}$
(9)	+0.936	+1.10	0.164
(10)	+1.05	+1.21	0.160
(11)	+1.39	+1.55	0.160
TTF	+0.342	+0.721	0.379

(a) Reversible oxidations in MeCN with added Et_4NClO_4 (0.05 m) vs. Ag/Ag$^+$ (0.1N in MeCN) with a glassy carbon electrode as the working electrode; the resulting values are given in V with respect to the saturated calomel electrode.

    In summary, we have attempted to incorporate molecular fea-
tures into the π-donors as suggested by present theoretical understanding of the macroscopic properties of conducting organic
charge-transfer salts. Several features are worth noting further.
(1) Even though the desired features of increased polarizability
and diminished charge transfer are present in (3), the fact that
the average conductivity of the TCNQ salt of (3) is lower by a factor of approximately 5×10^6 compared to those of (4a) and (4b)
stresses the importance of steric factors in achieving suitable
crystal packing in the salt (2). The instability of (7) is most
likely related to the unusually high electron density on the central
methylene carbon atom calculated for quinodimethane structures. This
high density is stabilized by the dithiole ring to a larger degree
than in the parent p-quinodimethane but not as much as in TCNQ itself where energetically low lying π* state are available in the CN
groups. Finally (3), in view of our demonstration that TTF can be
alternatively oxidized, namely, to TTF-S-oxide (9), it is tempting
to speculate that small amounts of (9) may exist as a hitherto un-
suspected impurity in TTF samples and may act as a dopant for the
solid state properties of TTF-TCNQ crystals.

    This work was supported by the National Science Foundation,
National Science Foundation-MRL Program, and NATO.

REFERENCES

1.   A.F. Garito and A.J. Heeger, Acc. Chem. Res., 7, 232 (1974).
2.   The discussion is limited to synthetic studies in our labora-
      tories due to space limitations, and the reader may wish to
      refer to recent reviews such as
      (a) E.M. Engler, Chem. Tech., 274 (1976).
      (b) M. Narita and C.U. Pittman Jr., Synthesis, 489 (1976).
3.   (a) H.K. Spencer, M.P. Cava and A.F. Garito, J. Org. Chem.
           41, 730 (1976).
      (b) M. Mizuno, A.F. Garito and M.P. Cava, Chem. Commun., 18
           (1978).
      (c) M.V. Lakshmikantham, M.P. Cava and A.F. Garito, Chem.
           Commun., 383 (1975).
      (d) H.K. Spencer, M.V. Lakshmikantham, M.P. Cava and A.F.
           Garito, Chem. Commun., 867 (1975).
      (e) M.V. Lakshmikantham and M.P. Cava, J. Org. Chem. 41, 882
           (1976).
4.   (a) H.K. Spencer, M.P. Cava, A.F. Garito, Chem. Commun., 966
           (1976).
      (b) N.C. Gionnella and M.P. Cava, J. Org. Chem. 43, 369 (1978).
      (c) M. Mizuno, M.P. Cava and A.F. Garito, J. Org. Chem. 41,
           1484 (1976).
5.   (a) M.V. Lakshmikantham and M.P. Cava, J. Org. Chem., 43,
           82 (1978).
      (b) M. Sato, M.V. Lakshmikantham, M.P. Cava and A.F. Garito,

J. Org. Chem., 43, 2084 (1978).

6. M. Mizuno, M.P. Cava and A.F. Garito, J. Org. Chem. 41, 1484 (1976).

7. D.C. Green, Chem. Commun., 161 (1977).

NEW DIRECTIONS IN ORGANIC ELECTRONIC MATERIALS BASED ON TETRATHIAFULVALENE

Edward M. Engler, Robert R. Schumaker and Frank B. Kaufman

IBM Thomas J. Watson Research Center

Yorktown Heights, New York 10598 USA

Widespread interest has been shown recently in the chemistry[1] and physics[2] of charge-transfer salts based on the organic π-donor, tetrathiafulvalene (TTF) since some of these materials were found to display unusual metal-like properties. This paper will describe two approaches that we have taken for preparing new organic electronic materials based on some of the insights and understanding gained from studying TTF. The first approach is essentially synthetic in orientation, and involves the preparation of extended π-systems based on the TTF ring system. The second approach involves the study of electron transport in less ordered systems, namely polymeric materials in which TTF has been incorporated as a pendant group.

EXTENDED π-SYSTEMS BASED ON TTF

The synthesis of TTF derivatives typically involves the coupling of 1,3-dithiole half-units.[1] Conceptually, one can imagine elaborating a variety of extended TTF derivatives based on bicyclic 1,3-dithiole derivative 1 (see Scheme 1). Recently, we discovered a simple, high yield synthesis[3] of this new ring system which is called 1,3,4,6-tetrathiapentalene or thiapen for short. Attempts to polymerize thiapendione (1,X=O) with trimethylphosphite led only to the formation of dimer (2). Dimer 2 could be further reacted with base and transition metal salts to give insoluble polymeric powders (3). Interestingly, the nickel compound (3 with M=Ni) was

Scheme 1

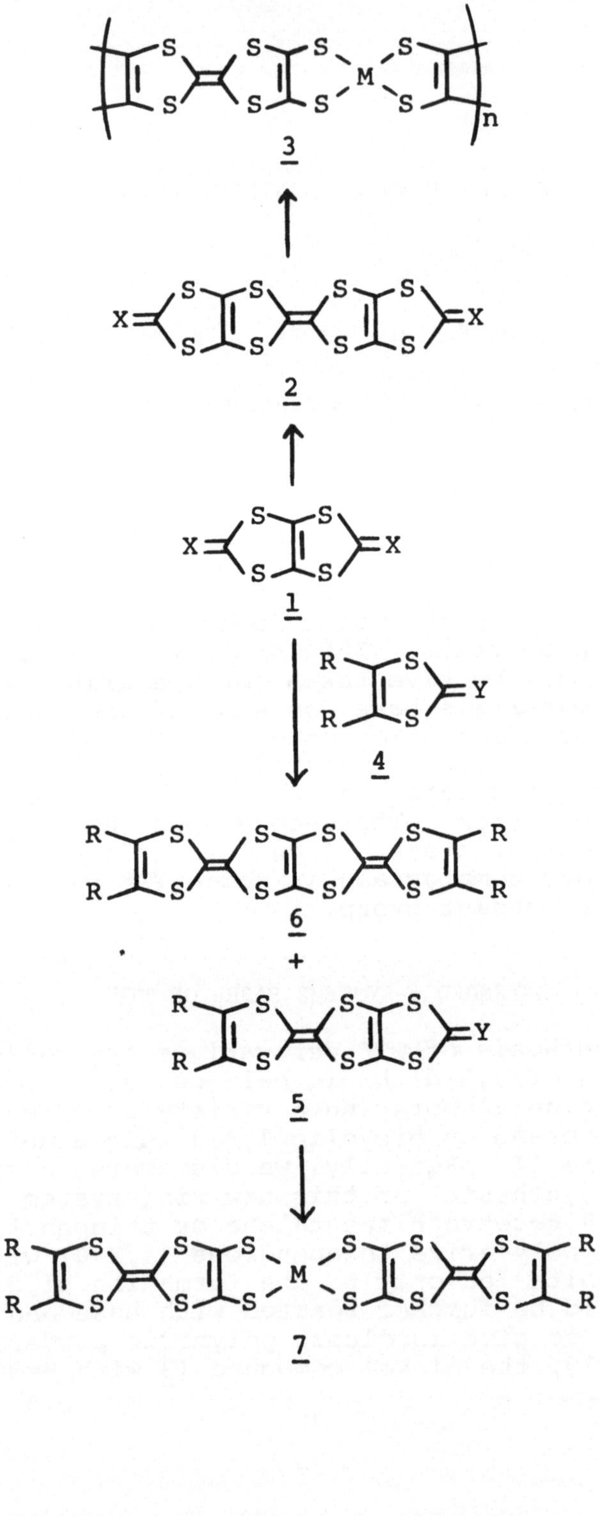

found to be highly conducting ($\sim$30/ohm-cm).[4]

Thiapendione (1,X=0) undergoes cross-coupling reactions with 1,3-dithiole derivatives (4) to yield mono-
and bis-capped products 5 and 6. The capping reaction
is very dependent on the nature of the R and Y substituents in 4 and succeeds for R=CO_2CH_3, CF_3, Y=S and R=CN,
Y=0. The half-capped material (5) can be further reacted
with base (NaOCH$_3$) and transition metal salts to provide
a multi-component π-system (7) which incorporates both
donor and acceptor units into a single conjugated molecular framework. Compound (7) is of particular interest
since one has the possibility of internal charge-transfer
to give a "single stack" conductor. Preliminary studies
have revealed high conductivity (σpowder$\sim$1/ohm-cm) in
one derivative of 7 (M=Ni, R=SCH$_3$).

Study of the properties of the symmetrical bis-TTF
systems 6 has been hampered by their poor solubility.
Recently, the unsymmetrically substituted bis-TTF (9)
has been synthesized which has shown improved solubility
characteristics.

$$\underline{9}$$

In comparison with a similarly substituted TTF derivative,
bis-TTF 9 was found electrochemically to be a slightly
better donor, and to display intense low energy electronic absorptions when dissolved in strong acid media.

POLYMER BOUND TTF

For a number of reasons, it would be desirable to
incorporate the interesting electronic properties of TTF
into polymer substrates. While a variety of approaches
to this problem are possible, we chose to attach TTF as a
pendant group along a cross-linked polymer matrix. By
varying the amount of donor coverage and degree of cross-
linking, the extent and nature of donor site-site interactions could be systematically probed.

Suitably monofunctionalized TTF derivatives have
been prepared[5,6] and attached in high yield (>90%) to
1-2% cross-linked chloromethylated polystyrene (see Eq.
1).

$$\text{(P)}-CH_2Cl \quad + \quad M^+\overline{O}RTTF \quad \longrightarrow \quad \text{(P)}-CH_2ORTTF \qquad (1)$$

$$-e \Big\updownarrow +e$$

$$R = CO, \ C_6H_4$$
$$M = \text{alkali metal} \qquad \text{(P)}-CH_2OR(TTF)^+$$

Although these TTF polymers are insoluble in common organic solvents by virtue of the internal cross-links present, it was discovered that the TTF polymer where $R=C_6H_4$ (Eq. 1) could be cast onto electrode surfaces to form porous high surface area films.[7] In contact with an organic solvent supporting electrolyte such as CH_3CN ($0.5Et_4N^+ClO_4^-$), the application of a voltage ($V>0.5$ Volts) causes these films to change color from yellow to green. Reversal of the voltage brings the films back to the original state. Optical studies on these films indicate that the applied voltage causes the formation of (polymer bound) colored donor ion radicals throughout the polymer film bulk. In contrast to most other organic electrochromic materials, free diffusion of the electroactive molecule to the electrode surface is limited in these TTF polymers. Therefore, in these polymers the bulk transport of holes (and ions) between donor sites must be involved in the observed optical changes.

REFERENCES

1.   For a review see: M. Narita and C. U. Pittman, Jr.,
     Synthesis, 489 (1976).
2.   For a recent review see: "Chemistry and Physics of
     One-Dimensional Metals", NATO Advanced Study In-
     stitutes Series (Physics), edited by H. J. Keller
     (Plenum, New York, 1977).
3.   R. R. Schumaker and E. M. Engler, J. Am. Chem. Soc.,
     99, 5519, 5521 (1977).
4.   N. Martinez-Rivera, E. M. Engler and R. R. Schumaker,
     to be published.
5.   D. C. Green, J.C.S. Chem. Comm., 161 (1977).
6.   E. M. Engler and V. V. Patel, to be published.
7.   F. B. Kaufman and E. M. Engler, to be published.

DESIGN AND SYNTHESIS OF A VARIABLE FILLED BAND MOLECULAR CONDUCTOR

Joel S. Miller* and Arthur J. Epstein[†]

Rockwell International Science Center*
Thousand Oaks, California 91360

Xerox Webster Research Center[†]
Rochester, New York 14618

New pseudo one-dimensional (1-D) substances are required to understand the salient features associated with unidimensionality. To date a variety of organic (e.g., tetrathiofulvalene (TTF), 7,7,8,8-tetracyano-p-quinodimethane (TCNQ)), inorganic (e.g., $K_2Pt(CN)_4Br_{0.3} \cdot 3H_2O$(KCP)), and covalent polymers (e.g., poly-(sulfurnitride)$(SN)_x$) have been studied.[1] However, due to significant differences in the types of bonding, resultant bandwidth, different stoichiometries (Tables I-II), and crystal structures detailed comparisons between these classes of materials are difficult to make. To help clarify the problem we focused our attention at approaching the situation from a different viewpoint, namely, studying the physical properties solely as a function of band filling, i.e., Fermi energy. In contrast with "classical"

TABLE I
Prototype 1-D Systems

Class of Substance (Conducting Chain)	Type of Bonding	Typical Bandwidth, eV
Organic (based on TCNQ)	p_z orbitals $(b_{2g}-b_{2g})$	1/3
Inorganic (based on TCP*)	d_{z^2} orbitals (a_g-a_g)	4
Polymer $((SN)_x)$	covalent bonds (mixture of $sp^2/sp^3/p_x/d$)	1

* TCP = tetracyanoplatinate

35

TABLE II
Representative 1-D Materials

One-Dimensional Substance[a]	Ground State[b]	Type[c]	$d\sigma/dT$[d]	Stoichiometric	Degree of Band Filling
$(SN)_x$	SC	P	+	Yes	semimetal
$Qn(TCNQ)_2$	I	O	+	Yes	0.25
$(NMP)_x(Phen)_{1-x}(TCNQ)$	I	O	+	No	0.25-0.5[f]
$(TTF)(TCNQ)$	I	O	+	Yes	0.30[g]
$(TTF)_x(TSeF)_{1-x}TCNQ$	I	O	+	No	0.30-0.32
$NH(CH_3)_3(I)(TCNQ)$	I	O,H	+	Yes	0.33
$(HMSeF)(TCNQ)$	HC	O	+	Yes	0.37[g]
$(NMP)(TCNQ)$	I	O	+	?	0.33-0.50
$(TTF)Br$	I	O,H	−	Yes	0.50
$(TTT)I_{2.7}$	I	O,H	+	Yes	0.55
$(TTF)Br_{0.76}$	I	O,H	+	No	0.62
$(TTF)(SeCN)_{0.58}$	I	O	+	No	0.71
$(TTT)I_{1.6}$	HC	O,H	+	Yes	0.73
$(TSeT)_2Cl$	HC	O,H	+	Yes	0.75
$K_{1/2}Ir(CO)_2Cl_2$	I	I	?	No	0.75
$Rb_2Pt(CN)_4(FHF)_{0.40}$	−	I	?	No	0.80
$K_{1.6}Pt(O_2C_2O_2)_2\cdot1.2H_2O$	−	I	?	No	0.80
Hg_3AsF_6	SC	I	+	No	0.83
$Ni(Pc)I$	I	I,H	+	No	0.84
$K_2Pt(CN)_4Br_{0.3}\cdot3H_2O$	I	I	+	No	0.85
$K_{1.75}Pt(CN)_4\cdot1.5H_2O$	I	I	−	No	0.875
$[C(NH_2)_3]_2Pt(CN)_4Br_{0.25}\cdot H_2O$	I	I	+	No	0.875
$Ni(HDPG)_2I$	I	I,H	−	No	0.90
$Cs_2Pt(CN)_4F_{0.19}$	−	I	?	No	0.91
$Ni(HBQD)_2I_{0.5}$	I	I,H	−	No	0.92
$[(SN)Br_{0.4}]_x$	ML	P,H(?)	+	No	0.93
$[(CH)I_{0.22}]_x$	ML	P,H(?)	+	No	0.96
$Ir(CO)_3Cl$	−	I	?	?	0.96-1.00
$Ni(HBQD)_2I_{0.02}$	−	I	?	No	0.99
$(CH)_x$	I	P	−	Yes	1.00

[a] H_2DPG = diphenylglyoxime, Pc = phthalocyanine, TTF = tetrathiofulvalene, NMP = N-methylphenazinium, Phen = phenazine, Qn = quinolinium, HMSeF = hexamethyleneselenafulvalene, TTT = tetrathiotetracene, TSeF = tetraselenofulvalene, H_2BQD = 1,2-benzoquinonedioxime, TSeT = tetraselenatetracene

[b] SC = superconductor, I = insulator, HC = highly conducting but $d\sigma/dT < 0$, ML = metal-like

[c] P = covalent polymer, O = organic, I = transition metal chain (inorganic), H = halogen chain.

[d] Room temperature, $d\sigma/dT < 0$ => semiconductor, $d\sigma/dT < 0$ => metal-like

[e] Calculated based on charge conservation and elemental composition assuming two charges per site and a uniform chain

[f] Degree of band filling depends on x

[g] From x-ray scattering

semiconductors, e.g., silicon, where doping with electron deficient, e.g., gallium, or electron rich, e.g., arsenic, atoms is
easily accomplished to vary the Fermi energy, wide variation of the
band filling of molecular based materials had not been previously
reported.

In order to synthesize such an isomorphous series of complexes
which differ to a first approximation only in the degree of band
filling, it is necessary to study the structural aspects of the
system. Consider the segregated 1-D anion (e.g., TCNQ⁻) and cation
chains illustrated in Fig. 1(a). Removal of an electron from the
anion chain (Fig. 1(b)) requires the removal of an electron from
the p_z based conduction band. Thus, due to the delocalized valence
band structure this implies that on the average less than one valence electron resides per TCNQ site. To conserve charge, however,
a cation must be removed from the cation chain (Fig. 1(b)). The
cation loss should result in a crystallographic void (Fig. 1(c)) in
the cation chain and the prediction that the unit cell would collapse. If one, however, were able to fill the cation void
(Fig. 1(d)) with a neutral molecular of similar size, shape, and
polarizability, it is feasible that the electron deficient chain
could be stabilized and through variation of the amount of electrons removed one could modulate the Fermi energy and concommitant-
ly vary the physical properties.

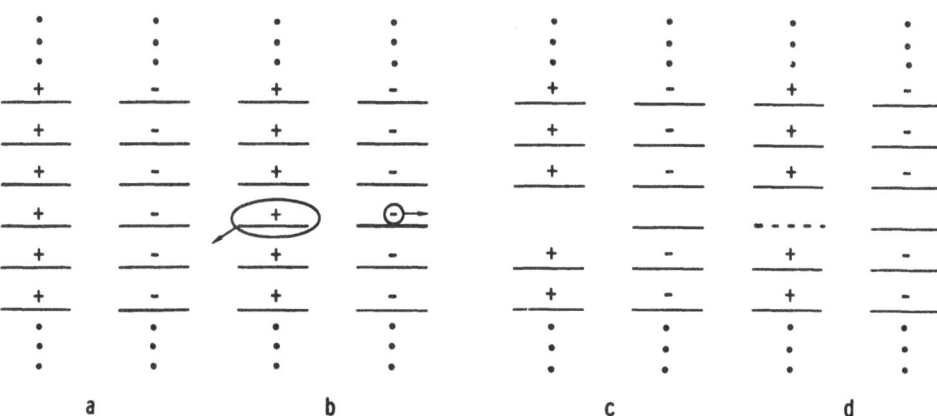

a   b   c   d

Fig. 1   (a) Illustration of segregated 1-D cation (e.g., NMP⁺) and
          anion (e.g., TCNQ⁻) chains. (b) Removal of an e⁻ from the
          anion chain and cation from the cation chain resulting in
          less than one valence electron in the conducting anion
          chain per anion and a void in the cation chain (c).
          (d) Filling the void with a neutral molecule (----) should
          stabilize the structure.

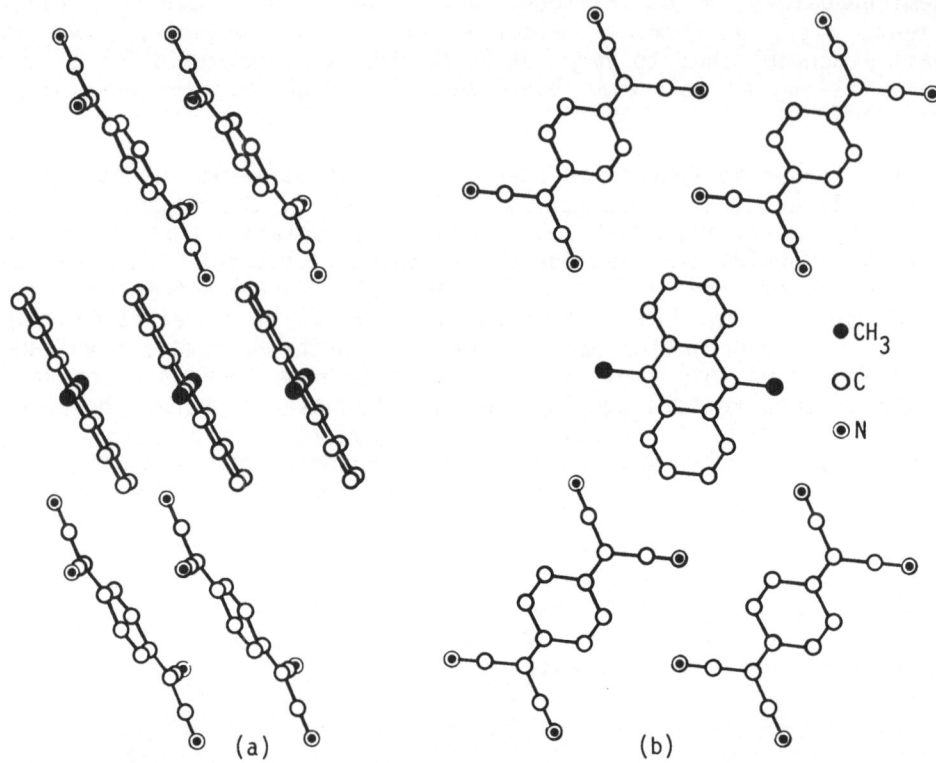

Fig. 2   Crystal structure of (NMP)(TCNQ). (a) Side view, (b) top
         view.   Random methyl groups (ref. 6a) are illustrated
         by ●.

With this concept in mind the (NMP)(TCNQ) (NMP^+=N-methylphena-
zinium, 1) system was chosen as the model system.[2-5] The structure
of (NMP)(TCNQ),[6] Fig. 2, shows that the methyl groups (●) are randomly orientated. Given the lack of positional preference of the
methyl group it seems reasonable to replace NMP^+ with phenazine,
phen, 2, which is of similar size, shape, and polarizability to
NMP^+.

NMP^+
1

phen
2

Upon reaction of phen, NMP^+, $TCNQ^0$, and $TCNQ^-$ in acetonitrile black needle crystals of $(NMP)(Phen)_{1-x}(TCNQ^-)_x(TCNQ)_{1-x} \equiv (NMP)_x-(Phen)_{1-x}TCNQ$ were collected and characterized by solution absorption spectroscopy and x-ray diffraction.[2] The unit cell parameters, Table III, show that for $x = 0.26$ and 0.46 cases the material is isomorphous to the well characterized authentic $(NMP)(TCNQ)$ (i.e., $x = 1$). Thus, we have prepared a series of 1-D materials which differ to a first approximation only in the degree of band filling, i.e., Fermi energy. Consequently, we are currently investigating the physical properties as a function of band filling.

TABLE III
Unit Cell Parameters for $(NMP)_x(Phen)_{1-x}TCNQ$

	$(NMP)(TCNQ)^{6a}$	$(NMP)_{0.74}(Phen)_{0.26}-(TCNQ)$	$(NMP)_{0.54}(Phen)_{0.46}-(TCNQ)$
x	1.00	0.74	0.54
$\underline{a}$, Å	3.8682 (4)	3.890 (8)	3.865 (7)
$\underline{b}$, Å	7.7807 (8)	7.799 (3)	7.611 (32)
$\underline{c}$, Å	15.735 (2)	15.706 (6)	16.329 (51)
α	91.67 (1)	91.75 (6)	93.73 (49)
β	92.67 (1)	92.96 (13)	91.53 (31)
γ	95.38 (1)	95.45 (2)	94.65 (20)
V, Å^3	470.7	473.4	477.4

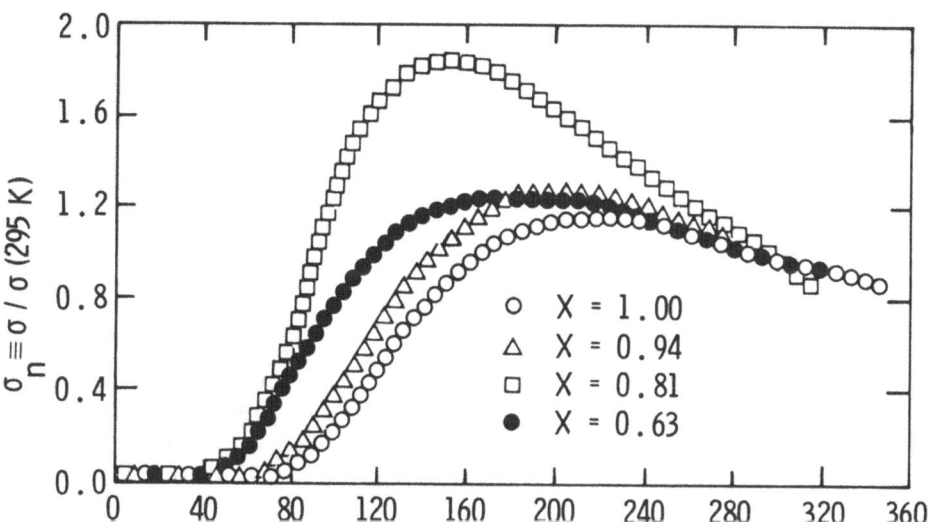

Fig. 3 Temperature dependence of the conductivity
for $(NMP)_x(Phen)_{1-x}TCNQ$ (Refs. 4,5).

The temperature dependence of the dc conductivity, $\sigma(T)$, for $(NMP)_x(Phen)_{1-x}TCNQ$ is shown in Fig. 3 for four representative samples. The first thing noticed is that the phenazine containing samples (i.e., $x < 1$) exhibit a similar $\sigma(T)$ to $(NMP)(TCNQ)$. The differences in $\sigma(295°K)$, Table IV, are not significant due to the errors involved in measuring the cross-sectional areas of the small samples. The temperature for the maximum conductivity, T_m, is lower for pheno replaced samples with respect to $(NMP)(TCNQ)$,[7] and the normalized conductivity, σ_n, increases with phenazine content.

TABLE IV
Conductivity Parameters for $(NMP)_x(Phen)_{1-x}(TCNQ)$

x	$\sigma(295°K)$, ohm^{-1}cm^{-1}	$\sigma_n(T_m)$	T_m, °K
1.00	200	1.17	220
0.94	100	1.27	205
0.81	100	1.85	155
0.63	70	1.26	175

The conductivity for $(NMP)_x(Phen)_{1-x}(TCNQ)$ for $60°K < T < 400°K$ data can fit[4,5] $\sigma_n(T)$ data by

$$\sigma_n(T) = A\,T^{-\alpha}\exp{-\Delta(x)/T}$$

with $\Delta(x)$ constant for all samples of the same phenazine content and α a sample dependent constant, Table IV. The constant A is fixed by $\sigma_n(295°K) \equiv 1$.

Thus, $(NMP)_x(Phen)_{1-x}TCNQ$ is a semiconductor at room temperature where it exhibits metallic-like conductivity, i.e., $d\sigma/dT < 0$. To confirm this it is desirable to determine the single crystal reflectivity at low frequency to determine the optical bandgap. Crystals of $(NMP)_x(Phen)_{1-x}TCNQ$ are too small for performing this measurement in the usual manner; however, large crystals of $(HNMe_3(I)(TCNQ))$, which exhibit a similar metal-like $\sigma(T)$ near room temperatures were grown.[8,9] The reflectivity data for $(HNMe_3(I)(TCNQ))$ confirm the presence of the bandgap at ~ 0.1 eV at room temperature where $d\sigma/dT < 0$.[10] Further systematic study of the physical properties as a function of band filling is in progress.

REFERENCES

1. J. S. Miller, Ann. N.Y. Acad. Sci., 313, 1 (1978) and references therein; "Synthesis and Properties of Low-Dimensional Materials," J. S. Miller and A. J. Epstein, Eds., N.Y. Acad. Sci., 313 (1978); A. F. Garito and A. J. Heeger, Acc. Chem. Res., 7, 232 (1974); M. Narita and C. U. Pittman, Synthesis, 8, 489 (1976); B. P. Bespalov and V. V. Titov, Russ. Chem. Rev., 44, 1091 (1975); J. J. Andre, A. Bieber and F. Gaultier, Ann. Phys. (Paris), 1, 145 (1976); L. Pál, G. Grüner, A. Jánossy, and J. Sólyom, Ed., Lect. Notes Phys., 65 (1977); H. J. Keller, Ed., Study Inst., Ser., Ser. B, 25 (1977); H. J. Keller, Ed., Study Inst. Ser., Ser. B, 7 (1975); J. S. Miller and A. J. Epstein, Prog. Inorg. Chem., 20, 1 (1976); G. B. Street and R. L. Greene, IBM J. Res. Dev., 21, 99 (1977).

2. J. S. Miller and A. J. Epstein, J. Am. Chem. Soc, 100, 1639 (1978).

3. J. S. Miller, Ann. N.Y. Acad. Sci., 313, 25 (1978).

4. A. J. Epstein and J. S. Miller, Solid State Commun., 27, 325 (1978).

5. A. J. Epstein, E. M. Conwell, and J. S. Miller, Ann. N.Y. Acad. Sci., 313, 183 (1978).

6. (a) C. J. Fritchie, Jr., Acta Cryst., 20, 892 (1966); (b) B. Morosin, Phys. Lett. A, 53, 455 (1975); (c) H. Kobayashi, Bull. Chem. Soc. Jap., 48, 1373 (1975).

7. A. J. Epstein, E. M. Conwell, D. J. Sandman, and J. S. Miller, Solid State Commun., 23, 355 (1977).

8. M. A. Abkowitz, A. J. Epstein, C. H. Griffiths, J. S. Miller, and M. L. Slade, J. Am. Chem. Soc., 99, 5304 (1977).

9. M. A. Abkowitz, J. W. Brill, P. M. Chaikin, A. J. Epstein, M. F. Froix, C. H. Griffiths, W. Gunning, A. J. Heeger, W. A. Little, J. S. Miller, M. Novatny, D. B. Tanner, and M. L. Slade, Ann. N.Y. Acad. Sci., 313, 459 (1978).

10. D. B. Tanner, J. E. Deis, A. J. Epstein, and J. S. Miller, submitted for publication.

DOPING ORGANIC SOLIDS – ITS USES TO PROBE AND TO

MODIFY ELECTRONIC PROPERTIES

Y. Tomkiewicz, E. M. Engler, B. A. Scott,
S. J. La Placa and H. Brom

*IBM Thomas J. Watson Research Center
Yorktown Heights, NY 10598, U. S. A.*

Organic compounds in general provide interesting systems for doping experiments. On the one hand, the versatility of the organic molecules in terms of chemical changes can be the source of a variety of isostructural dopants. On the other hand, the relatively low density of the structure is helpful in terms of interstitial doping. Thus, it is experimentally feasible to perform a variety of doping experiments on these compounds. Moreover, these experiments are of great interest for two different reasons: doping can be used to gain understanding of the multitude of phase transitions that occur in organic conductors, and dopants can be used to introduce charge carriers into organic insulators and thus increase their conductivity. The purpose of this article is to demonstrate the usefulness of doping in these two respects by exploring some examples. Let us start with the doping as a probe of phase transitions.

TTF-TCNQ was found to have three structural phase transitions,[1] at 52.6K, 49K and 38K. The highest-temperature transition was found[2] to be a metal-insulator transition. This multitude of phase transitions was explained[3] in terms of having different 3-D ordering temperatures on the donor and acceptor stacks respectively and in terms of interactions between distortions on the donor and acceptor stacks. The respective roles of the different kinds of stacks in the different phase transitions was found by decomposing the total magnetic susceptibility by the g-value decomposition technique[4] and by Knight-shift measurements.[5] Here we shall show how selective introduction of dopants into the donor and acceptor stacks led[6] to similar conclusions. Fig. 1 shows on a logarithmic scale the resistance of TTF-TCNQ *vs.* 1/T. The discontinuities in the slope, corresponding to the above-mentioned phase transitions, are clearly seen in the derivative plot, shown in Fig. 2. The effects of selective

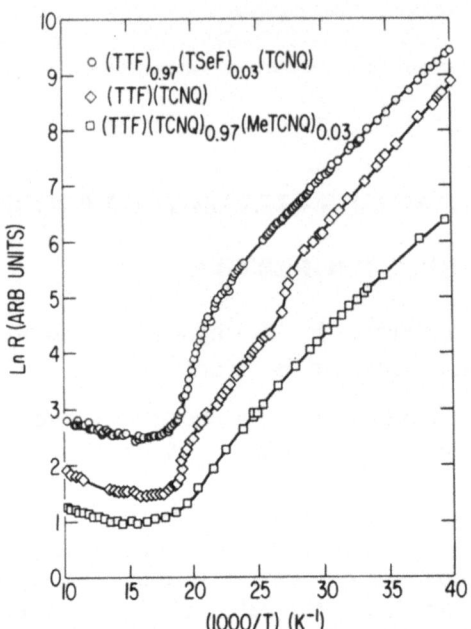

Figure 1.   The logarithm of the resistance as a function of reciprocal temperature of pure and doped (donor and acceptor stacks doping) TTF-TCNQ.

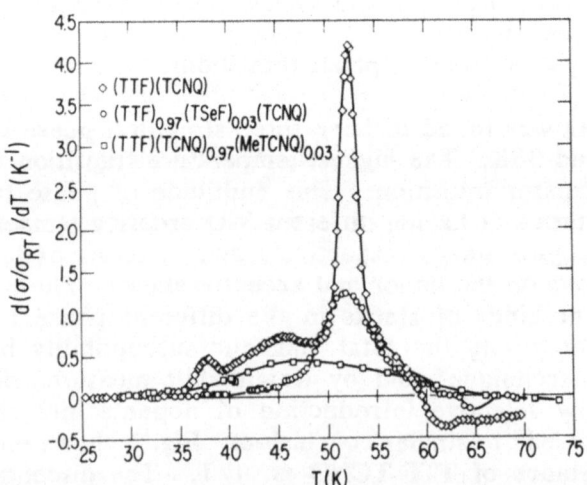

Figure 2.   The derivative of the conductivity with respect to T as a function of temperature for the three samples shown in Fig. 1.

introduction of dopants into donor (TSeF) and acceptor (MTCNQ) stacks are also shown. While donor-stack doping only broadens and smears the metal-insulator transition, acceptor-stack doping with a dopant producing a comparable perturbation,[6] also shifts the transition to 46K. Thus, the conclusion that had been previously inferred, that the acceptor stacks drive the metal-insulator transition, was confirmed. Moreover, both kinds of doping have similar effects on the transitions at 49K and 38K – they wipe them out completely. This result is in agreement with the interpretation[3] of the structural data, according to which both kinds of stacks are involved in these transitions: The 49K transition involves the donor stacks because the superlattice of the acceptor stacks is allowed to shear, thereby distorting the donor stacks. At 38K, the transverse periodicity of the superlattice jumps[1] discontinuously from 3.5a to 4a, which affects both kinds of stacks.

In the example described above, the doping experiments confirm conclusions obtained by other experimental techniques. However, in certain instances doping experiments provide results that cannot be obtained by any other means. An example of such a situation is the effect of doping on the phase transition of TSeF-TCNQ, the isostructural[7] analogy of TTF-TCNQ. The comparison of these two compounds demonstrates how sensitive are the quasi-1D organic metals to small changes such as the replacement of the sulfur atoms by selenium atoms in the TTF molecules. While TTF-TCNQ has three structural phase transitions, TSeF-TCNQ has only one,[8] in which both kinds of stacks undergo three-dimensional ordering. The temperature at which this ordering occurs, 29K, is lower than all the transition temperatures of TTF-TCNQ. Two different mechanisms have been suggested[9,10] as the possible source of the differences between TTF-TCNQ and TSeF-TCNQ. The first of these[9] relates to the experimental observation[8] that there are stronger $2k_F$ fluctuations in TSeF-TCNQ and that they definitely involve the TSeF stacks. Thus the situation with respect to the charge-density-wave amplitudes and correlation lengths is more symmetrical between the donor and acceptor stacks in TSeF-TCNQ than in TTF-TCNQ. Despite the transverse periodicity of the ordered state in TSeF-TCNQ being 2a, there is a competition *between* the interactions among fluctuations *within* and between the donor and acceptor subsystems leading to a lower transition temperature. A second possible source[10] of the lowering of the transition temperature is the greater banding in the transverse direction between donor and acceptor stacks in TSeF-TCNQ which has been demonstrated by measurements of electron-spin relaxation [10], nuclear-spin relaxation[11] and transverse conductivity.[12] This transverse banding, because of the unique band structure[13] of the TTF-TCNQ family of compounds, causes an opening of a semimetallic gap at the Fermi level, thereby providing competition for the Peierls-distortion mechanism. Thus larger transverse banding would mean a lower transition temperature. Both of these effects would imply that introduction of dopants into the donor stack would both raise the metal-insulator transition temperature and cause a splitting of the single transition of TSeF-TCNQ into two transitions, one of them corresponding to the donor-stack transition and the other to the acceptor-stack transition. In the first mechanism, the reason for this is that on the particular stacks into which the dopants are introduced, the charge-density waves will now have both smaller amplitudes and correlation lengths, so that the competition they provide to the ordering tendency that is solely within the other subsystem is now decreased. In the second mechanism, dopants affect transverse banding by smearing electronic k-values. The experimental consequence of the donor-stack doping of TTF-TCNQ with TSeF-TCNQ is shown in Fig. 3. The single transition of TSeF-TCNQ splits – the metal-insulator transition temperature increases as a function of increasing dopant concentration while the second transition is lowered until it is smeared out completely. The lower transition was identified[14] as the donor-stack ordering temperature. Thus the doping experiments confirmed our understanding of the difference between the phase transitions of TTF-TCNQ and TSeF-TCNQ and provide

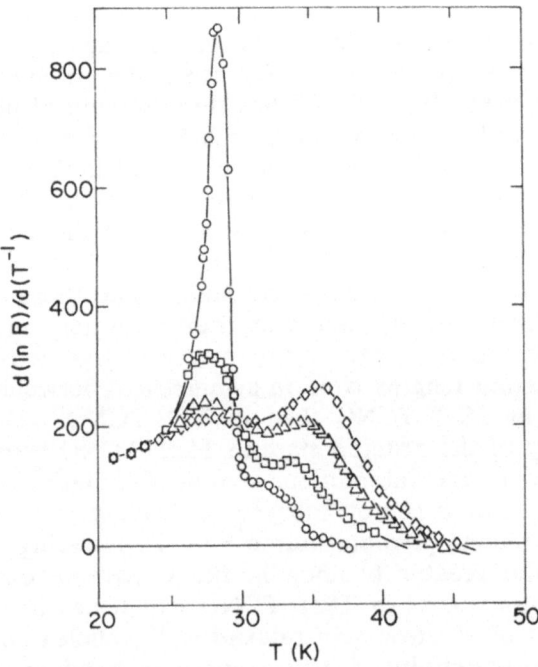

Figure 3. The logarithmic derivative of the resistivity with respect to T^{-1} as a function of temperature. The four curves shown are from samples of $(TSeF)_{1-x}(TTF)_x(TCNQ)$ with different impurity doping where $x = 0$ (circles), $x = 0.003$ (squares), $x = 0.0125$ (triangles), and $x = 0.025$ (diamonds).

a stimulus for additional theoretical work concerning the splitting of the single transition of TSeF-TCNQ.

Let us now turn to the second aspect of doping – introduction of charge carriers into organic insulators. The example that demonstrates the feasibility of these kinds of experiments is the doping of α-TTF single crystals with halogens like bromine. The TTF molecules in this particular structure[15] form stacks that are similar to the TTF stacks in TTF-TCNQ with a stacking distance of 4.023Å corresponding to a bandwidth[16] of 0.3eV. Thus, in principle, doping can introduce holes into this band. Fig. 4 shows the effect of bromination on the conductivity of such a TTF

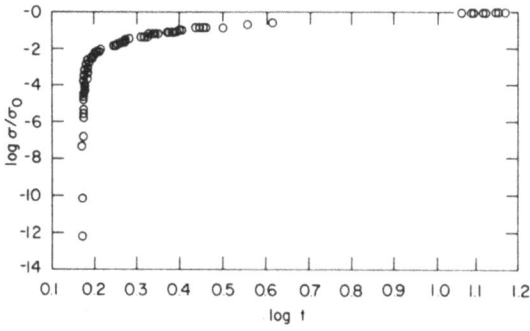

Figure 4.   The relative conductivity of TTF crystal as a function of exposure time to bromine gas.

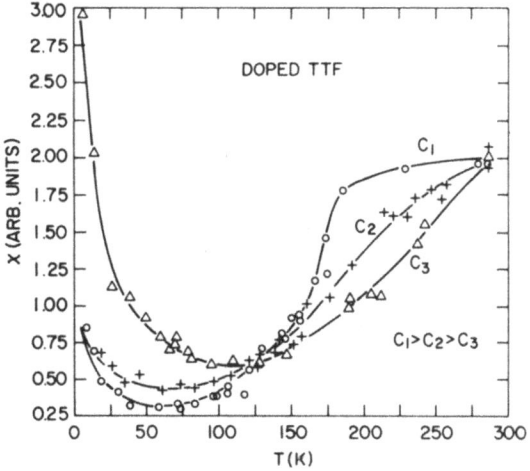

Figure 5.   The magnetic susceptibility of doped TTF as a function of temperature for different concentrations of bromine.

crystal. The relative conductivity is shown as a function of the exposure time to bromine at constant vapor pressure. The dramatic increase of the conductivity up to values of 10 $(\Omega\text{-cm})^{-1}$ is accompanied by a color change of the crystal from yellow to black. The color change originates in the absorption spectrum of TTF-ions which are formed as a consequence of doping. By varying the vapor pressure and/or exposure time, one can vary the concentration of charge carriers continuously.[17] The effect of this continuous change on the temperature dependence of the magnetic susceptibility is shown in Fig. 5. The values of the magnetic susceptibility for the three different concentrations of bromine (as determined by the microprobe technique) were normalized to agree at room temperature. It is clearly seen that the relative magnitude of the Curie tail in comparison to the room temperature value decreases with increasing dopant concentration. Since a Curie type of temperature dependence is typical of non-interacting spins, the following picture emerges: At low dopant concentrations, the holes introduced into the TTF band are localized, forming bound states with the halogen ions. As a consequence of this localization and the low density of dopants, a significant fraction of the spins are not interacting. This gives rise to the Curie type of temperature dependence of the magnetic susceptibility. At higher dopant concentrations, the bound states overlap, giving rise both to interacting-spin behavior and band-like conductivity.

The experiments described above were chosen to give a taste of the variety and versatility of doping experiments. The work on both of these aspects, particularly the effects of doping on conductivity of organic insulators, is now being carried further in our laboratory.

Acknowledgments — Helpful discussions with T. D. Schultz are greatly appreciated.

REFERENCES

1. See for example: S. Kagoshima, H. Anzai, K. Kajimura and T. Ishiguro, J. Phys. Soc. Japan **39**, 1143 (1975); S. K. Khanna, J. P. Pouget, R. Comes, A. F. Garito and A. J. Heeger, Phys. Rev. B **16**, 1468 (1977).

2. See for example: S. Etemad, Phys Rev. B **13** 2254 (1976).

3. P. Bak and V. J. Emery, Phys Rev. Lett. **36**, 751 (1976), T. D. Schultz and S. Etemad, Phys. Rev. B. **13**, 4928 (1976)

4. Y. Tomkiewicz, A. R. Taranko and J. B. Torrance, Phys. Rev. Lett. **36**, 751 (1976); Y. Tomkiewicz, A. R. Taranko and J. B. Torrance, Phys. Rev. B **15**, 1017 (1977).

5. E. F. Rybaczewski, A. F. Garito, A. J. Heeger and B. Silbernagel, Phys. Rev. B **14**, 2746 (1976).

6. Y. Tomkiewicz, R. A. Craven, T. D. Schultz, E. M. Engler and A. R. Taranko, Phys. Rev. B **15**, 3643 (1977).

7. S. Etemad, T. Penney, E. M. Engler, B. A. Scott and P. E. Seiden, Phys. Rev. Lett. **34**, 741 (1975).

8. C. Weyl, E. M. Engler, K. Bechgaard, G. Jehanno and S. Etemad Solid State Commun. **19**, 925 (1976).

9. T. D. Schultz, Solid State Commun. **22**, 229 (1977).

10. Y. Tomkiewicz, E. M. Engler and T. D. Schultz, Phys. Rev. Lett. **25**, 456 (1975).

11. G. W. Clark, Y. Tomkiewicz and E. M. Engler, unpublished.

12. D. Guidotti, P. M. Horn and E. M. Engler, Proceedings of the Conference on Organic Conductors and Semiconductors, Siofok, Hungary (1976).

13. See for example: M. H. Cohen, J. A. Hertz, P. M. Horn, and V. K. S. Shante, Int. J. Quantum Chem. Symp. **8**, 491 (1976); Bull. Am. Phys. Soc. **19**, 297 (1974).

14. R. A. Craven, Y. Tomkiewicz, E. M. Engler, and A. R. Taranko, Solid State Commun **23**, 429 (1977).

15. W. F. Cooper, N. C. Kenny, J. W. Edmonds, A. Nagel, F. Wudl and P. Coppens, Chem. Comm. 889 (1971); W. F. Cooper, J. W. Edmonds, F. Wudl and P. Coppens, Cryst. Struct. Comm. **3**, 23 (1974).

16. B. D. Silverman, private communication.

17. Overexposure leads to a formation of the stable phase of TTF bromide – TTF-Br$_{0.55}$.

NEW TYPE OF ORGANIC ALLOYS: MODIFICATION OF PHASE TRANSITION OF $(NPPy)(TCNQ)_n$

M. Murakami and S. Yoshimura

Matsushita Research Institute Tokyo, Inc.

Ikuta, Tama-ku, Kawasaki 214, Japan

INTRODUCTION

A wide range of electrical, magnetic and structural properties of the organic conductors based on 7,7,8,8,-tetracyanoquinodimethane, TCNQ, have been observed, and these have set up active arguments on their origin and how the physical properties can be designed. (1). We have suggested that substituted pyridinium and thiazolium salts of TCNQ form a unique class of materials serving as a good tool for studying the solid state chemical properties of organic conductors (2,3). In particular, N-n-propylpyridinium(TCNQ)$_n$ ((NPPY)(TCNQ)$_n$ or $\underline{A}$) shows a reversible first-order phase transition at 108°C accompamied with a sudden change in resistivity by a factor upto 20 (2). The high temperature phase is featured by both temperature-independent resistivity of 2 Ωcm and magnetic suscepti-bility of 6.6×10^{-4} emu/mol. The crystal structure change is supposed to dominate the phase transition, because its nature can be varied strikingly by modifying the structure of the cation molecule. Moreover, the reversible transition is found only for salts having specific cations; e.g., N-n-butylpyridinium(TCNQ)$_n$ ((NBPy)(TCNQ)$_n$ or $\underline{B}$), N-n-propylthiazolium(TCNQ)$_n$ ((NPTh)(TCNQ)$_n$ or $\underline{C}$) and N-n-butyl-thiazolium(TCNQ)$_n$ ((NBTh)(TCNQ)$_n$ or $\underline{D}$) (2,3). The butyl-substituted salts, $\underline{B}$ or $\underline{D}$, undergo a transition in which the resistivity increases towards high temperatures by a factor of ca. 2, and $\underline{C}$ exhibits a transition similar to $\underline{A}$ marked by a large temperature hysteresis.

This paper will report on the modification of the phase transi-tion of $(NPPy)(TCNQ)_n$, presenting a new family of organic alloys composed of above mentioned pyridinium and thiazolium TCNQ salts.

TABLE 1.   Characteristic data on the phase transitions in N-alkyl-
            substituted pyridinium and thiazolium TCNQ salts.

Compound	n	ρ_{RT} (Ωcm)	Temps. (°C) T_H	T_1	ΔT (°C)	ψ	$\Delta H*$ (cal/g)
(NPPy)(TCNQ)$_n$ (A)	1.9,2.5	3000	108	97	11	-0.62	4.5
(NBPy)(TCNQ)$_n$ (B)	1.8,1.9	700	134	111	23	+0.17	2.5
(NPTh)(TCNQ)$_n$ (C)	1.8	100	132	67	65	-1.49	5.4
(NBTh)(TCNQ)$_n$ (D)	1.8	40	98	78	20	+0.23	3.6

EXPERIMENTAL

Four basic components used for the preparation of alloys are
listed in Table 1, where ΔT designates the supercooling temperature,
$\Delta T = T_H - T_1$, and ψ the logarithm of the ratio of resistivity change
at T_H (on heating) defined as negative for the ordinary metal-
insulator transition.   Various approaches to obtaining alloys of
the formula (NPPy)$_{1-x}$M$_x$(TCNQ)$_n$ (M = NPTh, NBPy and NBTh) were survey-
ed (4), such as co-recrystallization in acetonitrile, rapid co-
precipitation in water, mixing and grinding in a mortar and hot
pressing.   Although all these methods were found essentially effective, we mainly employed the co-precipitation process on account of
its ease in preparing alloys with both the largest ψ and $\Delta H*$.   Since
the precipitation proceeded at a very high rate, leading to a high
yield of product (over 90%), we presumed that the composition of the
product was almost the same as that charged in the precipitating
solution.   The precipitated powders were finally heat-treated at
temperatures between 80 and 160°C, after which the formation of a
new modification (an alloy) was confirmed by means of X-ray and
differential scanning calorimetry measurements.

RESULTS

The resistivity-temperature, ρ-T, curves for A-C, A-B and A-D
are shown in Fig. 1.   The data were recorded by scanning the temperature at a constant rate (1°C/min).   The A-C system shows continuous shift of T_H from 108 to ca. 70°C along with decreased ψ and
temperature coefficient for ρ in the insulating phase.   The A-B and
A-D systems exhibit phase transitions accompanied with striking
increase in both ψ and ΔT.   Figure 2 summarizes the characteristic
temperatures obtained from the ρ-T curves as a function of the composition.   The heat associated with the transition, $\Delta H*$, was detected at T_H with all the compounds, and its dependence on the composi-

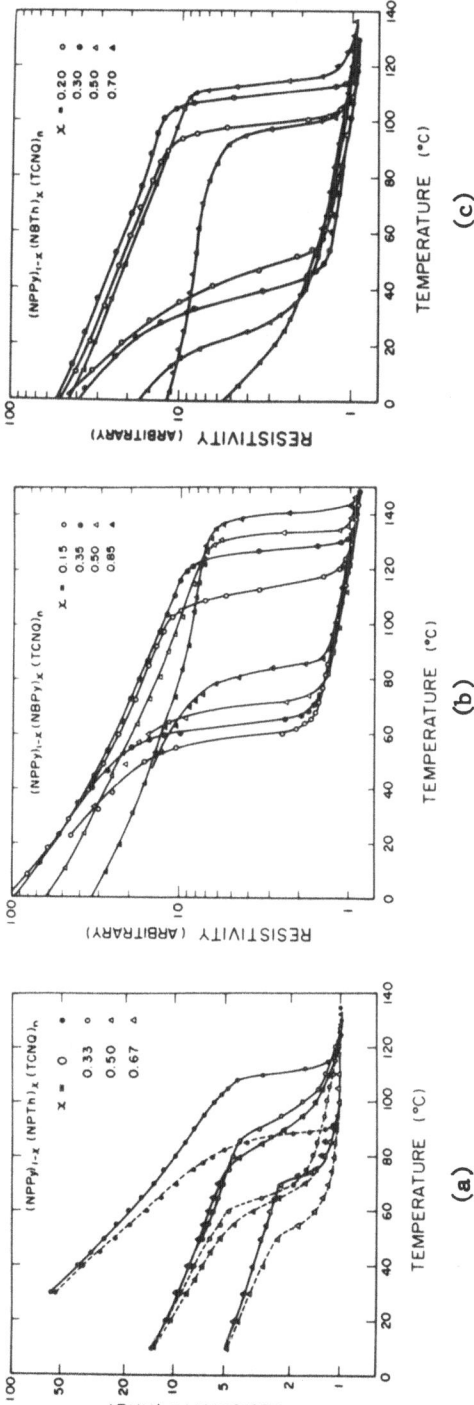

Fig. 1 Temperature dependence of relative resistivity for organic alloys; $\underline{A}$–$\underline{C}$ (a), $\underline{A}$–$\underline{B}$ (b) and $\underline{A}$–$\underline{D}$ (c).

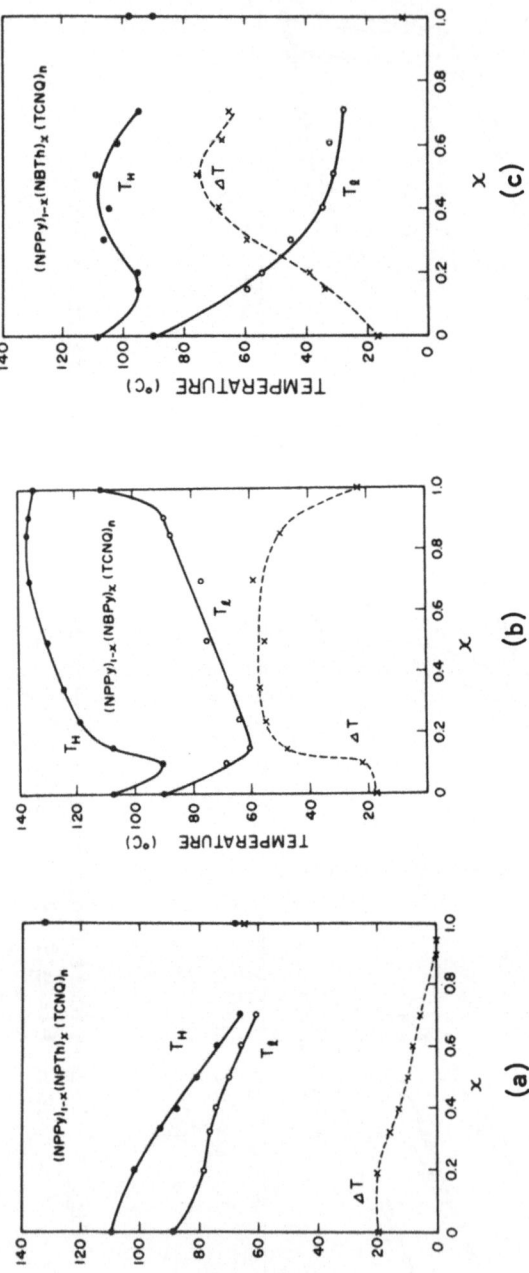

Fig. 2   Dependence of transition temperatures T_H, T_l and supercooling ΔT on alloy composition x.

tion is shown in Fig. 3.   For the A-C system, the changes in the
transition temperatures and ΔT are monotonous with the composition,
which is consistent with the fact that A and C are mutually isostructural (3).   In fact, the powder X-ray measurements suggested
that the A-C's were also isostructural with A or C, clearly indicating the solid solutions or alloys.   In the case of A-B or A-D,
however, a sharp phase change with composition was found to exist
at the concentration near 15% of B or D (see Figs. 2 and 3).   X-ray
patterns of A-B and A-D with compositions x > 0.2 showed no trace
of B and D components, respectively, and both suggested an isostruc-
ture with A or C.   A major differece in the nature of transitions
between A-B and A-D alloys was that the rate of the backward transition at T_1 was much lower for the latter; it took from 10 min to 10
hr to complete the transition, the time increasing with increasing
the concentration of D from x = 0.2 to 0.66.   Figure 4 shows X-ray
powder patterns of an A-D (1:1) alloy measured at 70°C after the
alloy had been stood at room temperature for a predetermined period.
The most interesting features of the results are that a strong peak
at θ = 6.23° gradually shifted to higher angles with time and that,
between 20 and 30° angles, several new peaks (e.g., 30.0°) developed
and original peaks (e.g., 27.3°) diminished.   The former variation
implies a homogeneous transformation, while the latter suggests a
heterogeneous one with the formation of domains of a new phase (5).

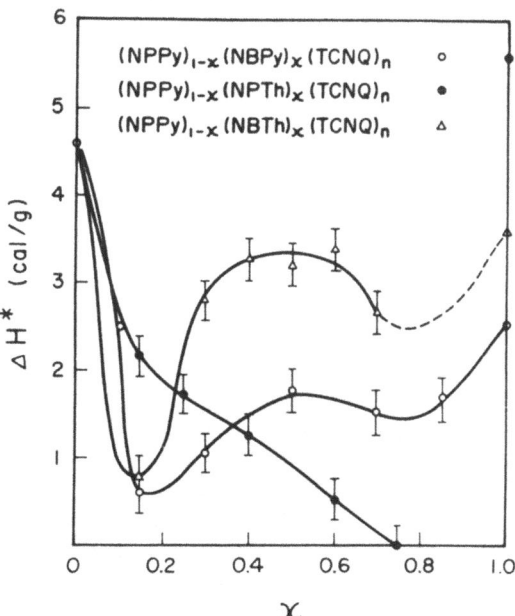

Fig. 3.   Heat associated with the phase transition of organic alloys
measured at T_H on heating.

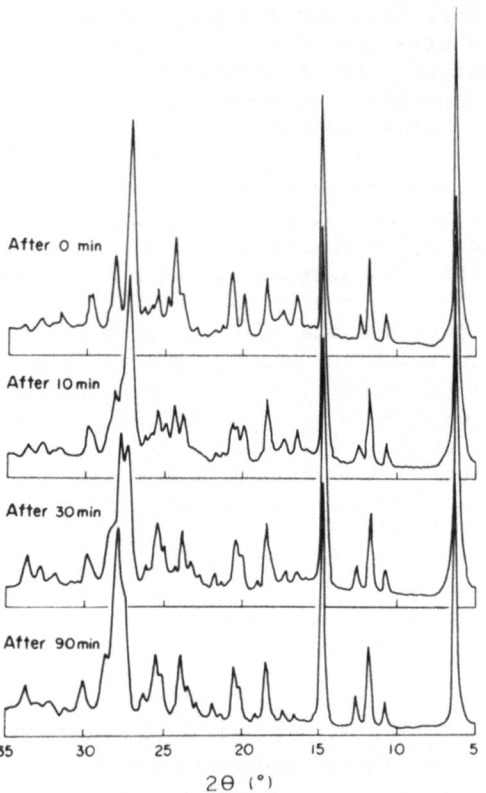

After 0 min

After 10 min

After 30 min

After 90 min

$$2\theta \ (°)$$

Fig. 4   Powder X-ray diagrams of $(NNPy)_{0.5}(NBTh)_{0.5}(TCNQ)_n$ during
the backward transition.   (Measured at 70°C after being
left at room temperature for described periods.)

DISCUSSIONS

It was found in this work that the organic alloys
$(NPPy)_{1-x}M_x(TCNQ)_n$ (M = NPTh, NBPy and NBTh) can be included in the
isostructural family of organic conductors, $(NPPy)(TCNQ)_n$ and
$(NPTh)(TCNQ)_n$, which undergo a sharp MI transition with a temperatur
hysteresis. The supercooling temperature, ΔT, qualitatively
corresponded to the heat of transition, ΔH^*, for all the alloys
(see Figs. 2 and 3).   If we assume that the phase transition on
cooling is driven by a chemical potential change retarded by the
formation of phase boundaries, ΔT is obtained from the following
equation (3):

$$3k(\Delta H)^2(\Delta T)^3 - 3kT_H(\Delta H)^2(\Delta T)^2 + 16\nu^2\sigma^3 T_H^2 = 0$$

where k is the Boltzmann constant, ΔH the molar heat of transition at T_H, ν the molar volume and σ the interfacial energy. This equation has a solution in the range $0 < \Delta T < 2T_H/3$ and ΔT is roughly proportional to $\sigma^{3/2} T_H^{1/2} (\Delta H)^{-1}$ for small ΔT. Comparing the experimental data with the theory, we conclude that the difference in ΔT between A-B and A-D can be viewed as arising from that in the interfacial energy.

The mode of the phase transition in the A-D alloy was speculated from the X-ray results (Fig. 4). The peak at 27.3° is supposed, without any knowledge of the crystal structure, to be that coming from the planes intersecting the linear chains of TCNQ, with a spacing nearly equal to the intermolecular distance between neighboring TCNQ molecules in the stack (5). So that the variation of this plane during the phase transition is indicative of a domain wall motion along the direction of TCNQ stacking. The continuous shift or no change in the low angular peaks, which can equally be assumed to be arising from the planes parallel to the stacking axes, implys that there may be no domain motion to the transverse direction. In addition, there may be some matching of the crystal structure between the parent and product phases resulting a semi-coherent interface (6) with respect to the planes parallel to the linear TCNQ chains.

The supposed difference in σ between A-B and A-D is accounted for by the effect of the sulphur-containing cation, as has been observed in C. So that the sulphur atom may work to retard the one-dimensional propagation of new phase, presumably by S-N (cation-TCNQ) interactions.

REFERENCES

1. (a) J. H. Perlstein, Angew. Chem. 16, 519 (L977); (b) J. J. Torrance, to be published in Annal N.Y.A.S. (1978); (c) J. S. Miller, to be published in Annal N.Y.A.S. (1978).
2. (a) M. Murakami and S. Yoshimura, Abstr. 7th Mol. Cryst. Symp. Nikko, Japan, Sept. 8 - 12, 1975, p.193; (b) M. Murakami and S. Yoshimura, Chem. Lett., 929 (1977).
3. S. Yoshimura and M. Murakami, to be published in Annal N.Y.A.S. (1978); M. Murakami and S. Yoshimura, to be published.
4. W. A. Barlow, G. R. Davis, E. P. Goodings, R. C. Hand, G. Owen, and M. Rhodes, Mol. Cryst. Liq. Cryst. 33, 132 (1976).
5. M. Murakami and S. Yoshimura, J. Phys. Soc. Japan 38, 488 (1975).
6. V. Raghavan and M. Cohen, in Treatise on Solid State Chemistry, vol 5, ed. N.B. Hannay (Plenum Press, N.Y. 1975) p. 67.

ELECTRONIC PROPERTIES OF A NEW RADICAL-CATION SALT, TETRA-
SELENOTETRACENE-IODIDE: EVIDENCE FOR QUASI METALLIC BEHAVIOR
AT VERY LOW TEMPERATURES

P. Delhaes, C. Coulon, J.P. Manceau, S. Flandrois

Centre de Recherche Paul Pascal CNRS
Domaine Universitaire, 33405 TALENCE (France)

B. Hilti, C. W. Mayer

Zentrale Forschungslaboratorien CIBA GEIGI A. G.
CH. 4002 BASEL (C. H.)

INTRODUCTION

The investigation of nearly one dimensional metals has reached
these last years a quite interesting stage of development. Among
the new materials which have been discovered, the organic charge
transfer complexes (CTC) belonging to the TTF (tetrathiofulvale-
nium)-TCNQ (tetracyanoquinodimethane) series present an outstanding feature. However, one of them, only, does not present a
metal-insulator transition due to an electronic PEIERLS distortion and shows therefore an appreciable electrical conductivity
at very low temperatures : this is the hexamethylene - tetrase-
lenofulvalenium HMTSeF-TCNQ (1). Furthermore, under pressure a
semi-metallic behavior has been found because of a large increase of the electrical conductivity along the cations and anions
chains (2).

Actually, in spite of ingenious chemistry synthesis, it seems
that no large improvements can be done on this series of compounds. The interest spreads out therefore to other materials.
Among them, the C.T.C. and radical-ions made with tetrathiotetra-

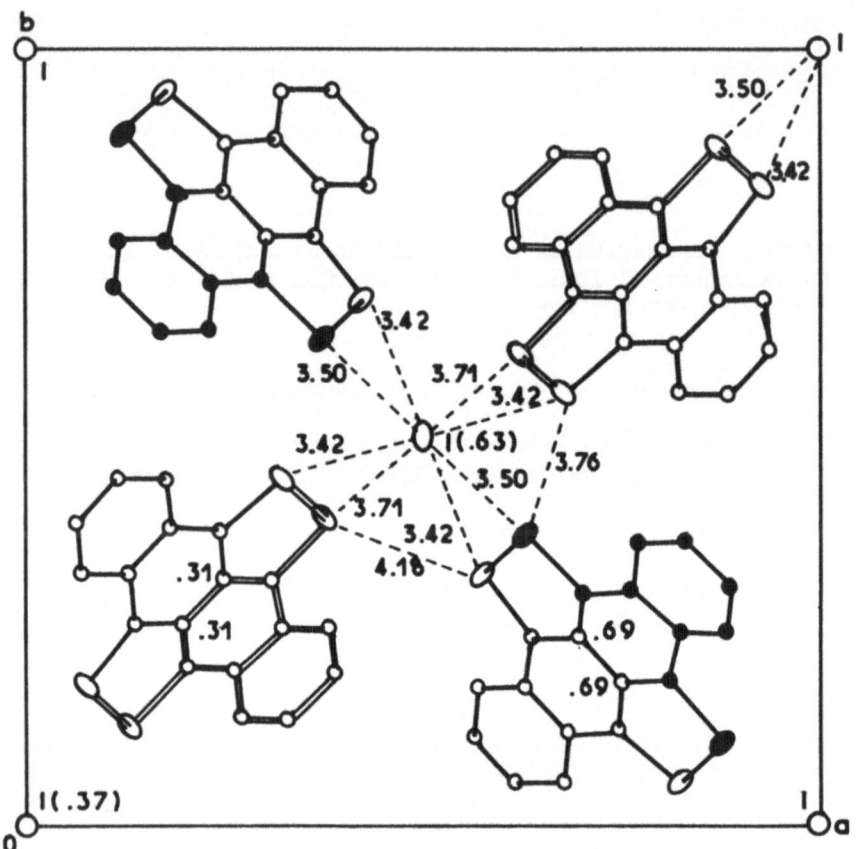

Figure 1 : Crystal structure of TSeT$_2$ - I (4) : Projection on
the a-b plane showing the different Selenium-Iodine interatomic
distances.

cene (TTT) and its selenium derivative (TSeT) appear noteworthy.
In the course of our investigations on them, we have synthetized
and studied a new radical cation with tetraselenotetracene and
iodide. We report here the first results which show that this
compound behaves almost like a metal at very low temperatures,
down to 0.1 K.

EXPERIMENTAL

In continuation of a previous work on TTT$_2$-I$_3$ (3) the new chalco-
genated metallic compound TSeT$_2$-I has been prepared by cosublima-
tion in vapour phase (4). Single crystals in the form of needles
in dimensions of about 5-10 mm x 0.02 mm x 0.2 mm have been obtained. These crystals with a silvery appearance exhibit a strong

electrical conductivity between 300 and 10 K. The X-rays crystal
structure has been determined at room temperature. This material
crystallizes in the orthorhombic space group $P2_4 2_4 2$; it consists
of   one set of crystallographically equivalent chains with an
interplanar stacking distance equal to 3.40 A and a good overlap
between neighbouring molecules (figure 1). It must be noticed
that the analogue compound $TSeT_2$-Cl reported by ZOLOTUKHIN et al.
(5) does not seem to be isomorphous in spite of obvious similarities.

2.1 – *Transport properties*

On a given batch good single crystals have been selected ; they
have been mounted on four suspended goldwires on which the electrical contacts are made with platinium paste. After a preliminary investigation of d.c. electrical conductivity at liquid nitrogen temperature the best samples have been chosen on the basis of
the largest increase of conductivity between 295 and 78 K (4).

The absolute values at room temperature range between 10^{+3} and
3.10^{+3} $\sim^{-1}$ cm $^{-1}$; this relatively wide spread of values appear to
be due to irregularities in the growth pattern of single crystals
which exhibit a tendency to form fibers.

On three single crystals, the thermal variation of d.c. conductivity has been undertaken down to 1.5 K for two of them and below
1 K for the last one with an He^3 - He^4 dilution refrigerator : these
results are presented on figure 2.

Every sample presents a metallic behavior between 300 and 50 K ;
around this temperature, a conductivity maximum occurs ($\frac{\sigma_{50K}}{\sigma_{R.T}}$) =
2.7 - 3.0) with at lower temperature a smooth decrease.
However, below 4 K a plateau of conductivity is detected, down to
0.1 K for the sample 7-5, which is at a larger value than the
room temperature figure

On single crystals 7-11 and 7-13, we have tried without any
success to detect a magnetoresistance effect. At 1.5 K, a magnetic field strength of 48 kgauss has been applied in order to detect a transverse magnetoresistance. Its absolute value must be
lower than 2 % to be undetectable with our experimental set up.

We have also measured the anisotropy in the conductivity at room
temperature by the MONTGOMERY's technique (6) and we have found
the $\sigma\| /\sigma\perp \simeq 130$ where $\sigma\|$ is the conductivity along the stacking
axis and $\sigma\perp$ is a transverse component.

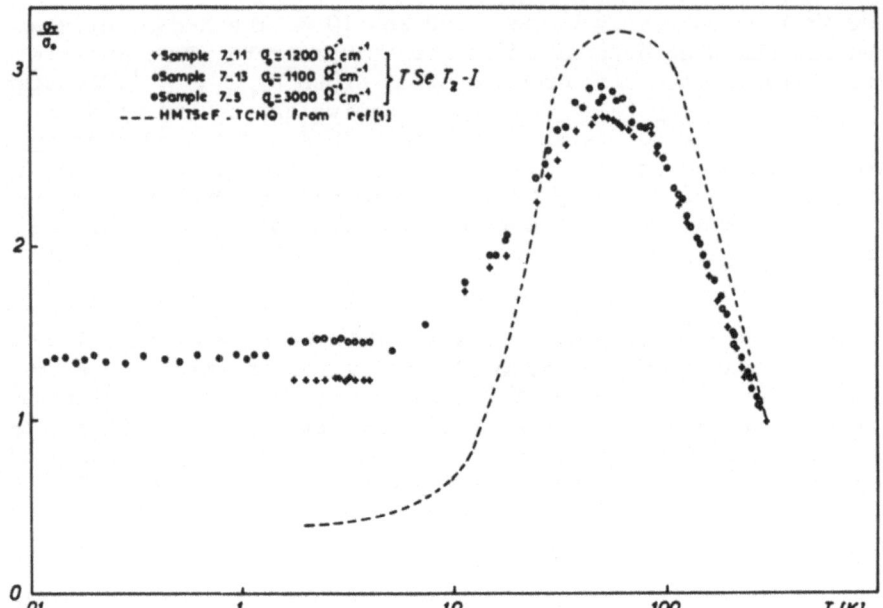

<u>Figure 2</u> : Normalized conductivity thermal variations observed
on three different single crystals of TSeT$_2$ - I. For comparison
the termal variation observed on HMTSeF-TCNQ by BLOCH et
al. (1) has been reported.

2.2 - *Magnetism*

On single crystals issued from the same batch, we have looked
for an EPR signal between 10 and 300 K with a standard X band
spectrometer. As for HMTSeF-TCNO (1) (2), the linewidth is so
large that any resonance cannot be detected.

In order to get some information, we have measured the static
susceptibility with a FARADAY's balance. Because a larger amount
is necessary (60-100 mg of compound) this experiment has been
done on another batch which furnishes the necessary quantity in
a polycrystalline form made under similar experimental conditions.

The constant diamagnetic contribution which is large has been
substracted following PASCAL's additivity rule ($\chi_d = -5.30$ x
10^{-4} e.m.u. CGS for a TSeT$_2$ - I mole) and the thermal variation
of paramagnetic susceptibility is presented on figure 3. At low
temperatures, the observed paramagnetism can be fitted with a

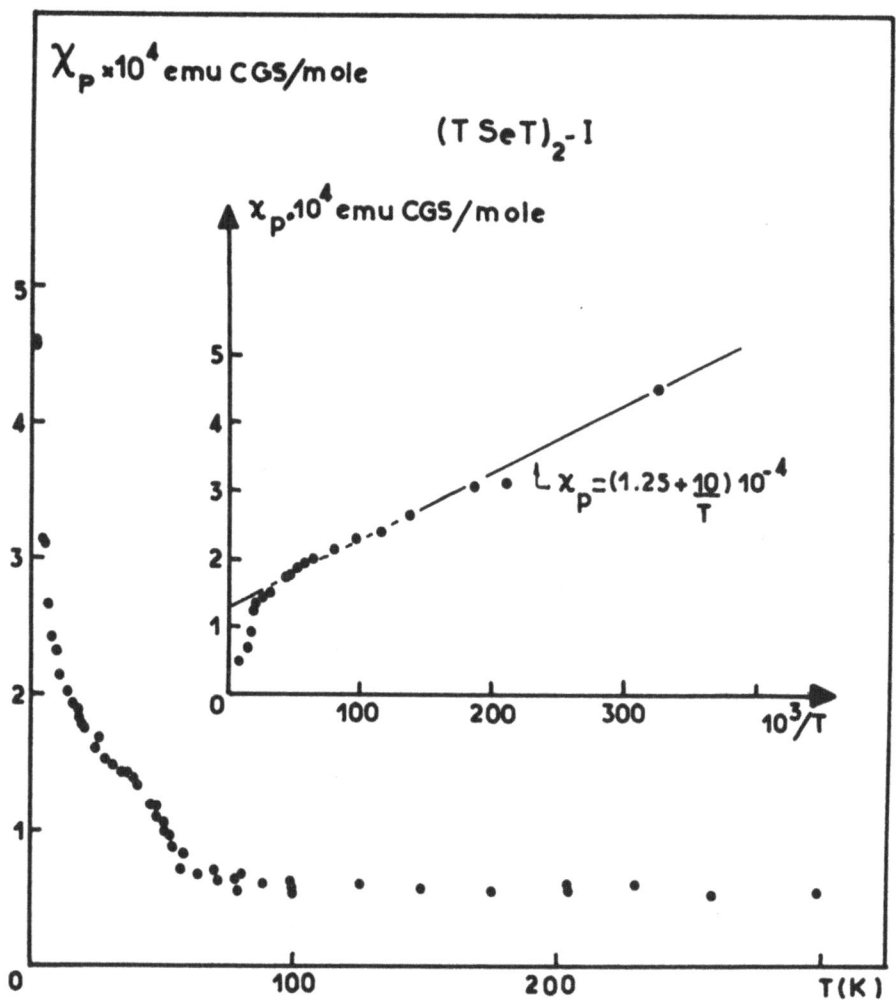

Figure 3 : Thermal variation of the paramagnetic susceptibility χ_p for a given batch of TSeT$_2$ - I ; the inset gives a plot of χ_p versus T^{-1} with the equation which fits this low temperature behavior.

CURIE law and a constant term (see insert of the figure 3). From the CURIE constant, we can see that the number of localized spin centers is less than 0.3 %, assuming g = 2 and S = 1/2. But the constant term appears to be larger, at very low temperature, than around room temperature with a small change near the temperature where the maximum of conductivity is. This result,

differs from that of ZOLOTUKHIN et al., and (5) does not allow us to
show the occurrence of a LANDAU diamagnetism.

DISCUSSION

The basic assumption is a full charge transfer between iodine and
TSeT which gives rise to a conduction band filled by three quar-
ters. We are therefore in front of one kind of efficient stacks
where the transport properties are governed by holes whereas the
counter ions (I^-) are diamagnetic.

3.1 - *Analysis of the experimental results*

At high temperatures (T > 10 K), we observe a conductivity maximum which is characteristic of 1 D electronic systems. To compare
with previous results, we have plotted our data in the form suggested by GROFF et al. (7). The basic assumption is that the resistivity can be written in the form of a MATTHIESSEN's rule :

$$\rho(T) = \rho_o + bT^\nu$$

where ρ_o is a constant which is a sample dependent quantity and
bT^ν furnishes the characteristic temperature variation for the
considered material. For the best sample (7-5) we have found :

$$\rho(T) = 2.52 \ 10^{-9} \ T^2 + 1.06 \ 10^{-4} \ \Omega\text{-cm}$$

A square law dependence is followed as for TTT_2-I_3 (8). Actually,
in several C.T.C. a similar power law is found but as proposed
by JEROME (9) the discrepancy from a linear temperature dependence may be due to a large thermal dilatation coefficient along the
stacking axis whatever the scattering mechanism is.

If we calculate the mean free path of the carriers from the 1 D
tight binding approximation we find around 50 Å at room temperature. Furthermore from the constant PAULI paramagnetism observed
at high temperatures the overlapping integral $t_{||}$ can be estimated
to equal 3100 K. These results indicate that the bandwidth is
quite large, around 1 eV, with a degenerate electronic gas without
electron-electron correlations. These results confirm the one
dimensional metallic character of this material in the high temperature range. Below 10 K a temperature independent conductivity
is detected which shows that a PEIERLS distortion does not occur
in this compound because of large enough 2D or 3D electronic
interactions. Indeed, around 50 K a small jump of paramagnetism is
observed (figure 3) which seems to indicate the occurrence of a
phase transition to another metallic state.

At low temperature, according to HOROVITZ, GUFTREUND and WEGER

(10), the ratio $(t\| /t\bot)$ $(t\bot$ is the transverse hopping integral) must be larger than a critical value which depends upon the electron-phonon coupling inside a chain. Besides, we must have a coherent transverse conductivity between chains as indicated by the weak conductivity anisotropy observed at room temperature. Thermal variations of the d.c. conductivity by using the MONTGOMERY's method are in progress to check this point. At the present step it is not possible to develop more quantitative arguments.

3.2. - *Comparison with other 3 D anisotropic materials*
- TTT_2-I_3 : this radical cation is not isomorphous with $TSeT_2-I$ Triodides (I_3) are present as non-stoechiometric counterions ; they introduce a disorder which should be able to prevent a PEIERLS distortion as indicated from specific heat measurements. However, transport properties, i.e. resistivity and strong positive magnetoresistance, show a phonon assisted hopping process which is characteristic of a disordered semi-conductor (11).

- HMTSeF-TCNQ : magnetoresistance and Hall effect measurements have demonstrated that this C.T.C. behaves at low temperature and under pressure as an anisotropic yet 3 D semimetal (12). For the present compound, we are just in presence of a one-carrier conduction mechanism which does not give rise to large magnetoresistance and Hall coefficient probably. If we compare the magnetic susceptibility studies we don't observe any predominant LANDAU diamagnetism at low temperature. This comparison indicates that the effective mass might be larger than in HMTSeF-TCNQ ($m^* \simeq 2m_e$) (12).

CONCLUSION
This material presents a quasi-metallic character at ambient pressure between 0.1 and 300 K. The interchain coupling is strong enough to prevent the PEIERLS distortion usually observed on linear chain systems.

More detailed physical investigations have to be done in connection with synthesis improvements concerning the knowledge of iodine stoichiometry and the quality of single crystals. Similarities with the problem encountered on the inorganic polymer $(SN)_x$ (13) seems to indicate a possible progress in this research field.

REFERENCES

(1)   A.N. Bloch, D.O. Cowan, K. Bechgaard, R.E. Pyle, R.H. Banks
      and T.O. Poehler, Physical Review Letters, $\underline{34}$ (25), 1561
      (1975).

(2)   J.R. Cooper, M. Weger, D. Jerome, G. Lefur, K. Bechgaard,
      A.N. Bloch and D.O. Cowan, Solid State Comm., $\underline{19}$, 1209 (1976).

(3)   B. Hilti and C.W. Mayer, Helvetica Chimica Acta, $\underline{61}$, 501
      (1978).

(4)   B. Hilti, C.W. Mayer and G. Rihs, Helvetica Chimica Acta, $\underline{61}$,
      1462 (1978).

(5)   S.P. Zolotukhin, V.F. Kaminskii, A.I. Kotow, R.G. Lyubovskii,
      M.L. Khidekel, R.P. Shibaeva, I.F. Shchgolev and E.B. Yagub-
      skii, Pisma Zh. Elisp. Theor. Fiz., $\underline{25}$ (10), 480 (1977).

(6)   H.C. Montgomery, J. Appl. Phys., $\underline{42}$, 7 (1971).

(7)   R.P. Groff, A. Suna and R.E. Merrifield, Phys. Rev. Letters,
      $\underline{33}$, 418 (1974).

(8)   I.C. Isett and E.A. Perez-Albuerne, Solid State Comm., $\underline{21}$,
      433 (1977).

(9)   D. Jerome, J. de Physique Lettres, $\underline{38}$, L 489 (1977).

(10) B. Horowitz, H. Guftreund and M. Weger, Phys. Rev. B, $\underline{12}$,
      3174 (1975).

(11) P. Delhaes, J.P. Manceau, C. Coulon, S. Flandrois, B. Hilti
      and C.W. Mayer, (Conference on One-Dimensional Conductors,
      Dubrovnik, September 1978).

(12) D. Jerome and M. Weger, Chemistry and Physics of one-dimen-
      sional metals, (Ed. H.J. Keller) NATO Advanced Study Insti-
      tutes series B, $\underline{25}$, 341 (1977).

(13) R.L. Greene and G.B. Street, Chemistry and Physics of one-
      dimensional metals (Ed. M.J. Keller) NATO Advanced Study
      Institutes Series B, $\underline{25}$, 167 (1977).

A SYSTEMATIC STUDY OF AN ISOMORPHOUS SERIES OF ORGANIC SOLID STATE CONDUCTORS BASED ON TCNQ AND IODINE

S. Flandrois, C. Coulon, P. Delhaès,

Centre de Recherche Paul Pascal, CNRS,
Domaine Universitaire, 33405 Talence (France)

P. Dupuis

Laboratoire de Chimie Physique Macromoléculaire,
1, rue Grandville, 54042 Nancy (France)

I - INTRODUCTION

The research of organic solid state conductors, as for any type of materials, can be made following two different approaches. The first one is the extensive study of all the physical and structural properties of one given compound exhibiting peculiarly interesting features : the best example of this kind of approach for recent years is certainly TTF-TCNQ (tetrathiofulvalenium-tetracyanoquinodimethanide). The second approach consists of systematically studying a series of compounds which obey some requirements such as isomorphism and isostoichiometry and allow one to obtain relationships between structure and physical properties.

Unfortunately, due to the character up to now fortuitous rather than predictable of the synthesis of organic conductors, comparative studies are scarce. Moreover they often are reduced to a comparison of only two members, for example TTF-TCNQ and TMTTF-TCNQ or TMSeF-TCNQ and TMSeF-DMTCNQ (1) ; it was pointed out that the methyl substitution causes a lowering of the transition temperature : from 53 K to 35 K in the first case and from 57 K to 42 K in the second one(1).Another attempt of systematic study to be mentionned is that concerning the salts $TTF(SCN)_{0.54}$, $TTF(SeCN)_{0.54}$ and $TTF-I_{0.71}$ (2). These salts are very nearly isomorphous and there is a direct relation between the transition temperature and the interplanar distance between TTF molecules within a stack. Other systematic studies have been recently undertaken on cation-radical salts, for example $TMTTF_2-X$ with $X = BF_4$, ClO_4, PF_6, SCN

and Br (3). However to our knowledge such a study has not appeared
with anion-radical salts and especially with TCNQ salts containing
diamagnetic cations. It is the purpose of this paper to present
a new series of TCNQ salts which seem the most convenient for the
research of general trends in organic conductors.

II - THE SERIES

During our study of TCNQ salts with ammonium cations, we
prepared a novel three-component material containing parallel
chains of both TCNQ and I_3^- (4,5). This salt, the cation of which
is trimethylammonium (TMA$^+$), exhibits a metal-like conductivity
at high temperature (6) and around 150 K undergoes a transition
to an insulating phase. We at first thought that it was an interesting but exceptional case ; however we recently found (7) this
is the first example of a new class of materials. All of them
contain regular TCNQ stacks separated from one another by I_3^- chains
on one side and the cations on the other side. The stoichiometry
(three TCNQ molecules for one I_3^- anion and three monopositive cations) is such as each TCNQ bears 2/3 of one electron, so that the
electronic band is one third filled.

To our knowledge they are the only metal-like compounds
with a 1/3 filled band. But their main interest lies on the possibility of a systematic study as a function of the counterion. The
members of this series are isostoichiometric, isomorphous, at
least for many of them, and the only structural variation is the
size of the cation. Finally they consist of only one stack efficient for electrical and magnetic properties and in that sense
they are simpler than charge transfer complexes such as TTF-TCNQ
with two types of stacks, the respective contribution of which is
difficult to ascertain.

III - SYNTHESIS AND MAIN STRUCTURAL AND PHYSICAL FEATURES

TCNQ anion-radical salts are generally obtained by TCNQ oxidation of iodide salts following the reaction :

$$3 M^+I^- + 2x \text{ TCNQ} \rightarrow 2 M^+(\text{TCNQ})_x^- + M^+I_3^-$$

where M^+ is an organic or inorganic cation and x most often is
equal to 1, 1.5 or 2 (8). In the case of trimethylammonium iodide
(TMA$^+$I$^-$) this oxidation leads to well-developed green crystals of
(TMA)$^+$(TCNQ)$^{2/3-}$(I$_3^-$)$_{1/3}$ instead of (TMA)$^+$(TCNQ)$_x^-$.

Numerous attempts with various substituted ammonium cations
have shown us that the occurence in the crystals of iodine together
with TCNQ is related to the structure of the cation which must be
of the type : R - Me$_2$N$^+$H. However although this structure is a
necessary condition, it is not a sufficient one, and the reason
why I_3^- ions are retained is as yet obscure. For example we succeeded

to prepare the salts of the cations shown in table I ; in particular in the case of dications of the type $HN^+Me_2 - (CH_2)_n - Me_2N^+H$
with $n = 2$, 6 and 9 iodine is retained, while with $n = 3$, 4 5, 7,
8 the reaction leads to blue crystals of formula $(M^{2+})(TCNQ)_4^{2-}$
which are semiconducting ($\sigma_{RT} \sim 10^{-3} \ \Omega^{-1} \ cm^{-1}$) and probably have
the diadic crystal structure of parent compounds :

$$\left[N^+Me_3 - (CH_2)_n - Me_3N^+\right]\left[TCNQ\right]_4^{2-} \quad (9)$$

The crystal structures at room temperature were determined
for two compounds with trimethylammonium (TMA) and tetramethyl-
hexamethylenediammonium (TMHDA) cations (4)(5)(10). Both compounds
have a similar structural organization, the main features of which
are the following (fig. 1) :

δ regular stacks of TCNQ molecules, equally distant from
3.23 A and showing the "ring-double bond" overlap characteristic
of conducting TCNQ salts,

- linear triiodide anions chains which alternate with TCNQ
stacks in the a-direction,

- alternative hydrogen bonds (N-H---N) between cations and
TCNQ.

Both d.c. electrical conductivity and magnetic susceptibility
measurements gave evidence for a phase transition at low temperature, which could not be detected by X-Ray diffraction investigations. For the entire series the room temperature values of the
conductivity parallel to the needle axis range from 10 to 60 Ω^{-1}
cm^{-1} depending on crystal quality. As the temperature is decreased,

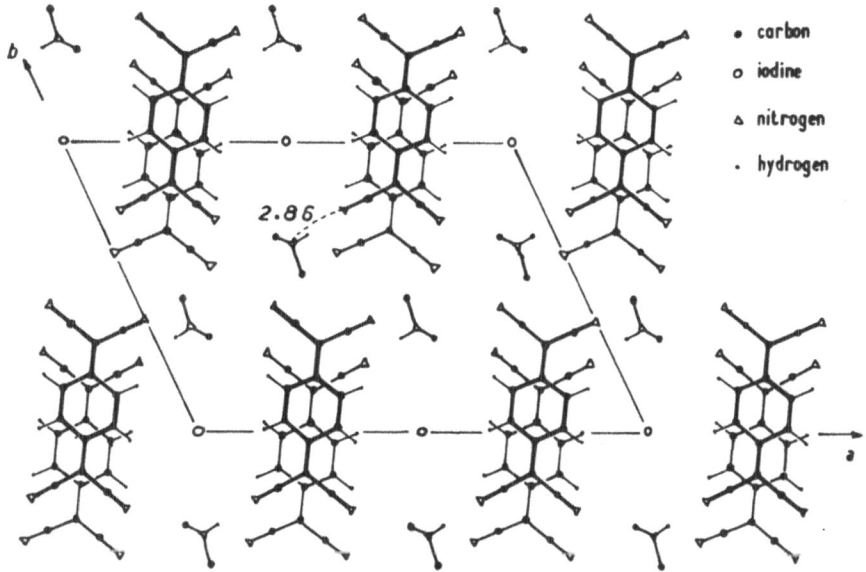

Fig. 1 : CRYSTAL STRUCTURE OF TMA SALT

the conductivity increases, reaches a rounded maximum ($\sigma_{max}/\sigma_{R,T} \sim$ 1.1-1.5) and then decreases continuously with a slight accident at a lower temperature clearly visible on a Log σ vs. 1/T plot.

The room temperature values of paramagnetic susceptibility are given in table I. For all the compounds the paramagnetism is almost constant at high temperatures , but again at low temperature the transition is evidenced by a sharp paramagnetic decrease, characteristic of an activated process. More details on these electrical and magnetic behaviours can be found in a recent paper relative to the TMHDA salt (11).

IV - DISCUSSION

Nature of the transition

For all the compounds of the series single crystal conductivities and magnetic susceptibilities measurements indicate the occurence of a transition. These transitions were reversible without apparent hysteresis. X-Ray diffraction experiments only showed a regular lattice contraction (10). A specific heat anomaly was detected on TMA salt at the transition temperature (5) but the phase transition order is not quite clear. According to an argument given by Mc MILLAN (12), using a LANDAU model, when the longitudinal distortion associated with 2 k_F is equal to the third of the associated reciprocal lattice vector, the phase transition must be of first order. At the transition temperature a three dimensional ordered state commensurate with the original lattice occurs and is responsible for such a first order transition.

In the high temperature phase the magnetic susceptibility is almost temperature independent with a rather high value, and the conductivity has a metallic behaviour , although a broad maximum is exhibited. After the transition the susceptibility becomes activated while the resistivity increases sharply . Thus the transition is a kind of metal-insulator transition.

However it must be pointed out that according to EPSTEIN (13) the transition observed in the TMA salt would be a semiconductor-to-semiconductor transition ; the apparent "metal-like" conductivity would result from a strongly temperature dependent mobility and an activated carrier concentration. Actually, with a number of carriers increasing sharply with temperature, the paramagnetic susceptibility should exhibit a similar variation. The quasi temperature independence of paramagnetism seems thus to be in contradiction with the EPSTEIN's assumption.

The question now arises : is the transition an electronic PEIERLS transition ? The best prove would be the crystallographic evidence of a TCNQ surstructure at low temperature. This surstructure would be reflected by the appearance of new diffuse lines on X-Ray diagrams. However the possible diffusion by TCNQ molecules could be masked by that of triiodide ions. Neutron diffusion experiments on deuterated samples appear therefore necessary. In the lack of such data we can only speculate from the behaviour of physical properties. The examination of the magnetic and electrical behaviour of compounds exhibiting a PEIERLS transition shows that our compounds behave similarly. We think therefore that the transitions observed are PEIERLS transitions.

Effect of counterion size : interchain interactions

These new compounds form an unique series for the study of interchain interactions. Their main advantages are based on their great number and the ability of varying only one parameter (e.g. the distance between TCNQ stacks) without modifying the others, like the distance and overlap between TCNQ molecules. Thus the crystal structure determinations of TMA and TMHDA salt show that only the distance between TCNQ stacks is modified in the direction of the long axis of TCNQ molecules.

Furthermore all members of the series possess the same electrical and magnetic features. The room temperature values of the conductivity have the same order of magnitude and, as seen in Table I, the magnetic susceptibility is generally between 2 and 3×10^{-4} e.m.u./mole. The only difference in behaviour is the position of the transition temperature.

From Table I it appears clearly that the increase of the cation size has the general effect of lowering the transition temperature. For the three salts where crystal data are up to now available, the larger the unit cell volume the lower is the transition temperature (table I). This result is in qualitative agreement with the predictions of HOROVITZ et al. (14) : the increase of one-dimensionality causes a lowering of the transition temperature. More X-Ray crystal studies are in progress which should lead to quantitative evaluations of interchain interactions effects.

V - CONCLUSION

Although this new isomorphous series of anion-radical salts appears very promising for the study of interchain couplings,some problems have yet to be resolved, in particular the disorder effects. X-Ray diagrams show that at room temperature either 2-D or 3-D ordering between iodine chains may exist in different compounds (10) or different batches of a same compound (TMA salt (15)). However the transition temperature is not sensitive to this disorder.

Another point to be cleared up is the influence of the hydrogen bond between each cation and each TCNQ. It is noteworthy that the replacement of the H atom by deuterium in the TMA salt (Table I) induces an increase of the transition temperature as large as 9 K.

TABLE I

Room temperature paramagnetic susceptibility and transition temperature for the new series $(Cation^+)$ $(TCNQ)^{2/3-}(I_3^-)_{1/3}$.

Cation		$X_p \cdot 10^4$ emu/mole of TCNQ site	Unit cell volume (Å^3)	T_{trans} (K)
(NDMA)	Neo-Menthyl-N^+Me_2H	0.99	1290	197
(TMA-d)	$CH_3-N^+Me_2D$	2.60		159
(TMA)	$CH_3-N^+Me_2H$	2.42	1650	150
(EDMA)	$CH_3-CH_2-N^+Me_2H$	2.41		143
(IPrDMA)	$\begin{matrix}CH_3 \\ CH_3\end{matrix}CH_2-N^+Me_2H$	3.09		139
(IPentDMA)	$\begin{matrix}CH_3 \\ CH_3\end{matrix}CH_2-CH_2-CH-N^+Me_2H$	5.04		136
(CMDMA)	$C_6H_{11}-CH_2-N^+Me_2H$	2.65		135
(DDMA)	$CH_3-(CH_2)_{11}-N^+Me_2H$	2.63		125
(TMDDA)	$HMe_2N^+-(CH_2)_2-N^+Me_2H$	2.82		156
(TMHDA)	$HMe_2N^+-(CH_2)_6-N^+Me_2H$	2.28	1780	120
(TMNDA)	$HMe_2N^+-(CH_2)_9-N^+Me_2H$	2.14		96

REFERENCES

(1) A.N. Bloch, Lecture Notes in Physics, vol. 65, p. 317, 1977. Y. Tomkiewicz, J.R. Andersen and A.R. Taranko, Phys. Rev. B 17, 1579, 1978.

(2) F. Wudl, D.E. Schafer, W.M. Walsh, L.W. Rupp, F.J. Disalvo, J.V. Waszczak, M.L. Kaplan and G.A. Thomas, J. Chem. Phys., 66, 377, 1977.

(3) P. Delhaès, C. Coulon, J. Amiell, S. Flandrois, E. Torreilles, J.M. Fabre and L. Giral, Molecular Cryst. Liq. Cryst., in press.

(4) A. Cougrand, S. Flandrois, P. Delhaès, P. Dupuis, D. Chasseau, J. Gaultier and J.L. Miane, Mol. Cryst. Liq. Cryst., 32, 165, 1976.

(5) P. Delhaès, A. Cougrand, S. Flandrois, D. Chasseau, J. Gaultier, C. Hauw and P. Dupuis, Lecture Notes in Physics, Springer Verlag, vol. 65, 1977, p. 493.

(6) M.A. Abkowitz, A.J. Epstein, C.H. Griffiths, J.S. Miller and M.L. Slade, J. Amer. Chem. Soc., 99, 5304, 1977.

(7) P. Dupuis, S. Flandrois, P. Delhaès and C. Coulon, J. Chem. Soc. Chem. Comm., 337, 1978.

(8) L.R. Melby, R.J. Harder, W.R. Hertler, W. Mahler, R.E. Benson and W.E. Mochel, J. Amer. Chem. Soc., 84, 3374, 1962.

(9) S. Flandrois, D. Chasseau, P. Delhaès, J. Gaultier, J. Amiell and C. Hauw, To be published.

(10) M. Rovira, Thesis, University of Bordeaux I, 1978. A. Filhol, M. Rovira, C. Hauw, J. Gaultier, D. Chasseau and P. Dupuis, Acta Cryst. B, in press.

(11) S. Flandrois, C. Coulon, P. Delhaès, J. Amiell and P. Dupuis, Proceedings of the Dubrovnik Conference, 1978, to be published.

(12) W.L. Mc Millan, Phys. Rev. B 14, 1496, 1976.

(13) A.J. Epstein, These proceedings.

(14) B. Horovitz, H. Gutfreund and M. Weger, Phys. Rev. B 12, 3174, 1975.

(15) A.J. Epstein, Private Communication.

CONDUCTING CHARGE-TRANSFER SALTS OF PHENAZINES AND QUINOLINES

H.J. Keller, W. Moroni, D. Nöthe, V. Seifried
and M. Werner

Anorganisch-Chemisches Institut der Universität
Heidelberg, Im Neuenheimer Feld 270
D-6900 Heidelberg 1/GFR

From the original work on TCNQ compounds (1) and
from several more recent publications dealing with this
matter (2,3) one has to conclude, that most crystalline
1:1 charge transfer adducts of planar nitrogen hetero-
cycles with the acceptor TCNQ are   n o t   highly conducting. Only a few exceptions have been reported so
far: the so-called "NMP-TCNQ" (5-Methylphenazinium-
7,7,8,8-tetracyanoquinodimethanide) (3,4), 5,8-Dihydro-
xyquinolinium-7,7,8,8-tetracyanoquinodimethanide (DHQn-
TCNQ) (1), 1-Amino-2,3,5,6-tetramethyl-4-hydroxybenze-
nium-7,7,8,8-tetracyanoquinodimethanide (AHB-TCNQ) (1)
and 1-Methylquinoxalinium-7,7,8,8-tetracyanoquinodime-
thanide (1-MQ-TCNQ) (2).

All the other so far reported highly conducting
1:1 adducts of TCNQ with planar heterocyclic donors contain one or more sulfur or selenium as heteroatoms (5).

The question arises whether the above cited conducting examples are just accidental exceptions or
whether there are directed routes to highly conducting
compounds of planar nitrogen heterocyles with TCNQ and
similar acceptors. (As proposed by J.B.Torrance e.g.
(5)).

This paper discusses two different routes to new
highly conducting charge transfer solids made up of
planar nitrogen heterocylic donors and TCNQ as an acceptor.

At the first sight the structure and stoichiometry of the above listed <u>conducting</u> compounds give no hint how to proceed in preparing new solids of this type. But a detailed study of some of the known materials reveals that the 1:1 1-MQ-TCNQ adduct mentioned first by Melby (2) is <u>not</u> highly conducting if purified. The pure 1:1 compound crystallizes presumably in a mixed stack arrangement. The crystals show the typical magnetic features of dimerized radical species (triplet excitons in the e.s.r. spectra) (6). The high conductivity mentioned earlier for this compound is an artifact caused by $(1-MQ)_2(TCNQ)_3$ impurities presumably.

Furthermore it has been shown that the so-called "NMP-TCNQ" easily crystallizes in the presence of appreciable amounts of phenazine and $NMPH^+$ only (3,4) and it can be assumed that <u>pure</u> NMP-TCNQ without phenazine and $NMPH^+$ impurities in the lattice has not been investigated in detail so far.

These results leave only two exceptions of conducting 1:1 adducts of nitrogen heterocyles with TCNQ: DHQn-TCNQ and AHB-TCNQ. The common feature of these planar donors is an additional -OH group attached to the aromatic ring. We therefore reinvestigate these compounds and additionally we are preparing a broad spectrum of 1:1 TCNQ adducts with OH-substituted aromatic molecules.

Here we present the results of a rather simple example: If equimolar amounts of LiTCNQ and 8-hydroxyquinolinium-perchlorate are dissolved in <u>hot</u> ethanol, poured together and quenched by cooling the mixture in a dry ice/methanol bath, a black microcrystalline precipitate is formed. The same compound can be obtained by mixing warm solutions of TCNQ and a tenfold excess of 8-hydroxyquinoline in acetonitrile. The combined solutions are cooled <u>rapidly</u>. The analytical data prove that a 1:1 adduct between TCNQ and 8-hydroxyquinolinium has been obtained. The compound loses neutral TCNQ on heating above 130° C. Small TCNQ crystals are growing at the surface on further heating. The remaining material melts at 173° C. Though many donor acceptor complexes of 8-hydroxyquinoline are known (7), a TCNQ compound of this stoichiometry has not been published to our best knowledge.

The pressed pellet conductivity is in the range of $2\text{-}4 \cdot 10^{-2}\Omega^{-1}cm^{-1}$ for different samples. Very characte-

ristic are the i.r. spectra. While TCNQ, TCNQ⁻, 8-hydro-
xyquinoline and its salts show highly resolved spectra
with the expected absorptions the i.r. bands of the 1:1
adduct are hidden under an intense and broad electronic
absorption. A very similar i.r. behaviour was found
earlier for different phenazinium-TCNQ compounds (8).
This intense electronic absorption strongly suggests a
"metallic" conductivity as indicated by the pressed
powder data which are in the same range as those for
highly conducting (impure) "NMP-TCNQ".

The e.s.r. spectra of polycrystalline samples show
an intense exchange narrowed but slightly anisotropic
absorption very similar to the one of "NMP-TCNQ". g_N and
$g_\perp$ as determined from the spectra are 2.0035 and 2.0026
respectively. The same values are found for the so-called
NMP-TCNQ under the same conditions. The g-value anisotropy slightly decreases with increasing temperature.
But the intensity of the signal remains constant on
changing the temperature from 170° K to 400° K which
indicates a Pauli susceptibility for this compound.

These observations suggest that 8-hydroxyquinoli-
nium-TCNQ opens a new route to highly conducting and
thermally stable 1:1 TCNQ CT solids.

There is another possibility to prepare highly conducting TCNQ adducts of nitrogen heterocylic donors by
starting from semiconducting or insulting phases. As
pointed out by Sandman (3) the doping of the mixed stack
phase of 5-ethylphenazinium-TCNQ (NEP-TCNQ) with about
15 % phenazine leads to the formation of a new phase
with a conductivity which is 9 orders of magnitude higher
than the one of the starting material. The consequences
of doping upon mixed stack phases of 5-alkylphenazinium-
TCNQ with neutral phenazine seems to be quite general.
Recently we observed (9), that doping of the "mixed
stack" compound 5,10-dihydro-5,10-dimethylphenazinium-
TCNQ (M_2P-TCNQ) (10) with 15 % neutral phenazine leads
to a fundamental structural change, and consequently different physical properties like high electrical conductivity for the new phase. Single crystal X-ray data
reveal that the new compound crystallizes in the triclinic space group with a = 7.75(1) Å, b = 3.850(7) Å, c =
18.09(3) Å, α = 88.99(9)°, β = 117.5(1)° and γ = 85.3(1)°.
These data are very different from those reported for
the pure mixed stack M_2P-TCNQ (10) and the similar mixed
stack NMPH-TCNQ (11) but are very similar to the unit
cell data of the so-called NMP-TCNQ (12) and to the doped

and conducting NEP-TCNQ prepared recently by Sandman
(3). These data might hint to a segregated structure
for the doped M_2P-TCNQ which would be in accordance
with the observed physical properties. A full structure
determination is in progress.

These results suggest that the "mixed stack" arrangement is typical for all 5-alkyl-phenazinium-TCNQ
and 5,10-dihydro-5,10-dialkylphenaziniumyl-TCNQ solids
(including pure 5-Methylphenazinium-TCNQ, the so-called
NMP-TCNQ) and that segregated phases can only be obtained after doping. This suggestion would imply that all
so far measured NMP-TCNQ compounds are doped with phenazine. Additional doping with phenazine therefore has
only minor and gradual effects (13).

Finally we would like to point out that the 5,10-
dihydro-5,10-dialkyl-phenazinium salts with acceptors
others than TCNQ exhibit quite high conductivities as
well (14).

This work was supported by Deutsche Forschungsge-
meinschaft, Bonn-Bad Godesberg, through grant No.
Ke 135/19. We would like to thank Dr.R.Martin and Dipl.-
Chem.W.Steiger for the X-ray data of "doped" M_2P-TCNQ.

References

1)   L.R.Melby, R.J.Harder, W.R.Hertler, W.Mahler, R.E.
     Benson, and W.E.Mochel, J.Amer.Chem.Soc. 84, 3374
     (1962)
2)   L.R.Melby, Can.J.Chem. 43, 1448 (1965)
3)   D.J.Sandman, J.Amer.Chem.Soc. 100, 5230 (1978) and
     references cited therein
4)   H.J.Keller, D.Nöthe, W.Moroni, and Z.G.Soos, J.Chem.
     Soc.Chem.Comm. 1978, 331
5)   See for example:
     J.B.Torrance, in "Synthesis and Properties of Low-
     Dimensional Materials", ed. J.S.Miller, Ann.N.Y.
     Acad.Sci 1978, in print and Accts.Chem.Res. in press
     (1978)
6)   H.J.Keller, D.Nöthe, and M.Werner, in preparation
7)   See for example:
     C.K.Prout and A.G.Wheeler, J.Chem.Soc. A1967, 469
8)   Z.G.Soos, H.J.Keller, W.Moroni, and D.Nöthe, J.Amer.
     Chem.Soc. 99, 5040 (1977)
9)   W.Moroni, Dissertation, University of Heidelberg,1977

10) I.Goldberg and U.Shmueli, Acta Cryst. B29,421(1973)
11) B.Morosin, Acta Cryst. B34, 1905 (1978)
12) C.J.Fritchie, Acta Cryst. 20, 892 (1966)
13) J.S.Miller and A.J.Epstein, J.Amer.Chem.Soc. 100, 1639 (1978)
14) H.Endres, H.J.Keller, W.Moroni, D.Nöthe, M.H.Vartanian, and Z.G.Soos, J.Chem.Phys.Solids, submitted

METAL VAPOR CHEMISTRY RELATED TO MOLECULAR METALS

Kenneth J. Klabunde, Robert G. Gastinger, Thomas J. Groshens, and Michael Brezinski

Department of Chemistry, University of North Dakota

Grand Forks, North Dakota 58202 U.S.A.

One of the apparent requirements for the production of new molecular metals is the use of "flat" molecules.[1] Bulky three dimensional substituents do not allow formation of stacked molecular layers as found in the noteworthy example TTF-TCNQ. By studying metal atom (vapor) reactions with flat ligands, we anticipated the production of new two dimensional organometallics in the absence of normal three dimensional stabilizing ligands. The following types of metal - "flat ligand" complexes have been considered, and successfully synthesized. So far our results have progressed to the

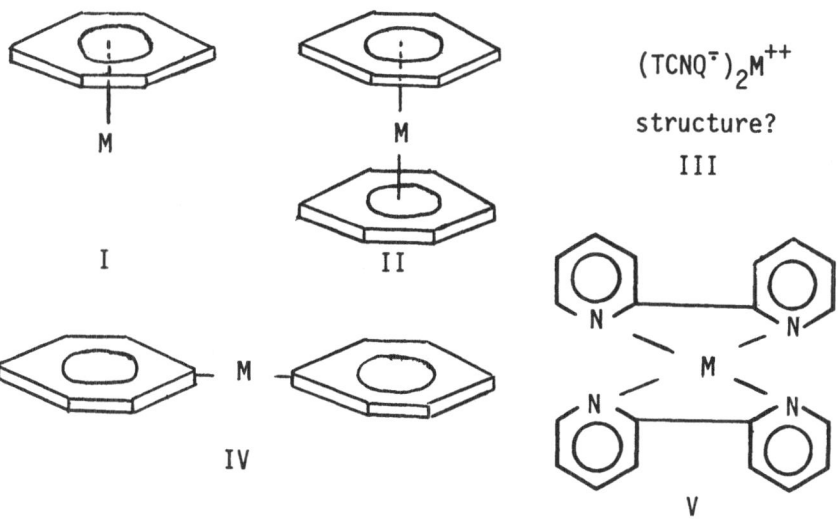

$(TCNQ^{\mp})_2 M^{++}$

structure?

III

stage of doing the syntheses and elucidating the structural properties and chemistry of these species. Very little has been done to
date regarding electrical properties of these molecules or their
charge transfer complexes.

π-COMPLEXES

For structures I and II we developed not only an interest in
the synthesis of such compounds, but also in their structural perm-
utations by electron donating and electron withdrawing substituents.
It of importance to understand in detail the bonding in such systems
because this knowledge may answer questions about arene complexation
to clean metal films. These films, vanadium in particular, have
changed transition temperatures for superconductivity when exposed
to arenes such as anthracene or polychloro analogs such as 1,4,5,8-
tetrachloroanthraquinone.[2] Thus, understanding electronegativity
effects on the M-arene interaction is important for understanding
electron movement on the surface of these metal films and an the
possibility of forming "molecular alloys" proposed by McConnell,
Gamble, and Hoffman.[2] In these M-organic films on metals, "conductive conjugation" is needed which means that the arene ligands donor-
acceptor properties must be carefully adjusted so that electron
transfer from M to M-complex is favorable. Thus, arene electronegativity would appear to be a very important variable to study,
and in particular it would be valuable to determine the limits
concerning covalent complexation (simple orbital mixing) vs. ionic
bonding (electron transfer to or from ligand). This is doubly
important with the recent discovery of partial charge transfer requirements for molecular metals.[3]

Several years ago we began our studies of π-arene-M complexes
with a study of electronegative arenes such as C_6F_6 compared with
C_6H_6.[4] It was found that V and Cr formed the very stable known
bis(arene)-sandwich compounds with C_6H_6, but extremely unstable explosive complexes with C_6F_6. On the other hand, the later transition
metals Fe, Co, and Ni formed more stable complexes with C_6F_6 than
C_6H_6, although none of these were stable enough for isolation and
purification. It was noted through chemical trapping studies that
the Ni-C_6F_6 complex was a 1:1 complex. Work by Efner, Tevault, Fox,
and Smardzewski[5] on system I has since shown that the Fe-arene,
Co-arene, and Ni-arene complexes are 1:1. It is likely that in solution additional arenes coordinate from the back side, but only
weakly in a solvating sense. Further work in our laboratory has
shown that other electron withdrawing substituents, such as CF_3,
serve to stabilize the Co-arene and Ni-arene complexes. It seems
likely that our initial conclusion is correct that good π-acceptor
arene ligands favor complexation to the later transition metals because of good d-orbital donation to π* orbitals of these arenes.
In general the Fe, Co, Ni-arene complexes are stable in solution in

the -20 to -80°C temperature range, are extremely reactive, and their main use for us has been as organometallic intermediates for further synthetic schemes.[6]

A series of polyfluoro arenes was studied, particularly with Cr and V. It was found that 1 or 2 F or CF_3 substituents could be tolerated, and stable bis(arene) complexes (system II) still produced. However, more than two such substituents caused the complexes to be quite unstable (sometimes explosive in the F substituted cases).[7]

The effects of these substituents on structure is interesting. In the case of $(1,4-C_6H_4F_2)_2V$, single crystal x-ray studies showed definite ring deformations due to the presence of F, the carbons bearing the F substituents bent up out of the plane of the arene p-orbitals.[8] However, this distortion is not found in similar CF_3 substituted V complexes or in F or CF_3 substituted Cr complexes.[9] It is not yet known how other halogens will affect these structures. So far, the V system with F substituents appears unique.

The presence of F and CF_3 substituents causes drastic changes in the ability of system II to be oxidized. Pulling one electron out is extremely difficult for $[m-(CF_3)_2C_6H_4]_2Cr$ (+0.39 ev) but quite facile for $(C_6H_6)_2Cr$ (-0.81 ev).

Detailed analysis of substituent effects on the oxidation potentials indicated that the bonding in the complexes must involve a great deal of the σ-framework as well as π-framework of the arene.[10,11] Also, studies in the IR indicate a slight M-ring bond weakening due to the presence of CF_3 substituents.[6]

Extension of these studies to ligands where complete electron transfer is expected to occur, has been carried out. For example, the non-arene ligand TCNQ has been codeposited as a vapor with metal vapors. Earlier work with Cs and Al yielded charge transfer salts $M^+ TCNQ^-$.[12] We extended this work to large scale with transition metals in order to prepare non-hydrated[13] transition metal-TCNQ films. Green films of $Ni^{++}(TCNQ^-)_2$ have been prepared by codeposition of Ni vapor and TCNQ vapor.[11] Studies of the product by IR under anaerobic conditions indicated the absence of $Ni^{++}TCNQ^-$. The dianion of TCNQ[14] apparently does not form under these conditions.

σ-COMPLEXES

Structures IV-V interested us for several reasons, one reason being the possible solid state structures with stacking of M over M, or in staggered arene-M arrangements:

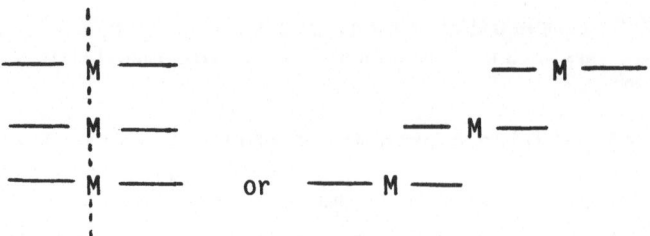

Transition metals again were of primary interest since soluble covalently bound materials were desired. The difficulty that immediately arises, however, is that coordinative unsaturation is present
in IV for all transition metals, and in V for many transition metals.
Thus, in the generation of Co and Ni analogs of IV (σ-bound arene =
C_6F_5), coordinative unsaturation is such a dominant feature that
the molecules coordinate a molecule of relatively inert toluene
solvent to form a new type of π-arene bis-σ-arene complex.[15] The
π-bound arene is extremely labile, however, and can be displaced by
a variety of relatively weak ligands. When M = Ni, dienes, acetylenes, phosphines, and ethers all ready displace the arene to form
new organometallic complexes.[16] Other arenes readily exchange with
the coordinated toluene at room temperature or below. When M = Co
however, the π-arene ligand is much less labile. X-ray structural
studies show that in the Co system the π-arene is much more tightly
bound than in the Ni system.[15,17] The Co system is paramagnetic
but the Ni system is not.

     Can the π-arene ligand be removed and structure IV be crystal-
lized by itself? The lability of the π-arene ligand would argue for
this. So far, however, we have not been successful in this endeavor.
Other similar coordinatively unsaturated systems have been isolated in
the absence of bound solvent however, such as C_6F_5PdBr[18] and CF_3PdI.[19]
Thus the possibility still exists for generating unusual solid state
materials form these reactive molecules.

     We have prepared structure V where M = Co, Ni, or Fe (further
characterization work is still needed, especially for the Fe system).
Metal vapor techniques were employed to prepare I, and this converted

IV →

to V by low temperature addition of bipyridine.

Are these zero-valent complexes (V) square planar D_4h, tetra-hedral Td, or in between D_2d? Currently, x-ray structural studies are underway. NMR studies imply that the V (M = Ni) is diamagnetic and probably square planar. When M = Co, there is currently some confusion as NMR indicates paramagnetism, but ESR does not confirm this. The dark green (Ni), blue-green (Co) and green (Fe) are quite interesting for these zero valent structures. Further work is underway.

ACKNOWLEDGEMENT:

The generous support of the National Science Foundation (CHE-7402713) is truly appreciated.

REFERENCES

(1)   A. F. Garito and A. J. Heeger, Accts. Chem. Res., 7, 232 (1974).

(2)   (a) H. M. McConnell, F. R. Gamble, and B. M. Hoffman, Proc. National Acad. Sci., 57 (5), 1131 (1967).

       (b) B. M. Hoffman, F. R. Gamble, and H. M. McConnell, J. Amer. Chem. Soc., 89, 27 (1967).

(3)   (a) J. B. Torrance, B. A. Scott, and F. B. Kaufman, Solid State Comm., 17, 1369 (1975).

       (b) L. J. LaPlaca, P. W. R. Corfield, R. Thomas, and B. A. Scott, Solid State Comm., 17, 635 (1975).

(4)   K. J. Klabunde and H. F. Efner, J. Fluorine Chem., 4, 115 (1974).

(5)   H. F. Efner, D. E. Tevault, W. B. Fox, and R. R. Smardzewski, J. Organomet. Chem., 146, 45 (1978).

(6)   K. J. Klabunde, H. F. Efner, T. O. Murdock, and R. Ropple, J. Amer. Chem. Soc., 98, 1021 (1976).

(7)   (a) K. J. Klabunde and H. F. Efner, Inorg. Chem., 14, 789 (1975).

       (b) Private communications with Professor P. L. Timms.

(8)   L. J. Radonovich, C. Zuerner, H. F. Efner, and K. J. Klabunde,
      Inorg. Chem., 15, 2976 (1976).

(9)   L. J. Radonovich and C. Zuerner, unpublished work.

(10) G. Essenmacher, P. Treichel, H. F. Efner, and K. J. Klabunde,
      unpublished work.

(11) K. J. Klabunde, Annals of the New York Acad. Sci., 295, 83
      (1977).

(12) (a) B. H. Schechtman, S. F. Lin, and W. E. Spicer, Phys. Rev.
      Lett., 34 (11), 667 (1975).

      (b) F. R. Gamble and H. M. McConnell, Phys. Letters, 26A (4),
      162 (1968).

(13) For hydrated systems see: L. R. Melby, R. J. Harder, W. R.
      Hertler, W. Mahler, R. E. Benson, and W. E. Mochel, J. Amer.
      Chem. Soc., 84, 3374 (1962).

(14) A. R. Sidle, J. Amer. Chem., Soc., 97, 5931 (1975).

(15) K. J. Klabunde, B. B. Anderson, M. Bader, and L. J. Radonovich,
      J. Amer. Chem. Soc., 100, 1313 (1978).

(16) R. G. Gastinger and K. J. Klabunde, unpublished work.

(17) B. B. Anderson, C. Behrens, L. J. Radonovich, and K. J.
      Klabunde, J. Amer. Chem. Soc., 98, 5390 (1976).

(18) B. B. Anderson and K. J. Klabunde, unpublished results.

(19) K. Neuenschwander and K. J. Klabunde, unpublished results.

(20) Original preparation of Ni analog (Co and Fe previously un-
      known) done by conventional methods:

      (a) H. Behrens and A. Muller, Z. Anorg. Allgem. Chems., 341,
      124 (1965);

      (b) H. Behrens and K. Meyer, Z. Naturforschung, 216, 489 (1966);

      (c) A. Misono, Y. Uchida, T. Yamagishi, and H. Kageyama, Bull.
      Chem. Soc. Jap., 45, 1438 (1972).

STRUCTURAL INVESTIGATIONS OF THE PEIERLS TRANSITIONS IN TTF-TCNQ

AND RELATED COMPOUNDS (TSeF-TCNQ, HMTTF-TCNQ, NMP-TCNQ)

S. Megtert, J.P. Pouget, R. Comès

Laboratoire de Physique du Solide associé au CNRS

Université Paris-Sud - 91405 ORSAY   (France)

I. - INTRODUCTION

The present Advanced Research Institute is the third meeting of this type dealing from slightly different points of view with one dimensional conductors. If we restrict ourselves to the structural aspects of the Peierls transitions in such 1-D system, a regular progression can be observed through the proceedings of these successive meetings : in 1974, the available results concerned exclusively the case of the platinum chain Krogmann salts (1) ; in 1976, a series of new investigations (2) allowed a relatively detailed description of the successive phase transitions of TTF-TCNQ (3), which contributed considerably to the understanding of this compound (4). Since then, a number of general reviews (5-7) and a book (8) including a chapter on the structural aspects (9) have been published, so that in the present paper we shall deal exclusively with the most recent results.

The latest structural investigations concerned mainly other compounds of the TTF-TCNQ family, but with the aim to try to answer some of the questions which were left open about TTF-TCNQ : essentially the nature of molecular motions involved in the Kohn anomaly, and the controversy about the origin of the $4k_F$ scattering. We shall successively examine the cases of TSeF-TCNQ, HMTTF-TCNQ, NMP-TCNQ and TTF-TCNQ.

II. - TSeF-TCNQ

TSeF-TCNQ is the most obvious compound of the family to compare with TTF-TCNQ. Both structures are isomorphous in the metallic state with the simplifying factors that TSeF-TCNQ only

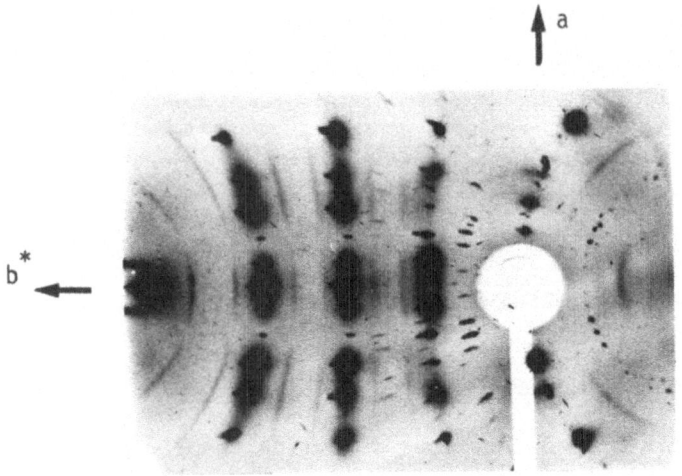

<u>Figure   1</u>

X-ray diffuse scattering from TSeF-TCNQ at 50°K showing
the $2k_F$ scattering at the wave vector 0.315 b^* in the
metallic phase (from 17).

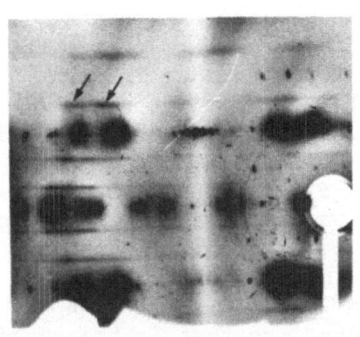
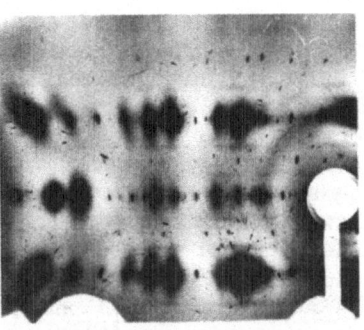

(a)   <u>Figure   2</u>   (b)

Patterns from TSeF-TCNQ at 50°K showing the Q dependence of the
$2k_F$ scattering. In both patterns, the b^* axis is vertical.

a)   Close to the b^*, c^* reciprocal plane (o k ℓ), note the
     increase of intensity of the $2k_F$ lines for both increasing k
     and increasing ℓ (towards the left), demonstrating the exis-
     tence of both b^* and c^* polarizations. Arrows show successive
     diffuse maxima on the $2k_F$ lines, demonstrating the existence
     of a weakly 2D regime.

b)   Close to the a^*,b^* reciprocal plane (h k o), note the absence
     of $2k_F$ scattering for increasing h (towards the left) in
     particular along the equatorial line (h, o±$2k_F$, o) where
     the contribution of the longitudinal polarization component
     is negligible (from ref 17).

undergoes one phase transition (10), and that the X-ray scattering
is so largely dominated by the selenium atoms that one can safely
neglect the contributions from the other atoms for weak effects
such as a Kohn anomaly or a Peierls distortion.

        In a first study, C. Weyl et al (11) could observe a
high temperature Kohn anomaly at the wave vector 0.315 b* attributed to $2k_F$ and condensing around 29°K into a 3-D superlattice
2a x 3.17b x c. This work however brought some confusion regarding (i) the polarization of the molecular motion responsible for
the Kohn anomaly, which was claimed to involve all three components a, b, c in contrast with the case of TTF-TCNQ which presented only such components along b, c, in consistence with a modulation of the intermolecular spacing (12) and (ii) the nature of
this molecular motion itself for which the possibility of an
important contribution of librational modes was left open. These
assumptions were of particular importance since they provided
some experimental support to a model theory for TTF-TCNQ by
Weger et al (13) based on a particular kind of librations (around
an axis perpendicular to the molecular plane) which was able
to explain a number of physical properties of TTF-TCNQ. More precisely for the present purpose, it could account for the occurence
of both the $4k_F$ scattering and the sequence of phase transitions
challenging other work which generally assumed important electron
correlations for the origin of the $4k_F$ scattering (14), and
successive ordering of both molecular species for the sequence
of phase transitions (15, 16).

        Two independent reinvestigations of TSeF-TCNQ by
Megtert et al (17) and Kagoshima et al (18) have now clarified
the case of TSeF-TCNQ. Both studies confirm the existence of a
single high temperature Kohn anomaly at the wave vector 0.315 b*
attributed to $2k_F$ with no detectable $4k_F$ effect (fig 1), and
the earlier reported low temperature 3-D ordering below 29 K.

        The detailed study of the temperature dependence of
the high temperature Kohn anomaly above 29 K has brought several
new observations :
a) at high temperature (up to 200°K) weak (extending only over
   2 to 3 lattice spacing in c direction) 2 dimensional coupling
   is found between stacks of identical molecules of the b, c
   planes ; at the present stage, it is even not sure that a
   truly one dimensional Kohn anomaly exists in TSeF-TCNQ (17)
   (fig 2) ;
b) closer towards the 29 K phase transition (18), short range
   3-D coupling is observed in a relatively large temperature
   domain (up to about 50 K) and of great interest is that this
   short range order corresponds to a different superlattice
   than that stabilized below 29 K ; a detailed description of

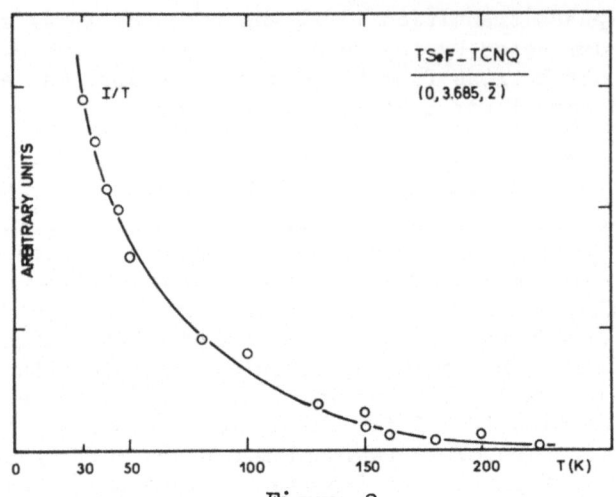

<u>Figure 3</u>

Temperature dependence of the intensity of the $2k_F$ X-ray scattering from TSeF-TCNQ, corrected for the Boltzman population
factor. As the diffuse intensity is almost exclusively due to
the Selenium atoms, the divergence around 30 K shows that the
TSeF molecules must contribute to the low temperature 3-D
distortion.

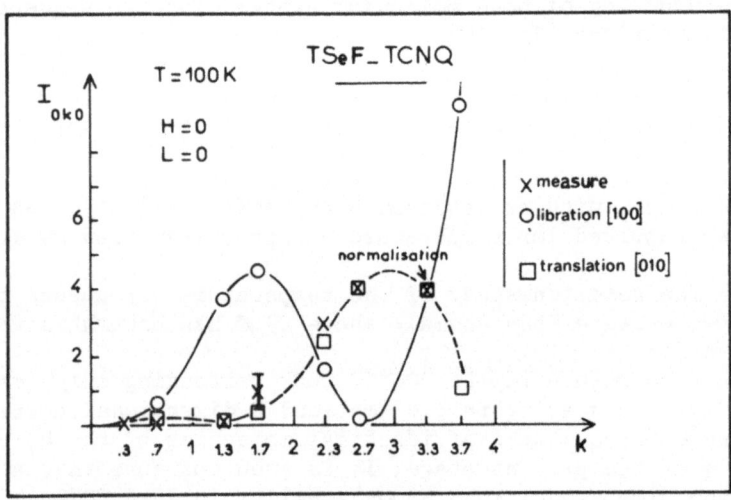

<u>Figure 4</u>

Observed intensity of the $2k_F$ scattering in the successive sheets
along the b^{x} direction (points (o, k_i ±$2k_F$, o), compared to the
intensity calculated for translations along b and c^{x}, and librations around the a axis which is the only simple movement of this
type with only b and c^{x} polarizations.

these results and of their implications can be found in the paper of Kagoshima et al (18) ;

c) the temperature dependence of the diffuse scattering intensity at the $2k_F$ wave vector (fig 3) reveals a divergent behaviour around 29-30°K. This is in principle expected for a precursor scattering leading to a phase transition ; however here in the case of TSeF-TCNQ, the 2a transverse multiplicity of the low temperature phase can lead to the assumption that only one molecular kind was undergoing a Peierls transition at 29 K believed to be the TCNQ stacks. As we have already mentioned above, the X-ray scattering from the TCNQ molecules can be completely neglected compared to the scattering from the Selenium atoms from the TSeF stacks, the divergent behaviour shown in fig 3 after correction for the temperature dependent population factor therefore implies that the TSeF stacks take part in the low temperature distortion.

The intensity distribution within the $2k_F$ diffuse sheets of the high temperature quasi one dimensional regime was investigated in some detail by Megtert et al (17) around 100°K regarding the polarization and the nature of the molecular movements.

Using the same scattering geometries as previously done by Khanna et al (12) for TTF-TCNQ, it can be shown unambiguously that only b and c^* polarization components contribute appreciably to the $2k_F$ scattering (fig 2). In this respect, TSeF-TCNQ and TTF-TCNQ are comparable. This is exclusively compatible with molecular translations along b and c^* and eventually molecular librations around the a axis, and therefore rules out, at least in the <u>high temperature quasi-one dimensional regime</u>, the type of libration assumed for the model of Weger et al (13).

Recording the intensity of the $2k_F$ scattering from TSeF-TCNQ along the b^* direction (points : $0, a^*$, $k_i \pm 2k_F$, $0c^*$), from a series of patterns taken for different sample orientations, one can further show (fig 4) that only translations can account for the observed intensity distribution. A more general survey of this intensity distribution within the $2k_F$ sheets reveals that both translations components are of comparable magnitude, and that the molecules along the stacks (TSeF) are modulated as rigid units (17).

As was the case for TTF-TCNQ, the experimental difficulties have prevented the collection of sufficient data in the low temperature distorted phase of TSeF-TCNQ below 29 K in order to make a more precise structure determination of the insulating state. In particular, in spite of the fact that it seems now esta-

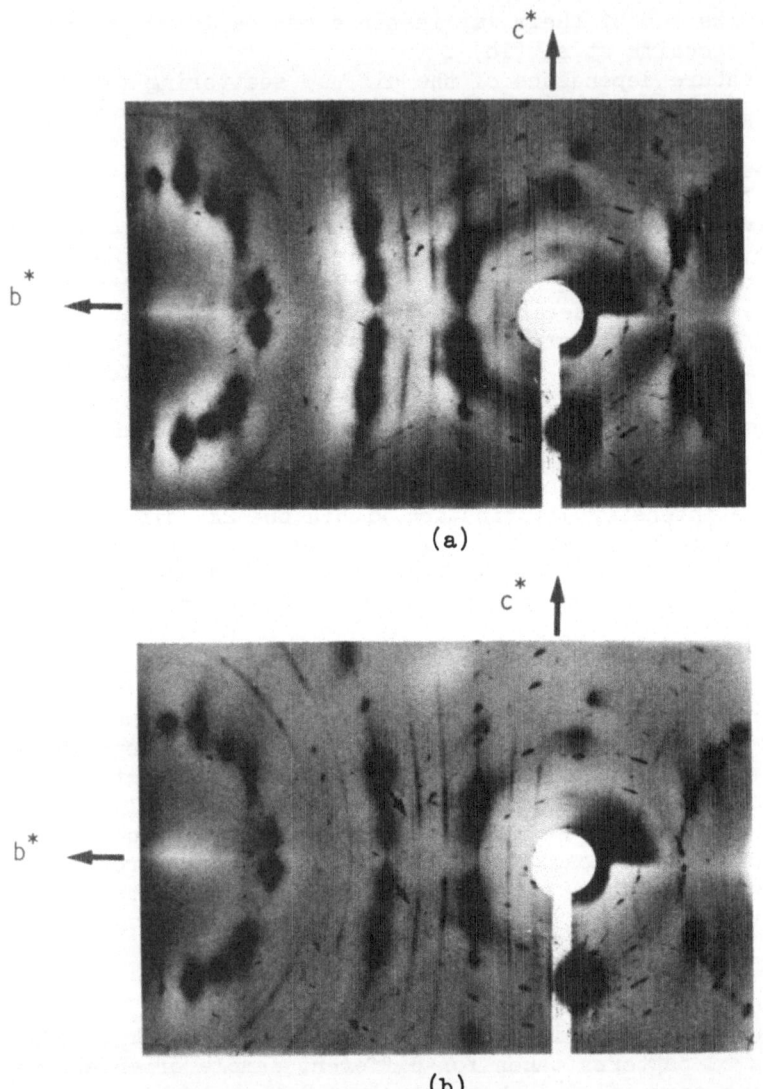

Figure 5

X-ray diffuse scattering patterns from HMTTF-TCNQ.

a) At 100 K : only the $2k_F$ = 0,36 b* scattering is detectable.

b) Around 45 K, additional 1-D $4k_F$ = 0.28 b* (=(1-2x0.36)b*) scattering is clearly observable (see arrows).

blished that librational movements don't play an appreciable
role in the high temperature quasi-one dimensional regime, the
possibility of such molecular movements in relation with the
onset of the lower temperature 3-D regime is still an open
question.

III. - HMTTF-TCNQ

Since, as shown above for TSeF-TCNQ and earlier for
TTF-TCNQ (13), the nature of the molecular movements related to
the 1-D Kohn anomalies seems to rule out the explanation of the
$4k_F$ scattering via a particular kind of librations (13), we are
left with the only other, but still controversial, explanations
in terms of strong intrasite repulsive coulomb interactions between electrons (14, 20, 21). From an experimental point of view,
one can try to check this kind of explanation by the investigation of other parent compounds of TTF-TCNQ. On the side of
(assumed) slightly weaker intrasite coulomb interactions are
TSeF-TCNQ and HMTTF-TCNQ (22). As shown in fig 1, TSeF-TCNQ
only exhibits the more usual $2k_F$ Kohn anomaly. HMTTF-TCNQ is
slightly more complicated than TSeF-TCNQ, but compared to
TTF-TCNQ, has still the advantage of a higher orthorhombic
symmetry (23) and only two conductivity anomalies at 49 and 43 K
(24).

A first structural investigation by Megtert et al (25)
revealed 1-D diffuse scattering with a wave vector of 0.36 b*
attributed to a $2k_F$ kohn anomaly (fig 5a) and visible up to room
temperature, confirming the analysis of the EPR work by Tomkiewicz et al (22) which concluded that the fluctuations in this
compound should be softer than in TTF-TCNQ and TSeF-TCNQ. In
contrast to the cases of this former compounds, the polarization
of the molecular movements responsible for the $2k_F$ scattering is
found to be strictly longitudinal to the chain direction which
is thought to be related to the higher symmetry of HMTTF-TCNQ.

The onset of short range 3-D coupling is observed at
48K where the first conductivity anomaly was earlier observed,
but the long range 3-D distortion is only established at 43 K
where the second conductivity anomaly takes place (fig 6). The
surprising result is that the corresponding superstructure is
not only incommensurate in chain direction ($2k_F$), but also in
the transverse a direction, but nevertheless, within experimental errors, temperature independent with a satellite reflexion
wave vector of : 0.42±03$\vec{a}$, 0.36±0.01$\vec{b}$, 0±0.1$\vec{c}$ (fig. 7).
A Ginzburg-Landau model worked out by Bjelis et al (26) seems
to account for the transverse incommensurability and its temperature independence on the basis of the higher symmetry of
HMTTF-TCNQ,with a dominant role of the distortion on the TCNQ
stacks between 48 and 43 K, and a sharp increase of the dis-

Microdensitometer scans
in the $k_i = 2 + 2k_F$
plane and along a^* showing the formation of
the ($\pm 0.58a^*$, $2.36b^*$, $\overline{2}c^*$)
satellites of HMTTF-TCNQ.
Note the temperature
independent value of the
position of the satellite
along the a^* direction
(from 25).

Figure 6

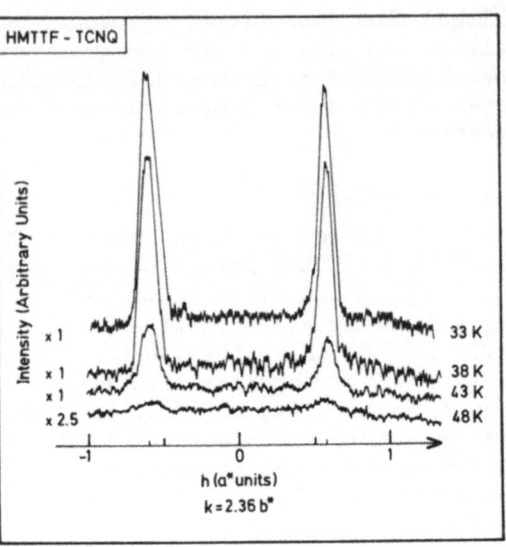

Figure 7

Temperature dependence
of the intensity of the
($0.58a^*$, $3.36b^*$, $5c^*$)
satellite. The insert
shows the HWHM tempera-
ture dependence, revea-
ling the short range
correlation in HMTTF-
TCNQ between 43 and
49 K.

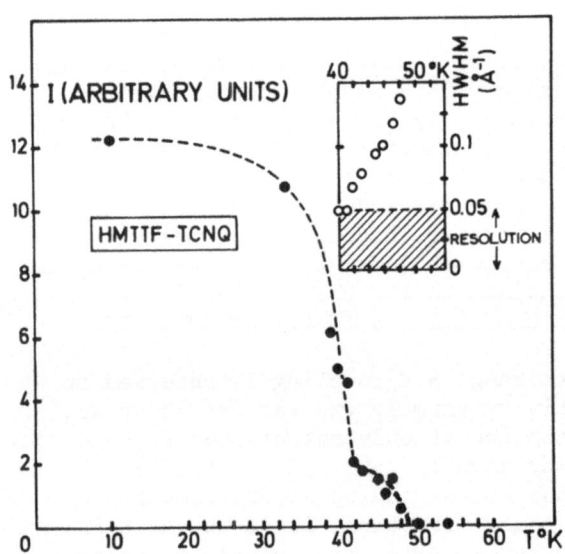

sion on the HMTTF stacks below 43 K.

In this initial study of HMTTF-TCNQ only one kind of higher temperature 1-D scattering could be detected and was therefore attributed to the $2k_F$ wave vector (0.36 b^*) implying a charge transfer of 0.72 electron ; the experimental conditions are here however much more difficult than in the case of TSeF-TCNQ because of much smaller crystal sizes and the absence of a strong X-ray scatterer. In fact, patterns obtained very recently below 100 K and in a different orientation, revealed a second 1-D scattering at the wave vector 0.28 a^* (= 0.72 a^* = 2 x 0.36 a^*) corresponding to $4k_F$ (fig 5b). This scattering does not have a dominant role in any temperature range, in contrast with the high temperature $4k_F$ scattering of TTF-TCNQ, but seems again too strong to be attributed to a second order diffraction effect. It can naturally be explained here by a second harmonic of a non sinusoidal distortion, but the origin of such a second harmonic still needs to be explained. If we assume increasing intrasite Coulomb interactions in the sequence TSeF-TCNQ → HMTTF-TCNQ → TTF-TCNQ, the $4k_F$ scattering of HMTTF-TCNQ shown in fig 5b could be of the same origin as the more striking effects observed in TTF-TCNQ, and support the interpretation of this scattering as reflecting intrasite Coulomb repulsion between electrons.

IV. - NMP-TCNQ

After the encouraging results obtained with HMTTF-TCNQ concerning the occurence of $4k_F$ scattering, the logical next compound to investigate is NMP-TCNQ. The magnetic properties of this compound indeed reveal strong Coulomb interactions between electrons, and it was even suggested that NMP-TCNQ could undergo a Mott-Hubard metal-insulator phase transition around 200 K (27). It is therefore widely agreed that even stronger intrasite Coulomb repulsion between electrons than in TTF-TCNQ should exist in NMP-TCNQ.

NMP-TCNQ is known to crystallize in several different forms :

a) a semiconducting monoclinic phase named NMP-TCNQ II not considered here, more likely to correspond to $NMPH^{\pm}$ $TCNQ^-$ (28)

b) two highly conducting triclinic forms which can be distinguished from a structural point of view by differences in the degree of orientational order of the NMP molecules due to the location of the methyl group (29). We shall only consider below the most studied of these two forms NMP-TCNQ (IA) which shows a metallic behaviour between room temperature and 200 K, temperature at which the electrical conductivity presents a broad maximum followed by a very

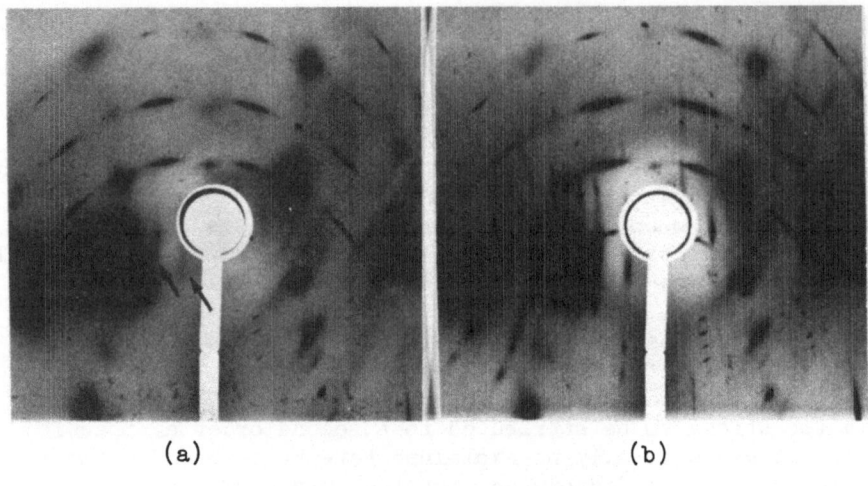

(a)   (b)

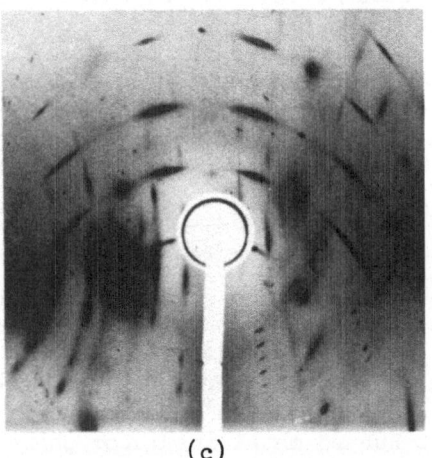

(c)

Figure 8

X-ray diffuse scattering from the metallic form of NMP-TCNQ (IA).
The stronger streaks visible on all three patterns are due to the
short range ordering of the methyl group described in ref 32 and
not considered here. On the 3 patterns the a^{x} axis is horizontal.

a) 300°K weak 1-D scattering at the wave vector $q_2 = \frac{1}{3} a^{x}$ (see
    arrows) demonstrates the existence of a 1-D lattice modula-
    tion in chain direction with a wave length of 3a.
b) 61°K the $q = 1/3 a^{x}$ scattering is partly condensed in broad
    spots showing the existence of short range 3-D ordering ;
    additional weak quasi 1-D scattering becomes visible around
    this temperature (see arrows) with wave vector $q_1 = 1/6 a^{x}$.
c) 20°K Both type of scattering have sharpened but only short
    range 3-D order exists.

rapid decrease for lower temperatures (27, 30), recalling
the temperature dependence of the conductivity of KCP (31).

The X-ray diffuse scattering investigation by Pouget
et al (32) dit not only contribute together with an earlier work
by Kobayashi et al (33) to a clear characterization of the structural differences between the two highly conducting forms of
NMP-TCNQ, but also revealed that the drop of conductivity
below 200 K of metallic NMP-TCNQ (IA) can be associated with the
onset of a 3D lattice distortion.

A first quasi-one dimensional diffuse scattering
already visible at room temperature (fig 8a) is observed at the
wave vector q_2 = 1/3 a^{*} (the high conduction and the stacking
are directed along the a axis in this compound). The intensity
of this scattering increases with decreasing temperature, and
starting around 200 K, broad diffuse spots begin to build up
reflecting a progressive growth of 3-D coupling between the
modulation waves. The precise transverse components (in b and
c directions) of the wave vector of these diffuse spots could
not be determined because of the combined effects of the
broadness of the diffuse spots which never condense into sharp
diffraction satellites (no long range order down to 20°K) and
the poor resolution of the experimental photographic technique.

For temperatures lower than about 70 K, a second type
of scattering located in diffuse sheets with a wave vector of
q_1 = 1/6 a^{*} becomes detectable (fig 8b). Besides the wave vector,
another noticeable difference between the two kinds of diffuse
scattering is that this last scattering never has a quasi 1-D
character ; although it is mainly located in reciprocal sheets,
broad maxima of intensity are already observable when this
scattering is detected, reflecting weak transverse coupling
between the modulation waves up to 70 K. As is the case for
the 1/3 a^{*} scattering, the intensity of the 1/6 a^{*} scattering
increases and the broad diffuse spots sharpen with decreasing
temperature but no long range order is achieved down to 20°K
(fig 8c).

At first sight and by analogy with the earlier
observations on TTF-TCNQ (19), one is tempted to assign the
q_1 = 1/6 a^{*} scattering to a $2k_F$ scattering and the q_2 = 1/3 a^{*}
scattering to a $4k_F$ scattering, result which would unambiguously
confirm the interpretation of the $4k_F$ scattering as arising from
strong intrasite Coulomb repulsion between electrons and close
the $4k_F$ controversy.

As shown in great detail by Pouget et al (32), the
situation is not that simple. Because of the particular value

of the wave vectors of the observed scatterings and the fact
that there is only one electron (and not two as in TTFTCNQ for
example) to share between the conduction bands of the two different molecular species different assignments are possible. If
most of the possible assignments listed by Pouget et al (32)
include a $4k_F$ type scattering, others do not : for example,
with 2/3 electron in the TCNQ conduction band the $2k_{F(TCNQ)}$
Kohn anomaly is expected at 1/3 a*, leaving 1/3 electron
in the NMP conduction band and a $2k_{F(NMP)}$ Kohn anomaly at 1/6 a*
which is also compatible with the experimental observations.
NMP-TCNQ can therefore not give the availed conclusive experimental demonstration of the origin of the $4k_F$ scattering !

        In another respect, the incomplete ordering of the
3-D distortions in an intrinsically disordered system as
NMP-TCNQ, confirms the earlier interpretations in terms of
pinning of the phase of modulation waves on defects or disorder
put forward for KCP (34).

        Before closing this section on NMP-TCNQ, it is
noteworthy to mention that the charge transfer deduced in this
compound (1/3 or 2/3 electrons) from the X-ray diffuse scattering study is much smaller than was usually assumed (close to
1 electron) from other investigations in particular NMR (35)
measurements. One should however note that in a careful
examination of bond lengths in a variety of 1-D conductors,
Flandrois et al. (36) concluded to a charge transfer of only
0.42 electron, and that a theoretical model for NMP-TCNQ from
Ovchinikov et al (37) also predicts a small charge transfer
(0.2 to 0.5). In this respect, small amounts of $NMPH^+$ molecules
on the NMP stacks could also influence the charge transfer.

V. - TTF-TCNQ

        Besides the work on parent compounds, TTF-TCNQ itself
has also been reinvestigated by Pouget et al (38) with inelastic and elastic neutron scattering. Regarding the nature of the
$4k_F$ scattering, in spite of several measurements around the $4k_F$
position in different Brillouin zones, this scattering could
not be detected ; from this work, it was concluded that a
strong $4k_F$ phonon anomaly in the longitudinal acoustic branch
can be eliminated, but that due respectively to the background
level and the sample size the existence of either a weak elastic
$4k_F$ scattering or a phonon anomaly in a higher frequency optic
phonon branch cannot be ruled out. Regarding the 3-D ordering,
this work clearly showed, as was earlier assumed from more
limited data (12), that the successive phase transitions at
54 K and 49 K correspond to successive $2k_F$ distortions respectively polarized along the transverse c* direction (at 54 K)
and along the longitudinal b direction (at 49 K).

The high resolution measurements performed on the satellite reflexions below 54 K, further clearly established that long range 3-D order in TTF-TCNQ only exists below 38 K. In the intermediate phases (54-49 K) and (49-38 K), the satellite reflexions are broadened in _a_ direction. This last result again reveals stronger coupling between the stacks of identical molecules in the b, c planes, as also mentioned above for TSeF-TCNQ with the observation of weak 2-D coupling of the same type. In fact, the $2k_F$ X-ray diffuse scattering from TTF-TCNQ well above the metal insulator phase transition and around 100 K reveals similar weak 2-D coupling in the b, c planes.

From all these newer studies on ordered systems, TSeF-TCNQ (17), HMTTF-TCNQ (25), TTF-TCNQ (38), it seems now clear that the higher temperature distorted "phases" are only short range ordered.

VI. - CONCLUSION

We have deliberately centered the present paper on structural aspects and work which aimed principally at solving the $4k_F$ dilemna. By doing so we have omitted to describe other very nice and recent structural studies dealing more generally with Peierls transitions in related systems. Among them, the work on systems with in particular only one kind of conducting chains, on alkali-TCNQ (Na, K, Rb) by Terauchi (39), or on TTF-SCN$_{.588}$ by Thomas et al (40), and other intriguing new investigations of high purity HMTSF-TCNQ where a phase transition has now been observed by Miljak et al (41) on transport measurements and as shown fig 9 confirmed in a preliminary X-ray study by Pouget et al (42).

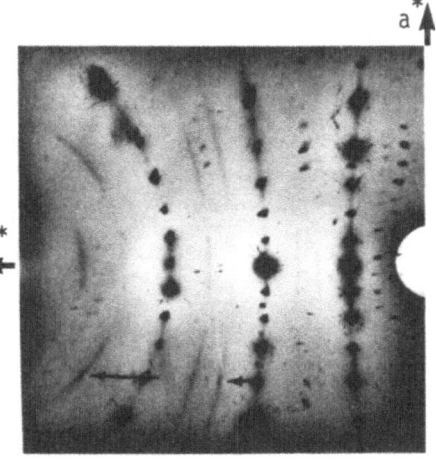

Figure 9

Pattern from high purity HMTSeF-TCNQ at 26°K (just above the 24°K phase transition (41)). Diffuse maxima at the positions $(0a^*, 0.37b^*, xc^*)$ give evidence of short range 3-D ordering. The higher temperature 1-D scattering at $2k_F = 0.37 b^*$ was earlier observed in ref (11).

This choice was motivated by the fact that experimental
information on the strength of Coulomb interactions between electrons in these one dimensional organic conductors is one of the
key parameters for a further progress in the understanding of
these systems. One of the most recent and striking illustration
of this, is the interpretation by Caron et al (43) of the new
high pressure conductivity measurements from Friend et al (44)
on TTF-TCNQ. If the strength of the $4k_F$ scattering now observed
in TTF-TCNQ, HMTTF-TCNQ and perhaps NMP-TCNQ even only gives
an estimation of the relative magnitude of such interactions
between electrons for different compounds, such structural
investigations provide a unique direct information. The cases
which we have examined in some detail above do not completely
establish that the $4k_F$ scattering really reflects strong intrasite Coulomb repulsion between electrons, but they tend
to eliminate other possible explanations for the occurrence
of such scattering and as a whole accumulate evidence in that
direction.

As is the case for detailed studies of the dynamical properties close to the phase transitions, further structural
investigations regarding the $4k_F$ scattering, for example
TTF-TCNQ under high pressure, entirely depend on the availability
of larger size deuterated single crystals.

*We are grateful to our colleagues and collaborators
A.F. GARITO, A.J. EPSTEIN, K. BECHGAARD, L. GIRAL, S.M. SHAPIRO,
G. SHIRANE who made the different studies mentioned above possible.*

BIBLIOGRAPHY

(1) In :
     Low dimensional cooperative phenomena, H.J. Keller (Ed),
     Plenum, New York, (1975).

     See also :
     Lecture notes in physics, Vol 34, Springer, New York,(1975).

(2) R. COMES, G. SHIRANE, S.M. SHAPIRO, A.F. GARITO, A.J. HEEGER-
     Phys. Rev., B 14, 2376, (1976).

     G. SHIRANE, S.M. SHAPIRO, R. COMES, A.F. GARITO, A.J. HEEGER-
     Phys. Rev., B 15, 2413, (1976).

     W.D. ELLENSON, S.M. SHAPIRO, G. SHIRANE, A.F. GARITO - Phys.
     Rev., B 16, 3244, (1977).

(2) J.P. POUGET, S.K. KHANNA, F. DENOYER, R. COMES, A.F.GARITO,
     A.J. HEEGER - Phys. Rev. Lett., 37, 437, (1976).

     S. KAGOSHIMA, T. ISHIGURO, H. ANZAI - J. Phys. Soc. Jap.,
     41, 2061, (1976).

(3) In Chemistry and Physics of one dimensional metals,
         H.J. Keller (Ed), Plenum, New York, (1977).
     See also Lecture notes in physics, Vol 65, Springer,
         New York, (1977).

(4) P. BAK, V.J. EMERY - Phys. Rev. Lett., 36, 978, (1976).
     T.D. SCHULTZ, S. ETEMAD - Phys. Rev., B 13, 4928, (1976).
     K. SAUB, S. BARISIC, J. FRIEDEL - Phys. Lett., 56A,302,(1976).
     A. BJELIS, S. BARISIC - Phys. Rev. Lett., 37, 1517, (1976).
     B. HOROWITZ, D. MUKAMEL - Sol. St. Comm., 23, 285, (1977).
     E. ABRAHAMS, J. SOLYOM, F. WOYNAROVICH - Phys. Rev., B 16,
     5238, (1977).

(5) A.M. BERLINSKY - Contemp. Physics, 17, 331, (1976).

(6) J.S. MILLER - Progress in org. chem., Vol 20, Wiley (1976).

(7) G.A. TOOMBS - Physics reports, 40 c, 182, (1978).

(8) Highly conducting one dimensional solids, J. Devreese (Ed),
     Plenum - to be published (1978-79).

(9) R. COMES and G. SHIRANE in (8).

(10) S. ETEMAD, T. PENNEY, E.M. ENGLER, B.A. SCOTT, P.E. SEIDEN -
     Phys. Rev. Lett., 34, 741, (1975).

     S. ETEMAD, E.M. ENGLER, T.D. SCHULTZ, T. PENNEY, B.A. SCOTT,
     Phys. Rev., B 17, 513, (1978).

(11) C. WEYL, E.M. ENGLER, K. BECHGAARD, G. JEHANNO, S. ETEMAD -
     Sol. St. Comm., 19, 925, (1976).

(12) S.K. KHANNA, J.P. POUGET, R. COMES, A.F. GARITO, A.J. HEEGER
     Phys. Rev., B 16, 1468, (1977).

(13) M. WEGER, J. FRIEDEL - J. Phys. France, 38, 241, (1977).

     See also :
     H. MORAWITZ - Phys. Rev. Lett., 34, 1096, (1975).

(14) V.J. EMERY - Phys. Rev. Lett., 37, 1227, (1976).

     J.B. TORRANCE in ref (3) and Phys. Rev., B 17, 3099,(1978).

     P.A. LEE, T.M. RICE, R.A.KLEMM - Phys. Rev., B 15, 2984,(1977).

     J.R. FLETCHER and G.A. TOOMBS - Sol.St.Comm., 22, 555, (1977).

     J. KONDO, K. YAMAJI - J. Phys. Soc. Jap., 43, 424, (1977).

     J. HUBBARD - Phys. Rev., B 17, 494, (1978).

(15) See ref (4) for theories.

(16) Y. TOMKIEWICZ, A.R. TARANKO, J.B. TORRANCE – Phys. Rev.
Lett., 36, 751, (1976).

E. RUBACZEWSKI, S. SMITH, A.F. GARITO, A.J. HEEGER,
B. SILBERNAGEL – Phys. Rev., B 14, 2746, (1976).

(17) S. MEGTERT – Thèse 3ème Cycle, Orsay, (1977).

S. MEGTERT, A.F. GARITO, J.P. POUGET, R. COMES – to be
published.

(18) S. KAGOSHIMA et al – to be published.

(19) J.P. POUGET et al in ref (2).

(20) A.A. OVCHINIKOV – Sov. Phys. JETP, 37, 176, (1973).

(21) J. BERNASCONI, M.J. RICE, W.R. SCHNEIDER, S. STRASSLER –
Phys. Rev., B 12, 1090, (1975).

(22) Y. TOMKIEWICZ, A.R. TARANKO, R. SCHUMAKER – Phys. Rev.,
B 16, 1380, (1977).

Y. TOMKIEWICZ, B. WELBER, P.E. SEIDEN, R. SCHUMAKER –
Sol. St. Comm., 23, 471, (1977).

(23) D. CHASSEAU, G. COMBERTON, J. GAULTIER, C. HAUW – Acta
Cryst., B 34, 689, (1978).

(24) R.L. GREENE, J.J. MAYERLE, R. SCHUMAKER, G. CASTRO,
G. CHAIKIN, S. ETEMAD, S.L. LAPLACA – Sol. St. Comm.,
20, 943, (1976).

R.M. FRIEND, D. JEROME, J.M. FABRE, L. GIRAL, K. BECHGAARD–
J. Phys., C 11, 263, (1978).

(25) S. MEGTERT, J.P. POUGET, R. COMES, A.F. GARITO,
K. BECHGAARD, J.M. FABRE, L. GIRAL – J. Phys. Lett. (France)
39, L 118, (1978).

(26) A. BJELIS, S. BARISIC – to be published.

(27) A.J. EPSTEIN, S. ETEMAD, A.F. GARITO, A.J. HEEGER – Phys.
Rev., B 5, 952, (1972).

(28) B. MOROSIN – Acta Cryst., B 32, 1176, (1976) and private
communication.

(29) C.J. FRITCHIE – Acta Cryst., 20, 892, (1966).

B. MOROSIN – Phys. Lett., 53 A, 455, (1975).

(30) A.J. EPSTEIN, E.M. CONWELL, D.J. SANDMAN, J.S. MILLER –
Sol. St. Comm.,23, 355, (1977).

(31) M. THIELMANS, R. DELTOUR, D. JEROME, J.R. COOPER –
Sol. St. Comm., 19, 21, (1976).

(32) J.P. POUGET, S. MEGTERT, R. COMES, A.J. EPSTEIN - to be published.

(33) J. KOBAYASHI - Bull. Chem. Soc. Jap., 48, 1373, (1975).

(34) B. RENKER, L. PINTSCHOVIUS, W. GLASER, H. RIETSCHEL, R. COMES, L. LIEBERT, W. DREXEL - Phys. Rev. Lett., 32, 836, (1974).

J.W. LYNN, M. IZUMI, G. SHIRANE, S.A. WERNER, R.B. SAILLANT- Phys. Rev., B 12, 1154, (1975).

(35) M.A. BUTLER, F. WUDL, Z.G.SOOS - Phys. Rev., B 12, 4708, (1975).

F. DEVREUX, M. GUGLIEMI, N. NETSCHEIN - J. Phys. France, 39, 541, (1978).

(36) S. FLANDROIS, D. CHASSEAU - Acta Cryst., B 33, 2744, (1977).

(37) A.A. OVCHINIKOV, V. YA. KRIVNOV, V.E. KLIMENCO, I.I. UKRAINSKI in Lecture notes in physics, 65, p. 103, (Springer, 1977).

(38) J.P. POUGET, S.M. SHAPIRO, G. SHIRANE, A.F. GARITO, A.J. HEEGER - to be published.

(39) H. TERAUCHI - Phys. Rev., B 17, 2446, (1978).

(40) G.A. THOMAS, D.E. MONCTON, F. WUDL, M. KAPLAN, P.A. LEE - Phys. Rev. Lett., 41, 487, (1978).

(41) M. MILJAK, D. JEROME - to be published.

(42) J.P. POUGET et al - unpublished.

(43) L.G. CARON, M. MILJAK and D. JEROME - to be published.

(44) R.H. FRIEND, M. MILJAK, D. JEROME - Phys. Rev. Lett., 40, 1048, (1978).

ELECTRONIC PROPERTIES OF TTF-TCNQ AND DERIVATIVES

D.Jérome

Laboratoire de Physique des Solides

Université Paris-Sud - 91405 ORSAY (France)

I. - COMMENSURATE CDW STATE UNDER PRESSURE

Extensive high pressure studies of the phase diagram have been performed in TTF-TCNQ and TSeF-TCNQ using the resistivity change occuring at phase transitions (1,2).

At atmospheric pressure, three transitions are detected. The "53 K" transition is attributed to the onset of three dimensional ordering in the TCNQ chain, figure 1(3). The transition at 48 K corresponds to three dimensional ordering occuring on the TTF-chain. The temperature interval 48-38 K corresponds to a region in which both TTF and TCNQ are three dimensionaly ordered but as a result of interchain interactions the transverse period of the distorsion along the a-axis is drifting continuously from 2a at 48 K to 3.3 a at 38 K (4,5). A lock-in of the transverse period along the a-axis occurs at 38 K through a first order transition (6). The period stays 4 a at low temperature. Hysteresis is observed at the 38 K transition and also in the 38-48 K domain. This fact can possibly indicate the existence of domain walls(discommensurations) in the sliding period region. One can also notice from figure 1 that a fairly large conductivity remains in the 53-48 K temperature interval. This is a consequence of the existence of non-activated current carriers on TTF-chains.

The 4 Kbar pressure run establishes quite clearly that the onset temperature of 3-D ordering on TTF chains is rapidly suppressed by pressure.

The phase diagram up to 33 Kbar is displayed on figure 2. The important features are the followings:

105

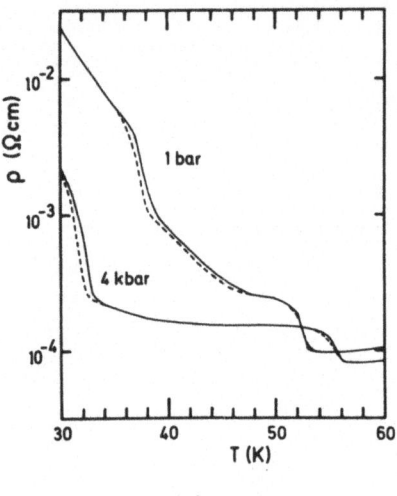

Figure 1

Cooling (dashed) and warming
(solid) resistivity curves
for TTF-TCNQ at 1 bar and
4 Kbar (N^{15} substituted samples) $\sigma(300 \text{ K})$ normalized to
400 Ωcm^{-1}, $\sigma(\text{peak})/\sigma(300K)$
= 25.

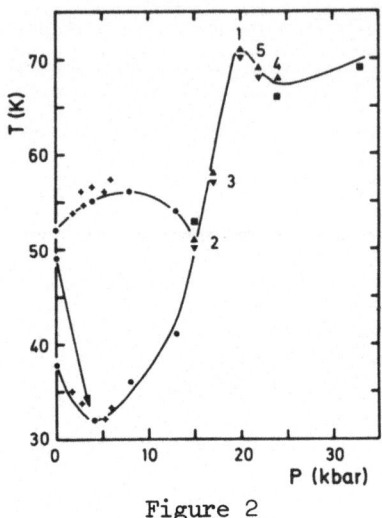

Figure 2

Phase diagram of TTF-TCNQ.

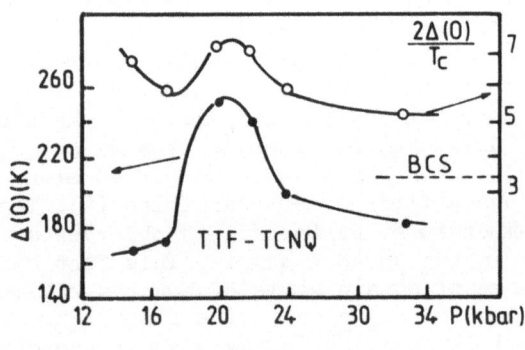

Figure 3

Activation energy $\Delta(0)$ derived from the relation $\rho(T) = \rho_{min}$
exp $\Delta(T)/kT$ and $2\Delta(0)/T_c$ in TTF-TCNQ versus pressure. The commensurability domain is characterized by a first order transition at
T_c, and a large value of $2\Delta(0)/T_c \sim 7$.

a) A minimum of the TTF-Chain transition temperature is situated at
32 Kbar around 5 Kbar. The 38 K lock-in transition is certainly suppressed by high pressure.Actually in order to know whether it merges
or not with the TTF-transition more detailed pressure work is needed.
b) Initially, the TCNQ transition at 53 K increases under pressure
and then merges into the TTF transition at 15 Kbar.
c) All onset transitions are second order up to 15 Kbar, as seen
from the absence of hysteresis on the resistivity.
d) The single transition temperature peaks sharply at 71 K and 20
Kbar.
e) The transition towards a strongly insulating low temperature state
becomes first order (1 K hysteresis) in the pressure domain of 6Kbar
wide centered at 20 Kbar.
f) In this narrow pressure domain the low temperature order parame-
ter, $\Delta(0)$, defined from the resistivity data by $\rho(T)=\rho_{min} \exp \frac{\Delta(T)}{kT}$
exhibits a sharp peak. Similarly does the ratio $2\Delta/T_c$,figure 3.

The experimental features d-f, around 20 Kbar can be understood
by the existence in the Landau free energy expansion of a cubic term.
This term could explain both the peaking and the first order character of the transition (7). Such a cubic term can exist if the spanning vector of the one dimensional Fermi surface, $2 k_F$, becomes equal
to the third of a reciprocal lattice vector, $1/3 b^*$. Under such cir-
cumstances the wavelength of the periodic lattice distorsion along
the b direction in the 3 dimensionaly ordered state becomes 3 b. We
are therefore in the presence of a commensurate state, similar to
commensurate states occuring in layer materials (8) for which Mc
Millan (9) has developed a theory. The pressure domain 20 ± 3 Kbar
in which the low temperature longitudinal distorsion is commensurate
with the underlying lattice is connected with the adjustment of the
charge transfer to the value $2/3 = 0.66$ (10).

The free energy expansion of the commensurate state, with Mc-
Millan's notations, becomes neglecting fluctuations:

$$F_c=(\frac{a_o}{2}+e_o(2k_F-\frac{b^*}{3})^2)\Delta^2 - \frac{1}{4} b_1\Delta^3 \delta (2k_F-\frac{b^*}{3}) + \frac{3}{8} C_o \Delta^4 +\ldots \quad (1)$$

The cubic term in (1) corresponds to the energy gained adjusting the
wavelength of the distorsion to a multiple of the intermolecular spacing. The harmonic contribution e_o represents the energy loss associated with the necessary lattice deformation as the spanning vector
is in the vicinity of $b^*/3$.

The present resistivity investigation does not provide any reliable information about the relative phasing of distorsions belonging to neighbouring chains. However considering the strongly insulating character of TTF-TCNQ at 20 Kbar below $\sim$ 71 K and the results
on figure 3 we can forsee that distorsions are large on both chains
and that possibly the most likely period along the a-axis is the one

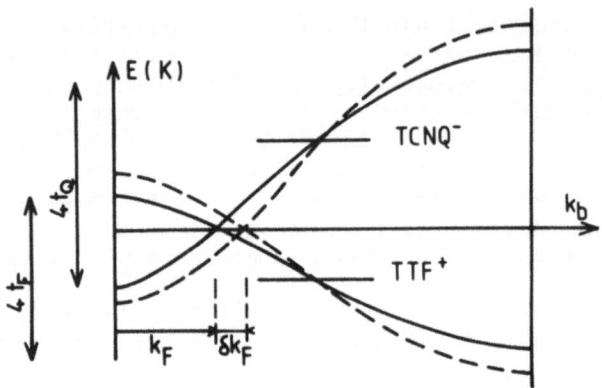

<u>Figure 4</u>

Schematic effect of high pressure on the TTF-TCNQ bands.

corresponding to the out of phase stacking, namely a. According to
figures 2 and 3, the width of the commensurability range is $\approx \pm 3$
Kbar, corresponding to a variation $\pm 0.5\%$ of 2 k_F, namely
$2\ k_F = b^*/3(1 \pm \delta_c)$ with $\delta_c = 0.005$. The Landau theory(1) provides
a relation between the width δ_c, the kinetic energy gain at the
transition and the additional lock-in energy gain.

$$2\ \delta_c \approx \frac{\text{lock-in energy}}{\text{kinetic energy}} \frac{b}{\xi} \qquad (2)$$

From the value $2\ \delta_c = 1\%$ and fig 3, an estimate $\xi/b \sim 60$ is derived
for the coherence length, which compares favourably with the experimental determination (11).

 A major consequence of the longitudinal lock-in effect is the
existence of a charge transfer increasing from 0.59 electrons/mole-
cule at 1 bar to 0.66 electrons/molecule at 20 Kbar (4). Such an
increase, when scaled with compressibility data (12) corresponds to
a pressure coefficient $\partial \ell n k_F/\partial P \approx 1\%$ Kbar^{-1} at atmospheric pressure,
in good agreement with sound velocity measurement experiments(13).

 With the assumption of energy difference between molecular levels TCNQ$^-$ and TTF$^+$ being independent of pressure, and of similar
bandwidth pressure dependence, the relation

$$\frac{\Delta t_{\text{\tiny \textbar\textbar}}}{t_{\text{\tiny \textbar\textbar}}} = \frac{\pi}{2}\ \Delta\rho\ \text{tang}\ \frac{\pi\rho}{2}$$

holds between the relative change $\dfrac{\Delta t_{\text{\tiny \textbar\textbar}}}{t_{\text{\tiny \textbar\textbar}}}$ of the tight binding bandwidth
and the change $\Delta\rho$ of the charge transfer, figure 4.

Following this procedure a bandwidth pressure coefficient $\frac{\partial \ell n t_{\shortparallel}}{\partial P} \sim 1.75\%$ Kbar^{-1} is derived for TTF-TCNQ at ambient pressure.

Tight binding calculations provide an averaged coefficient of $\sim 2\%$ Kbar^{-1}(14). Thus, we feel that a tight binding approach is quite appropriate for the description of the bare bands of TTF-TCNQ. Admittedly, a direct observation of the charge transfer increase and of the commensurate state would be one of the most useful experiments for future work. This could be achieved for instance, using the diffuse scattering of X-rays or neutrons under pressure in the temperature range 120-77 K. Correlatively the b parameter is expected to remain constant at low temperature in the commensurability range and a very slight discontinuity, $\Delta b/b \sim 0.06\%$ may be visible at the limits 17 and 23 Kbar (2).

Extensive high pressure experiments up to 33 Kbar have also been performed on other charge transfer salts, for example TSeF-TCNQ, figure 5, whose charge transfer is 0.63 at ambient pressure (15). In this compound a single second order phase transition is observed up to 32 Kbar and Δ and $2\Delta/T_c$ peak at 8 Kbar, figure 6. A slight peaking of T_p at 8 Kbar is not inconsistent with data of figure 5. We believe it is obscured by an underlying strong pressure enhancement related to the increase of the electron-phonon coupling constant.

The high value of the transition temperature found at 81 K in the new compound t-TTF-TCNQ(16,17) and pressure dependences of Δ and $2\Delta/T_c$ could also indicate the existence of a commensurate state at ambient pressure in this compound. The structure of t-TTF is hybrid between that of TTF and HMTTF, therefore a 2/3 charge transfer might result from the average between charge transfer of TTF and HMTTF donors compounded with TCNQ.

If the relation between plasma frequency and free carrier density is accepted, namely $\omega_P^2 = ne^2/m^*$ the tight binding approximation provides $\omega_P^2 \sim b^2 t_{\shortparallel} k_F$. Hence the plasma frequency pressure dependence becomes

$$\frac{\partial \ell n \omega_P^2}{\partial P} = 2 \frac{\partial \ell n b}{\partial P} + \frac{\partial \ell n t_{\shortparallel}}{\partial P} + \frac{\partial \ell n k_F}{\partial P} \tag{3}$$

Pressure coefficients for $t_{\shortparallel}$ and k_F of 1.75% Kbar^{-1} and 1% Kbar^{-1} respectively have been derived from the commensurability phenomenon. With the measured value of the b-axis compressibility 0.47% Kbar^{-1} (12), we derive from (3)

$$\frac{\partial \ell n \omega_P^2}{\partial P} = 1.7\% \text{ Kbar}^{-1}$$

This value is consistent with the directly measured pressure

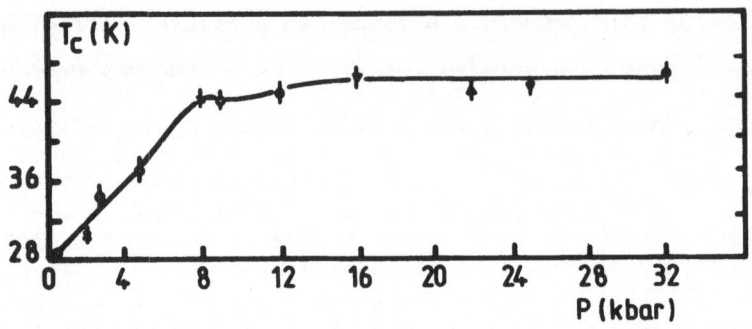

<u>Figure 5</u>

Pressure dependence of the single phase transition of
TSF-TCNQ, J.R.Cooper and D.Jérome unpublished work, Orsay.

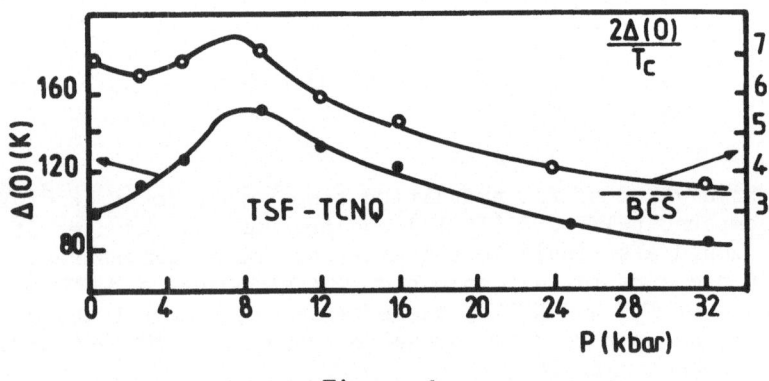

<u>Figure 6</u>

Activation energy $\Delta(0)$ and $2\Delta(0)/T_c$ in TSF-TCNQ.

dependence of ω_P in TTF-TCNQ and TSeF-TCNQ (14).

In conclusion, the message of the high pressure investigation
of TTF-TCNQ is that electronic properties connected with the bare
bandwidth: charge transfer, plasma frequency, etc... are adequately
described by a tight binding approach. The phase diagram of
TTF-TCNQ suggests that the charge transfer is related to the overlap
integral $t_{\parallel}$ and that a commensurate CDW state occurs at 20 Kbar with
a 3 b superlattice. Similarly, commensurability effects may occur
in TSeF-TCNQ at 8 Kbar.

II. - RESISTIVITY UNDER PRESSURE

The striking feature of the conductivity of charge transfer salts belonging to the TTF-TCNQ family is the metallic behavior of the resistivity ($d\rho/d^T > 0$) above the metal to insulator transition.

High pressure enhances significantly the conductivity above the room temperature value, figure 7. However, the asymptotic value of the conductivity of all two-chain conductors looks quite similar under very high pressure, ~ 6-$7 \times 10^3 \, (\Omega cm)^{-1}$.

This observation tends to suggest the existence of strongly pressure dependent scattering governing the resistivity of TTF-TCNQ (18) and to a lesser extent the conductivity of the selenium analogues, HMTSF-TCNQ and HMTSF-TNAP.

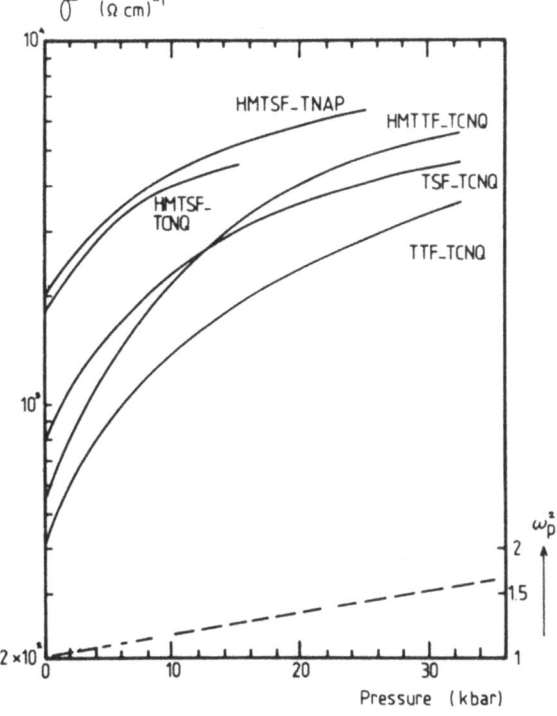

Figure 7. Conductivity versus pressure for various charge transfer compounds at room temperature. The pressure dependence of the plasma frequency is given for comparison.

The fact that the maximum conductivity at low temperature is also
very close to the asymptotic high pressure conductivity of figure 7,
leads to the following remark: the scattering mechanism operating at
ambient temperature can be removed either by lowering temperature or
by the application of high pressure.

The DC conductivity can be related to the plasma frequency by
the Drude relation $\sigma_{DC} = \omega_p^2 \tau_{\shortparallel}$ (although the interpretation of optical
properties is still subject to questions in TTF-TCNQ)and at room tem-
perature the pressure dependence can come either from ω_p^2 or $\tau_{\shortparallel}$ (or
from both). The pressure dependence of ω_p is approximately known,
from the analysis of the previous section and from direct measurements (14). It amounts to $\gtrsim 2\%$ Kbar^{-1}, far too small to account for
the 30% Kbar^{-1} change of the conductivity. The pressure dependence
of the scattering time $\tau_{\shortparallel}$ must be large $\sim 28\%$ Kbar^{-1}, or equivalently

$$\partial\ln \tau_{\shortparallel}/\partial\ln V = - 30$$

Such an unusually large volume dependence of the scattering time
implies significant consequences for the temperature dependence of
the resistivity. Most theories provide constant volume results, whereas observations are performed at constant pressure. The correction
can be made if the pressure coefficient of the resistivity is known
(19). Such a correction should be small whenever the pressure coefficient is small, but it can become an extremely important effect
for the pressure coefficient observed in TTF-TCNQ at high temperature. Cooper (20) has drawn attention on the effect of thermal expansion on the resistivity which was not properly taken into account
in earlier works. A careful reinvestigation of the TTF-TCNQ resisti-
vity(21,22) has been performed, figure 8, and has allowed a derivation of the constant volume curve, figure 9. The procedure taken for
the derivation of figure 9 is the following. On one hand X-ray measurements (23) have shown a thermal expansion effect of 1%, 2.3%
and 0.5% for a, b and c axis respectively between 60 and 295 K. On
the other hand, axial compressibility experiments(12) have revealed
that 5 Kbar amounts to a contraction of 1.3%, 2.3% and 1.6% for a,
b and c axis respectively. Thus, one can crudely say that a hydrostatic pressure of 5 Kbar is equivalent to the thermal contraction
between 300 and 60 K. Such equivalence between pressure and temperature effects is certainly not fullfilled as far as the c-axis is
concerned, since high pressure is about 3 times more efficient than
thermal expansion in reducing the c parameter.

Recent elasto resistance measurements (24) have shown that
strains along a and c directions are nearly as important as the
strains along the b direction. Therefore, as a first approximation
we estimate that constant volume and constant b-axis reductions are
nearly equivalent as far as the intrinsic temperature dependence
of the resistivity is concerned.

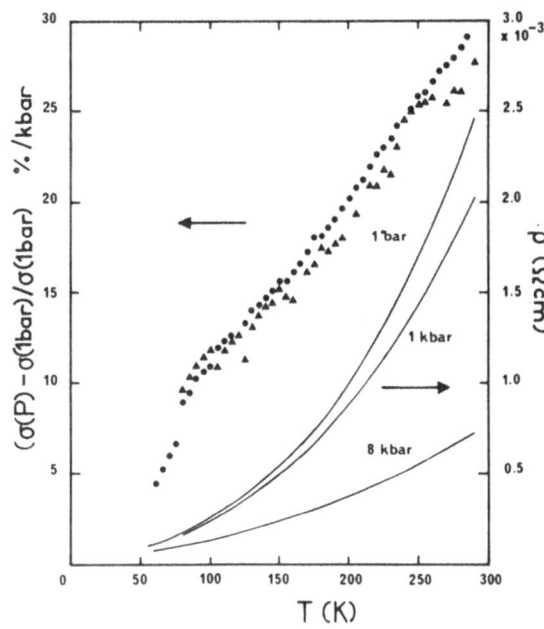

Figure 8

Right hand scale: resistivity versus temperature at constant pressure for TTF-TCNQ with $\rho(300 \text{ K}, 1\text{bar})$ normalized to $2.5\times10^{-3}\,\Omega\text{cm}$. Left hand scale $\partial\ln\sigma/\partial P$ versus temperature.

Figure 9

Resistivity versus temperature for TTF-TCNQ with b lattice parameter constant at its 60 K value. Data for two samples exhibiting $\sigma(\text{peak})/\sigma(300 \text{ K}) = 15$ and 25, $\rho(300 \text{ K}, 1 \text{ bar})$ is normalized to $2.5\times10^{-3}\,\Omega\text{cm}$ for both samples.

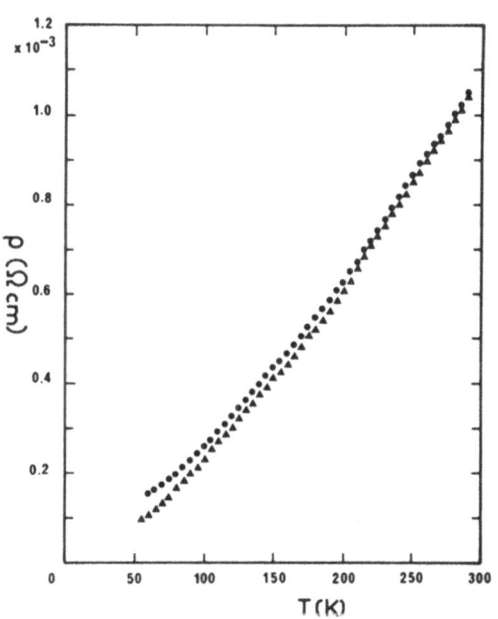

At each temperature the pressure required to bring the b parameter back to its value at 60 K has been calculated and then the
measured value of the resistivity has been corrected to its value
at that pressure. The temperature dependence at constant b parameter for the resistivity is shown on figure 9 (22). The temperature
dependence is close to a linear law above 140 K, contrasting significantly with the $T^{2.3}$ law observed under ambient pressure (25). If
one assumes the Boltzmann equation formalism to be an adequate description for one dimensional conduction (26) scattering by acoustic
phonons could lead to a linear T dependence for $T > \theta_D$ and one would
expect a pressure dependence going like $\partial \ln \tau_{\shortparallel} / \partial \ln V = -2\gamma_G$. The
volume Gruneisen constant $\gamma_G \simeq 1.4$ can be derived from the pressure
dependence of the compressibility (12,27). It is thus obvious that
acoustic phonon scattering are unable to explain a pressure coefficient of -30. Other phonon scattering processes have been claimed
to explain the resistivity, optical phonons, second order scattering
with librations. All these approaches are based on the assumption of
a strong pressure coefficient for the characteristic frequency.

The resistivity can be decomposed into three components

$$\rho (T) = \rho_o + \rho_{ph} (T) + \rho_i (T)$$

where ρ_o and $\rho_{ph}(T)$ are the residual and phonon (Bloch-Gruneisen)
contributions respectively. In the low pressure domain around room
temperature, the resistivity of TTF-TCNQ is dominated by $\rho_i(T)$ such
that $\rho_i \gg \rho_{ph} > \rho_o$. The component ρ_i is strongly pressure dependent
and at high pressure the situation can become different with
$|\partial \ln \sigma / \partial \ln V| < 8$ and $\rho_i \sim \rho_{ph} \sim \rho_o$.

The temperature dependence of the resistivity approaches linearity at high pressure in TTF-TCNQ (even at constant pressure). Such
a tendency results from the effects of weaker pressure coefficients
and thermal expansion at high pressure.

In all selenium derivatives a linear law is actually observed
at high pressure ($P \sim 32$ Kbar). For TSeF-TCNQ, figure 10, we assume
that at 32 Kbar $\rho_{ph} \gtrsim \rho_i$, therefore

$$\frac{\partial \rho_{ph}}{\partial T} \lesssim 10^{-6} \ \Omega cm \ K^{-1}$$

Since $\omega_p \sim 1.4 eV(14)$, and using the Hopfield relation (28)

$$\omega_P^2 \ \frac{\partial \rho_{ph}}{\partial T} = \frac{2\pi \ k_B}{\hbar} \lambda \quad \text{an upper value of 0.35 is deri-}$$

ved for the electron-phonon coupling constant at 32 Kbar. This corresponds to an upper estimate of 0.26 for λ at ambient pressure with
the tight binding model. The temperature dependence of the pressure

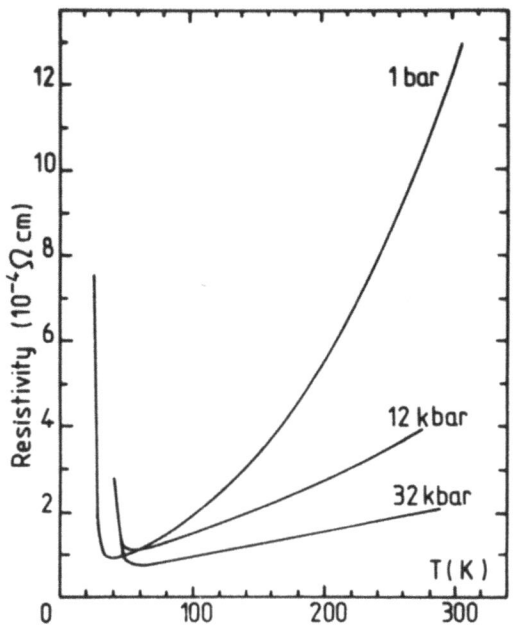

Figure 10

Resistivity versus temperature in TSF-TCNQ, showing
the quasi linear law <u>under
pressure</u> at constant pressure.

Figure 11

Temperature dependence of

$\dfrac{\partial \ln \sigma}{\partial \ln b}$ in two samples of TTF-TCNQ
showing a maximum around 300 K,
continuous line.
Calculated temperature depen-
dence of $\dfrac{\partial \ln \text{ Im } \chi_{SDW}(2k_F)}{\partial \ln b}$
(dashed line) for TTF-TCNQ,
according to the theory of
reference 29.

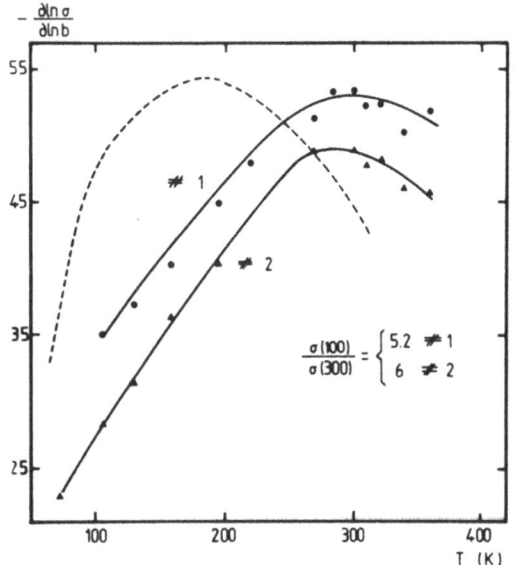

coefficient $\partial\ln\sigma/\partial\ln V$ is quasi linear below room temperature (22), figure 8. At low temperature ($T \sim 60$ K) values $\partial\ln\sigma/\partial\ln V \sim -6$ can be explained by a Bloch-Gruneisen contribution to the resistivity as $T \sim \theta_D$. Recent high temperature investigation of the pressure coefficient has demonstrated the existence of a maximum of $\partial\ln\sigma/\partial\ln b$ (or $\partial\ln\sigma/\partial\ln V$) around room temperature, figure 11 (29). Experiments lead to a phenomenological relation between resistivity and magnetic response function, and suggest the relation $\rho/T \sim \text{Im } \chi_{SDW}(2 k_F)$.

i) The TCNQ chain in TTF-TCNQ is carrying most of the current. Figure 12 shows that a relation $\rho/T \sim \text{Im } \chi_{SDW}^{TCNQ}(2 k_F)$ is well obeyed in temperature, a similar law follows from the pressure dependence around atmospheric pressure.

ii) A comparative study of ρ_i and $\text{Im } \chi_{SDW}(2k_F)$ of various charge transfer salts is also suggestive of the relation $\rho_i \sim \text{Im } \chi(2k_F)$. Taking TTF-TCNQ and HMTSF-TCNQ for example; the ratio $\rho_i(TQ)/\rho_i(HM-Q) \sim 8$ is in good agreement with the NMR result(30) $\text{Im } \chi^{TQ}/\text{Im } \chi^{HM-Q} \sim 10$. We must actually notice that the relation $\rho_i > \rho_{ph}$ does not hold in HMTSF-TCNQ since $\rho_i/\rho_{ph} \sim 3/2$ at room temperature and ambient pressure.

The inability of a Bloch-Gruneisen theory to account for the pressure dependence of the resistivity and the close analogy existing between spin response function and resistivity suggested a magnetic origin to the intra-chain scattering time (21). It has been recognized that electron-electron scattering leads to a $\rho \sim T$ law in a one dimensional band at $T \ll T_F$ only for the half-filled situation (U-processes)(31).

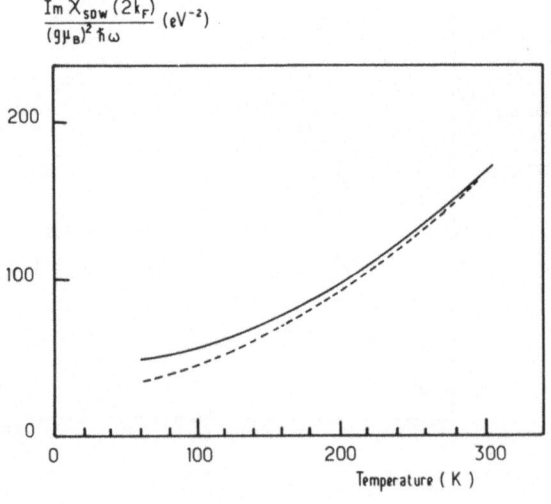

Figure 12

Temperature dependence of $\text{Im } \chi_{SDW}(2k_F)$ for the TCNQ chain in TTF-TCNQ (29), and temperature dependence of ρ/T normalized at room temperature (dashed line), $\sigma(\text{peak})/\sigma(300 \text{ K}) = 25$.

In TTF-TCNQ however, the filling of the band is not one half(ρ=0.55 at 300 K)(4) and consequently we believe electron-electron intrachain scatterings cannot explain the resistivity.

Interchain electron-hole scattering could in principle lead to a contribution to resistivity(32,33). We find this possibility unlikely since materials in which current is carried by one chain only, TTT_2-I_3(34) $TTF-Br_x$(35) or $TMTSF-DMTCNQ_{75}-MTCNQ_{25}$(36) exhibit temperature and pressure dependences similar to those currently observed in TTF-TCNQ. The model that we propose is based on the coexistence in correlated one dimensional conductors of two divergent modes at $2 k_F$ characterized by the response functions χ_{SDW} and N_{CDW} (37). When $\ln T_F/T \sim 2$, namely, around room temperature the divergence is actually much more pronounced on χ_{SDW} than on N_{CDW}. Because of interchain couplings the divergence towards infinity of the coupled chain response occurs only in the CDW channel (38,39).

Fluctuations towards an antiferromagnetic state of wave vector $2 k_F$ are thus strongly enhanced in a correlated one dimensional electron gas. The life time of these fluctuations (paramagnons) is long $\tau_S \sim 10^{-13}$ s at room temperature, with $\hbar/\tau_S \sim E_F/S$ where S is the enhancement factor of Im χ_{SDW} ($2 k_F$), $E_F \sim 0.1$ eV (30).

It is well established that paramagnons in nearly magnetic metals scatter current carriers and lead to an important contribution to the resistivity (40).

If paramagnon-modes play the role of heavy carriers, the scattering amplitude of current carriers off paramagnons becomes Im χ_{SDW} ($2k_F$) and the diffusion time can be written

$$1/\tau_{\shortparallel} \sim \frac{T}{T_J} \text{ Im } \chi_{SDW} (2 k_F)$$

where T_J is a short range ordering temperature of $\sim$ 250 K in TTF-TCNQ. If this picture for the resistivity is accepted, the main experimental features can be explained easily:

i) Proportionality between ρ/T and Im χ_{SDW} ($2 k_F$) is an inherent property of our picture.

ii) The value of the room temperature resistivity of TCNQ is related to the large measured enhancement of Im χ_{SDW}($2 k_F$). An order of magnitude estimate has been performed for ρ(300 K) when Im χ_{SDW} is approximated by its RPA value (21).

iii) The deviation from a linear T law below 140 K is related to the decrease of Im χ_{SDW} ($2 k_F$) (even at constant volume) due to pseudo gap effects (29).

iv) The magnitude and the temperature dependence of $\partial\ln\sigma/\partial\ln V$ can be explained by the calculated $\partial\ln$ Im $\chi_{SDW}/\partial\ln V$, (figure 11). In particular this latter quantity exhibits a maximum in temperature around 300 K (29).

v) The spin-phonon coupling is large in TTF-TCNQ (probably due to the strong volume dependence of Coulomb interaction screening), therefore momentum 2 k_F can be transferred to the lattice, without bottleneck effect, through a paramagnon-3 dimensional phonon scattering.

In conclusion the proposed resistivity channel is a natural consequence of a point of view based on the importance of intrachain Coulomb correlations in TTF-TCNQ.

"Non magnetic"theories have been proposed for the resistivity:

i) Collective mode theory for the conductivity, proposed by Allender, Bray, Bardeen above a mean field (41) transition temperature.

ii) Froehlich sliding mode below a mean field transition temperature (42).

iii) Intra molecular phonon scattering (43).

iv) Second order scattering of electrons against librations (44).

v) Interchain electron-electron scattering (45).

vi) Variable range hopping conductivity (46).

We believe that none of the previous models can explain correctly the very large variety of peculiar behaviours observed in TTF-TCNQ, such as, magnetic enhancement, temperature dependence of χ_S pressure dependences, phase diagram, constant volume resistivity, thermopower and sample dependence.

BIBLIOGRAPHY

(1)   D.JEROME and M.WEGER, Chemistry and Physics of One-Dimensional Metals, H.J.KELLER editor, Plenum Press, New York 1977.

(2)   R.H.FRIEND, M.MILJAK and D.JEROME, Phys.Rev.Lett.40, 1048,1978.

(3)   E.F.RYBACZEWSKI, L.S.SMITH, A.F.GARITO, A.J.HEEGER and B. SILBERNAGEL, Phys.Rev.B 14, 2746, 1976.

(4)   R.COMES, page 315, in reference 1.

(5)   S.KAGOSHIMA, T.ISHIGURO and H.ANZAI, J.Phys.Soc.Japan 41,2061, 1976.

(6)   D.JEROME, W.MULLER and M.WEGER, J.Physique Lett.35, L-77,1974.

(7)   L.LANDAU et E.LIFCHITZ, Physique Statistique Editions Mir,
      Moscou, 1967.

(8)   J.A.WILSON, F.J.Di SALVO and S.MAHAJAN, Adv.in Physics $\underline{24}$,117
      1975.
      P.M.WILLIAMS, Crystallography and Crystal Chemistry of Mate-
      rials with layered Structures, F.LEVY Editor, D.REIDEL publis-
      hing Company, Holland, 1976.

(9)   W.L.Mc MILLAN, Phys.Rev.B $\underline{12}$, 1187, 1975.

(10) J.FRIEDEL, Electron-Phonon interactions and Phase transitions,
      T.RISTE editor, Plenum Press, New-York 1977.

(11) S.K.KHANNA, J.P.POUGET, R.COMES, A.F.GARITO, and A.J.HEEGER,
      Phys.Rev.B $\underline{16}$, 1468, 1977.

(12) D.DEBRAY, R.MILLET, D.JEROME, S.BARISIC, L.GIRAL and J.M.FABRE,
      J.Physique Lett.$\underline{38}$, L-227, 1977.

(13) T.TIEDJE, R.R.HAERING, M.H.JERICHO, W.A.ROGER and A.SIMPSON,
      Solid State Comm.$\underline{23}$, 713, 1977.

(14) B.WELBER, P.E.SEIDEN and P.M.GRANT, preprint 1978.

(15) C.WEYL, E.M.ENGLER, K.BECHGAARD, G.JEHANNO and S.ETEMAD, Solid
      State Comm.$\underline{19}$, 925, 1976.

(16) D.CHASSEAU, J.GAULTIER, C.HAUW, J.M.FABRE, L.GIRAL, E.TORREILLES,
      Acta Cryst.(in press).

(17) G.KERYER, J.AMIELL, S.FLANDROIS, P.DELHAES, E.TORREILLES,J.M.
      FABRE, L.GIRAL, Solid State Comm.$\underline{26}$, 541, 1978.

(18) D.JEROME, G.SODA, J.R.COOPER, J.M.FABRE and L.GIRAL, Solid State
      Comm.$\underline{22}$, 319, 1977.

(19) J.M.ZIMAN, Electrons and Phonons p.367.Clarendon Press,Oxford
      1960.

(20) J.R.COOPER, Phys.Rev.B, Comments in press.

(21) D.JEROME, J.Physique Lett.$\underline{38}$, L-489,1977.

(22) R.H.FRIEND, M.MILJAK, D.JEROME, D.L.DECKER and D.DEBRAY, J.Phy-
      sique Lett.$\underline{39}$, L-134,1978.

(23) A.J.SCHULTZ, G.D.STUCKY, R.H.BLESSING and P.COPPENS, J.Am.Chem.
      Soc.$\underline{98}$, 3194, 1976.

(24) S.BOUFFARD and L.ZUPPIROLI, Solid State Comm.,in press.

(25) J.P.FERRARIS, T.F.FINNEGAN, Solid State Comm.18, 1169, 1976.

(26) This problem has been raised by L.P.GORKOV. The problem of ap-
      plicability of the Boltzmann equation in Quasi-One-Dimensional
      conductors at high temperature is still not settled.

(27) J.C.SLATER, Phys.Rev.57, 744, 1940.

(28) J.J.HOPFIELD, Superconductivity in d and f Band Metals p 358,
      D.H.DOUGLAS AIP N Y 1972.

(29) L.CARON, MILJAK and D.JEROME, J.Physique, December 1978.

(30) The magnetic properties of TTF-TCNQ and HMTSF-TCNQ are treated
      extensively in the article of L.G.CARON and D.JEROME.Proceedings
      of the Conference on One-Dimensional Conductors Dubrovnick 1978.

(31) L.P.GORKOV,I.E.DZYALOSHINSKII, JETP Lett 18, 401, 1974.

(32) P.E.SEIDEN proposed such a mechanism at the Dubrovnik Conferen-
      ce 1978.

(33) This possibility has also been suggested by L.P.GORKOV, private
      communication.

(34) V.F.KAMINSKII, M.L.KHIDEKEL, R.B.LYBOVSKII, I.F.SHCHEGOLEV, R.
      P.SHIBAEVA, E.B.YAGUBSKII, A.V.ZVARYKINA, G.L.ZVEREVA, Phys.
      Stat.Sol (a) 44, 77, 1977.

(35) Unpublished results, Orsay.

(36) C.S.JACOBSEN, K.MORTENSEN, J.R.ANDERSEN, K.BECHGAARD, Phys.Rev.
      B, 18,905,1978.

(37) P.A.LEE, T.M.RICE, R.A.KLEMM, Phys.Rev.B 15 2984, 1977.

(38) S.BARISIC, J.Physique C 2, 262, 1978.

(39) N.MENYHARD, Sol.State Comm 21, 495, 1977.

(40) D.L.MILLS and P.LEDERER, J.Phys.Chem.Solids 27, 1805, 1966.

(41) D.A.ALLENDER, J.W.BRAY, J.BARDEEN, Phys.Rev.B 9, 119, 1974.

(42) A.J.HEEGER, Chemistry and Physics of One Dimensional Metals
      p 87, H.J.KELLER, Plenum Press 1977.

(43) E.M.CONWELL, Phys.Rev.Lett, $\underline{39}$, 777, 1977 and Proceedings of the Dubrovnik Conference, 1978.

(44) H.GUTFREUND and M.WEGER, Phys.Rev.B $\underline{16}$, 1753, 1977.

(45) P.E.SEIDEN, Proceedings of the Dubrovnik Conference 1978.

(46) See for example section III of Organic Conductors and Semi-conductors, Springer Verlag 1976.

THE ELECTRICAL CONDUCTIVITY OF TTF-TCNQ AND RELATED COMPOUNDS

Meir Weger

Hebrew University

Jerusalem, Israel

Charge-transfer complexes such as TTF-TCNQ have been the center of intensive research because of their high electrical conductivity, which is close to that of metals. Therefore, any understanding of these interesting materials, as well as a meaningful discussion about future progress in their research, requires some understanding of the mechanism giving rise to the electrical resistivity. Nevertheless, the understanding of the resistivity has not been in the center of research in this field. Still, several conflicting theories have been proposed to account for the resistivity, and caused quite a measure of controversy. In contrast, the experimental situation is far more satisfying. The electrical resistivity, its temperature and pressure dependence, is by now well established. Experiments on the frequency dependence and the effect of radiation damage exist. The Hall mobility and magnetoresistance have been investigated; related transport phenomena, such as thermoelectric power, heat conductivity, and spin diffusion are known. A rather large number of high-conductivity charge transfer complexes are by now known. Recently Shubnikov-de Haas oscillations were seen in HMTSF-TCNQ, providing the most striking evidence for the metallic nature of this substance.

The enormous extent of experimental data available, raises the question, how is it possible that so many mutually-inconsistent theories can be proposed. In the past, the temperature-dependence of the resistivity in ordinary metals, perhaps supplemented by thermal conductivity and TEP, was sufficient to establish the Bloch theory of metallic resistivity. Moreover, solid state physics is by now better understood than 40 years ago. In this work, I shall give a short resume of the experimental data, of the salient features of the various theories, and show how the libron theory, developed mainly in Jerusalem, accounts for all the experimental observations.

EXPERIMENTAL FEATURES OF THE RESISTIVITY

1)   Absolute Value

The conductivity of TTF-TCNQ at ambient temperature is about (1) 500-800 $(\Omega\text{-cm})^{-1}$. Probably the higher value represents an "ideal" material (2) (large, in particular long, crystals to ensure a uniform current flow in spite of the high anisotropy).

2)   Temperature Dependence

At ambient pressure, the resistivity follows the law (3) $\rho=\rho_0+BT^n$ with $n\approx2.3$. For ideal samples, ρ_0 is close to zero, and B is sample-independent for good samples (4), i.e. Matthiessen's law is obeyed. At pressures in the range 2-8 kbar, n falls to 2 (5); at higher pressures ($\sim$30 kbar), n falls to close to 1 (6).

3)   Pressure Dependence

At ambient temperature, $d\ln\rho/d\ln b\approx56$ for good samples. This value is extremely large - values of order 6 are typical for ordinary metals. The logarithmic derivative falls to about 32 at 100 K; below 80 K, it falls very rapidly (7). Above ambient, it falls slightly. The logari-thmic derivative increases slighly with increasing pressure.

This large pressure dependence creates a qualitative difference from ordinary metals. Virtually all electronic properties are strongly pressure dependent (8) (susceptibility, NMR relaxation time, EPR line-width, etc.) As a result, pressure plays an extremely important role in the investigation of organic metals - analogous to the role of a magnetic field, at low temperatures, for ordinary metals.

4)   Temperature Dependence at Constant Volume

In principle, this is not an independent experimental property, since it follows from the temperature dependence at constant pressure, and pressure dependence at constant temperature. Nevertheless, because of the emphasis put on this quantity by Cooper and Jerome (9), it deserves a separate discussion. Above 150 K, $\rho_V=C(T-T_0)$, with $T_0\approx60$ K. Around 120 K, $\rho=-a+bT^{n'}$, with $n'\approx1.4$ (7).

This pseudo-linear law with negative intercept, is characteristic of S-shaped curves; at low temperature, $\partial^2\rho/\partial T^2>0$; if at high temperature $\partial^2\rho/\partial T^2<0$, then the second derivative must vanish in between; thus the ρ vs. T curve must be straight there; and since $\rho(0)=0$, this straight line must have a negative intercept. Here, the straight section is particularly long (at least up to 300 K, probably up to 400 K as well. The material is not stable above 400 K).

5)   Frequency Dependence

The conductivity drops rapidly at frequencies of order 10 cm^{-1}, particularly at low temperatures ($\sim$60 K), according to Heeger(2). It starts to rise slowly around 100 cm^{-1} until 1000 cm^{-1}, from where it drops again, following the Drude law, up to the plasma frequency. Unfortunately, the striking drop at low frequencies has not been verified by other groups.

6.   Radiation Damage Dependence

Radiation damage (10) changes the temperature dependence at ambient pressure to $\rho=\rho_0'+AT+BT^n$, i.e. introduces a linear term of up to $0.15\mu\Omega$-cm/deg, while B approximately doubles for heavy radiation damage (11). In HMTSF-TCNQ at low temperatures, each radiation-induced defect renders a region with radius of order 30 Å, insulating   (12).

7.   Dependence Upon Chemistry

Selenium compounds (TSF-TCNQ, HMTSF-TCNQ) have a higher conductivity than the analogous sulphur compounds; attaching tri-methylene bridges also increases the conductivity.The conductivities of the individual stacks are approximately given by $\sigma(HMTSF)>\sigma(TSF)>\sigma(HMTTF)>\sigma(TCNQ)>\sigma(TTF)$. The conductivity of a given stack may be slightly different in different materials (13). Attaching CH_3 groups to the TCNQ ring reduces the conductivity. The donor/acceptor ratio of the conductivities can be estimated from the thermoelectric power (14) or the Hall Effect (15).

8.   Dielectric Constant

The dielectric constant is the imaginary part of the conductivity $\varepsilon=(4\pi/\omega)$ $Im(\sigma)$. It is positive at ambient temperature, and about 2000-3000 in TTF-TCNQ, and about 6000 in TSF-TCNQ (16). At low temperatures it has not yet been properly measured, because the microwaves do not penetrate into the high-conductivity material.

9.   Thermal Conductivity

The electronic part of the thermal conductivity (about 10% of the total around 60 K) obeys the Wiedemann-Franz law approximately(17).

10.   Transverse Conductivity

The anisotropy in TTF-TCNQ and TSF-TCNQ at ambient is about 300-600 (2). In HMTTF-TCNQ and HMTSF-TCNQ it is about 30 in one direction, and 500 in the other (18). It increases somewhat at low temperatures (19) and is nearly pressure independent (20).

11.   Relation with Escape Time from a Chain

The escape time of an electron from a chain, $\tau\perp$, can be measured by NMR (21), from the frequency dependence of the relaxation time. This time is related to the transverse conductivity by the Einstein relation, $\sigma\perp =[n(\varepsilon_F)\ k_BT]\ e\mu\perp$; $\mu\perp=eD/k_BT$; $D\perp=d^2/\tau\perp$, where d is the interchain distance, $\tau\perp\approx10^{-11}$ sec $\approx$ 3000 $\tau\parallel$ at ambient temperature.

12.   Spin Diffusion

The spin-diffusion constant is approximately equal to the momentum-diffusion constant, $D_\parallel^{spin}(T,P)\approx v_F^2\ \tau_\mu= D_\parallel^{momentum}(T,P)$.

It is measured by NMR from the frequency dependence of the relaxation time (21).

13.   Magnetoresistance

HMTSF-TCNQ possesses a large, anisotropic magnetoresistance at low temperatures (below 50 K). A field of 3T doubles the resistance (22). The resistance starts to increase quadratically with the magnetic field, but soon the increase becomes linear.

14. Hall Mobility

The Hall mobility of TTF-TCNQ is rather low (15). $R_H \simeq 1/nec$, where n is the density of conduction electrons (~ 0.6 per molecule), leading to $\mu \simeq 2$ cm^2/Vs at ambient, increasing to about 50 at 60 K. In HMTSF-TCNQ, μ increases to 40000 cm^2/Vs at helium temperatures, under pressure (18). This increase is associated with a decrease in carrier concentration to about 1 per 1000 molecules.

15. Relation with Diamagnetism

In HMTSF-TCNQ at low temperatures, where the magnetoresistance and Hall mobility build up, a diamagnetic susceptibility appears, which overwhelms the spin paramagnetism (23).

16. Relation with Spin Susceptibility

The spin susceptibility follows the behaviour of the mean-free-path ℓ; when ℓ^{-1} increases, so does χ_s, and when it decreases, say by the application of pressure, so does χ_s (24,25).

17. Relation with $T_1 (H \rightarrow \infty)$

At high temperatures, the NMR relaxation time in the limit of high field is proportional to the electrical conductivity. At low temperatures, $T_1(H \rightarrow \infty)T \simeq$ const., but the value of T_1 is considerably shorter than predicted by the Korringa relation (21). In HMTSF-TCNQ, T_1 is closer to the Korringa value (23).

18. Broadening of Diffuse X-Ray Diffraction Lines

Diffuse reflection lines are observed at $2k_F$ and $4k_F$ (26). They are thermally broadened. Naively, one might expect a broadening due to the finite mean-free-path of the electrons. However, the $4k_F$ reflections in particular, are much narrower than the inverse mean free path, at ambient temperature.

19. Non Linearity

There is a slight non-linearity of the I-V characteristics at low temperatures (around 100 K). This non-linearity has been measured up to now only at microwave frequencies (27).

PROPOSED THEORIES

Theories can be classified as electron-phonon and electron-electron ones. In the first, resistivity is due to scattering of electrons by phonons, and in the second, by other electrons. Also, theories cna be classified as single-particle and collective ones.

Electron-Phonon Theories
Single-Particle Theories

i) The most obvious theory is scattering by ordinary translational phonons, which is the mechanism responsible for resistance in ordinary metals. Difficulties with this mechanism are, the T^2 dependence (at constant pressure); the very strong pressure dependence ($\rho \propto 1/\omega^2$ for this mechanism, and $-\partial \ln \omega / \partial \ln b = 10$; thus it is difficult to account for $\partial \ln \rho / \partial \ln b = 50$. Even if we postulate an unreasonably large effect due to change in charge-transfer (28), this does not account for TTF-Br$_x$, where the charge-transfer does not change under pressure (9)).

With the constant volume curve, it is hard to account for the negative intercept at T=0, and the $T^{1.4}$ law around 120 K, as well as the high value of ρ , which requires $\lambda>1$ or more, a value consistent with the Peierls transition at 53 K (yielding $\lambda\approx0.2$).

ii) Molecular librations interact with electrons in second order only, by symmetry (transverse phonons, in the a-direction, do so as well). As a result, $\rho\propto T^2/\omega^4$. Thus, they account for all experimental properties, as will be discussed in detail in the next section.

iii) Intra-molecular vibrations interacting with the electrons, were invoked as a source of the Peierls transition already in 1973 (29). They were invoked by Conwell (30) to account for the resistivity. The distribution of frequencies may perhaps account for the non-linear temperature dependence at P=0, however, the pressure dependence cannot be explained by this mechanism (9).

Collective Mode Theories

iv) Allender, Bray, and Bardeen suggested in 1974 (31) fluctuations into a Peierls state (with mobile charge density waves) as a mechanism for enhanced conductivity. The lack of critical enhancement of the conductivity near T_p, as well as the pressure dependence, seem to rule against this mechanism.

v)   Lee, Rice and Anderson suggested (32) that the mean-field Peierls temperature is around 500 K, and thus only the phase of the CDW fluctuates (phasons). This was suggested to account for an exponential increase in the conductivity suggested by Heeger in 1973 (and since, discarded).

vi) Abrikosov (33), Rashba (34), and Cohen (35) suggest localisation of electrons by phonons , the localisation being characteristic to 1-D systems. The Russians suggest this mechanism for low-conductivity materials, such as NMP-TCNQ. Cohen suggests this mechanism may apply to TTF-TCNQ as well. The explanation of the T^2 law by this mechanism is tenous, since it invokes diffusive motion ($\ell<b$), while the T^2 law is observed for coherent motion ($\ell>b$); the pressure dependence is not accounted for at all. Nor is this, as well as the other collective mechanisms, consistent with the anisotropic magnetoresistance which proves a single-particle motion, and the increase in the Hall mobility to 40000, which proves the 1-electron Bloch picture (36). In materials such as NMP-TCNQ, where the conduction is diffusive, there is no T^2 law. ρ actually increases when the temperature falls.

Electron-Electron Theories

vii) Seiden and Cabib (37) suggested Baber scattering between electrons on the same molecule as a source of resistivity, to account for the T^2 law. Actually, in 1 dimension, Baber scattering gives rise to a T law; moreover, electron-electron collisions do not serve to transfer momentum to the lattice. Nor is there a strong pressure dependence for this mechanism.

viii) Seiden (38) suggested electron-electron scattering between
electrons on neighbouring chains as a source for the resistivity.
He found a scattering proportional to ω_p^{-16} ; thus the small pressure
dependence of the plasma frequency ($-\partial \ell n \omega_p / \partial \ell nb \approx 2-3$) may account
for the large pressure dependence of the resistivity.   Still,
there is no mechanism for momentum transfer to the lattice. Also, the
mobilities on both donor and acceptor stacks should be about the same
for donor-acceptor scattering, while experimentally the mobility of
the TCNQ stack is 3-4 times (at least) as big as the TTF stack mobility.

ix)   Jerome (9) suggested scattering of electrons by paramagnons.
This accounts for the correlation between the pressure dependence of
the susceptibility, and that of the resistivity. The pseudo-linear
constant volume resistivity is invoked to justify this mechanism.
However, it is not clear where the negative intercept comes from.Also,
it is not clear how the momentum is transfered to the lattice. Also,
the use of the RPA when $n(\varepsilon_F)U \approx 1$ is very questionable. Also, U has to
change by a factor of 2 for a 2% expansion of the lattice. Nor is
there any evidence for the existence of paramagnons.

LIBRON THEORY

Interaction of electrons with librons was first proposed by
Morawitz (39). A Linear Interaction between CDW's and librons on
Neighbouring Chains (LINC) was proposed to account for the cascade of
phase transitions (at 53 K, 49 K, 38 K), the $4k_F$ reflections, and the
transverse period of the CDW (40). For interaction between librons
with rotation axis in the bc* plane, and electrons on the same chain,
$\partial t/\partial \theta = 0$ by symmetry. Thus, $t = t^\circ + \frac{1}{2} \partial^2 t/\partial \theta^2)\theta^2$ where t is the
transfer integral, and we have a Quadratic Interaction (QISC). By the
Golden Rule, $$1/\tau_\parallel = (2\pi/\hbar)[\tfrac{1}{2}\partial^2 t/\partial \theta^2 <\theta^2>]^2 n(\varepsilon_F)$$
for a libron frequency $\hbar \omega_L < k_B T$, $<\theta^2> = k_B T/I \omega_L^2$
thus, $\rho = (m^*/ne^2)(1/\tau_\parallel) = (2\pi/\hbar)(\tfrac{1}{2}\partial^2 t/\partial \theta^2)^2 (k_B T/I \omega^2)^2 n(\varepsilon_F)$ and
$\rho \propto T^2/\omega^4$. Because the molecules are long, a rotation by about 5°
is sufficient to decrease the transfer integral by a factor of 2 or
so; this accounts for the absolute value of the resistivity; the T^2
law is obvious.

The pressure dependence is accounted for by using a Lennard-
Jones potential $V = a/r^{12} - b/r^6$. For such a potential, the force
constant is given by $K = 72 a/r_0^{14}$, and $d\ell nK/d\ell nr_0 = -21$. Thus, for
$\delta r/r = \delta b/b$ we have $\partial \ell n\omega/\partial \ell nb = -10.5$, and $\partial \ell n\rho/\partial \ell nb = 42$, which is
close to the experimental value which varies between about 35 and 55
for various materials (see Fig. 1).

Including thermal vibrations, $\delta \omega/\omega_0 = -10.5 <\delta r>/r_0 + 93<\delta r^2>$
$/r_0^2$. At constant volume, $<\delta r> = 0$, but $<\delta r^2> \propto T$, and therefore
$\delta \omega/\omega_0 \approx 0.2(T/300)$ (41).   Thus, $\rho_V \propto T^2/(1+0.2T/300)^4$. This relation

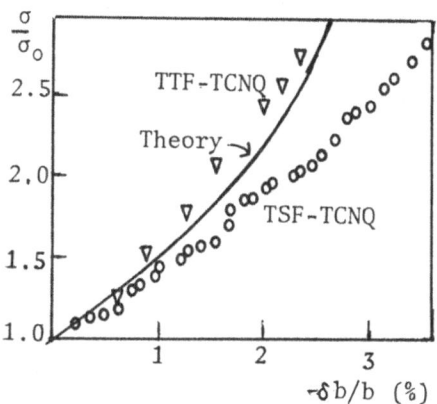

Figure 1

Pressure dependence of
the electrical conductivity
(ref. 41)

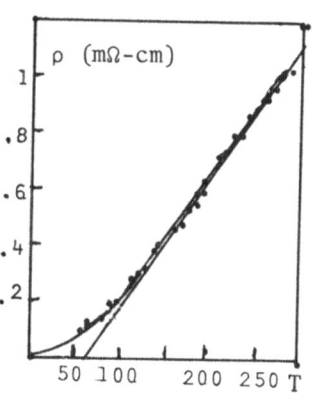

Figure 2

Temperature dependence of
the resistivity at constant b.
Experiment: ref. 7.
Theory: $\rho \propto T^2/(1+_{0.2} T/300)^4$
(ref. 41)

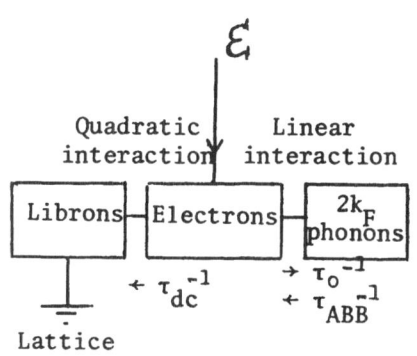

Figure 3

Momentum balance among
electrons, $2k_F$ phonons, and
general-q librons (ref. 42).
The $2k_F$ phonons are bottle-
necked at dc, but contribute
to the resistivity at high
frequencies.

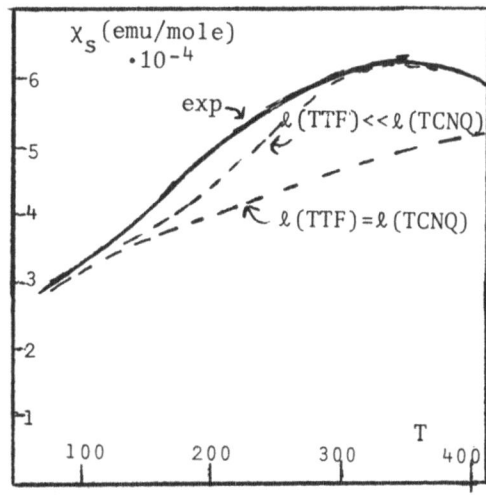

Figure 4

Enhancement of the spin
susceptibility due to localisation
by small mean free path (ref. 48).

yields the observed pseudo-linear law with astonishing accuracy (Fig. 2).

The frequency dependence is accounted for by phonon-drag (Fig.3). At dc, the $2k_F$ phonons coupled linearly to the electrons are dragged along without contributing to the resistivity (42). At $\omega \approx 10$ cm^{-1}, equilibrium with these phonons has no time to establish, and the heavy $2k_F$ phonons ($M \approx 100$ m_e according to ABB(31)) rest in place and contribute to the resistivity.

This phonon-drag effect should be quenched by defects that scatter $2k_F$ phonons, such as defects produced by irradiation. Thus, such defects should give rise to the T-law of the resistivity, which is eliminated by the phonon drag. This accounts for Heeger's observations (10,11). Also, it is clear why radiation damage affects the dc conductivity, and not the optical conductivity.

Phonon drag also accounts naturally for the peak observed in the TEP of TSF-TCNQ (43). In "ordinary" metals such peaks are the fingerprint of the phonon-drag effect.

The difference in the conductivity of different materials, is accounted for by the different rigidity towards librations, σ being proportional to K^2, where $K = I\omega_L^2$ is the force constant. K is particularly weak for TTF in TTF-TCNQ (as evidenced by the Debye-Waller factor (49) for librations in the molecular plane) and it is strong for HMTSF where the large seleniums, and the trimethylene bridges, contribute significantly to the rigidity. In TTF-TCNQ, the force constant for TCNQ is about twice that of TTF, accounting for the electron-like TEP and R_H .

Thus, for these properties (1-7 of the first section) the second-order electron-libron coupling is essential.

The positive dielectric constant of about 2000-3000 has been used as an argument against free-electron theory (2) since for free electrons, obeying the Drude equation, $\epsilon \approx -(\omega_p \tau)^2$ and is negative. However, recently Pohl and Pollak (44) pointed out, that a large positive dielectric constant is characteristic of insulating and semiconducting chain compounds. In TTF-TCNQ, when the chains are broken by some defect, the capacitance of the break gives rise to a large dielectric constant, of order 10^4. Near the surface, the chains are certainly broken, and because the dielectric constant there is so large, even a damaged volume of 1% is sufficient to overcome the small ($\epsilon \approx -10$) negative dielectric constant of the bulk (45).

Properties 9-15 point out to the one-electron nature of the conductivity (to be distinct from a collective effective).

The thermal conductivity obeys the Wiedemann-Franz law (17). For a collective motion, the electrical conductivity of the electrons would be thousands of times larger than their thermal conductivity.

The transverse conductivity was shown to be proportional to $\tau_\perp^{-1}$. The escape rate $\tau_\perp^{-1}$ is given by: $1/\tau_\perp = (2\pi/\hbar) |t_\perp|^2 \tau_\parallel / h$ (21), where $t_\perp$ is the interchain hopping matrix element and $\tau_\parallel$ the longitudinal collision time. Since $\sigma_\| = ne^2\tau_\|/m^*$, the anisotropy is independent of $\tau_\|$ and thus nearly constant. Small deviations occur when $\hbar/\tau_\| < k_BT$, a situation that exist at low temperatures (46).

The spin-diffusion constant equals the free-electron value. Electron-electron collisions should cause spin flips, without delivering momentum to the lattice, and thus invalidate this relationship. Since this relationship holds, these collisions (as described by the Hubbard hamiltonian, say) are not very strong.

The galvanomagnetic properties of HMTSF-TCNQ require a one-electron picture. The anisotropic magnetoresistance, and Landau-Peierls diamagnetism, prove the 3-dimensional motion of the electrons. The increase of 3 orders in the Hall constant at low temperatures, shows that the volumes of the electron and hole pockets at those temperatures are small, and thus the small transverse hopping integral $t_\perp$ is sufficient to overcome thermal smearing, and all other possible effects, such as CDW motion, strongly correlated electrons ("spinless fermions") characteristic of "Big U" theories, etc. The Bloch band formalism is absolutely essential to establish these small electron and hole pockets (8,36).

The spin susceptibility has been a subject of extensive study (2,47). A Curie-like χ_s is characteristic of a localised state, while a Pauli χ_s characterises an itinerant state. Thus, the changeover of χ_s from a Pauli-like behaviour at low temperatures to a Curie-like one near ambient, indicates localisation. However, any localisation can account for it. The small mean-free-path ($\ell < b$ for TTF) clearly indicates localisation, and a quantitative calculation of the increase in χ_s due to "small ℓ" exists (48)(see Fig.4). Localisation by other factors, such as "Big U", also gives rise to an enhancement of χ_s, but cannot account for the temperature and pressure dependence, since U is practically constant.

The NMR relaxation time at infinite magnetic field is related to $Im(\chi_s)$. (At finite fields, effects characterised by D^{spin} and τ enter). For $\ell k_F \gg 1$, the Korringa relation may apply (as in ordinary metals). For much smaller ℓ, the relaxation rate should diverge, since for a localized state, the spin precesses at the hyperfine frequency (a few hundred MHz). For $\ell < b$, we may estimate T_1 by the standard BPP expression; $1/T_1 = (\gamma_N H_{hf})^2 \tau_{hop}/(1+\omega_N^2 \tau_{hop}^2) \propto \tau_{hop}$, since $\omega_N \tau_{hop} \ll 1$. The conductivity is proportional to the diffusion constant $D = b^2/\tau_{hop}$; thus T_1 is proportional to σ in this region, in accord with experiment. Again, the source of localisation is irrelevant; "small ℓ" is as good as "Big U"; however, "small ℓ" accounts automatically for the T and P dependence, while "Big U" does not.

The narrow width of the $4k_F$ diffuse reflection lines (in particular) is not yet properly accounted for, since $\ell \approx b$, and the Fermi

surface should be smeared out. Apparently, part of the effect of
"small ℓ" is to renormalise the band structure (qualitatively, like
in Holstein's polaron theory), and only a part contributes to the
width of the diffraction lines (48).

Acknowledgements

The work on the libron theory started during a sabbatical in
Orsay in 1976, with Prof. Friedel. It was applied to the metallic
state of TTF-TCNQ in Jerusalem, with Prof. Gutfreund, and the phonon-
drag work was carried out with Prof. Gutfreund and Dr. Kaveh.
Support for the work is provided by a grant from the National
Research Council for Research and Development, Israel, and the KfK
(Kernforschungszentrum, Karlsruhe, GmbH), Germany.

REFERENCES

1.   G.A. Thomas et al, Phys. Rev. B13, 5105 (1976).
2.   A.J. Heeger in "Chemistry and Physics of One Dimensional Metals",
     H.J. Keller Editor, NATO ASI Series B25, Plenum Press, New York,
     1977.
3.   R.P. Groff, A. Suna, and R.E. Merrifield, Phys. Rev. Lett., 33,
     418 (1974).
4.   J.P. Ferraris and T.F. Finnegan, Solid State Commun., 18, 1169
     (1976).
5.   J.R. Cooper, D. Jerome, M. Weger, and S. Etemad, J. Physique
     Lett., 36, L-219 (1975).
6.   C.W. Chu, J.M. Harper, T.H. Geballe, and R.L. Greene, Phys. Rev.
     Lett., 31, 1431 (1973).
7.   R.H. Friend, M. Miljak, D. Jerome, D.L. Decker and D. Debray,
     J. Physique Lett., 39, L-134 (1978).
8.   D. Jerome and M. Weger in Ref. 2.
9.   J.R. Cooper, Phys. Rev. B Comments (in press); D. Jerome, J.
     Physique Lett., 38, L-489 (1977).
10. C.K. Chiang, M.J. Cohen, P.R. Newman, and A.J. Heeger, Phys. Rev.
     B16, 3163 (1977).
11. M. Kaveh, H. Gutfreund, and M. Weger, Phys. Rev. B (in press).
12. L. Zuppiroli, J. Arionceau, M. Weger, K. Bechgaard, and C. Weyl,
     J. Physique Lett., 39, L-170 (1978).
13. C.S. Jacobsen, K. Mortensen, J.R. Andersen, and K. Bechgaard,
     Phys. Rev. B18, 905 (1978).
14. P.M. Chaikin, R.L. Greene, S. Etemad, and E.M. Engler, Phys.
     Rev. B13, 1627 (1976).
15. J.R. Cooper, M. Miljak, G. Delplanque, D. Jerome, M. Weger, J.
     M. Fabre, and L. Giral, J. Physique, 38, 1097 (1977).
16. W.J. Gunning, A.J. Heeger, I.F. Shchegolev, and S.P. Zolotukia,
     Solid State Commun., 26, 981 (1978).
17. M.B. Salomon, J.W. Bray, G. de Pasquali, R.A. Craven, G. Stucky,
     and A. Schultz, Phys. Rev. B11, 619 (1975).

18. J.R. Cooper, M. Weger, G. Delplanque, D. Jerome, and K. Bech-
    gaard, J. Physique Lett., 37, L-349 (1976).
19. L. Coleman, Thesis, University of Pennsylvania, 1975 (unpublished).
20. J.R. Cooper, S. Etemad, D. Jerome, and E.M. Engler, Solid State
    Commun., 22, 257 (1977).
21. G. Soda, D. Jerome, M. Weger, J. Alizon, J. Gallice, H. Robert,
    J.M. Fabre, and L. Giral, J. Physique, 38, 931 (1977).
22. J.R. Cooper, M. Weger, D. Jerome, D. Lefur, K. Bechgaard, D.O.
    Cowan, and A.N. Bloch, Solid State Commun., 19, 749 (1976).
23. G. Soda, D. Jerome, M. Weger, K. Bechgaard, and E. Pedersen,
    Solid State Commun., 20, 107 (1976).
24. J.C. Scott, A.F. Garito, and A.J. Heeger, Phys. Rev. B10, 3131
    (1974).
25. C. Bertier, J.R. Cooper, D. Jerome, G. Soda, C. Weyl, J.M.
    Fabre, and L. Giral, Mol. Cryst. Liq. Cryst., 32, 267 (1976).
26. S. Kagoshima, H. Anzai, K. Kajimura, and T. Ishiguro, J. Phys.
    Soc. Japan, 39, 1143 (1975); J.P. Pouget, S.K. Khana, F.
    Denoyer, R. Comes, A.F. Garito and A.J. Heeger, Phys. Rev. Lett.,
    37, 437 (1976).
27. K. Seeger and A. Maurer, Solid State Commun. (in press).
28. E.M. Conwell, Phys. Rev. B Comments (in press).
29. H. Gutfreund, B. Horovitz, and M. Weger, J. Phys. C7, 383
    (1974).
30. E.M. Conwell, Phys. Rev. Lett., 39, 777 (1977).
31. D. Allender, J.W. Bray, and J. Bardeen, Phys. Rev. B9, 119 (1974).
32. P.A. Lee, T.M. Rice, and P.W. Anderson, Phys. Rev. Lett., 31,
    462 (1973).
33. A.A. Abrikosov and I.A. Ryzhkin, Solid State Commun., 24, 319
    (1977).
34. E.I. Rashba, A.A. Gogolin, and V.I. Melnikov, in "Organic Con-
    ductors and Semiconductors", conference proceedings, Siofok
    Hungary 1976, Lecture Notes in Physics, 65, Springer Verlag
    (1977), p. 265.
35. M.H. Cohen, ibid., p. 225.
36. M. Weger, Solid State Commun., 19, 1149 (1976).
37. P.E. Seiden and D. Cabib, Phys. Rev. B13, 1846 (1976).
38. P.E. Seiden (unpublished).
39. H. Morawitz, Phys. Rev. Lett., 34, 1096 (1975); H. Morawitz, ref.
    34, p. 303; H. Morawitz, in "Synthesis and Structure of Low
    Dimensional Systems", Proceedings of a conference, New York
    Academy of Sciences, New York, 1977.
40. M. Weger and J. Friedel, J. Physique, 38, 241 (1977); ibid 38,
    881 (1977).
41. H. Gutfreund and M. Weger, Phys. Rev. B16, 1753 (1977); H. Gut-
    freund, M. Weger, and M. Kaveh, Solid State Commun., 27, 53
    (1978).
42. M. Weger and H. Gutfreund, Comments on Solid State Physics, 8,
    135 (1978).
43. P.M. Chaikin, R.L. Greene, and E.M. Engler, Solid State Commun.,
    25, 1009 (1978).

44. H.A. Pohl and M. Pollak, J. Chem. Phys., 66, 4031 (1977).
45. M. Weger and M. Kaveh (unpublished).
46. M. Weger, in proceedings of LT15 conference, J. Physique (in press).
47. J.B. Torrance, Y. Tomkievicz, and B.S. Silverman, Phys. Rev. B15, 4738 (1977).
48. S. Marianer, M. Weger, and H. Gutfreund (unpublished).
49. A.J. Schultz, G.D. Stucky, R.H. Blessing, and P. Coppens, J. Am. Chem. Soc., 98, 3194 (1976).

EFFECTS OF IMPURITIES ON THE ORDERED PHASES

OF ONE-DIMENSIONAL SYSTEMS

H. Gutfreund and W. A. Little*

The Racah Institute of Physics, The Hebrew U., Israel

Physics Department, Stanford U., Stanford, CA*

It has long been known that disorder or the presence of impurities cause localization of the single-electron states in one-
dimensional systems (1). In addition to this we shall show that
such impurities, in general, can have a large effect upon the
degree of order which tends to appear as T→0 and upon the mean
field transition temperature T_c of any such ordered phases. However, this effect is selective and depending upon the nature of
the scattering introduced by the impurities can severely suppress
one type of order while leaving another virtually unaffected. The
result which we obtain is a generalization to all types of order
in one-dimensional systems, of the Anderson-Maki theorem(2,3) on
the insensitivity of the singlet superconducting transition temperature to the presence of non-magnetic impurities, while showing a marked depression of T_c with increasing number of mangetic
impurities. We show that to each type of ordered state in one
dimension, whether it be a charge density wave (CDW), spin density wave (SDW), triplet superconductivity (TS) or singlet superconductivity (SS) there exists a type of impurity scattering
mechanism which has a negligible effect upon the mean field transition temperature of this state. Impurities whose scattering is
not of this type have a severe effect upon T_c.

The physical reason for this behaviour is as follows. In the
pure state each of the ordered states is characterized by the
pairing of two electron states or an electron and a hole state,
each of well defined momentum and spin orientation. These pairs
define the order parameter of each of the phases. Upon the introduction of the impurity these momentum and spin states become
mixed. However, certain types of mixing leave the gap equation,
which determines the order parameter for each of the phases,

essentially unchanged. The best known example of this is the
effect of magnetic and non-magnetic impurities upon singlet superconductivity in three dimensional metals. As shown first by
Anderson non-magnetic impurities mix states of different momenta
but leave the gap equation for singlet superconductivity virtually
unchanged. This follows because the gap equation results from the
pairing of time reversed states and the time reversed nature of
the paired states is preserved even in the presence of non-
magnetic impurities. Magnetic impurities on the other hand mix in
states of a different symmetry and this results in a destructive
interference between terms in the gap equation and a drastic
reduction in T_c.

For our work in order to describe the Peierls state, the
superconducting states and the effects of both magnetic and non-
magnetic scattering it is convenient to use a Nambu representation
similar to that used by Horovitz (4). We describe the system in
terms of an 8-component vector,

$$\Psi^\dagger(k) \equiv \left(\psi^\dagger_{1\uparrow}(k), \ \psi^\dagger_{1\downarrow}(k), \ \psi^\dagger_{2\uparrow}(k), \ \psi^\dagger_{2\downarrow}(k), \ \psi_{1\uparrow}(-k), \ \psi_{1\downarrow}(-k), \right.$$

$$\left. \psi_{2\uparrow}(-k), \ \psi_{2\downarrow}(-k) \right) \qquad\qquad (1)$$

in which $\quad \psi^\dagger_{1s}(k) \equiv \psi^\dagger_s (k_f + k)$

and $\quad \psi^\dagger_{2s}(k) \equiv \psi^\dagger_s (-k_f + k)$, where k is small and s is the spin
state. Then each of the different order parameters can be des-
cribed by a characteristic 8 x 8 matrix operating on this vector
space. This 8 x 8 matrix can be decomposed into a direct product
of three 2 x 2 Pauli matrices or the 2 x 2 unit matrix (e.g.,
$\sigma_i \cdot \tau_j \cdot \rho_k$).

The first matrix σ_i describes the type of spin pairing, the
second τ_j the pairing of states on one side or the other of the
Fermi surface and the third ρ_k the pairing of particles or of
particles and holes. As an example, singlet superconductivity
which is described by the order parameter
$< \psi^\dagger_{1\uparrow}(k) \ \psi^\dagger_{2\downarrow}(-k) \ - \ \psi^\dagger_{1\downarrow}(k) \ \psi^\dagger_{2\uparrow}(-k)>$ would be described by the
matrix $(i\sigma_2 \cdot \tau_1 \cdot \rho^\dagger)$ where $\rho^\dagger = 1/\sqrt{2} \ (\rho_1 + i\rho_2)$.

Similarly, terms in the Hamiltonian which represent the various scattering mechanisms can likewise be given a matrix representation. Using methods similar to that used by Maki (3) one can

calculate the gap equation for each of the ordered states. Then
one can show that depending on the commutation or anti-commutation
properties of the order parameter matrix with the particular impurity matrix, T_c is either strongly suppressed (commutes) or is
left virtually unchanged (anti commutes). The matrix method is
simply a compact way of describing this calculation.

We have considered four types of impurity scattering mechanisms:
(a) forward scattering with no spin flip, (b) forward scattering
with spin flip, (c) backward scattering with no spin flip and
(d) backward scattering with spin flip. If the scattering matrix
commutes with a particular order parameter matrix then the effect
upon T_c can be described in each case by a parameter

$$\alpha_c \equiv \frac{\hbar}{2\pi k T_c \tau}$$ where τ is now the scattering time due to the

particular type of scattering processes a, b, c or d above. If
T_{c0} is the mean field transition temperature in the absence of
impurities and T_c the actual transition temperature then in each
case one finds

$$\log (T_c/T_{c0}) = \psi(\tfrac{1}{2}) - \psi(\tfrac{1}{2} + \alpha_c), \qquad (2)$$

$$\text{where} \quad \psi(z) \equiv \operatorname*{Lt}_{n\to\infty} \left[\log n - \sum_0^n \frac{1}{m+z} \right] \qquad (3)$$

is the digamma function (5). A strong depression in T_c then occurs
for increasing impurity concentration.

On the other hand if the scattering matrix anti commutes with
the order parameter matrix then the gap equation is essentially
unchanged and T_c remains virtually imdependent of impurity concentration. It should be noted that we are treating the impurities
in the Born approximation and have used the mean field approximation for T_c. The latter implies that we have assumed sufficient
interchain coupling that a phase transition in the one-dimensional
system can occur. Furthermore, such interchain coupling allows one
to neglect pathological effects of a rigorously one-dimensional
system which would lead to strict localization and other multiple
scattering effects.

In Table I we show the commutation (C) and anti-commutation
behaviour of the different order parameters with the various scattering mechanisms. Note that an entry with anti commutation (A)
property causes no change of T_c while the commutation (C) property

results in a depression in T_c.

TABLE I

State / Scattering Mechanism	SS	CDW	TS			SDW		
	0	0	0	1	-1	0	1	-1
Forward No Spin Flip	A	C	A	A	A	C	C	C
Forward Spin Flip	C	C	A	A	C	A	C	A
Backward No Spin Flip	A	C	C	C	C	C	C	C
Backward Spin Flip	C	C	C	C	A	A	C	A

Discussion. Previously Zavadovskii (6) had considered the effects of impurities on the superconducting and charge density wave states and showed that in the quasi classical limit where $|V| \ll E_f$, impurities do not affect the transition temperature of either. Our result however shows that when this approximation is relaxed the two phases are affected differently.

We wish to acknowledge support for this work from the Binational Fund, National Aeronautical and Space Agency, Contract JPL 953752 and National Science Foundation, Grant DMR 76-82087-A02.

REFERENCES

1. D. J. Thouless, Physics Reports 13 (3) 93 (1974).

2. P. W. Anderson, J. Phys. Chem Solids 11, 26 (1959).

3. K. Maki, "Superconductivity" Ed. R. D. Parks, Marcel Dekker, Inc., New York (1969) p 1035.

4. B. Horovitz, Solid State Commun. (USA) 18, 4, 445 (1976).

5. H. Jeffreys and B. Jeffreys, Methods of Mathematical Physics, Cambridge University Press, London (1972), p 465.

6. A. Zavadovskii, Soviet Physics JETP $\underline{27}$, 5, 767 (1968).

VALENCE BOND THEORY OF PARTLY-FILLED CONDUCTORS

Z. G. Soos, S. Mazumdar, and T.T.P. Cheung

Department of Chemistry, Princeton University
Princeton, New Jersey, 08540, U.S.A.

I. MODELING ORGANIC CONDUCTORS

The presently known[1,2] π-molecular conductors share important common features. They all contain segregated stacks ... $A^{-\gamma}A^{-\gamma}A^{-\gamma}$... of anion radicals and/or ...$D^{+\gamma}D^{+\gamma}D^{+\gamma}$... of cation radicals, as illustrated by A = TCNQ (tetracyanoquinodimethane) and D = TTF (tetrathiofulvalene). They all crystallize with a single molecule per unit cell along the stack, or high conductivity, axis. They all have partial charge transfer, or $\gamma < 1$, in contrast[2] to the semiconducting simple salts based on D^+ or A^- stacks. They are narrow-band systems, with $4t \sim 0.5$ eV and π-electron overlaps along the stack of 3.1 - 3.5 Å, or just shorter than the van der Waals separation. These common features, rather than differences in low-temperature phases, must guide the development of theoretical models.

The classification[2] of different types of π-molecular solids is naturally based on a site representation[3] involving diamagnetic D, D^{+2} and A, A^{-2} sites along with paramagnetic $D^+\alpha, D^+\beta$ and $A^-\alpha, A^-\beta$ sites arranged according to various observed structures. The good conductors have partly-filled regular segregated stacks. Site representations are exact in the limit of zero overlap (when $t \to 0$) and generalize the Heitler-London, or valence bond (VB), treatment of H_2. Simple VB theory overcorrelates electrons, just as simple molecular-orbital (MO) or band theory misses correlations by focusing on one-electron states. The demonstrated convergence of MO and VB for small systems like H_2 is irrelevant in practice for solids, where even the starting point involves approximations.

Since organic conductors have rather narrow bands, VB may well prove superior to the more commonly used and highly developed MO (or band) techniques. The need for improved computational methods, which may not necessarily require improved models, is illustrated by the static susceptibility $\chi(T)$ of TTF-TCNQ shown in Fig. 1. The magnitude of $\chi(300)$ suggests[4] strong correlations $(U > 4t)$ rather than a band $(U = 0)$ picture, since $4t \sim 0.5$ eV is needed for other data. Other organic conductors[5] show similar $\chi(T)$ behavior, in spite of having different molecules and packing. It is disconcerting that the available model solutions miss the temperature dependence in the conducting regime of $T > 60°K$. After all, $\chi(T)$ is a static property that depends only on the energy spectrum, whereas the transport and relaxation properties debated by competing models[1] also involve the dynamics.

The generalization of VB theory to partly-filled regular segregated stacks, as required for organic conductors, offers an improved treatment of various Hubbard models. The simplest case

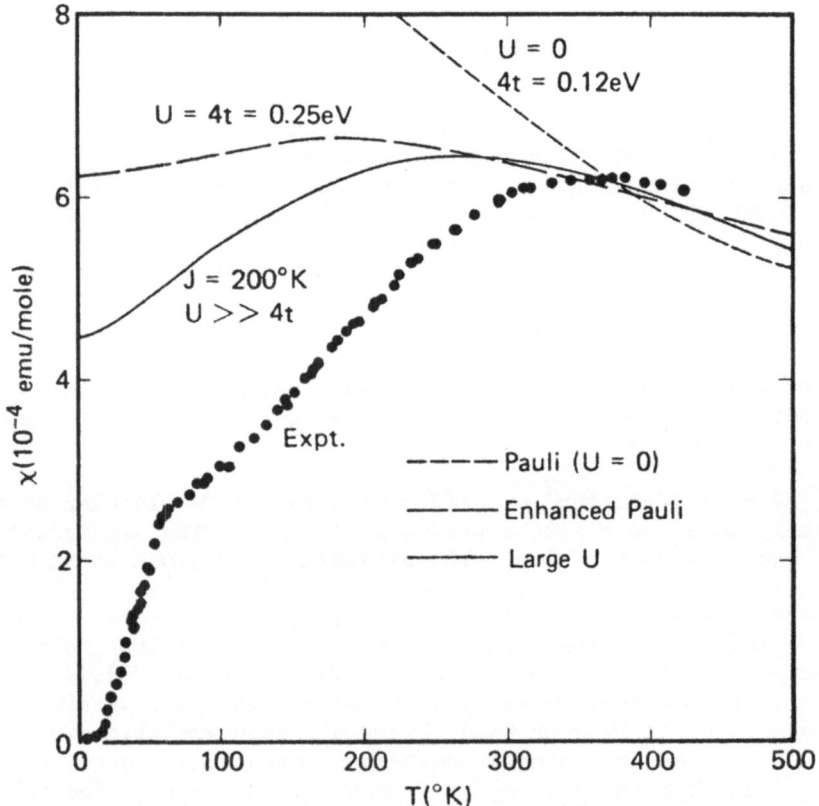

Fig. 1   Three attempts at fitting the experimental $\chi(T)$ for TTF-TCNQ at high temperature (Ref. 4).

of only on-site correlations $U > 0$ is

$$\mathcal{H}_o = t \sum_{n\sigma} (a^+_{n\sigma} a_{n+1\sigma} + a^+_{n+1\sigma} a_{n\sigma}) + U \sum_n a^+_{n\alpha} a^+_{n\beta} a_{n\beta} a_{n\alpha} \tag{1}$$

$\mathcal{H}_o$ is to be solved for γN electrons on N equivalent sites, as discussed elsewhere.[2,4] The inclusion of coulomb interactions M_n
between singly-charged sites separated by n lattice spacings is
straightforward for small n and undoubtedly leads to physically
more realistic models. The essential feature of $\mathcal{H}_o$ for VB computations is the nearest-neighbor transfer of electrons, with spin
conservation, by the t term. This connects different possible
assignments of electrons to molecular sites.

II. VB FOR SOLID STATE MODELS

VB methods[6] for molecular computations are an old, and not
entirely successful, generalization of the Heitler-London wave
function for H_2. The idea is to consider singlet-correlated
electrons, or covalent bonds, in many-electron molecules. The six
π electrons in benzene form three bonds, which can be represented
by lines, and the VB basis consists of the various ways of spin-
pairing the six electrons into three bonds. The most serious problem is that the atomic orbitals for each atom allow a potentially
infinite number of VB functions, with only "chemical intuition" to
guide efforts at minimizing the energy. Furthermore, the evaluation
of molecular integrals over these many-electron functions is difficult. The VB basis is not even orthonormal. Thus MO methods are
now overwhelmingly used for molecular computations, and correlations
are occasionally approximated by configuration interactions (CI).

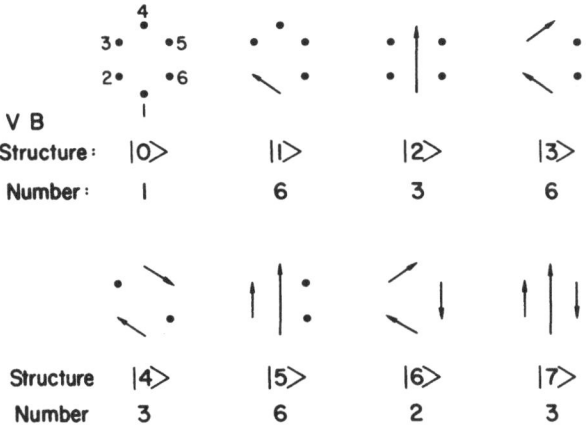

Fig. 2   VB diagrams for the singlet state of a mixed regular
stack of N = 6, with donors at 1, 3, 5, and acceptors
at 2, 4, 6 (Ref. 7).

Solid-state models typically avoid these complications. Thus the parameters t and U in Eq. (1) are given, rather than evaluated, and their magnitude is a separate question. Hubbard models are limited to four states per site, such as A, A^-_α, A^-_β, A^{-2}, even though ultimately the higher electronic states may have to be included. There are, at most, 4^N basis states for N sites and the VB description is potentially complete. The emphasis in solid-state models is on treating $N \to \infty$ sites. The lack of orthonormality is relatively unimportant. VB analyses of solid-state models, with their high symmetry, can routinely include over 10^4 states, or many more than the 10 - 100 used in molecular VB computations.

These VB ideas for ground electronic states are illustrated in Fig. 2 for a <u>mixed</u> •••DADA••• stack of N = 6 sites arranged in a ring. Diamagnetic D or A sites are represented by dots, singlet electron-hole pairs $^1|D^+A^-\rangle$ by arrows, thereby fixing phases. The 30 possible singlet VB functions in Fig. 2 reduce to at most 7 on account of the 6-fold rotational symmetry. The transfer (t) term in Eq. (1) interconnects these VB states, which consequently generalizes[2,7] the Mulliken DA dimer to one-dimensional arrays. Similar VB diagrams can be developed[7] for triplets, with $^3|D^+A^-\rangle$ represented by a notched arrow, or for higher spin states. The analysis of ground-state energies for N = 10 sites starts with almost 16,000 VB diagrams, before block diagonalizing according to spin and rotational symmetry.

As in computations on spin systems,[8] the mixed chain for N = 4, 6, 8, and 10 sites gives reliable $N \to \infty$ extrapolations. VB methods thus provide[7] the first rigorous computation for the charge-transfer γ in •••$D^{+\gamma} A^{-\gamma} D^{+\gamma} A^{-\gamma}$••• stacks and for the magnetic gap $\Delta E_m(z)$ to the lowest triplet. The model parameter $z = \delta/\sqrt{2}|t|$ relates the energy -2δ for the excitation DA $\to D^+ A^-$ to the bandwidth. The neutral ($\gamma \approx 0$, or $z \to -\infty$) stack is diamagnetic, with $\Delta E_m \sim 2|\delta|$, while the ionic ($\gamma \simeq 1$, $z \to \infty$) stack is paramagnetic, with $\Delta E_m = 0$. The neutral-ionic interface at $\gamma_c = 0.68 \pm 0.01$ for $N \to \infty$ is signaled by a change in the spatial symmetry of the ground singlet state and by the vanishing of ΔE_m. The small $\Delta E_m \sim 0.07$ eV in the 1:1 complex of TMPD (NNN′N′ tetramethyl-p-phenylenediamine) and TCNQ, which crystallizes in a mixed regular stack, then requires incomplete charge-transfer $0.6 < \gamma < \gamma_c$, in excellent agreement[9,7] with recent resonance Raman results of $\gamma \sim 0.7$ based on calibrating a $TCNQ^{-\gamma}$ vibration against γ in systems with known ionicity.

III.   PARTLY-FILLED SEGREGATED STACKS

Modeling $\chi(T)$ for organic conductors requires the complete energy spectrum for partly-filled ($\gamma < 1$) regular segregated stacks. This entails considerable modification of VB methods summarized above for ground state properties of mixed stacks. We only consider ••• $A^{-\gamma} A^{-\gamma} A^{-\gamma}$•••, since the corresponding results for

cationic systems merely involve interchanging electrons and holes.
The VB prescription is to assign γN electrons to N sites like
A, $A^-\alpha$, $A^-\beta$, or A^{-2}, and then to solve Eq. (1) or some generalization involving intersite interactions M_n. The atomic limit
$(U \to \infty)$ precludes doubly-occupied A^{-2} sites and is readily soluble,[9]
but it leads to the unsatisfactory $\chi(T) \alpha T^{-1}$ dependence because the
spins are completely decoupled. We thus consider $N_d = 0, 1,$
$\cdots \gamma N/2$ (γN even) nondegenerate doubly-occupied sites, $\gamma N - 2N_d$
doubly degenerate A_σ^- sites, and $N - \gamma N + N_d$ nondegenerate A sites.
The total number N_T of linearly independent VB functions for γN
even is consequently

$$N_T = \sum_{N_d=0}^{\gamma N/2} \frac{N! \, 2^{(\gamma N - 2N)}}{N_d! \, (\gamma N - 2N_d)! \, (N - \gamma N + N_d)!} \tag{2}$$

N_T increases rapidly with the ring size N. There are N_T eigen-
energies, many of which are degenerate, that must be included in
the partition function for $\chi(T)$ computations.

Since organic molecules have small spin-orbit coupling, the
total spin S is a good quantum number even in far more complete
model Hamiltonians than Eq. (1). In addition, the equivalence of
molecular sites in the conducting phase can be modeled by forming
a ring of N sites and classifying the exact eigenstates by the
wavevector

$$k = 0, \, \pm \frac{2\pi}{N}, \, \cdots, \, \pm \frac{(N-1)\pi}{N}; \text{ or } \cdots, \pi \tag{3}$$

for N odd and even, respectively. Thus, exact $\{S,k\}$ subspaces
occur in general for organic conductors. These block diagonalize
the N_T VB functions, as illustrated[7] in mixed stacks. Quite
generally, the ground state for open spin chains[11] is a singlet
$(S = 0)$ when the coupling is antiferromagnetic.

The construction of VB functions that are symmetry-adapted
to $\{S,k\}$ subspaces is greatly facilitated by diagrammatic methods.
The transfer term in Eq. (1) conserves spins and produces singlet-
correlated electrons when acting on $A^{-2}A$. We define the operator

$$(n,n+1)^+ \equiv (a_{n+1\alpha}^+ a_{n\alpha} + a_{n+1\beta}^+ a_{n\beta})/\sqrt{2} \tag{4}$$

which moves an electron to the right, while its adjoint moves an
electron to the left. The reduced hamiltonian $h = -H_0/\sqrt{2} \, |t|$, with
$t = -|t|$ in Eq. (1), then describes all partly-filled regular
Hubbard models in terms of the single parameter $z \equiv U/\sqrt{2}|t|$. The
atomic limit $z \to \infty$ excludes A^{-2} and sharply reduces N_T in Eq. (2)
to the $N_d = 0$ term. The band limit $z = 0$ is also exactly soluble,
but Fig. 1 shows that neither limit fits $\chi(T)$.

We define the vacuum state $|o\rangle = |\cdots AAA\cdots\rangle$ and represent empty sites by dots. The creation and annihilation operators $a^+_{n\sigma}$, $a_{n\sigma}$ in Eq. (1) automatically guarantee the proper antisymmetrization. The problem is to define phases and to interrelate a diagram like Fig. 3 with operators acting on $|o\rangle$. This can be accomplished[12] by taking crosses at doubly-occupied A^{-2} sites to be $a^+_{n\alpha} a^+_{n\beta} |o\rangle$. A normalized singlet between A^- sites at n and m is a line corresponding to $2^{-1/2} (a^+_{n\alpha} a^+_{m\beta} - a^+_{n\beta} a^+_{m\alpha})|o\rangle$. Triplets are always taken with $S_z = 1$ and are represented by an arrow from n to m for $a^+_{n\alpha} a^+_{m\alpha}|o\rangle$. States with odd numbers of electrons, or with higher spin, present no difficulties. VB diagrams like Fig. 3 are far simpler to manipulate than the earlier[6] explicit representations of VB functions.

Once the eigenergies $\lambda_n^{S,k}$ in $\{S,k\}$ for $h = -\mathcal{H}_0/\sqrt{2}|t|$ are obtained for γN electrons on N sites, we construct the partition function Z_S

$$Z_S (k_BT/\sqrt{2}|t|) = \sum_k \sum_n (2S+1) \exp\left(\frac{k_BT}{\sqrt{2}|t|} \lambda_h^{S,k}\right) . \tag{5}$$

The k values are given in Eq. (3), the n roots in $\{S,k\}$ depend on γ and on N, and $k_BT/\sqrt{2}|t|$ is a reduced temperature. Z_S represents a total spin S, which for even γN can be $S = 0, 1, \cdots \gamma N/2$. The result for $\chi N (k_BT/\sqrt{2}|t|)$ is simply

$$\chi_N = \frac{g^2 \mu_B^2}{3k_BT} Z^{-1} \sum_S S (S+1)Z_S \tag{6}$$

where $Z = \sum_S Z_S$ is the complete partition function, $\mu\beta$ is the Bohr magneton, and $g \sim 2.003$ is the g-value for the ion radicals. Although χ_N depends on the chain length N and the filling γ, extrapolations of χ_N/N give the desired $N \to \infty$ behavior of $\chi(T)$.

IV.   LOCAL NEUTRALITY AND PRELIMINARY RESULTS

The block diagonalization of N_T VB diagrams into exact $\{S,k\}$

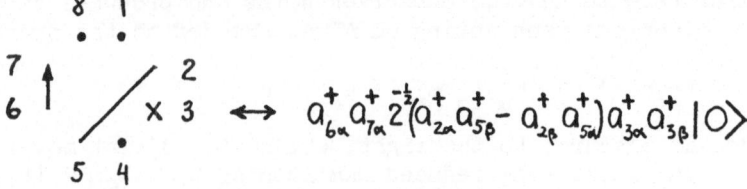

Fig. 3   Representative VB diagram for 6 electrons on 8 sites, with double occupancy at Site 3. The overall spin is S = 1, since Sites 6 and 7 are triplet correlated. The ordering of electron pairs in the explicit representation of the normalized VB ket is arbitrary.

subspaces still leads to rather large secular determinants for
$N \sim 10$. Before undertaking such computations, it is worth examin-
ing the validity of H_9. Narrow-band, quasi-one-dimensional systems
probably do not shield coulomb interactions, so that intersite
terms M_n may be important. Indeed, face-to-face stacks of planar
A or D molecules have $M_1 \sim 3$ eV, which is only slightly smaller
than the bare on-site repulsion $U_0 \sim 4$ eV. It follows that
adjacent $A^{-2} A^{-2}$ or $A^{-2} A^{-1}$ sites involve repulsions of $4M_1$ and
$2M_1$, respectively, that exceed U_0. All intersite interactions are
diagonal in the site, or VB, representation. Their inclusion is
limited only by practical considerations for the ring size N. In
this context, VB diagrams emphasize that simple solid state models
often contain severe, or even untenable, approximations.

Our current studies of partly-filled segregated[12] stacks consequently invoke a "local neutrality" condition that excludes all VB
diagrams, like $A^{-2} A^{-2}$ or $A^{-2} A^{-1}$, with more than two electrons on
any two adjacent sites. Local neutrality further reduces N_T in
Eq. (2) by eliminating some high-energy states. It automatically
retains the atomic limit $z \rightarrow \infty$, which serves as a simple check on
computations. The parameter U in Eq. (1) is then approximately
given by $U_0 - M_1 \sim 1$ eV, the excitation energy for $A^{-1} A^{-1} \rightarrow A A^{-2}$.
Such charge-transfer excitations around 1 eV characterize[2,1] both
semiconducting and conducting segregated stacks. The explicit
inclusion of isolated doubly-occupied sites with energy $z = U/\sqrt{2}|t|$
guarantees the preferential stabilization of singlet states, thus
leading to antiferromagnetic $\chi(T)$. Local neutrality is a physi-
cally-motivated improvement to H_0 that also simplifies the VB
analysis.

Although VB results for fixed N are exact, $N \rightarrow \infty$ extrapolations introduce approximations. Fortunately, $N \sim 10$ suffices[8,13]
for $\chi(T)$ extrapolations at temperatures of the order of $J \sim t^2/U \sim$
100 K, while $\chi(0)$ is exactly known[14] for any γ. Such $N \sim 10$ ring
sizes are manageable. VB analysis readily generates[12] the $\lambda_n^{S,k}(z)$
eigenvalues for any z in such simple cases as: $\gamma N = 2$, $N = 3,4$ etc;
$\gamma N = 3$, $N = 6$; and $\gamma N = 4$, $N = 6,8$. The $z \rightarrow \infty$ limit can be checked,
as can the local-neutrality condition (by restoring the VB diagrams
with $A^{-2} A^{-2}$ or $A^{-2} A^{-1}$ neighbors). We are especially interested
in $z \sim 5$, as suggested by $4t \sim 0.5$ eV and charge transfer excitations $U \sim 1$ eV.

These preliminary VB computations[12] have led to some surprises:
(1) In the special case $\gamma N = 2$, the factorization of space and spin
functions spoils the atomic limit $z \rightarrow \infty$ and produces gaps of order
$|t|$ (rather than t^2/U) between the $S = 0$ ground state and the $S = 1$
excited state. (2) While the $\gamma N = 4$, $N = 6$ open chain has the
expected[11] $S = 0$ ground state for any z in $0 \leq z < \infty$, the closed
ring has an $S = 1$ ground state for $3 < z < \infty$, presumably on account
of end effects. These small systems, as well as the mixed stack,[9]

demonstrate the computational advantages of VB. Typical band theory approaches to $\mathcal{H}_0$ for U $\gtrsim$ 4t would require prohibitively many configuration interactions to be exact. The possibility of more realistic models, as illustrated by the local-neutrality condition, shows that VB analysis of narrow-band solids offers a pwerful new approach to correlation problems. The special features of solid-state models, as opposed to molecular computations, are well suited to VB techniques. We anticipate many further developments and applications of diagrammatic VB methods to π-molecular solids.

The financial support of NSF-CHE76-07377 is gratefully acknow-ledged.

REFERENCES

1. Chemistry and Physics of One-Dimensional Metals, ed. H. J. Keller, (NATO Advanced Institute Series B25, Plenum, NY 1977). These proceedings survey current results and provide a guide to the original literature on conductors.
2. Z. G. Soos and D. J. Klein, in Molecular Association, V. 1, ed. R. Foster, (Academic, NY 1975), pp. 1-109; Z. G. Soos, Ann. Rev. Phys. Chem. 25, 121 (1974). Contains a survey and classification of π-molecular charge transfer solids.
3. D. J. Klein and Z. G. Soos, Mol. Phys. 20, 1013 (1971).
4. J. B. Torrance, pp. 137-166 in Ref. 1.
5. A. N. Bloch, T. F. Carruthers, T. O. Poehler, and D. O. Cowan, pp. 47-86 in Ref. 1.
6. L. Pauling, J. Chem. Phys. 1, 280 (1933); H. Eyring, T. Walter, and G. E. Kimball in Quantum Chemistry, (Wiley, NY 1944), Chap. 13.
7. Z. G. Soos and S. Mazumdar, Phys. Rev. B (in press); Chem. Phys. Letters, 56, 515 (1978).
8. J. C. Bonner and M. E. Fisher, Phys. Rev. 135, A640 (1964).
9. Z. G. Soos, S. Mazumdar, and T.T.P. Cheung, Fifth International Symposium on Chemistry of the Organic Solid State, Brandeis, 1978, and Mol. Cryst. (in press).
10. G. Beni, T. Holstein, and P. Pincus, Phys. Rev. B8, 312 (1973); D. J. Klein, ibid B8, 3452 (1973).
11. E. H. Lieb and D. C. Mattis, Phys. Rev. 125, 164 (1962).
12. S. Mazumdar, T.T.P. Cheung, and Z. G. Soos (unpublished results).
13. W. A. Seitz and D. J. Klein, Phys. Rev. B9, 2159 (1974).
14. H. Shiba, Phys. Rev. B6, 930 (1972).

ROLES OF CATION STACKS IN ORDERING OF CHARGE-DENSITY WAVES IN

TSeF-TCNQ: COMPARISON WITH TTF-TCNQ

S. Kagoshima*

Electrotechnical Laboratory

Mukodai 5-4-1, Tanashi, Tokyo 188, Japan

In the organic quasi-one-dimensional conductor TTF-TCNQ, the successive structural phase transitions have been studied in relation to the ordering of the charge-density waves (CDWs), which are the coupled modes of lattice distortions and one-dimensional electrons. The origin of the successive occurrence of the transitions is in the relative independence of the CDWs on each kind of molecular stack from those on the other. In fact, above the metal-insulator transition temperature 53K, the CDW on each stack is independent and has only one-dimensional correlations, and its correlation length is short above 53K [1]. In a system with two relatively independent kinds of CDWs, independent ordering of each kind of CDW at a different temperature may be possible. Experimentally and theoretically, there have been identified three transition temperatures, T_1 (53K), T_2 (49K) and T_3 (38K) in TTF-TCNQ [2,3]. T_1 is characterized as the three-dimensional long-range ordering temperature of the so-called $2k_F$-CDWs on the TCNQ stacks. At T_2, the long-range order of the TTF-CDWs begins to develop, although it is considered that this is an ordering induced by the ordered TCNQ-CDWs. Once two kinds of CDWs have long-range order, the coupling between them must be taken into account, since the magnitude of both order parameters becomes appreciable. Thus, the interactions among TCNQ- and TTF-CDWs brings about the decrease of the transverse wave number Q_a from 0.5a* with decreasing temperature [2] due to the competition of TCNQ-TCNQ and TCNQ-TTF interactions. The value of Q_a stays at 0.25a* below T_3 [2]. The presence of two kinds of molecular stacks

* Results reported here are part of an active collaboration with Drs. E. M. Engler, T. D. Schultz and Y. Tomkiewicz (IBM Watson Research Center) and Dr. T. Ishiguro (Electrotechnical Laboratory).

can give rise to a wide variety of structural features in two-chain systems like TTF-TCNQ. Theoretically, the successive phase transitions in TTF-TCNQ have been explained successfully by the so-called two-chain model [3,4].

The purpose of the present work is to clarify the CDW-orderings in TSeF-TCNQ, the isostructural analogue of TTF-TCNQ, using X-ray scattering, and thereby to expand our understanding of the general properties of the CDWs in two-chain systems.

According to the study of Weyl, *et al.* [5], only one kind of CDW (called the $2k_F$-CDW tentatively) occurs in TSeF-TCNQ, and only one transition temperature (29K) is seen, corresponding to the metal-insulator transition [6]. We have obtained similar results in our more detailed study. However, we have found two kinds of short-range order between CDWs beginning to develop at ~100K and at ~50K, the former having first been reported by Garito, *et al.* [7]. The experimental method was essentially the same as that used in the study of TTF-TCNQ [8]. Measurements were made on two samples selected from crystals prepared by Engler.

In Fig. 1 is shown the wave-number dependence of the diffuse scattering intensities measured at several temperatures. An anomaly appears below ~240K at the wave number $Q_b = 0.315b^*$, in agreement with the results of Weyl, *et al.* [5]. This implies that phonons with wave number ~$0.315b^*$ begin to soften and/or a CDW of this wave number begins to grow with decreasing

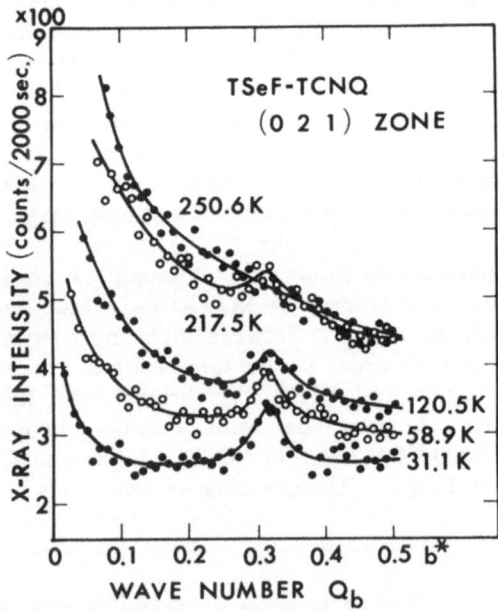

Fig. 1. The dependence of the X-ray diffuse scattering intensities on wave number Q_b as measured in sample #1. The solid lines are guides for the eye.

temperature below ~240K. Above ~100K, the anomaly shown in Fig. 1 was found at all points measured on the so-called diffuse sheet. Thus, the CDWs are one-dimensional and are independent of each other above 100K.

At ~100K, order along the c-direction begins to develop among the CDWs although its correlation length is short. Fig. 2(a) shows the development of the order with the wave number $Q_c=0c^*$ (*i.e.* period $c'=c$) along the c-axis. The correlation length ξ_c in this direction increases with decreasing temperature and reaches ~c at 70K. Below 70K, ξ_c does not change noticeably with temperature. Short-range order in the a-direction begins to appear at ~50K. Fig. 2(b) shows the ordering with wave number $Q_a=0a^*$ (*i.e.* period $a'=a$). The correlation length ξ_a along the a-axis reaches ~0.6a at 44K. As shown in Fig. 2(b), diffuse spots at $Q_a=\pm0.5a^*$ ($Q_b=0.315b^*$, $Q_c=0$) were found just above the metal-insulator transition temperature 29K. Thus, a crossover occurs from the state with $Q_a=0a^*$ to one with $Q_a=0.5a^*$ in the range ~40–29K.

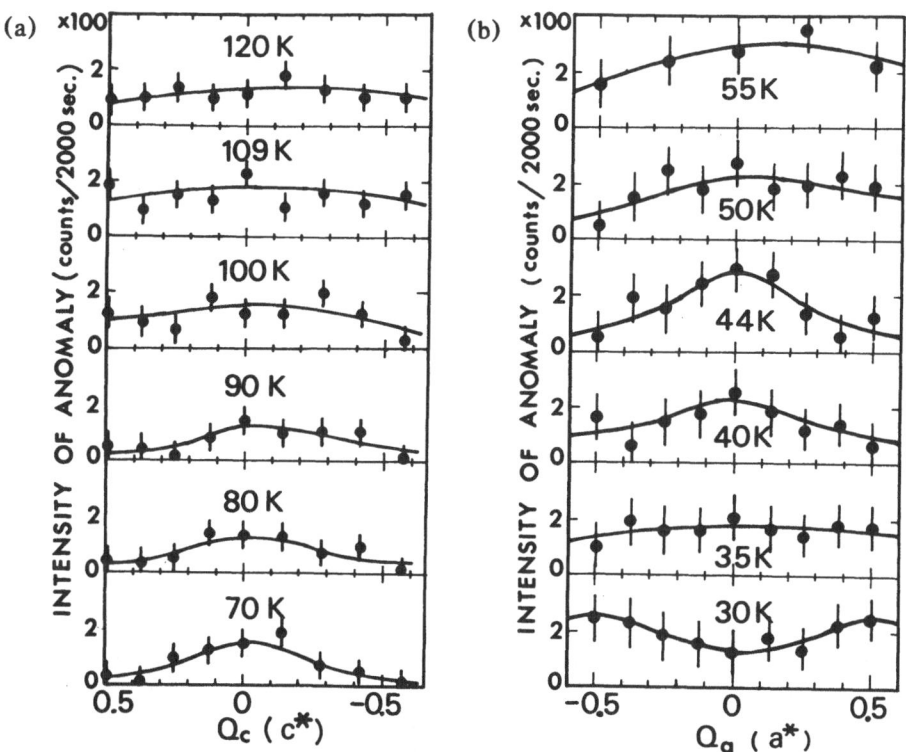

Fig. 2. (a) The dependence of the anomalous intensity on the wave number Q_c in the diffuse sheet measured in sample #2. (b) The dependence of the anomalous intensity on the wave number Q_a in the diffuse sheet measured in sample #1. The normal intensities have been subtracted. The solid lines are guides for the eye.

Below 29K, satellite spots were found, indicative of a superlattice with wave-vector components $Q_a=0.5a^*$, $Q_b=0.315b^*$ and $Q_c=0c^*$. These are independent of temperature down at least to 8K. The widths of the satellites are the same as the experimental resolutions, so the correlation lenghts are estimated to be larger than ~10a, ~30b and ~25c.

In TTF-TCNQ, no transverse short-range order has been found above T_1 (53K) [1,2] except for the appearance of precursors of the satellite spots [9]. Thus, we consider that the origin of the short-range order above 29K in TSeF-TCNQ must be in the CDWs on the TSeF-stack. In TTF-TCNQ, the TCNQ-CDWs play the primary role in the $2k_F$-CDW ordering [3] whereas in TSeF-TCNQ, both stacks play important roles [10], a difference that may be due to the TSeF bandwidth being larger than the TTF bandwidth. Thus, we expect that the main difference between CDWs in TTF-TCNQ and in TSeF-TCNQ is the larger amplitude of the cation CDWs in the latter system. The X-ray results for TSeF-TCNQ support this expectation as discussed below.

The periodicity of the short-range order at ~100K along the c-axis can be explained by the simple Coulomb interaction among neighbouring CDWs. Short-range order along the c-axis has not been found above 53K in TTF-TCNQ [1,2] although the interaction among TCNQ-CDWs must also be present in TTF-TCNQ. Therefore we expect that the short-range order at ~100K in TSeF-TCNQ occurs in the TSeF-CDWs. It is not clear at present, however, whether the short-range order along the c-axis is present in only the TSeF-CDWs or in both the TCNQ- and TSeF-CDWs.

In TTF-TCNQ, it is believed that the ordering of the CDWs along the a-axis at the first ordering temperature (53K) is dominated by the TCNQ-TCNQ interaction [4], since the amplitude of the TTF-CDW is expected to be smaller than that of the TCNQ-CDW. This interaction stabilizes the transverse wave-number at $Q_a=0.5a^*$. In TSeF-TCNQ, the pediodicity of the short-range order along the a-axis at ~50K can be explained by the Coulomb interaction among the TCNQ- and TSeF-CDWs. The dominance of the interactions between TSeF- and TCNQ-CDWs can be understood on two grounds: the larger amplitude expected of the TSeF-CDWs and the shorter distance along a between TCNQ- and TSeF-stacks than between TCNQ-stacks (or between TSeF-stacks).

The short correlation lengths along the a- and c-axes above 29K may be due to the shortness of the correlation lengths of the CDWs along the b-axis. The values of ξ_b at 70K and 44K are 6b and 9b, respectively, although ξ_b becomes larger than ~30b below 29K. Qualitatively, the Coulomb interaction among CDWs will be proportional to $\xi_{1d}^2\psi^2$, where ξ_{1d} and ψ denote the correlation length along the one-dimensional axis and the amplitude of the CDW, respectively. Thus, order along the transverse directions can occur with small ξ_{1d} when ψ is large enough. However, the transverse correlation lengths will be short, since a short correlation-length ξ_{1d} will inhibit the cooperative development of long-range order among the CDWs along the transverse directions.

If, in TSeF-TCNQ, the amplitudes of the TCNQ- and TSeF-CDWs are comparable and increase by similar amounts with decreasing temperature, then the higher-order interactions among the TCNQ- and TSeF-CDWs may become important at sufficiently low temperatures. It has been shown [10] that the fourth-order interaction $\psi_Q^2\psi_F^{*2}$+c.c., if strong enough, can stabilize the transverse periodicity at $Q_a=0.5a^*$, where ψ_Q and ψ_F are the order parameters of the TCNQ- and TSeF-CDWs, respectively. This interaction was proposed to explain why $Q_a\equiv0.5a^*$, independent of temperature, as observed in TSeF-TCNQ below 29K. If this is indeed what happens, then a crossover from $Q_a=0a^*$ to $Q_a=0.5a^*$ can be expected near the transition temperature when, as the order parameters grow, the fourth-order terms in the interaction become comparable to the second-order terms. The observed crossover can therefore be considered as additional support for the importance of fourth-order interaction terms.

In summary, the results of the present study indicate that the CDW amplitudes on the TSeF-stacks are comparable with those on the TCNQ-stacks in TSeF-TCNQ. We have argued that both the transverse short-range order at ~50 and the transverse long-range order at 29K are due to the interactions between the TCNQ- and TSeF-CDWs, in contrast to the dominant role of the TCNQ-TCNQ interactions at the transition temperature 53K of TTF-TCNQ. In Fig. 3 we contrast the behavior of these two systems and their CDWs schematically.

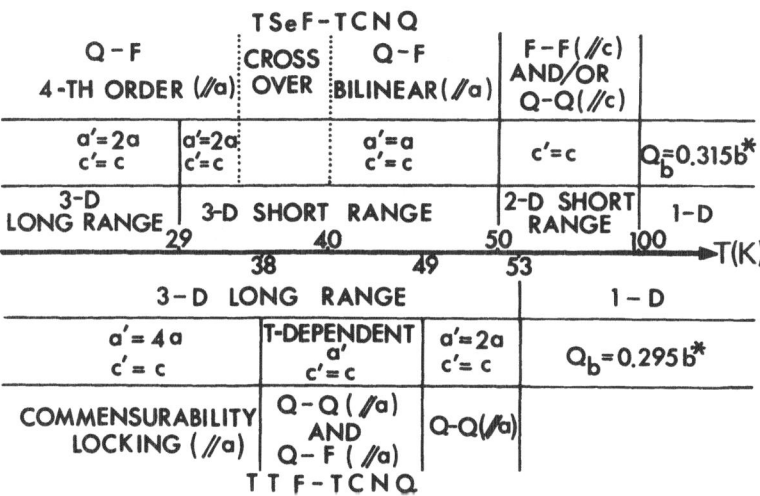

Fig. 3. Properties of the CDW-ordering and the expected dominant interactions in TSeF-TCNQ and TTF-TCNQ.

REFERENCES

1. F. Denoyer, R. Comès, A. F. Garito and A. J. Heeger, Phys. Rev. Letts. **35**, 445 (1975); S. Kagoshima, H. Anzai, K. Kajimura and T. Ishiguro, J. Phys. Soc. Japan **39**, 1143 (1975).

2. R. Comès, S. M. Shapiro, G. Shirane, A. F. Garito and A. J. Heeger, Phys. Rev. Letts. **35**, 1518 (1975); R. Comès, G. Shirane, S. M. Shapiro, A. F. Garito and A. J. Heeger, Phys. Rev. B **14**, 2376 (1976); W. D. Ellenson, R. Comès, S. M. Shapiro, G. Shirane, A. F. Garito and A. J. Heeger, Solid State Commun. **20**, 53 (1976); S. Kagoshima, T. Ishiguro and H. Anzai, J. Phys. Soc. Japan **41**, 2061 (1976); and W. D. Ellenson, S. M. Shapiro, G. Shirane and A. F. Garito; Phys. Rev. B **16**, 3244 (1977).

3. P. Bak and V. J. Emery, Phys. Rev. Letts. **36**, 978 (1976).

4. T. D. Schultz and S. Etemad, Phys. Rev. B **13**, 4928 (1976).

5. C. Weyl, E. M. Engler, K. Bechgaard, G. Jehanno and S. Etemad, Solid State Commun. **19**, 925 (1976).

6. S. Etemad, Phys. Rev. B **13**, 2254 (1976).

7. A. F. Garito, S. Megtert, J. P. Pouget and R. Comès, Bull. Am. Phys. Soc. **23**, 381 (1978).

8. S. Kagoshima, T. Ishiguro and H. Anzai, same as listed in Ref. 2

9. S. K. Khanna, J. P. Pouget, R. Comès, A. F. Garito and A. J. Heeger, Phys. Rev. B **16**, 1468 (1977).

10. T. D. Schultz, Solid State Commun. **22**, 289 (1977).

'METAL-LIKE' SEMICONDUCTORS - ROLE OF MOBILITY IN MOLECULAR
CONDUCTORS

Arthur J. Epstein

Xerox Webster Research Center

Rochester, New York 14644

In recent years, much work has focused on understanding charge
transport in molecular conductors.[1-4] There are many similarities
in the temperature (T) dependence of the conductivity (σ) of many
quasi-one-dimensional (1-D) materials which contain parallel segregated chains of large, planar, open-shell molecules, such as $TCNQ^-$
($TCNQ \equiv$ 7,7,8,8-tetracyano-p-quinodimethane).[5] The $\sigma(T)$ behavior
for these molecular conductors falls into three general classes:

I. Materials with strongly activated σ, such as $(alkali^+)$-
$(TCNQ^-)$.[6] In general, $\sigma(295K)$ is in the range of 10^{-6}-10^0 $\Omega^{-1}cm^{-1}$.
These materials are usually characterized as "semiconductors".

II. Those systems with a broad weak maximum, σ_m, in the tem-
perature dependence of their dc conductivity at a temperature T_m.
For these systems usually $\sigma_m/\sigma(295K) \stackrel{<}{\sim} 2$ and $\sigma(295K) \sim 100$ $\Omega^{-1}cm^{-1}$.
The σ versus T curves for these systems generally show negative
curvature ($d^2\sigma/dT^2 < 0$) over a large temperature range around T_m.
This class is typified by the σ versus T behavior of (NMP)(TCNQ)[7]
(NMP $\equiv$ N-methylphenazinium), Figure 1a. Numerous prior models have
been used to characterize $\sigma(T)$ for this class of materials (see
below). It is shown below how this broad group of materials can be
understood in terms of 'metal-like' semiconductors with a large
strongly T-dependent mobility.

III. Those systems with metallic σ and a sharp maximum in
their temperature-dependent dc conductivity. This class is typified by the σ versus T behavior of (TTF)(TCNQ) (TTF $\equiv$ tetrathiofulvalene).[8,9] Generally, for these systems $\sigma(T_m)/\sigma(295K) >> 2$
and σ versus T has positive curvature ($d^2\sigma/dT^2 > 0$) for $T > T_m$.
The room temperature conductivity is generally $\sim$500-1000 $\Omega^{-1}cm^{-1}$.

155

Considerable effort has focused on understanding Class II
materials. Their 'metal-like' behavior for $T > T_m$ and high room
temperature conductivity has led to early identification of two
distinct regions for charge transport, $T < T_m$ and $T > T_m$. An early
assignment was that a transition occured from a metallic to a Mott-
Hubbard[10] or Peierls[11] semiconductor as the temperature was lowered
below T_m. However, there was no quantitative modeling for $\sigma(T)$
within this model. A second school of thought focused on the crystallographic disorder that exists in many Class II materials,[12,13]
and the eigenfunction localization[14] this implies. However, explanations of $\sigma(T)$ as arising from hopping among disorder localized
states either did not simultaneously quantitatively fit data for
$\sigma(T)$ above and below T_m[15,16] or lead to inconsistent microscopic
parameters for these materials.[17]

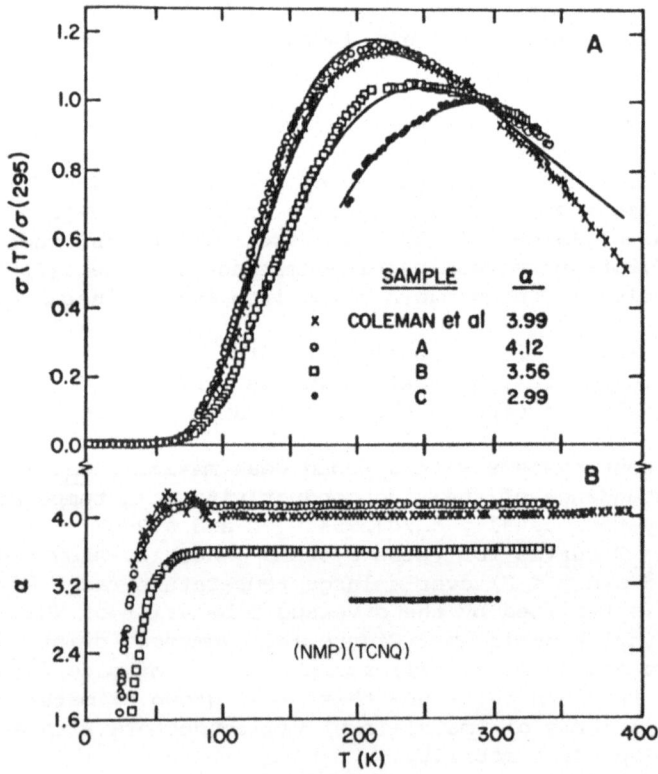

Figure 1
(a) Normalized four-probe a-axis conductivity, $\sigma(T)/\sigma(295K)$ versus
temperature for some representative (NMP)(TCNQ) samples. The 'x'
data is from ref. 8. The solid lines are computer fits to Equation
[1] with α indicated; (b) Calculated α from Equation [1] with experimental $\sigma(T)$ and $\Delta = 900K$.

We have shown that T dependence of σ for a large number of Class II materials can be readily understood in terms of the roles of a large, strongly T-dependent mobility, $\mu(T) \simeq \mu_0 T^{-\alpha}$ for T > 65K with 2 < α < 4.5 and μ(295K) $\sim$ 5 cm^2/Vsec, and an activated charge carrier concentration, n, with a system-specific activation energy, Δ[5,7,18], $n \propto \exp(-\Delta/T)$. That is,

$$\sigma(T) = \sigma_o \, T^{-\alpha} \exp(-\Delta/T) \qquad\qquad (1)$$

where σ_o is a constant. Equation (1) leads to a maximum in σ(T) at $T_m = \Delta/\alpha$, Fig. 2. The fit of this expression to several samples of (NMP)(TCNQ)[7] as well as $(NMP)_x(Phen)_{1-x}(TCNQ)$[19] is shown by the solid curves in Figs. la and 3a, respectively. The activation energy is found to be the same for all samples of the same material while α is a sample-dependent constant related to sample quality. To show even more convincingly that there is no change in charge transport mechanisms at T_m, or elsewhere in the range 65K < T < 400K, a plot of α as a function of T calculated from Equation (1) is shown in Figs. lb and 3b, using for σ the experimental values, and the appropriate system dependent Δ.

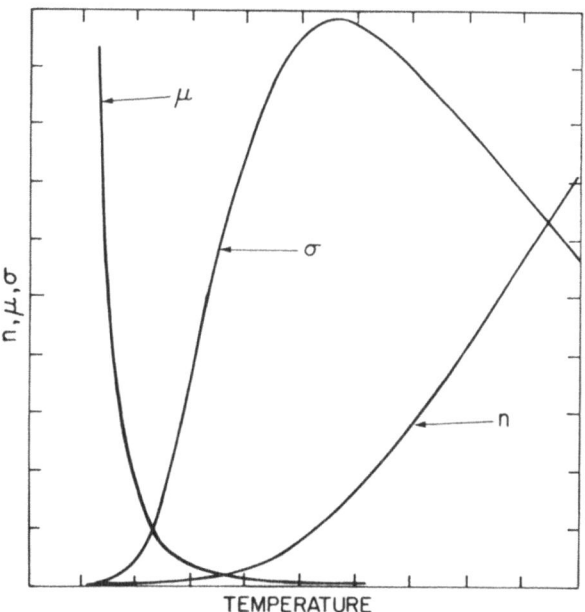

TEMPERATURE

Figure 2
Schematic temperature dependence of charge carrier concentration, $n \propto \exp(-\Delta/T)$, mobility $\mu \propto T^{-\alpha}$, and conductivity, $\sigma = ne\mu$. A maximum in σ occurs at $T_m = \Delta/\alpha$.

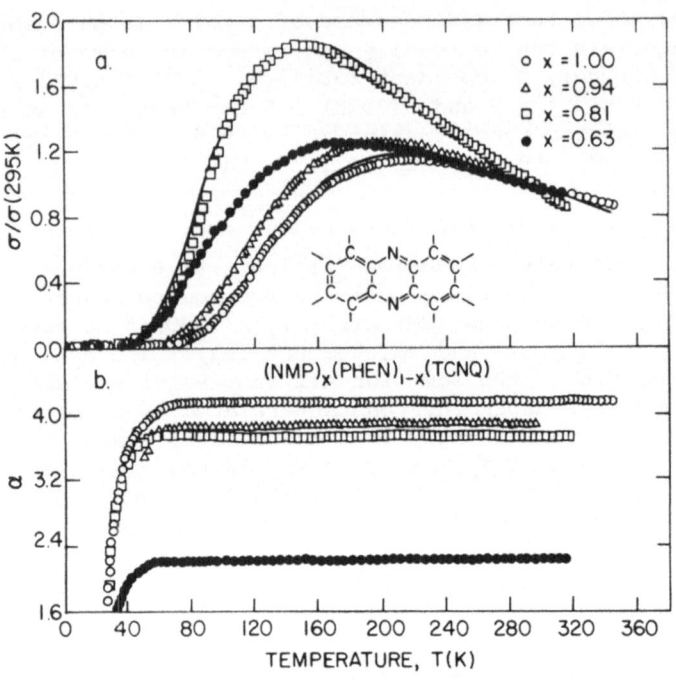

Figure 3

(a)   Normalized four-probe a-axis conductivity versus temperature for some representative $(NMP)_x(Phen)_{1-x}(TCNQ)$ samples.   The solid lines are computer fits to Eq. (1) with values given in Table I. The phenazine molecule is illustrated here.   (b) $\alpha(T)$ calculated from Eq. (1) with experimental $\sigma(T)$ and $\Delta(x)$ from Table I.

Table I
Parameterization of Class II and Class III Conductors

$$\sigma(T) = \sigma_o T^{-\alpha} \exp(-\frac{\Delta}{T}) = n(T)e\mu(T)$$

	α	Δ, °K	Ref.
(BTP)(TCNQ)	2.6	1400	22
(NMP)(TCNQ)	3-4	900	7
$(NMP)_{.94}(Phen)_{.06}(TNCQ)$	3.9	800	19
$(NMP)_{.81}(Phen)_{.19}(TCNQ)$	3.5	550	19
$(NMP)_{.63}(Phen)_{.37}(TCNQ)$	2.2	400	19
$(NMe_3H)(I)(TCNQ)$	3.3	800	23
$Qn(TNCQ)_2$	2.7	650	18
$Adz(TCNQ)_2$	3.0	450	18
$Adn(TCNQ)_2$	2.3	350	18
(TTF)(TCNQ)	2.3	0	5,9

The mobility obtained by this approach is large enough (> 1 cm^2/volt-sec) so that the band approach is self consistently valid. However, the very strong T-dependence is unusual, implying a strongly T-dependent scattering mechanism. Assuming that molecular vibrations are of central importance, we have been able to quantitatively calculate[5,7,18] this mobility utilizing the known molecular vibration frequencies[20] and electron-phonon coupling constants.[21] Experiments in controlling the band filling and disorder[19] have shown that disorder has a secondary role T > 65K in these systems. Table I summarizes the parameters in applying Equation (1) to Class II and Class III conductors.

The model of Class II materials as metal-like semiconductors with a strongly T-dependent mobility has been supported through the direct optical observation of the semiconducting gap in (NMe$_3$H)(I)-(TCNQ),[24] as well as thermoelectric power studies of (NMe$_3$H)(I)-(TCNQ)[25] and (NMP)$_x$(Phen)$_{1-x}$(TCNQ).[26] The latter studies have shown many of these Class II systems to have large on-site Coulomb repulsions (large U).[27,28] Studies are in progress to determine the source of these semiconducting gaps.

Extension of these ideas for Class II systems to Classes I and III is now apparent. Class I materials have high activation energies so that T_m is outside the region of measurement.[5] In contrast, Class III materials have little or no activation energy, and $\sigma(T) \propto \mu(T)$.[9]

Many stimulating discussion with P.M. Chaikin, E.M. Conwell, J.S. Miller, M.J. Rice, D.J. Sandman, and D.B. Tanner are acknowledged.

REFERENCES

1. J.S. Miller and A.J. Epstein, Eds., Ann. N.Y. Acad. Sci. 313 (1978).

2. H.J. Keller, Ed., NATO Adv. Study Inst. Ser., Series B, 25 (1977); 7 (1975).

3. L. Pál, G. Grüner, A. Jánossy, and J. Sólyom, Eds., Lecture Notes in Physics 65 (1977).

4. J.S. Miller and A.J. Epstein, Prog. Inorg. Chem. 20, 1 (1976).

5. A.J. Epstein, E.M. Conwell and J.S. Miller, Ann. N.Y. Acad. Sci. 313, 183 (1978).

6. R.G. Kepler, P.E. Bierstadt, and R.E. Merrifield, Phys. Rev. Lett. 5, 503 (1960).

7. A.J. Epstein, E.M. Conwell, D.J. Sandman, and J.S. Miller, Solid State Comm. 23, 355 (1978).

8. L.B. Coleman, J.A. Cohen, A.F. Garito, and A.J. Heeger, Phys. Rev. B7, 2122 (1973).

9. L.B. Coleman, M.J. Cohen, D.J. Sandman, F.G. Yamagishi, A.F. Garito, and A.J. Heeger, Solid State Comm. 12, 1125 (1973).

10. E.M. Conwell, Phys. Rev. Lett. 39, 777 (1977).
11. A.J. Epstein, S. Etemad, A.F. Garito and A.J. Heeger, Solid
     State Comm. 9, 1803 (1971); Phys. Rev. B 5, 952 (1972).
12. C.S. Jacobsen, J.R. Anderson, K. Bechgaard, and C. Berg,
     Solid State Comm. 19, 1209 (1976).
13. C.J. Fritchie, Jr., Acta. Cryst. 20, 892 (1966); H. Kobayashi,
     F. Marumo, and Y. Saito, Acta. Cryst. B 27, 373 (1971).
14. N.F. Mott and W.D. Twose, Adv. Phys. 10, 107 (1961).
15. A.N. Bloch, R.B. Weisman, and C.M. Varma, Phys. Rev. Lett.
     28, 753 (1972).
16. V.K.S. Shante, Phys. Rev. B 16, 2597 (1977).
17. A.A. Gogolin, S.P. Zolotukhin, V.I. Melnikov, E.I. Rashba,
     and I.F. Shchegolev, JETP Lett. 22, 278 (1975).
18. A.J. Epstein and E.M. Conwell, Solid State Comm. 24, 627
     (1977).
19. A.J. Epstein and J.S. Miller, Solid State Comm., in press.
20. R. Bozio, A. Girlando, and C. Pecile, J. Chem. Soc. Faraday
     Trans. II, 1237 (1975).
21. N.O. Lipari, C.B. Duke, R. Bozio, A. Girlando, C. Pecile,
     and A. Padva, Chem. Phys. Lett. 44, 236 (1976); C.B. Duke
     N.O. Lipari, L. Pietronero, Chem. Phys. Lett. 30, 415 (1975);
     M.J. Rice, L. Pietronero, and P. Bruesch, Solid State Comm.
     21, 757 (1977).
22. A.J. Epstein and D.J. Sandman, to be published.
23. A.J. Epstein and J.S. Miller, to be published.
24. D.B. Tanner, J.E. Deis, A.J. Epstein and J.S. Miller, to be
     published.
25. P.M. Chaikin, A.J. Epstein and J.S. Miller, to be published.
26. A.J. Epstein, P.M. Chaikin and J.S. Miller, to be published.
27. P.M. Chaikin, Proc. Int. Conf. on Thermoelectricity in Metal-
     lic Conductors, Aug. 1977 (Plenum Press 1978), in press.
28. E.M. Conwell, Phys. Rev. B, 18, 1818 (1978).

SYNTHESIS AND PROPERTIES OF SEMICONDUCTING

AND METALLIC DERIVATIVES OF POLYACETYLENE, $(CH)_x$

Alan G. MacDiarmid and Alan J. Heeger*

Department of Chemistry and Department
of Physics,* University of Pennsylvania
Philadelphia, PA 19104, U.S.A.

ABSTRACT

Polyacetylene, $(CH)_x$ is the simplest organic polymer. Through chemical doping, the electrical conductivity of films of $(CH)_x$ can be varied over twelve orders of magnitude with properties ranging from insulator ($\sigma < 10^{-10}$ ohm^{-1}cm^{-1}) to semiconductor to metal ($\sigma > 10^3$ ohm^{-1}cm^{-1}). Both donors and acceptors can be used with these flexible, free-standing polycrystalline polymer films (thickness 10^{-5}cm to 0.5 cm) to yield n-type or p-type material. In this review we summarize some of the more important chemical and physical properties of $(CH)_x$ and its doped derivatives.

Polyacetylene, $(CH)_x$ is the simplest possible conjugated organic polymer and is therefore of special fundamental interest. Early studies on this polymer, which was known only as a dark-colored insoluble powder, concentrated on the production of pure material. Hatano et al.[1] found the electrical conductivity depended on the crystallinity with higher crystallinity giving higher conductivity. Berets and Smith[2] studied the effect of oxygen content on polycrystalline powder and found that oxygen in the polyacetylene affected its conductivity; the lowest oxygen content yielded the highest conductivity. Their best samples had oxygen content as low as 0.7%.

In a series of studies Shirakawa et al.[3-6] succeeded in synthesizing high quality polycrystalline films of $(CH)_x$ and developed techniques for controlling the cis-trans content. (See Figure 1).

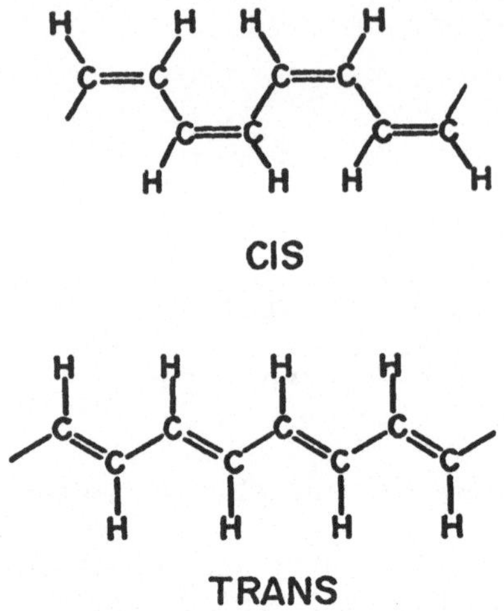

CIS

TRANS

Figure 1: Cis- and trans- polyacetylene, $(CH)_x$

The $(CH)_x$ films have a lustrous silvery appearance; they are
flexible and appear to have excellent mecanical properties. Films
can be made free standing, or on substrates such as glass or metal.
Films have been made with thickness varying from 10^{-5} cm to 0.5 cm.

(1) SYNTHESIS OF $(CH)_x$ FILMS

Polyacetylene films may be prepared by simply wetting the inside walls of a glass reactor vessel with a toluene solution of
$(C_2H_5)_3Al$ and $(n-C_4H_9O)_4Ti$ Ziegler catalyst and then immediately
admitting acetylene gas at any pressure from a few centimeters up to
ca 1 atmosphere pressure. The cohesive film grow on all surfaces
which have been wet by the catalyst solution during a few seconds to
1 hour depending on the pressure of acetylene and temperature employed. If a polymerization temperature of ca-78°C is used, the
film is formed almost completely as the cis-isomer; if a temperature
of 150°C is used (decane solvent) the film is formed as the trans
isomer. With room temperature polymerization the film is approx-
mately 80% cis-and 20% trans-isomer. If the film is carefuly
washed, analytically pure $(CH)_x$ is obtained (See Table 1). The
cis-isomer may be conveniently converted to the trans-isomer (the
thermodynamically stable form) by heating at ca 200° for ca 1 hour.

(2) STRUCTURAL PROPERTIES OF $(CH)_x$ FILMS

Electron microscopy studies show that the as-formed $(CH)_x$ films
consist of randomly oriented fibrils (typical fibril diameter of a
few hundred angstroms). The bulk density is ca 0.4 gm/cm^3 compared
with 1.2 gm/cm^3 as obtained by flotation techniques. This shows

TABLE 1

Chemical Analysis of Pure and Doped $(CH)_x$ [a]

		C%	H%	Halogen %	Total
1) $(CH)_x$	calculated	92.26	7.74		100.00
<u>cis</u>-$(CH)_x$	found	92.16	7.81		99.97
<u>trans</u>-$(CH)_x$	found	92.13	7.75		99.88
2) <u>trans</u>-$(CHI_{0.22})_x$ [b]	calculated	29.34	2.46	68.20	100.00
	found	29.14	2.62	68.26	100.02
3) <u>trans</u>-$(CHBr_{0.224})_x$ [b]	calculated	38.85	3.26	57.89	100.00
	found	38.89	3.05	58.16	100.10
3) <u>cis</u>-$[CH(AsF_5)_{0.099}]_x$ [b,c]	calculated	40.15	3.61	31.44	100.00
	found	39.86	3.75	31.48	99.78

[a] Galbraith Laboratories, Inc.

[b] The designation "<u>cis</u>" or "<u>trans</u>" refers to the isomeric form of the $(CH)_x$ employed in the doping experiment. It does not necessarily imply that the doped material has the same isomeric composition as the original $(CH)_x$.

[c] Arsenic: Calcd., 24.80%; Found, 24.69%.

that the polymer fibrils fill only about one-third of the total volume. X-ray studies show that the films are polycrystalline with interchain spacing of approximately 3.8 Å.[3-6]

Characteristic infrared absorption bands of thin films can be conveniently used to distinguish the <u>cis</u> and <u>trans</u> isomers and to estimate the relative amounts of each isomer in a partly isomerized film.[3] Solid state ^{13}C nmr studies[7] have been used to confirm the assignment of a given type of film as either <u>cis</u> or <u>trans</u>. Such studies also show that isomerically pure films (within the limits of experimental error) are obtained although there is a suggestion that both types of film may contain up to <u>ca</u> 5% of sp^3 -hybridized carbon atoms which might be acting as cross-linking atoms between adjacent $(CH)_x$ chains.

(3) MECHANICAL PROPERTIES OF $(CH)_x$ FILMS

Fresh films of both <u>cis</u>- and <u>trans</u>-$(CH)_x$ are flexible and can be easily stretched. These properties are more pronounced with <u>cis</u> isomer films which can be stretched in a few minutes (with partial

alignment of fibers) at room temperature up to 3 times their
original length. Reasonably good tensile strengths (up to ca 3.8
kg/mm^2) are routinely obtained.

(4) STABILITY OF (CH)$_x$ FILMS

The parent (CH)$_x$ films have good thermal stability when heated
in vacuum.[6] Thus, thermograms of both the cis and trans isomers
show an exothermic peak at 325°C. Decomposition is rapid at this
temperature. At 420°C volatile decomposition products are formed
in an endothermic reaction. The parent (CH)$_x$ films slowly become
brittle in air during several days and their resistance increases.
However, when coated with a thin plastic film, or wax, they are
stable for many weeks.

(5) ELECTRICAL PROPERTIES OF (CH)$_x$ FILMS

Shirakawa et al.[8] pointed out that the room temperature conductivity of crystalline films of polyacetylene depended on the cis-
trans content, varying from 10^{-5} ohm^{-1}cm^{-1} for the trans material
to 10^{-9} ohm^{-1}cm^{-1} for the cis-isomer. In view of the sensitivity
of polyacetylene to impurities and/or defects as demonstrated by our
doping studies,[9-16] it appears likely that the intrinsic conductivity of pure polyacetylene is even lower. This is supported by the
observation that exposure of trans-(CH)$_x$ to the vapor of the
donor, NH$_3$, causes the conductivity to fall more than four orders
of magnitude (to $< 10^{-9}$ ohm^{-1}cm^{-1}) without detectable weight increase. This may be due to the coordination of the NH$_3$ to traces
of the catalyst (which acted as a dopant). Subsequent reaction of
the film with a dopant such as AsF$_5$, which is described in following
sections, increases the conductivity many orders of magnitude to
metallic levels.

From theoretical and spectroscopic studies of short chain
polymers, the π-system transfer integral of (CH)$_x$ can be estimated
as $\beta \sim 2 - 2.5$ eV.[17] Thus the overall bandwidth would be of order
8-10 eV; $W = 2z\beta$, where z is the number of nearest neighbors, β is
the transfer integral and W is the bandwidth in the tight-binding
approximation. The electrons from the unsaturated π-system are
therefore delocalized along the polymer chains. However, because
of the combined effects of bond alternation and Coulomb correlation,
there is an energy gap in the excitation spectrum leading to semiconducting behavior. As a result of the large overall band width
and unsaturated π-system, (CH)$_x$ is fundamentally different from
either the traditional organic semiconductors made up of weakly
interacting molecules (e.g., anthracene, etc.), or from other
saturated polymers with monomeric units of the form R   R' where

$$(-\overset{|}{\underset{|}{C}}-)$$

there are no π-electrons (e.g., polyethylene, etc.). <u>Polyacetylene</u>
<u>is therefore more nearly analogous to the traditional inorganic</u>
<u>semiconductors; and indeed we have shown recently that (CH) can be</u>
<u>chemically doped with a variety of donors or acceptors to give n-</u>
<u>type or p-type semiconductors.</u>[9-16]

(6) DOPING OF $(CH)_x$ FILMS

We have developed methods for doping either <u>cis-</u>or <u>trans-</u>$(CH)_x$
to <u>p</u>- or <u>n</u>-type semiconductors or metals. These methods fall into
three chief categories. The silvery $(CH)_x$ films undergo very
little, if any change in appearance upon doping.

(i) Exposure of the $(CH)_x$ films to a known vapor pressure of a
volatile dopant e.g. I_2, AsF_5 etc. until a desired conductivity is
obtained; removal of the dopant vapor at that stage then essentially
"freezes" the conductivity at that value.

(ii) Treatment of the $(CH)_x$ film with a solution of the dopant
in an appropriate solvent (e.g. I_2 in pentane; sodium naphthalide in
THF, etc.) for a given time period.

(iii) Treatment of the $(CH)_x$ films with liquid Na/K alloy at
room temperature for a few minutes to give an <u>n</u>-type Na/K-doped
film.

A list of some doped $(CH)_x$ species is given in Table II. In
addition, we have found that the following compounds also increase
the conductivity of $(CH)_x$ films to either a good semiconducting or
metallic range: SbF_5, SiF_4, PCl_5, PF_5, and ICN. Exact compositions
and conductivities have yet to be determined.

Films doped with e.g. I_2 or AsF_5 retain their high metallic
conductivity after several days exposure to air and show only a
small decrease in conductivity. These films, and also those doped
with e.g. Na appear to be stable in air for many weeks when coated
with a protective film. In this respect it might be noted that un-
coated films doped with I_2 to the metallic regime undergo essentially no change in conductivity or composition when held under
water over night.

A few preliminary studies at elevated temperatures have been
carried out on I_2-doped and AsF_5-doped films. Films containing the
former dopant appear to retain their semiconducting properties up to
at least <u>ca</u> 100°C whereas films involving AsF_5 remain semiconducting
up to at least 150°-200°C.

TABLE 2

Conductivity of Polycrystalline Polyacetylene and Derivatives (As-Grown Films)[a]

Material	Conductivity σ(ohm^{-1}cm^{-1})(25°C)
cis-(CH)$_x$ [b]	1.7×10^{-9}
trans-(CH)$_x$ [b]	4.4×10^{-5}
trans-[(CH)(HBr)$_{0.04}$]$_x$	7×10^{-4}
trans-(CHCl$_{0.02}$)$_x$	1×10^{-4}
trans-(CHBr$_{0.05}$)$_x$	5×10^{-1}
trans-(CHBr$_{0.23}$)$_x$ [b]	4×10^{-1}
cis-[CH(ICl)$_{0.14}$]$_x$	5.0×10^{1}
cis-(CHI$_{0.25}$)$_x$	3.6×10^{2}
trans-(CHI$_{0.22}$)$_x$ [b]	3.0×10^{1}
trans-(CHI$_{0.20}$)$_x$ [b]	1.6×10^{2}
cis-(CHI$_{0.28}$)$_x$	5.0×10^{2}
cis-[CH(IBr)$_{0.15}$]$_x$	4.0×10^{2}
trans-[CH(IBr)$_{0.12}$]$_x$	1.2×10^{2}
trans-[CH(AsF$_5$)$_{0.03}$]$_x$	7×10^{1}
trans-[CH(AsF$_5$)$_{0.10}$]$_x$ [b]	4.0×10^{2}
cis-[CH(AsF$_5$)$_{0.14}$]$_x$	5.6×10^{2}
trans-[Na$_{0.28}$(CH)]$_x$	8×10^{1}

(a) The prefix "cis" or "trans" refers to the isomeric composition of the (CH)$_x$ which was used in a given doping experiment.

(b) Composition obtained by chemical analysis from Galbraith Laboratories, Inc. (Sum of all elements $\sim$ 99.8-100.1%).

(7) ELECTRICAL CONDUCTIVITY OF DOPED (CH)$_x$ FILMS AND SEMICONDUCTOR-METAL TRANSITIONS

When pure polyacetylene is doped with a donor or an acceptor, the electrical conductivity increases sharply over many orders of magnitude at low concentration, then saturates at higher dopant levels, above approximately 1%.[9-16] The maximum conductivity we have reported to date at room temperature for nonaligned cis-[CH(AsF$_5$)$_{0.14}$]$_x$ was 560 ohm^{-1}cm^{-1}. The typical behavior for the

conductivity as a function of dopant concentration (y) is shown in
Figure 2. The general features appear to be the same for the various donor and acceptor dopants, but with detailed differences in the
saturation values and the critical concentration at the "knee" in
the curve (above which σ is only weakly dependent on y). These
transport studies suggest a change in behavior near 1% dopant concentration; a semiconductor-to-metal transition.

To verify the existence of the semiconductor-to-metal transition, far infrared transmission data were taken[10] on samples of varying concentrations of iodine and AsF$_5$ (with qualitatively similar
results). The data for a series of iodinated samples are shown
in Figure 3 for y = 0.0, 0.9, 2.0 and 6.0 at%. In the case of the
6% sample, there is no observable transmission throughout the ir
down to 20 cm^{-1} implying a continuous excitation spectrum; i.e.,

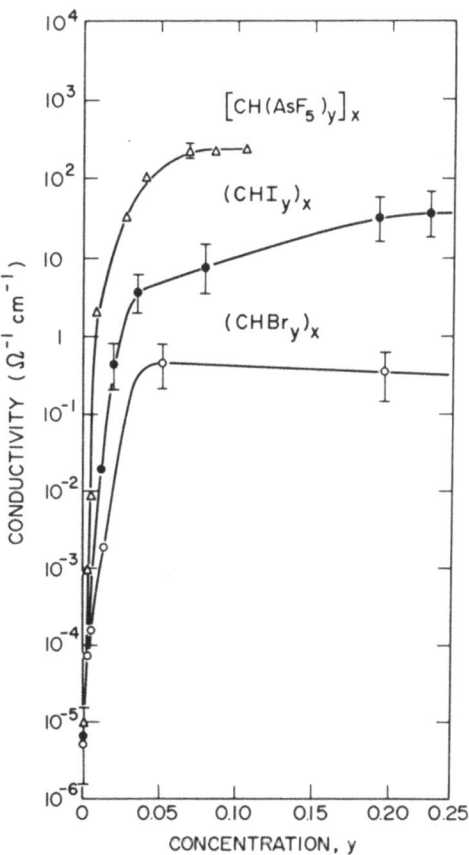

<u>Figure 2</u>: Electrical conductivity (room temperature)
as a function of dopant concentration.

metallic. For the 2% sample, the transmission was zero at the high
end of the spectrum (4000 cm^{-1} to 300 cm^{-1}), but increases below
300 cm^{-1} to about 60% by 40 cm^{-1} implying an energy gap at low frequencies. The far ir transmission through the 0.9% sample is near
90% with no significant change from an undoped sample. The inset
to Figure 3 shows the absorption coefficient, α, (uncorrected for
reflection) at 40 cm^{-1} as a function of dopant concentration. The
transition is sharp with a critical concentration (n_c) in the range
2-3%. Similar results have been obtained with AsF$_5$ although n_c appears to be slightly smaller. The values for n_c as inferred from
the ir and dc transport measurements are in agreement.

   As an initial point of view we treated this transition as
similar to that seen in heavily doped semiconductors. In this case,
one expects the halogen and AsF$_5$ dopants to act as acceptors with
localized hole states in the gap, with the hole bound to the acceptor
in a hydrogen-like fashion. For low concentrations, one expects the
combination of impurity ionization and variable range hopping to
lead to a combination of activated processes as observed experimentally. However, as extensively discussed by Mott[18] and others,[19]
if the concentration is increased to a critical level, then the
screening from carriers will destroy the bound states giving an in-
sulator-to-metal transition. This will occur when the screening
length becomes less than the radius of the most tightly bound Bohr
orbit of the hole and acceptor in the bulk dielectric;

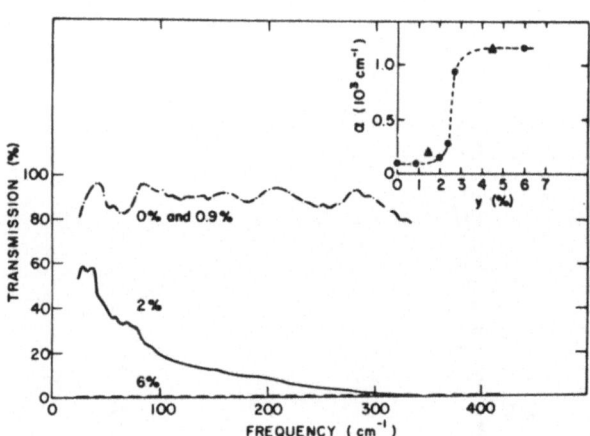

Figure 3:   Far infrared (20 cm^{-1} to 400 cm^{-1}) transmission of I$_2$ doped
            polyacetylene (sample temperature was 77 K). The inset
            shows the absorption coefficient (uncorrected for re-
            flection) at 40 cm^{-1} vs. concentration (solid points,
            trans-polymer; triangles, cis-polymer).

$$n_c^{1/3} \simeq (4a_H)^{-1} (\frac{m^*}{m_\epsilon})$$

where a_H is the Bohr radius, ϵ is the dielectric constant of the medium and m^*/m is the ratio of the band mass to the free electron mass. Assuming $m^*/m \sim 1$ and using $\epsilon \simeq 8$ from ir reflection measurements, we estimate $n_c \sim 10^{20} - 10^{21}$ cm^{-3}. Since the density of carbon atoms is about 2×10^{22} cm^{-3}, n_c would be in the range of a few percent assuming one carrier per dopant. The good agreement with our experimental results is probably fortuitous in view of the much over-simplified model. However, the overall features of the semiconductor-metal transition at high doping levels are very similar to those observed in traditional inorganic semiconductors.

(8) THE ENERGY GAP AND ABSORPTION SPECTRUM

For the undoped polymer, the absorption edge (see Figure 4) is quite sharp, rising much more rapidly than the typical three-dimensional (3d) joint density of states, which increases from the gap edge as $(\epsilon - E_g)^{\frac{1}{2}}$. In contrast, if we assume weak interchain coupling as suggested above, the 1d joint density of states has the well-known $(\epsilon - E_g)^{-\frac{1}{2}}$ singularity at the gap edge with a correspondingly steep absorption edge. Actual attempts to fit with a Gaussian broadened 1d density of states resulted in good agreement with the data near the gap edge. The magnitude of the absorption maximum $(\alpha \sim 3 \times 10^5$ cm$^{-1})$ is comparable to that for interband transitions in more common direct gap semiconductors.[20] A precise value for the energy gap requires a detailed theoretical model. If we assume a 1d band structure with broadened 1d density of states, $E_g = E_{max} = 1.9$ eV. Using the more conventional definition of the onset of absorption, one estimates $E_g \simeq 1.4 - 1.6$ eV.

These values are considerably larger than the activation energy obtained from temperature dependent resistivity studies of trans-$(CH)_x$ (0.3 eV).[8] This is consistent with the transport in undoped $(CH)_x$ being dominated by trace impurities or defects.[9-15]

Ovchinnikov[17] has argued that the energy gap extrapolated to infinite chain polyenes is too large to be accounted for by simple band theory of the bond-alternated chain, and he therefore concluded that Coulomb correlations play an important role. In a tight binding calculation, the band gap due to bond alternation would be

$\delta\beta = \delta x \frac{\partial\beta}{\partial x}\big|_{x_0} = \frac{\delta x}{a} \beta(x_0)$ where δx is the difference in bond lengths

and x_0 the average bond length; $\beta(x) = \beta_0 \exp(-x/a)$ is the transfer integral where $\underline{a}$ is a characteristic atomic distance ($a \sim 0.7$ Å) describing the fall-off of the carbon $2p_\pi$ wavefunction ($\beta(x_0) \simeq 2.5$ eV). If we assume δx takes the maximum value, equal to the dif-

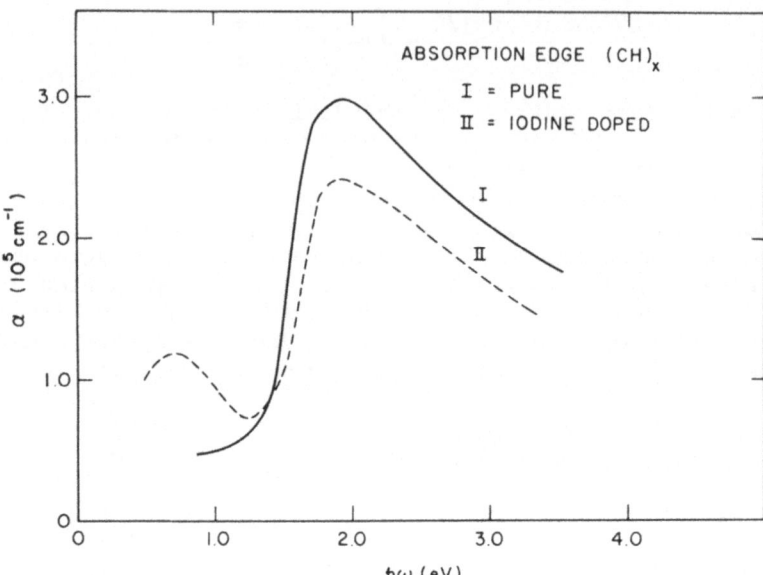

<u>Figure 4</u>:   Absorption coefficient of <u>trans</u>-(CH)$_x$ before and after
           doping to saturation as a function of frequency; film
           thickness 0.1 μm.

ference in bond lengths between a single bond (1.51 Å as in ethane)
and a double bond (1.34 Å as in ethylene) we estimate E$_g$ ~ 0.6 eV.
Certainly this question must be resolved with more detailed band
calculations.[21] However, based on results obtained thus far in our
transport[9-15] and magnetic[22] studies of (CH)$_x$, we see no experimental evidence suggesting strong Coulomb correlations.

    The absorption spectra after doping with iodine is shown in
Figure 4, similar results are obtained after doping with AsF$_5$.
There is relatively strong absorption at low frequencies within the
gap as would be expected for a heavily doped semiconductor. Detailed studies of the onset of absorption within the gap at lightest
doping levels are in progress. More important in the context of
this study is the observation that the strong interband transition
persists even at the highest doping levels. This result suggests
that at least for these dopants the basic π-electron band structure
of (CH)$_x$ remains intact in the doping process, consistent with the
charge transfer doping model[11,12] with A$^-$ species between chains.
An uninterrupted π-system is consistent with the excellent transport
properties of metallic doped (CH)$_x$. On the other hand recent studies[23] with bromine doping have revealed a major change in the absorption spectrum after doping, consistent with earlier observations[12]

that bromine tends to add to the double bond (at least at high concentrations) with a corresponding decrease in conductivity. Studies
must be carried out with a variety of dopants before general conclusions can be drawn.

Throughout the above discussion we have assumed that the absorption edge results from an interband transition with the creation of
electron-hole pairs. The possibility of electron-hole bound states
(excitons) on the chain must be considered since such exciton transitions can lead to a sharp absorption edge below the gap. However,
the observation of this absorption even at the highest doping levels
argues against a transition to an exciton bound state, which would
be screened in the metal. Moreover, early photoconductivity measurements[24] on powder samples suggest a photo-conductive edge consistent with the absorption edge.

(9) ORIENTED FILMS: ANISOTROPIC ELECTRICAL PROPERTIES[14]

The high conductivity in the metallic state above n_c (see
Figure 2) is particularly interesting since electron microscopy
studies[5,6] show that the $(CH)_x$ films consist of tangled randomly
oriented fibrils (typical fibril diameter of a few hundred angstroms). The bulk density[6] is 0.4 gm/cm^3 compared with 1.2 gm/cm^3
as obtained by flotation techniques, indicating that the polymer
fibrils fill only about one-third of the total volume. X-ray
studies[5] show that the $(CH)_x$ films are polycrystalline with interchain spacing of approximately 3.8 Å. Consequently we expect the
interchain electronic transfer integrals to be small, ~ 0.1 eV, i.e.,
less than or comparable to the intermolecular transfer integrals
along the b-axis in TTF-TCNQ where the intermolecular spacing is
3.6 Å. On the other hand, molecular spectroscopic studies of short
chain polymers lead to the conclusion that the intrachain transfer
integrals for carbon atoms separated by ~ 1.4 Å are of order 2 -
2.5 eV. Thus we anticipate a highly anisotropic band structure with
correspondingly anisotropic transport in $(CH)_x$. Indirect evidence
of this was obtained from the temperature dependence of the conductivity in $(CH)_x$ doped to concentrations above the semiconductor-
to-metal transition.[9-15] The conductivity was found to decrease on
lowering the temperature in a manner similar to that observed in
polycrystalline $(SN)_x$[25], sublimed films of $(SN)_x$[25] or polycrystalline
TTF-TCNQ where the transport is limited by a combination of anisotropy and interparticle contact. In these cases, the conductivity
decreases even though the single crystal transport measurements along
the principal conducting axis clearly imply metallic behavior.

It is well known that many polymers can be stretch-oriented by
mechanical elongation. Shirakawa and Ikeda have recently reported[27]
significant orientation of $(CH)_x$ after stretch elongation; they have
been able to vary the amount of orientation by combined mechanical

and thermal treatment resulting in elongation with $\ell/\ell_o \sim 1$ to 3 where ℓ is the final stretched length and ℓ_o is the unstretched length.

The room temperature results on partially oriented films are summarized in Figures 5 and 6.[14] The anisotropy is plotted in Figure 5 as a function of dopant concentration for $\ell/\ell_o = 2.1$ and $\ell/\ell_o = 2.9$. The induced anistropy of the undoped oriented films appears to increase approximately as the square of the elongation (prior to stretching the non-aligned samples are isotropic both before and after doping with AsF_5). The anisotropy remains after doping, increasing modestly with iodine and more steeply with AsF_5. The effects of elongation (alignment) on the absolute values of $\sigma_\parallel$ and $\sigma_\perp$ are shown on Figure 6 for the heavily doped metallic polymer $[CH(AsF_5)_{0.10}]_x$. The parallel conductivity increases dramatically with alignment; the solid curve follows $\sigma = \sigma_o \, (\ell/\ell_o)^2$ where $\sigma_o = 300$ $ohm^{-1}cm^{-1}$.

Scanning electron microscope pictures of the films as grown and after stretch alignment are shown in Figures 7 and 8 respectively. The characteristic branched and twisted fibrils of the unstretched polymer discussed earlier by Ito et al.[5,6] are clearly visible in Figure 7. Elongation results in partial alignment as shown in Figure 8. However comparison of the two shows that the fractional alignment is only modest.

The temperature dependence of the parallel and perpendicular conductivities and the anisotropy for an oriented film ($\ell/\ell_o = 2.9$) doped into the metallic regime with AsF_5 are shown in Figure 9. The solid points result from four probe measurements on two separate ($\parallel$ and $\perp$) films; the x points result from the Montgomery measurements. The three samples were taken from the same initial film and doped simultaneously to a final composition $[CH(AsF_5)_{0.10}]_x$. We expect the Montgomery technique to give the more reliable data since the measurements were taken on a single sample. The four-probe data come from two separate samples ($\parallel$ and $\perp$), so that slightly different final compositions are possible. Nevertheless, the results from the two independent sets of measurements are consistent, and the general agreement is excellent. The room temperature parallel conductivity is in excess of 2000 $ohm^{-1}cm^{-1}$; the average of the two measurements yields 2150 $ohm^{-1}cm^{-1}$. On cooling, $\sigma_\parallel$ and $\sigma_\perp$ decrease slowly; however, the conductivity remains high even at the lowest temperatures, consistent with metallic behavior. A more detailed examination of the data shows that $\sigma_\parallel$ remains approximately constant, increasing slightly ($\sim 0.5\%$) down to 260 K, whereas $\sigma_\perp$ decreases monotonically.

The results of these initial studies on oriented $(CH)_x$[14] must be compared with earlier results on the random polymer.[9-13] The general conclusion is that the transport is indeed limited by a

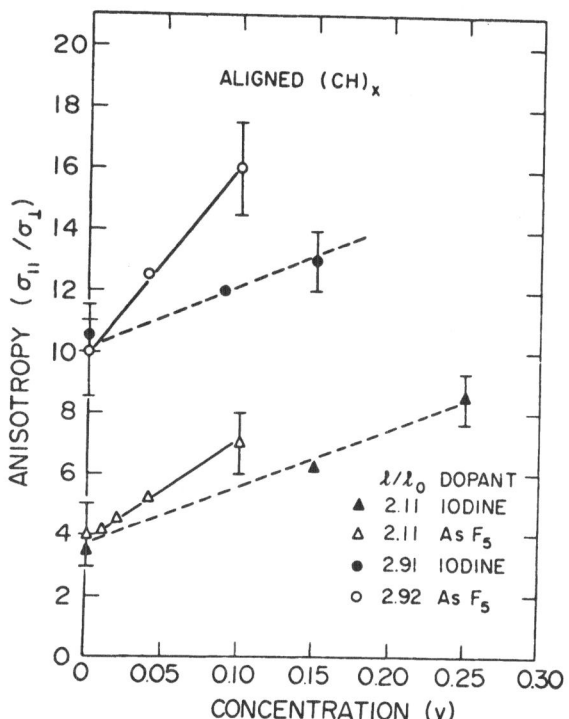

Figure 5: The electrical anisotropy, $\sigma_\parallel / \sigma_\perp$ as a function of dopant concentration for partially aligned $(CH)_x$ films at different values of elongation (ℓ/ℓ_0).

combination of interparticle contact and anisotropy even in the partially oriented films; the intrinsic conductivity along the $(CH)_x$ chain direction in the doped metallic polymer is much higher than the measured value. The trends in the data together with the electron microscope photographs suggest that better orientation will lead to considerable enhancement of the anisotropy and the absolute room temperature conductivity (Figure 5) with $\sigma_\parallel$ vs. (T) probably increasing substantially on cooling.

(10) ORIENTED FILMS: ANISOTROPIC OPTICAL PROPERTIES[15]

Direct visual inspection of oriented $(CH)_x$ films reveals a silvery reflection, similar to Al foil, but somewhat darker. Through a polarizer, the reflection polarized parallel to the fiber and polymer chain orientation direction is silvery, but the reflection polarized perpendicular is pastel orange. Doping with AsF_5 $(\sigma_\parallel \geq 10^3 \ ohm^{-1}cm^{-1})$ produced no obvious change by direct vision on parallel polarization, but the reflection polarized perpendicular became much darker indicative of increased anisotropy on doping.

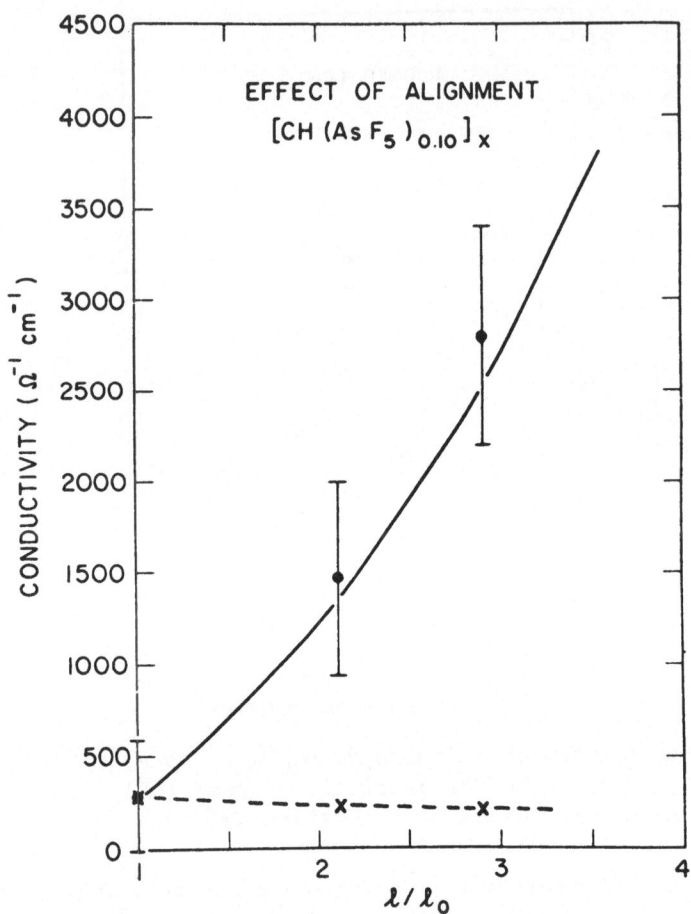

Figure 6:   Conductivity of metallic, heavily-doped $[CH(AsF_5)_{0.10}]_x$
as a function of elongation (ℓ/ℓ_o) as obtained by
Montgomery method.
            • • • parallel to alignment direction
            x x x perpendicular to alignment direction

        The reflectance results are shown in Figures 10 and 11. For
the pristine sample, $R_\parallel$ data show a broad maximum that corresponds
to the absorption peak near 2 eV. $R_\parallel$ decreases in the infrared
consistent with a semiconductor picture; extrapolation of the low
energy data suggests a low frequency reflectance of 12 to 18% imply-
ing a dielectric constant, $\varepsilon_\parallel (o) \simeq 5$. The perpendicular reflectance
is flat ($\sim 4\%$, $\varepsilon_\perp (o) \simeq 2$) at low frequencies, with a weak maximum
centered at 1.7 eV. At higher frequencies $R_\perp$ falls proportionately

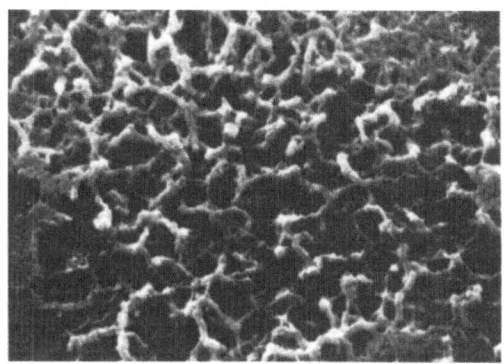

Figure 7:   Scanning electron microscope picture of as-grown $(CH)_x$.
(The average fibril diameter is approximately 200 Å.)

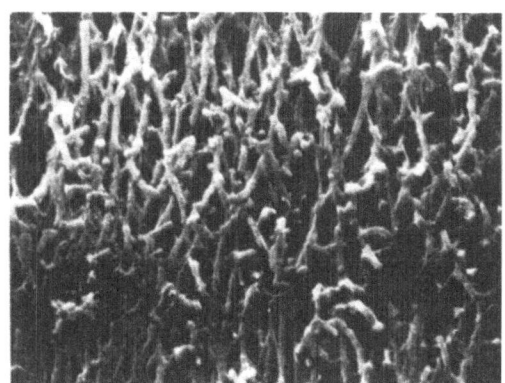

Figure 8:   Scanning electron microscope picture of stretch oriented $(CH)_x$. (The average fibril diameter is approximately 200 Å.)

faster than $R_\parallel$ suggesting that the observed structure in $R_\perp$ is intrinsic and not the result of incomplete orientation. The optical
anisotropy goes through a minimum of $(R_\parallel/R_\perp) = 4.7$ at 1.65 eV,
increases to $(R_\parallel/R_\perp) \simeq 10$ at 2.5 eV, then decreases at higher energy.

Heavy doping of the sample with AsF_5 ($\sigma_\parallel \geq 10^3$ ohm^{-1}cm^{-1})
increases $R_\parallel$ below 1.4 eV (Figure 11); the low frequency results
are similar to the free carrier reflectance in heavily doped semiconductors. The trans-$(CH)_x$ maximum near 2 eV remains after doping,
consistent with the absorption results described earlier. The overall result is an increase of the optical anisotropy at all energies
below 2.5 eV.

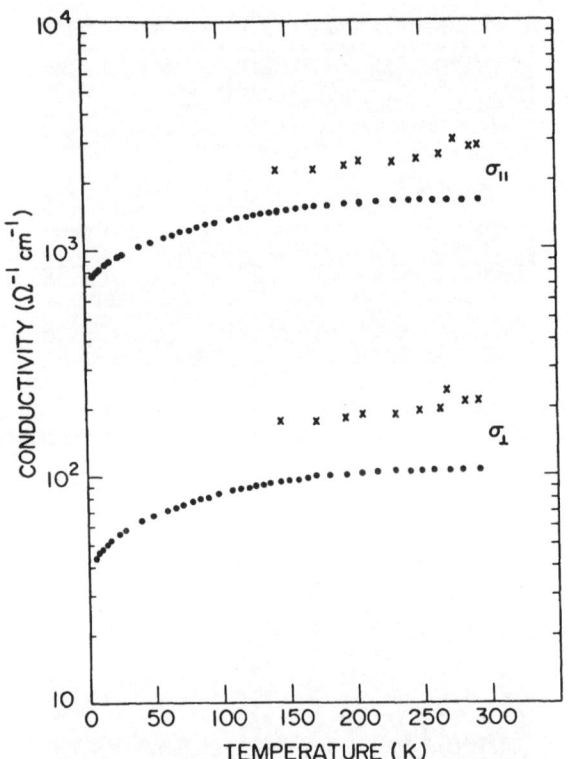

<u>Figure 9</u>:   Conductivity <u>vs</u>. temperature for oriented
            $[CH(AsF_5)_{0.10}]_x$. The film was stretch-oriented
            $(\ell/\ell_o = 2.9)$ prior to doping.

                    • • • four-probe measurements
                    x x x Montgomery measurements

        A somewhat more quantitative comparison of the absorption and
$R_{||}$ reflectance of the undoped polymer can be made by modeling the
interband transition as a single Lorentz oscillator centered at 2 eV
Thus, assuming

$$\varepsilon(\omega) = 1 + \frac{\Omega_0^2}{\omega_g^2 - \omega^2 - i\omega\Gamma}$$

leads to $E_g = \hbar\omega_g = 2$ eV, $\hbar\Omega_0 \simeq 4.0$ eV and $\hbar\Gamma \simeq 0.54$ eV; the
latter two parameters being determined by fitting to $\varepsilon_{||}(o)$ and
$R_{||}$ (2 eV) . From these one estimates a peak absorption at 2 eV of
$\alpha \simeq 5 \times 10^5$ cm^{-1} in good agreement with Figure 4. More detailed
comparison must await a Kramers-Kronig analysis of the full reflect-

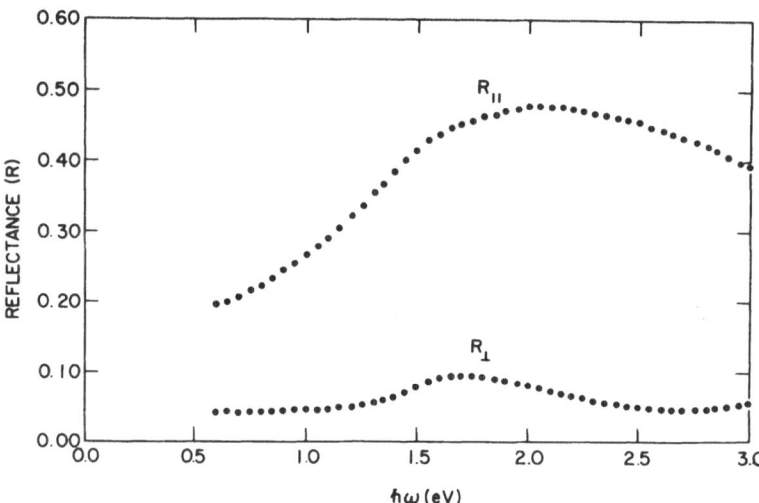

Figure 10: Anisotropic reflectance from partially oriented film of $(CH)_x$ ($\ell/\ell_o = 2.94$).

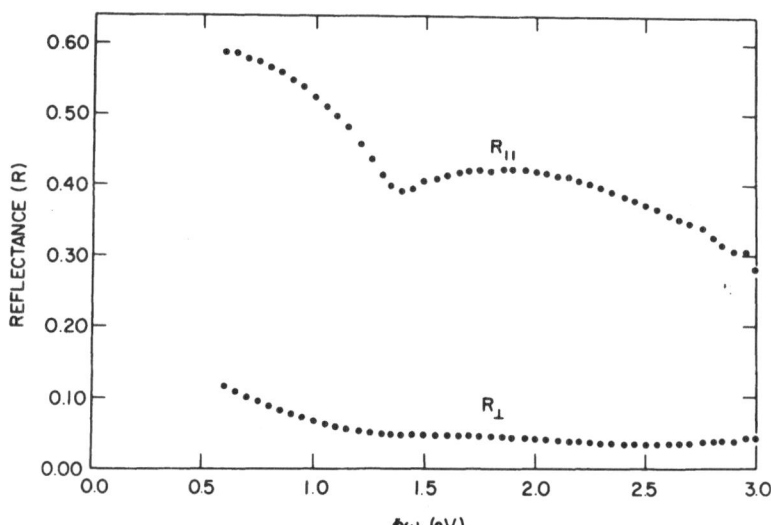

Figure 11: Anisotropic reflectance from partially oriented film of $(CH)_x$ ($\ell/\ell_o = 2.94$) after doping with AsF_5 ($\sigma_{\parallel} = 10^3$ ohm^{-1}cm^{-1}).

ance spectrum. Note, however, that the implied quantitative agreement between the $R_\parallel$ data from partially aligned films of undoped
$(CH)_x$ (Figure 10) and the absorption by non-aligned films (Figure 4)
implies that the strong absorption is polarized along the chain direction. We therefore conclude that the anisotropy is intrinsic and
is present on a single fiber scale in the non-oriented polymer. The
large optical anisotropy is consistent with a quasi-(1d) band
structure as described above.

The effect of doping on the reflectance is entirely consistent
with the previous interpretation of a metal-semiconductor transition
in doped polyacetylene.[9-15] The interband transition remains visible,
but the reflectance begins to rise at lower frequencies due to the
free carrier contribution to the dielectric function. Detailed
Drude fits require extension of the data into the far ir. These
studies are presently being carried out.

(11) SEMICONDUCTOR PHYSICS OF $(CH)_x$ FILMS: COMPENSATION AND JUNCTION FORMATION[13]

A series of experiments have been reported which demonstrate
that donors or acceptors can dope polyacetylene to n-type or p-type
respectively, and that the two kinds of dopants can compensate one
another.[13] The formation of a rectifying p-n junction as well as
Schottky barrier junctions have been demonstrated. These results
suggest the possibility of utilizing doped polyacetylene in a
variety of potential semiconductor device applications; in particular those involving solar cell applications.

Compensation of n-type material by subsequent acceptor doping
has been successfully demonstrated using Na (donor) and iodine or
AsF_5 (acceptors). Figure 12 shows the compensation of Na-doped
polyacetylene by iodine. The Na-doped films were prepared by treat-
ing the polymers with a solution of sodium naphthalide, $Na^+(C_{10}H_8)^{\frac{-}{\cdot}}$,
in THF whereupon electron transfer from the naphthalide radical
anion to the $(CH)_x$ occurred. In each case the pure cis-polyacetylene was first doped with sodium until the conductivity was in the
saturation range. Subsequent exposure to iodine vapor resulted in
the compensation curve plotted in Figure 12. The compensation
proceeds more slowly than the original doping; the electrical conductivity of the n-type sample gradually decreases and reaches a
minimum. Continued doping with iodine results in conversion to p-
type material with an associated increase in conductivity. Similar
compensation has been achieved with AsF_5 as the acceptor.

Starting with an initial composition $(CHNa_{0.27})_x$, the compensation point with iodine occurred at $(CHNa_{0.27}I_{0.28})_x$; all compositions
being determined by measurement of weight increase. Thus the compensation point corresponds approximately to a stoichiometric

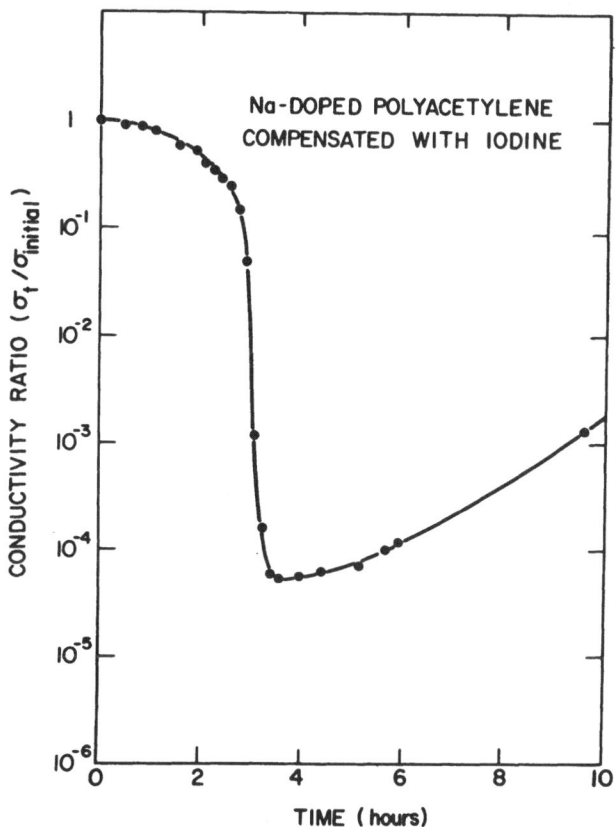

<u>Figure 12</u>:   Compensation curve for Na-doped polyacetylene; conduc-
tivity ratio $(\sigma_{(t)}/\sigma_{initial})$ <u>vs</u>. time. The sample
was initially doped <u>n-type</u> and subsequently exposed
to iodine vapor.

sodium to iodine ratio consistent (within the limits of error) to the
presence of equal concentrations of Na^+ and I^- in the compensated
polymer. Continued doping leads to p-type material, where the iodine
is known to be present as I_3^- from Raman studies.[26,27]

     The assignment of donor doped material as <u>n</u>-type and acceptor
doped material as <u>p</u>-type follows from the chemical properties of the
donor and acceptor dopants. Moreover, the assignments are consistent
with the results obtained from studies of graphite intercalated with
alkali metals and iodine or AsF_5 respectively.[28] Finally, thermoelectric power measurements on acceptor doped $(CH)_x$ yield a positive
Seebeck coefficient consistent with <u>p</u>-type material. Moreover, the
value of +15 $\mu V/K$ found at room temperature for $(CH)_x$ heavily doped
with AsF_5 is consistent with metallic behavior. Initial experiments

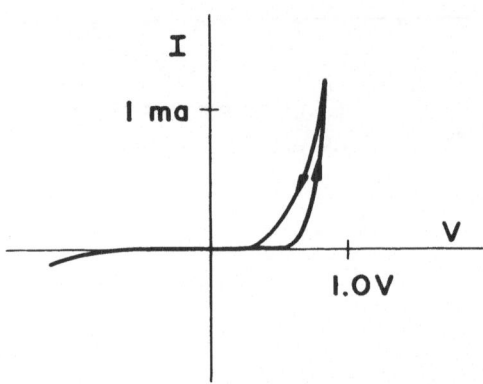

<u>Figure 13</u>: I-V curve for a doped polyacetylene <u>p-n</u> junction.

directed toward fabrication of <u>p-n</u> junctions are encouraging as
shown in Figure 13. The junction was made by mechanically pressing
together <u>n</u>-type (Na-doped) and <u>p</u>-type (AsF$_5$ - doped) strips of
polymer film. Although some hysteresis is evident, a typical diode
characteristic is seen in the I-V curve. Junctions have also been
made using a single polymer strip doped <u>n</u>-type on one-half and <u>p</u>-
type on the other half. Note that in all cases the forward bias
direction was consistent with the <u>p</u>-type and <u>n</u>-type character of the
acceptor and donor doped material.

Initial experiments directed toward fabrication of Schottky
barrier rectifying diodes have also provided encouraging results.
Both <u>n</u>-type material (e.g. Pt metal in contact with $[CH(Na)_y]_x$)
and <u>p</u>-type material (e.g. Na metal in contact with $[CH(AsF_5)_y]_x$)
can be used to obtain typical diode characteristics such as that
shown in Figure 14. Experiments to date have utilized point contact
geometry. More work is needed in order to elucidate the nature of
the interface, since, as shown in Figure 12, compensation has been
demonstrated and a possible chemical reaction may occur at the
interface between Na and the $[CH(AsF_5)_y]_x$.

(12) ELECTRICAL PROPERTIES OF DOPED (CH)$_x$ FILMS

At relatively low doping levels, the conductivity is activated
as shown for example in Figure 15 for (CHI$_y$)$_x$ (similar data have
been obtained for AsF$_5$, Br etc.) where the conductivity is plotted
<u>vs</u>. 1/T on a semilog scale. In general, we find that the conducti-
vity of doped polyacetylene decreases with decreasing temperature.
However, the plot of $\ell n\sigma$ <u>vs</u>. 1/T do not give straight line behavior
as seen in Figure 15. Plotting the data as $\ell n\sigma$ <u>vs</u>. T$^{-\frac{1}{4}}$ (or T$^{-\frac{1}{2}}$)

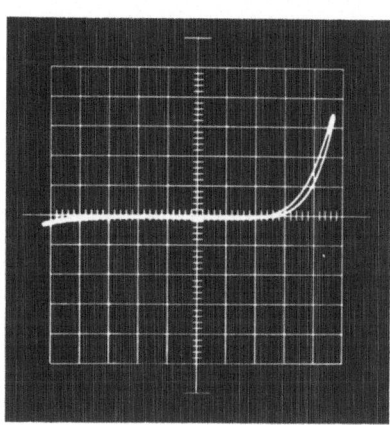

<u>Figure 14:</u>   Oscilloscope picture of I <u>vs.</u> V for Schottky barrier
          diode; Na contact on AsF_5-doped $(CH)_x$.

          x = 0.5 V/division; y = 1 μA/division at 60 Hz.

tends to give more nearly straight line behavior as shown in
Figure 16. Again this behavior is typical of that observed in non-
crystalline inorganic semiconductors such as amorphous Si, although
$(CH)_x$ films are at least partially crystalline as demonstrated by
X-ray diffraction. The twisted fibril structure of the films, (see
Figure.7) indicates the presence of significant disorder.

     The general behavior shown on Figures 15 and 16 is toward
smaller activation energy as the dopant concentration increases.
We use the initial slope of the 1/T plots to determine the approximate thermal activation energy, E_o, which serves as a simple index
of the conductivity behavior. The resulting activation energies
are shown in Figure 17 as a function of concentration y for both
$(CHBr_y)_x$ and $(CHI_y)_x$. Undoped polyacetylene has an activation
energy in the range from 0.3 eV (<u>trans</u>) to 0.5 eV (<u>cis</u>).[6] However,
the compensation experiments indicate that the conductivity in the
undoped polymer results from residual defects and/or impurities.[9-11]
Thus the intrinsic $(CH)_x$ activation energy is significantly higher,
in agreement with the optical studies (Figure 4). On doping with
halogen, the activation energy drops rapidly reaching a value as low
as 0.018 eV at about 20 mole % iodine. Similar results are obtained
from the bromine doping.

     The sudden change in the concentration dependance of the con-
ductivity and the activation energy near y = 0.02 is consistent
with a semiconductor-to-metal transition near the 2% dopant level,
in agreement with earlier far infrared and transport studies.[10] The
temperature dependence studies indicate that samples with y < 0.02

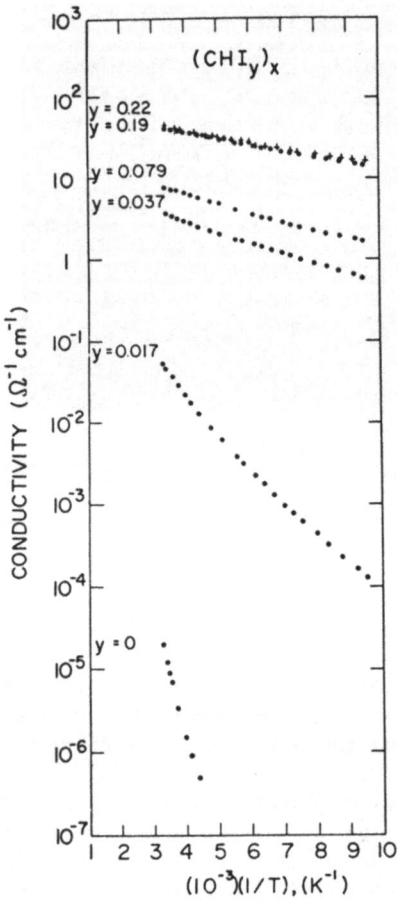

<u>Figure 15</u>: $\ell n\sigma$ <u>vs</u>. $1/T$ for $(CHI_y)_x$ for various concentrations (y) of iodine.

show an activated conductivity with the activation energy being a strong function of dopant concentration. For y > 0.02, the activation energy is sufficiently small that interfibril contacts in the polycrystalline polymer films are playing a limiting role.

The mobility of the heavily doped non-oriented polymer in the metallic regime, was estimated to be about 1 cm^2/Volt-sec.[10] This value was subsequently confirmed by Hall effect measurements.[29] However as demonstrated above, the transport is limited by the interfibril contact; the intrinsic mobility is undoubtedly considerably higher. Initial utilization of polymer processing techniques to orient the polymer fibrils have resulted in significant improvement of the conductivity as shown in Figure 6.[14] Using the higher conductivity of these partially oriented films to estimate the mobility

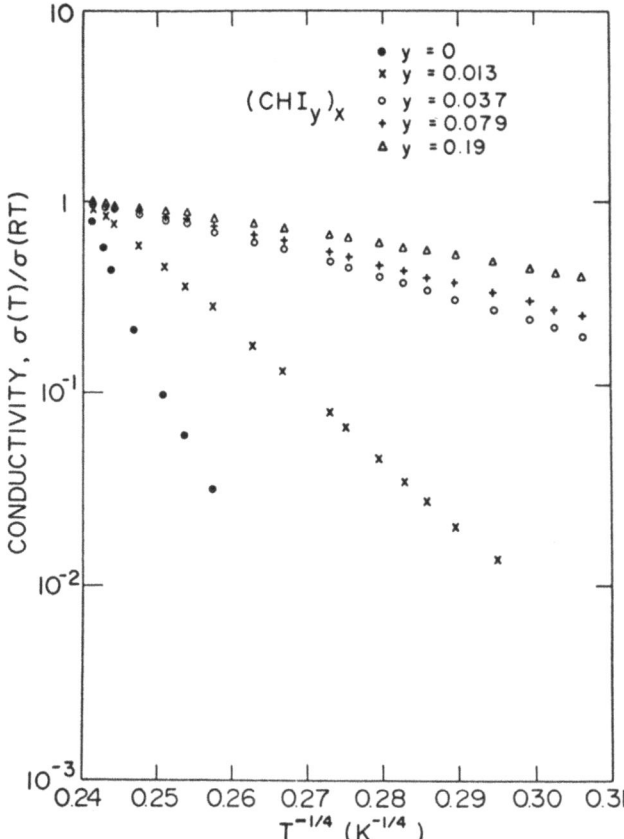

Figure 16: ℓnσ **vs.** $1/T^{\frac{1}{4}}$ for $(CHI_y)_x$ with various concentrations of iodine.

leads to a value of order 5-10 cm^2/Volt-sec; this clearly is a <u>lower</u> limit.

In conclusion, as can be seen from the following list of conductivities of common substances, $(CH)_x$ is quite remarkable in that its conductivity can be readily modified to span an extraordinarily large range. Considering possible polyacetylene derivatives, replacement of some or all of the hydrogen atoms in $(CH)_x$ with organic or inorganic groups, copolymerization of acetylene with other actylenes or olefins, and the use of different dopants should lead to the development of a large new class of conducting organic polymers with electrical properties that can be controlled over the full range from insulator to semiconductor to metal.

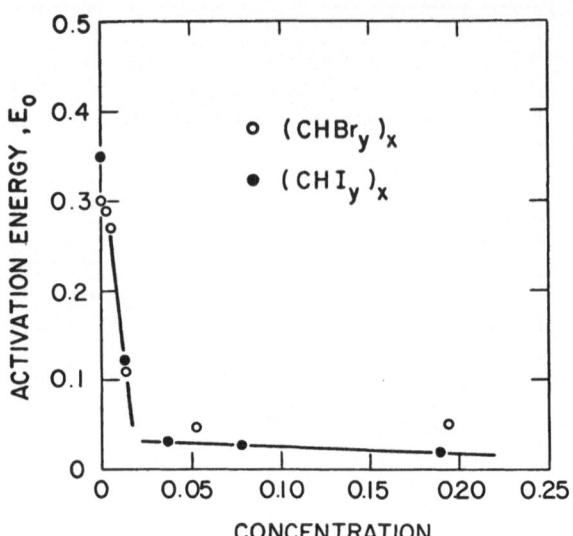

<u>Figure 17</u>: Activation energy of halogen-doped polyacetylene as a
              function of concentration. The activation energies
              were obtained from the slopes of the curves in Figure
              15.

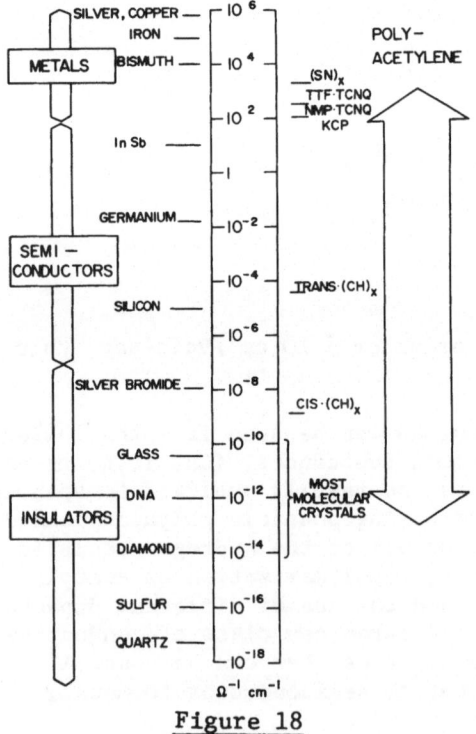

<u>Figure 18</u>

ACKNOWLEDGMENTS

The authors are indebted to the following persons for their important contributions to the work described in this review: Dr. Hideki Shirakawa (Tokyo Institute of Technology); C. K. Chiang, M. A. Druy, C. R. Fincher, Jr., S. C. Gau, E. J. Louis,Y. Matsumura, Y. W. Park, D. L. Peebles and A. Pron. This work was supported by the Office of Naval Research.

REFERENCES

1. M. Hatano, S. Kambara, S. Okamoto, J. Polym. Sci. 51, 526 (1961).
2. D. J. Berets and D. S. Smith, Trans. Faraday Soc. 64, 823 (1968).
3. H. Shirakawa and S. Ikeda, Polym. J. 2, 231 (1971).
4. H. Shirakwa, T. Ito and S. Ikeda, Polym. J. 4, 460 (1973).
5. T. Ito, H. Shirakawa and S. Ikeda, J. Polym. Sci. Polym. Chem. Ed. 12, 11 (1974).
6. T. Ito, H. Shirakawa and S. Ikeda, J. Polym. Sci. Polym. Chem. Ed. 13, 1943 (1975).
7. M. M. Maricq, J. S. Waugh, A. G. MacDiarmid, H. Shirakawa and A. J. Heeger, J. Amer. Chem. Soc., (in press) (1978).
8. H. Shirakawa, T. Ito, S. Ikeda, Die Macromoleculare Chemie, (in press) (1978).
9. H. Shirakawa, E. J. Louis, A. G. MacDiarmid, C. K. Chiang and A. J. Heeger, Chem. Comm. 578 (1978).
10. C. K. Chiang, C. R. Fincher, Jr., Y. W. Park, A. J. Heeger, H. Shirakawa, E. J. Louis, S. C. Gau and A. G. MacDiarmid, Phys. Rev. Lett. 39, 1098 (1977).
11. C. K. Chiang, M. A. Druy, S. C. Gau, A. J. Heeger, E. J. Louis, A. G. MacDiarmid, Y. W. Park, J. Amer. Chem. Soc. 100 (1013) (1978).
12. C. K. Chiang, Y. W. Park, A. J. Heeger, H. Shirakawa, E. J. Louis, and A. G. MacDiarmid, J. Chem. Phys. (in press) (1978).
13. C. K. Chiang, S. C. Gau, C. R. Fincher, Jr., Y. W. Park, A. G. MacDiarmid, and A. J. Heeger, Appl. Phys. Lett. (in press) (1978)
14. Y. W. Park, M. A. Druy, C. K. Chiang, A. G. MacDiarmid, A. J. Heeger, H. Shirakawa, and S. Ikeda, Phys. Rev. Lett. (Submitted).
15. C. R. Fincher, Jr., D. L. Peebles, A. J. Heeger, M. A. Druy, Y. Matsumura, A. G. MacDiarmid, H. Shirakawa and S. Ikeda, Solid State Commun. (in press) (1978).
16. The successful doping and resulting control of electrical properties over a wide range, including high conductivity (in the metallic range) at heavy doping levels have now been reproduced in many laboratories throughout the world. See, for example, J. F. Kwak, T. C. Clarke, R. L. Greene, and G. B. Street, Bull. Am. Phys. Soc. 23, 56 (1978).
17. For a summary and detailed references see A. A. Ovchinnikov, Soviet Phys. Uspekhi 15, 575 (1973).

18. N. F. Mott, Advances in Physics 21, 785 (1972).
19. See, for example, J. M. Ziman, Principles of the Theory of Solids, (Cambridge Univ. Press, 1972) p. 168-170.
20. F. Wooten, Optical Properties of Solids (Academic Press, New York (1972)), p. 116.
21. P. M. Grant, Bull. Am. Phys. Soc. 23, No. 3, 305 (1978).
22. B. R. Weinberger, A. J. Heeger, A. Pron, and A. G. MacDiarmid, (to be published).
23. H. Shirakawa, T. Sasaki, and S. Ikeda (to be published).
24. A. Matsui, K. Nakamura, J. Appl. Phys. (Jpn.) 6, 1468 (1967).
25. A. A. Bright, M. J. Cohen, A. F. Garito and A. J. Heeger, Appl. Phys. Lett. 26, 612 (1975); F. de la Cruz and H. J. Stoltz, Solid State Commun. 20, 241 (1976); R. J. Soulen and D. B. Utton, Solid State Commun. 21, 105 (1977); M. M. Labes, Pure Appl. Chem. 12, 275 (1966).
26. S. L. Hsu, A. J. Signorelli, G. P. Pez and R. H. Baughman, J. Chem. Phys. (in press) (1978).
27. H. Shirakawa and S. Ikeda, (to be published).
28. Proceedings of the International Conference on Intercalation Compounds of Graphite, Mat. Sci. Engineering 31, (December 1977).
29. K. Seeger, T. C. Clarke, W. D. Gill and G. B. Street, Bull. Am. Phys. Soc. 23, 56 (1978).

STRUCTURAL PERSPECTIVES FOR POLYMERIC METALS

R. H. Baughman, S. L. Hsu, L. R. Anderson,
G. P. Pez and A. J. Signorelli

Corporate Research Center
Allied Chemical Corporation
Morristown, New Jersey 07960, U.S.A.

INTRODUCTION

A major amount of research effort is now being devoted to the synthesis and properties of highly conducting derivatives of poly-acetylene. While the electrical conductivity of pure cis or pure trans polyacetylene is probably lower than $10^{-9}\Omega^{-1}$,[1] a variety of dopants are now known which increase the conductivity of polyacetylene at least nine orders of magnitude above this value. These include I_2, IBr, ICl, AsF_5, Na,[1-4] $AgBF_4$, $AgClO_4$,[5] and $(FSO_2O)_2$.[6] For the AsF_5 and $(FSO_2O-)_2$ dopants, conductivities of about $10^3\Omega^{-1}$ cm^{-1} or higher have been observed[2,5].

The absence of crucial structural information for these polymers presently hinders progress towards understanding electronic proper-ties. Until the present series of investigations, the crystal structure was unknown for not only all forms of polyacetylene, but also for any polymer having the polyacetylene type of backbone $(-C{=}C-)_x$. The limited amount of diffraction data obtainable for the available polycrystalline or more poorly ordered polymers dis-couraged structural investigations.

The present work evaluates the structure of cis and trans polyacetylene using x-ray diffraction, lattice packing calculations, and analysis of molecular packing in model compounds. A variety of property aspects are evaluated using as a basis these derived structures. Finally, structural aspects of the highly conducting iodine and $(FSO_2O-)_2$ doped polymers are derived using structural data for the parent polymers and x-ray diffraction, Raman, and x-ray photoemission results for the doped polymers.

II CIS-POLYACETYLENE

Basic aspects of the structure of cis-polyacetylene have been determined, using both diffraction intensity and crystal packing analyses.   Since only ten diffraction lines are observed for the polycrystalline, nonoriented cis-polymer, the molecular geometry used in these analyses is obtained from low molecular weight, conjugated olefins[7,8].   The C-C, C=C, and C-H bond lengths are 1.46, 1.35, and 1.11 Å, respectively.   The nominal C=C bond is parallel to the chain axis, the C=C-H bond angle is 120°, and the C-C=C bond angle (127.3°) is chosen to be consistent with the experimentally determined chain-axis periodicity.

The observed diffraction spacings are consistent with the orthorhombic unit cell Pnma, where a = 7.61, b = 4.47 (chain axis), and c = 4.39 Å at 300°K.   With one CH per asymmetric unit, there are eight CH units per unit cell, corresponding to a calculated density of 1.16 gm/cm^3 (1.165 gm/cm^3 observed density)[9].

The parameters determined from the intensity analysis are ϕ, the setting angle of the polymer chains with respect to the (001) plane, and B, an anisotropic temperature factor which is assumed identical for carbon and hydrogen.   The intensity analysis provides ϕ = 59° and B =12 Å^2.[9]

The basic structure determined by x-ray diffraction analysis is consistent with results obtained from lattice energy calculations[10].   These calculations predict the a-axis, c-axis, and setting angle parameters at 0°K via minimization of lattice energy, neglecting vibrational contributions.   The lattice energy is provided by the summation of the Williams IV intermolecular potentials for CH, CC, and HH[11] out to a distance of 15 Å.   Two minima are obtained from the packing analysis, which correspond to lattice energies of -1.675 kcal/mole (for a_1 = 7.14 Å, c_1 = 4.25 Å, and ϕ_1 = 51.7°) and -1.650 kcal/mole (for a_2 = 4.44 Å, c_2 = 6.92 Å, and ϕ_2 = 33.8°).   Note for these two solutions that $a_1 \simeq c_2$, $c_1 \simeq a_2$, and $\phi_1 \simeq$ 90-ϕ_2.   Consequently, as shown schematically in Fig.1, the chain-axis projections of these two packing solutions are essentially identical.   However, as also shown in Fig.1, the chain in the center of the unit cell is effectively rotated by 180° about the chain axis in going from the first solution (structure I) to the second solution (structure II). This effective rotation arises because the glide plane in structure I is normal to the short crystal axis, while this symmetry element is normal to the long crystal axis in structure II.   Structure I, with the calculated 1.5% lower energy and 1.3% higher density than for structure II, is the only solution consistent with the observed systematic absences in the x-ray diffraction data[9].   Approximating the thermal expansion of polyacetylene by that of polyethylene,[12,13] for reasons which will be discussed later, the discrepancy at 300°C between the calculated unit cell parameters for structure I and the experimental unit cell parameters is -2.7% for the a-axis and -1.1% for the c-axis. The slightly smaller predicted parameters are expected, since lattice expansions due to zero point vibrations are neglected in the packing calculations.[14]

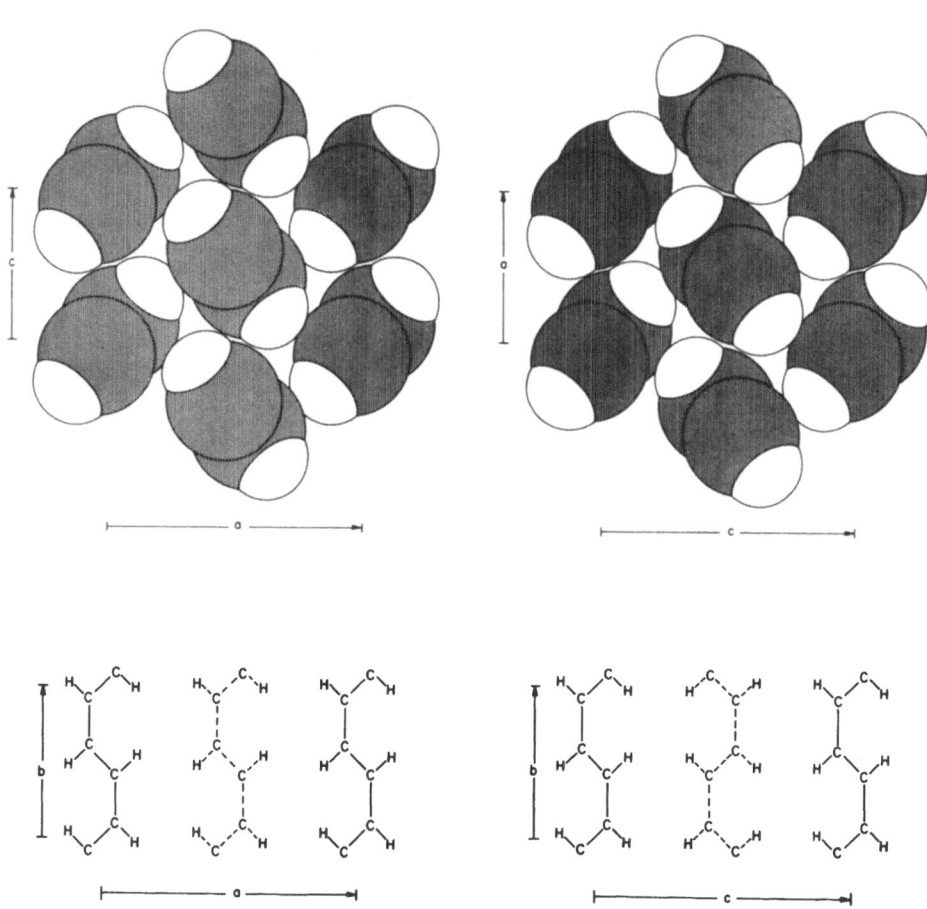

STRUCTURE I: Glide plane normal to short axis   STRUCTURE II: Glide plane normal to long axis

Fig. 1. Crystal structures corresponding to the two calculated energy minima for cis-polyacetylene. In the chain-axis projections (top) the white balls are hydrogens and the shaded balls are carbons. In the orthogonal projections (bottom), the dashed molecules are centered at ($\frac{1}{2}$, 0, $\frac{1}{2}$) with respect to the non-dashed molecules.

III TRANS-POLYACETYLENE

Since for trans-polyacetylene even fewer powder diffraction
lines were observed than for the cis polymer, guidance for the
structural analysis was sought from the crystal packing of model
compounds, the diphenylpolyenes $C_6H_5\text{-}(C=C)_n\text{-}C_6H_5$ [15-18]. As shown
by the crystal structure of diphenyloctatetraene in Fig.2,[15] such
polyenes are almost ideally suited for this purpose. The crystals
consist of layers of molecules with essentially parallel chain
axes having centers at $z=0$ for the first layer and at $z = c/2$ for
the second layer. The packing of molecules within the layers is
essentially the packing of all-trans $(CH)_{14}$ chains, with no dominant end effect, since the short intermolecular distances within
the molecular layer all involve the former atoms. The atoms in
the phenyls in diphenyloctatetraene largely determine the interlayer spacing and a rotation of the plane of the phenyl groups by
$\sim 5.4°$ with respect to the mean plane of the rest of the molecule.
Excluding the phenyl atoms, the remaining carbon atoms in the
molecule are coplanar to within a maximum deviation of 0.007 Å.[15]

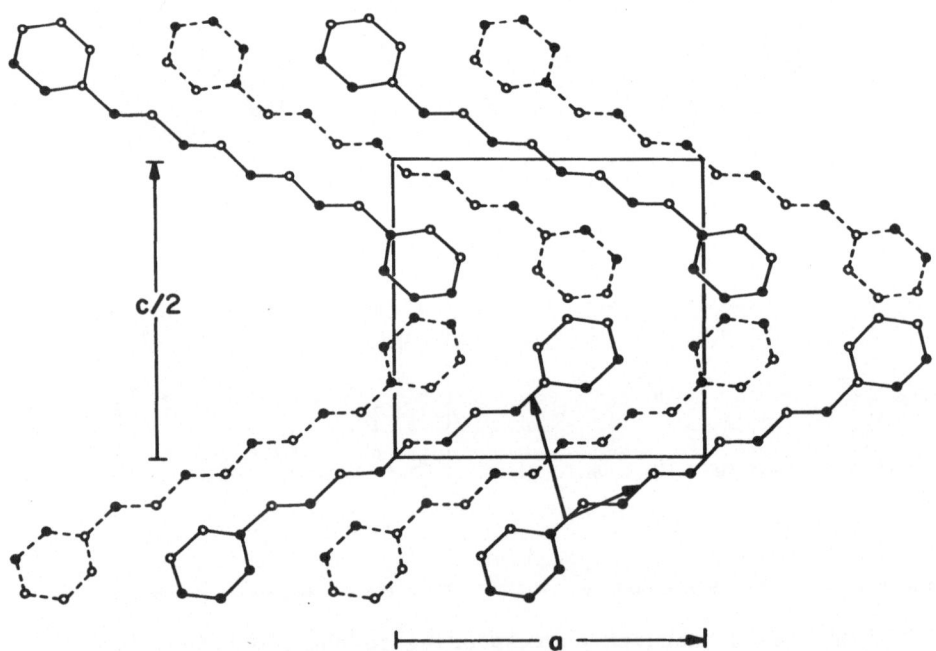

Fig. 2. The crystal structure of orthorhombic diphenyloctatetraene[15]
projected onto the (010) plane. The dashed and non-dashed molecules are related by the a-glide plane which is normal to the b-axis.
The unlabeled arrows indicate two axes of the derived cell for trans-
polyacetylene.

The symmetry obtained by projecting the atom coordinates of the polyene segments down the molecular axis direction of a layer is nearly Pgg. This result arises because the molecular axis is very nearly parallel to the (010), there exists a (010) glide plane, and each molecule sits on a center of symmetry which does not coincide with this glide plane.   This glide plane relates molecules having centers at y = 0 to those at y = b/2, so that the molecular-axis projection of the polyene segments look quite similar to that for cis-polyacetylene (Fig.1).   This result is encouraging, since the strong similarities in diffraction spacings and relative intensities observed for the cis and trans polymers suggest similar chain-axis projections.

The unit cell derived from the diphenylpolyene data is monoclinic, since the vector relating C-C≡C segments in different molecules (which can be nearly superimposed by a translation) is inclined to the chain-axis direction.   The only monoclinic space group having Pgg projection symmetry is P2$_1$/a.   This space group is not unexpected for trans-polyacetylene, since no orthorhombic space group is consistent with both the expected bond-alternate chain geometry, where alternate bonds are equivalent, and a chain projection symmetry Pgg. If trans-polyacetylene with this bond-alternate geometry and this projection symmetry had an orthorhombic space group, the location of the double and single bonds would have to be disordered.

The unit cell parameters calculated for trans-polyacetylene using the crystallographic data for the diphenylpolyenes are shown in Table I.   The results are compared with a unit cell which had been previously calculated on the basis of the seven observed diffraction spacings for trans-polyacetylene.   The observed spacings are consistent with an orthorhombic unit cell.   However, since all lines can be indexed as hk0 reflections, this analysis is equally consistent with a monoclinic cell with $\beta \neq$ 90 and where the calculated a-axis dimension is in reality a(sinβ) .

The crystal parameters derived from the diphenyloctatetraene and diphenyldecapentaene crystal structures are reasonably consistent (Table I).   For the longer polyene in the orthorhombic phase, the calculated parameters a(sinβ) and c are only 4.2% and 2.1% longer than the respective x-ray parameters for trans-polyacetylene.   Furthermore, as shown in Table I, the setting angle calculated for trans-polyacetylene (using the type of packing analysis described earlier for cis-polyacetylene) is nearly identical to the setting angle observed in the orthorhombic diphenylpolyene structures.

Table I   Structural Results for Trans-Polyacetylene

Origin	a (Å)	b (Å)	c (Å)	β (°)	ϕ (°)[a]
(CH)$_x$	4.08[b]	7.41			47.8[c]
C$_6$H$_5$(C≡C)$_4$C$_6$H$_5$	4.41[d]	7.50[d]	2.48[d]	99.1[d]	50.8[d]
		7.62[e]			
		7.39[f]			
C$_6$H$_5$(C≡C)$_5$C$_6$H$_5$	4.32[e]	7.57[e]	2.47[e]	100.1[e]	48.5[e]
		7.45[f]			

[a] Angle between chain backbone and (100).

[b] $a(\sin\beta)$ for space group $P2_1/a$.

[c] Packing calculation using unit cell parameters on left.

[d] Orthorhombic phase at -100°K.

[e] Orthorhombic phase at room temperature.

[f] Monoclinic phase at room temperature.

IV   COMPARISON OF CRYSTAL STRUCTURES AND RELATED PROPERTIES

It is useful at this point to compare structural aspects of cis and trans polyacetylene, (SN)$_x$, and more conventional polymers, such as polyethylene.

The chain symmetries retained on crystallization are $(m\bar{1})_t$ for cis-polyacetylene,[9,10] $(\bar{1})_t$ for trans-polyacetylene and 2_1 for (SN)$_x$[19-21]. Despite these differences, both of the polyacetylenes and the orthorhombic phase of (SN)$_x$ have the same chain-projection symmetry (Pgg) as orthorhombic polyethylene[22]. In fact, the setting angles for cis-polyacetylene, trans-polyacetylene, and polyethylene are close enough that the chain-axis projections of these polymer structures look nearly indistinguishable. Despite the differences in backbone structures, the chain symmetries of cis-polyacetylene and polyethylene are identical. The mirror plane normal to the chain axis in cis-polyacetylene relates neighboring CH units in the chain direction, while this symmetry element in polyethylene takes a CH$_2$ unit into itself. Hence, the chain repeat length (b-axis dimensions) in cis-polyacetylene is approximately twice as long as in polyethylene and the unit cell in cis-polyacetylene contains twice as many carbon atoms as in polyethylene. The axial lengths normal to the chain-axis direction are nearly

equal for cis-polyacetylene and trans-polyacetylene, and quite
similar to those for orthorhombic polyethylene.   However, the
glide plane is normal to the short axis (c) for cis-polyacetylene
and for polyethylene, while this symmetry element is normal to the
long axis (b) for trans-polyacetylene.   This difference is precisely the difference between the structures corresponding to the
two energy minima calculated for cis-polyacetylene (Fig.1), where
the trans-polyacetylene type of packing arrangement (structure II)
is calculated to have a slightly higher energy and lower density
than the observed packing arrangement for the cis-polymer
(structure I).

    The packing calculations described here for cis-polyacetylene
have been used to predict a variety of crystal properties for cis-
polyacetylene.   These calculations indicate that nonelectronic
solid-state properties related to intermolecular interactions for
polyacetylene are well approximated by those of polyethylene.
This point is useful, since it will probably be a long time before
the physical properties of polyacetylene are known to the same
extent as is now true for polyethylene.   Specifically, the calculated lattice energy per CH group in polyacetylene (-1.68 kcal/mole)[10]
is approximately equal to the experimentally-derived lattice energy
per CH_2 group in polyethylene (-1.84 kcal)[23].   Likewise, the low
temperature C_{11} and C_{33} elastic coefficients calculated for cis-
polyacetylene differ by only -3 and +7 percent from the corresponding polyethylene parameters calculated by Wobser and Blasenbrey[24].
Also, the calculated lattice Grüneisen coefficient for cis-polyacetylene (3.37)[10] nearly equals the value determined for polyethylene from bulk compressibility measurements (3.4 and 3.64)[25,26].
From these results, one can conclude that the lattice Debye
frequencies and low temperature specific heat of polyethylene
should be a good approximation for that of cis-polyacetylene.
These parameters could be of value, for example, for evaluating
phonon-electron interactions or the electronic contributions to
specific heat in the weakly-doped polyacetylenes.

    Because the packing of CH chains in the diphenylpolyenes
provides a good approximation for chain packing in trans-polyacetylene, these compounds provide a route for deriving properties
which cannot be directly measured for this polymer.   Specifically,
the microfibular morphology of polyacetylene, as well as present
inability to obtain fully oriented polymer, effectively precludes
reliable measurements of electrical or optical properties normal
to the chain direction.   Such information can be derived using
the diphenylpolyenes, which are obtainable as large single crystals.
For example, we can use refractive index data which is available
for the diphenylpolyenes[17-18] to examine the importance of interchain interactions and chain length upon optical properties orthogonal to the chain direction.   As shown in Table II, the

Table II   Refractive Indices of $H_5C_6(-C\equiv C-)_n C_6 H_5$ and $(CH)_x$

Polarization	n=3	n=4	n=5	trans-$(CH)_x$	cis-$(CH)_x$
Normal to glide plane	1.523^a	1.531^a	1.539^a	1.54^c	1.51^c
	1.537^b	1.537^b	1.543^b		
		1.513^c	1.521^c		
Glide direction				1.57^c	1.55^c

[a] Observed[15,17] for monoclinic phase.

[b] Observed[15,17] for orthorhombic phase.

[c] Calculated from bond polarizability tensors (no exaltation normal to chain) for orthorhombic diphenylpolyene or for polyacetylene.

refractive index (5893 Å) for light polarization normal to the chain direction shows little or no variation with increasing chain length of the diphenylpolyenes.   These refractive indices for the mono-clinic phases correspond to molar polarizabilities, normalized to the total number of carbon atoms in the molecule, which range from 0.1367 to 0.1402 (10^{-23} cm^3) in going from diphenylhexatriene to diphenyldecapentaene.   For the orthorhombic phases of these compounds, identical values are obtained for n=3 and n=5 (0.1434 x 10^{-23} cm^3) and a slightly lower value is obtained for n=4 (0.1428 x 10^{-23} cm^3).   In sharp contrast with these results, the solution measurements of Bramley and Le Fèvre[27] are consistent with a normalized molar polarizability in the maximum polarization direction which increases from 0.35 for n=3 to 0.53 for n=5.   The results for a polarization orthogonal to the chain direction suggest that neither strong interchain interactions nor bond exaltations of the isolated chain significantly affects the optical properties for this direction.   Hence, refractive indices were calculated using empirically determined polarizability tensors for CH, C=C, and C-C bonds[27] - neglecting any possible exaltations for directions normal to the chain direction.   Using the known orthorhombic structure of diphenyloctatetraene and diphenyldecapentaene[15,18], calculated refractive indices of 1.513 and 1.521, respectively, were obtained for the b-axis direction, which are in good agreement with the observed values (1.537 and 1.543, respectively)[17,18].   Using the setting angle for trans-polyacetylene deduced using the crystal structure of orthorhombic diphenyldecapentaene (Table I), the molar polarizabilities (per CH) calculated are 0.146 (a-axis) and 0.141 (b-axis), in units 10^{-23} cm^3.   Using the setting angle predicted at 300°K from packing analysis (49.8°), the corresponding quantities for the cis polymer (cis-transoid geometry) are 0.130 (a-axis) and 0.133 (c-axis).   The refractive indices obtained from these polar-izabilities and a density of 1.16gm/cm^3 are provided in Table II.

The structure of trans-polyacetylene can be used to deduce further information on the optical properties of this polymer. Since trans-polyacetylene is a bond-alternate polymer, the transition moment for the $\pi-\pi^*$ transition is expected to diverge from the chain-axis direction.   Consequently, as shown in Fig. 3, the polarization direction for maximum absorption would coincide with the chain-axis direction for light propagating normal to the (100) plane, but not for light propagating normal to the (010) plane.   In the former direction, since the transition moments of the two molecules in the unit cell are nonparallel, Davydov splitting should exist.   Because of the mirror plane normal to the chain-axis direction of the cis polymer, the transition moment for a chain must be parallel to the chain-axis direction.

Further examination of the diphenylpolyenes as model compounds for trans-polyacetylene is likely to yield useful insights. Diphenylpolyenes having as many as 16 conjugated double bonds have been synthesized[28].   As is true for polyacetylene, the cis forms readily isomerize to the all-trans.   With respect to the effect of chain length on light absorption, solution data is available in the literature[28].   The crystalline-state spectroscopy should be characterized for the entire series in order to predict the optical absorption spectra of a single crystal of trans-polyacety-lene.   Relevant to the effect of incorporating heteroatoms in the

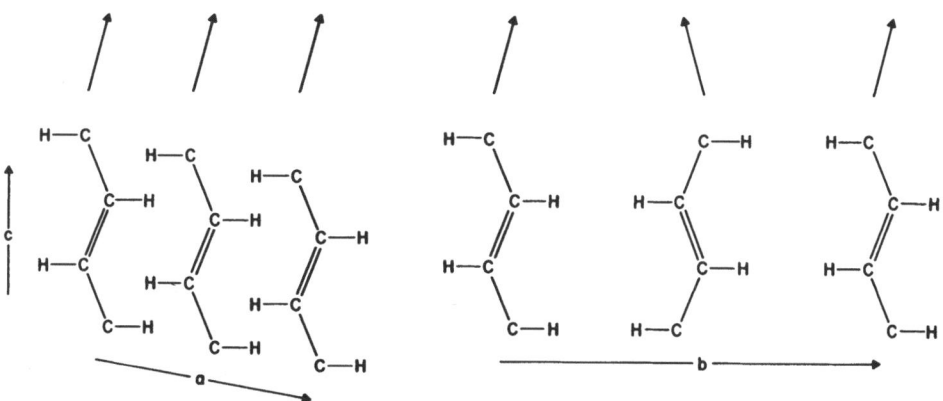

Fig. 3.   Projections of the trans-polyacetylene structure normal to the (010) plane (left) and normal to the (100) plane (right). The central molecule in each projection is related to the adjacent molecules by an a-glide plane.   The vectors above each chain represent schematically the transition moment direction of that chain.

polyacetylene backbone, it is interesting to note that replacement
of the first polyene CH in diphenyloctatetraene with N does not
appear to disrupt the 'polyacetylene-like' packing of chains.
Crystals of the substituted molecule appear to be isomorphous with
the octatetraene, having the same monoclinic space group and quite
similar unit cell parameters[29].   Apparently, the -N= atom in the
crystal is statistically disordered so as to preserve on the
average the diphenylpolyene symmetries and basic structure.   In
another vein, it is quite possible that the diphenylpolyenes will
form charge-transfer complexes having strong structural similarities
to those of the polyacetylenes, but with the advantage of being
obtainable as large single crystals.

V HIGHLY CONDUCTING DERIVATIVES OF POLYACETYLENE

The structure is not well understood for any of the doped
polyacetylenes, but the greatest amount of information is available
for the iodine complexes.   Compositions as high as $(CHI_{0.31})_x$ have
been observed in the pioneering research of Chiang, MacDiarmid,
Heeger and coworkers[1,3,30].   However, the major conductivity
increase (over many decades) occurs at less than a tenth of this
molar concentration, followed by about a decade conductivity
increase for higher concentrations[4].

Since the average fiber diameter of the undoped polymer is
typically about 200Å,[31] a monolayer deposition on these fibers
(with iodine-iodine bonds approximately parallel to the surface)
would result in a composition of about $(CHI_{0.03})_x$.   In order to
accommodate in this way the highest observed iodine concentrations,
the effective surface area must be increased about an order of
magnitude, corresponding to decreasing the average fiber diameter
by a factor of about three.   We cannot exclude the possibility
that a minor portion of the iodine sits on the fiber surface, but
can conclude that for the highest doping levels most of the iodine
either enters the polyacetylene phase or resides on newly created
surfaces.

The appearance of a new diffraction spacing, at roughly the
sum of the separation of close-packed planes in the parent polymer
and the effective van der Waals thickness of a dopant layer, is
an important feature of the iodine,[9,10,32] bromine, AsF$_5$, and
$(FSO_2O-)_2$ complexes.   These results suggest that these dopants
intercalate between close-packed planes of polyacetylene.   For
iodine complexes of cis and trans polyacetylene, the new diffraction spacing occurs at 7.6 - 7.9Å, which is reasonably close to
the sum of the spacings of the most intense diffraction lines in
the precursor polymers (3.7 - 3.8Å) and the van der Waals diameter

of iodine (3.96Å)[9,10,32].   Assuming that linear arrays of iodine ions (having a linear density well approximated by that for an I_3^- array)[33] are incorporated into one-half of the chain sites (such as by intercalation between adjacent close-packed planes) a composition of $(CHI_{0.33})_x$ is calculated for the fully-intercalated complex.   This composition is in good agreement with the highest observed composition, $(CHI_{0.31})_x$.[30]

Raman and x-ray photoemission measurements on the polyacetylene-iodine complex indicate the presence of two different halide species.[32,34]   A Raman line at 105-107 cm^{-1} and harmonics up to fourth order are observed, which can be assigned to the symmetric stretch of I_3^-.   In addition, a second strong Raman line is observed at 160 cm^{-1} for heavily doped polymer samples.   This second line is assigned to I_5^- or to strongly electronically perturbed I_2, rather than to the asymmetric stretch of I_3^-.   The relative intensity of the second Raman line (for 5145Å excitation) increases from 0.14 to 1.33 in going from $(CHI_{0.07})_x$ to $(CHI_{0.25})_x$, and decreases irreversibly upon thermal annealing $(50°C)$ on the application of mechanical stress (> 1 kbar).   The x-ray photoemission measurements[32] indicate a splitting of the iodine $3d_{3/2}$ and $3d_{5/2}$ core level lines, which is consistent with the presence of two iodine species.   The iodine $3d_{5/2}$ levels can be deconvoluted into two components at 620.6 eV and 619.0 eV, whereas only single, sharp 3d core level lines are observed for complexes known to contain only I_3^- (at 619.5 eV for CsI_3 and at 619.0 eV for a cobalt phthalocyanine-iodine complex)[35,36].   Also consistent with the Raman results, the higher binding energy component (less negatively-charged iodine species) decreases in relative intensity upon thermal annealing or upon the application of high mechanical stresses.   Valence band measurements on an iodinated cis-polyacetylene film, $(CHI_{0.22})_x$, are consistent with a finite density of states at the Fermi level, as expected for a metal.[32]

Complexes obtained by Anderson et al[6] via slow addition of $(FSO_2O-)_2$ to unoriented cis-polyacetylene (in SO_2F_2 solvent at low temperatures) show quite remarkable properties.   In contrast with the three-fold or higher monotonic decrease in conductivity reported for other polyacetylene complexes[2,3] in going from 300°K to 4°K, the conductivity of the peroxide-doped polyacetylene increases from 300°K $(700\,\Omega^{-1}cm^{-1})$ up to a maximum at 150°K $(800\,\Omega^{-1}cm^{-1})$.   Most interestingly, the conductivity at 4°K is about 0.90 of the 300°K value.

The form of the dopant is as yet unclear.   While one would expect SO_3F^- ions to predominate, the observed S=O stretching frequencies suggest that the fluorosulfonate might be covalently bonded to the chain or present as $(FSO_2O-)_2$.[6]   Further studies are required to resolve this important point.

The x-ray diffraction pattern for the doped cis polymer (21.3 weight percent dopant) shows no measurable shifts and no appreciable broadening of lines present in the parent polymer. However, a new and more diffuse diffraction line appears at a spacing of 8.37 $\pm$ 0.10Å.   Assuming the same dimensional increase for the intercalation of the cis polymer as for the stage 1 intercalation of graphite (4.46 $\pm$ 0.08Å)[37], the calculated spacing for the new diffraction line is 8.31 $\pm$ 0.08Å, which is the sum of the effective van der Waals thickness of the SO_3F^- layer in $C_{12}SO_3F$ and the separation of $(CH)_x$ planes normal to the (100).   The diffraction spacing calculated for the intercalated cis polymer by applying these same arguments to intercalated boron nitride, $(BN)_4SO_3F$,[37] is 8.49 $\pm$ 0.05Å.   Note that the observed spacing (8.37Å) is intermediate between these two calculated values. Assuming that the SO_3F^- intercalates between every (100) molecular plane of the cis polymer, having the same area per SO_3F^- as in $C_{12}SO_3F$ (31.3Å^2) or in $(BN)_4SO_3F$ (21.7Å^2), limiting compositions of $[CH(FSO_3)_{0.16}]_x$ and $[CH(FSO_3)_{0.23}]_x$, respectively, are calculated. Since the investigated composition, $[CH(FSO_3)_{0.036}]_x$, has a much lower dopant level than either of these estimates of the limiting composition, it seems clear why no measurable changes occur in the diffraction lines associated with the parent polymer. Apparently, in this sample the dopant is preferentially intercalating between the $(CH)_x$ planes in only limited regions of the polymer, leaving the remainder essentially untouched. The basic conclusions of these dimensional calculations are not dependent upon whether FSO_3 is present as an ion or as a dimer species.

VI CONCLUSION

We have pieced together structural models and related properties aspects for the polyacetylenes and the polyacetylene derivatives on the basis of the quite limited concrete data obtainable at the present time.

Further work is required to establish the validity of these models and to provide necessary refinements. The most important contribution towards both structure and properties characterization would be the development of methods for producing macroscopic single crystals of the polyacetylenes. Without this advance, uncertainties related to the structure and properties of these interesting materials will probably remain for a long time.

ACKNOWLEDGEMENTS

The authors thank A.G. MacDiarmid and coworkers for gifts of polyacetylene and R.R. Chance, A.G. MacDiarmid, R. Silby, and G.B. Street for helpful discussions concerning this work.   We

are especially indebted to J.O. Williams and J.M. Thomas for
valuable comments and for support when this work was completed at
the University College of Wales at Aberystwyth.

REFERENCES

1. C.K. Chiang, M.A. Druy, S.C. Gau, A.J. Heeger, E.J. Louis,
    A.G. MacDiarmid, Y.W. Park, and H. Shirakawa, J.Am.Chem.Soc.,
    100, 1013 (1978).

2. Y.W. Park, M.A. Druy, C.K. Chiang, A.G. MacDiarmid,
    A.J. Heeger, H. Shirakawa, and S. Ikeda, Phys.Rev.Lett.,
    in press.

3. C.K. Chiang, C.R. Fincher,Jr., Y.W. Park, A.J. Heeger,
    H. Shirakawa, E.J. Louis, S.C. Gau and A.G. MacDiarmid,
    Phys.Rev.Lett., 39, 1098 (1977).

4. C.K. Chiang, S.C. Gau, C.R. Fincher, Jr., Y.W. Park,
    A.G. MacDiarmid, and A.J. Heeger, Appl.Phys.Lett., 33, 18
    (1978).

5. T.C. Clarke, R.H. Geiss, J.F. Kwak, and G.B. Street, J.Chem.
    Soc.Chem.Commun., in press.

6. L.R. Anderson, G.P. Pez, and S.L. Hsu, J.Chem.Soc.Chem.Commun.,
    in press.

7. M. Traetteberg, Acta.Chem.Scand., 22, 628, 2297 (1968).

8. J.C.J. Bart and C.H. MacGillavry, Acta. Crystallogr. Sect.
    B24, 1569, 1587 (1968).

9. R.H. Baughman, S.L. Hsu, G.P. Pez and A.J. Signorelli,
    J.Chem.Phys., 68, 5405 (1978).

10. R.H. Baughman and S.L. Hsu, J.Polym.Sci., Polym. Lett.,
    in press.

11. D.E. Williams, J.Chem.Phys., 47, 4680 (1967).

12. G.T. Davis, R.K. Eby, and J.P. Colson, J.Appl.Phys. 41, 4316
    (1970).

13. G. Avitabile, R. Napolitano, B. Pirozzi, K.D. Rouse,
    M.W. Thomas, and B.T.M. Willis, J.Polym.Sci., Polym.Lett.,
    13, 351 (1975).

14. A. Warshel and S. Lifson, J.Chem.Phys., 53, 582 (1970).

15. W. Drenth and E.H. Wiebenga, Acta.Cryst., 8, 755 (1955).

16. R. Olthof, A. Vos, and D. Kracht, Rec.Trav.chim. Pays-Bas,
     87, 1295 (1968).

17. W. Drenth and E.H. Wiebenga, Rec.Trav.chim. Pays-Bas, 72, 39
     (1953).

18. W. Drenth and E.H. Wiebenga, Rec.Trav.chim. Pays-Bas, 73, 218
     (1954).

19. M. Boudeulle, Cryst. Struct. Commun., 4, 9 (1975).

20. A.G. MacDiarmid, C.M. Mikulski, P.J. Russo, M.S. Saran,
     A.F. Garito, and A.J. Heeger, J.Chem.Soc.Chem.Commun.
     1975, 476.

21. R.H. Baughman, P.A. Apgar, R.R. Chance, A.G. MacDiarmid,
     and A.F. Garito, J.Chem.Phys. 66, 401 (1977).

22. P.E. Teare, Acta Cryst., 12, 294 (1959).

23. F.W. Billmeyer, Jr., J.Appl.Phys., 28, 1114 (1957).

24. V.G. Wobser and S. Blasenbrey, Kolloid-Z.u.Z. Polymere,
     241, 985 (1970).

25. T. Ito and H. Marui, Polym.J. 2, 768 (1971).

26. T.P. Sham, B.A. Newman, and K.D. Pae, J.Matls. Sci., 12,
     771 (1977).

27. R. Bramley and R.J.W. LeFe'vre, J.Chem.Soc. 1960, 1820.

28. S.H. Harper, 'Rodd's Chemistry of Carbon Compounds' (Second
     Edition), S. Coffey, Ed. (Elsevier Scientific Pub.Co.,
     Amsterdam, 1974), Chapter 22, pp.216-222.

29. E.H. Wiebenga in 'Crystal Data Determination Tables' (Second
     Edition), J.D.H. Donnay, General Ed. (Amer. Crystallog.
     Assoc., 1963),p.233.

30. A.G. MacDiarmid, unpublished.

31. I. Ito, H. Shirakawa, and S. Ikeda, J.Polym.Sci., Polym.
     Chem.Ed., 12, 11 (1974).

32. S.L. Hsu, A.J. Signorelli, G.P. Pez, and R.H. Baughman,
     J.Chem.Phys., 69, 106 (1978).

33. J.S. Miller and C.H. Griffiths, J.Am.Chem.Soc. 99, 749 (1977).

34. R.H. Baughman, S.L. Hsu, and A.J. Signorelli, Proceedings of
     the Fifth International Conference on Organic Solid State
     Chemistry, submitted.

35. A.G. Maki and R. Forneris, Spectrochimica Acta, 23A, 867 (1967).

36. J.L. Peterson, C.S. Schram, D.R. Stojakovic, B.M. Hoffman, and
     T.J. Marks, J.Amer.Chem.Soc., 99, 286 (1977).

37. N. Bartlett, R.N. Biagioni, B.W. McQuillan, A.S. Robertson,
     and A.C. Thompson, J.Chem.Soc.Chem.Comm. 1978, 200.

PROPERTIES OF DOPED POLYACETYLENE, $(CH)_x$

R. L. Greene, T. C. Clarke, W. D. Gill, P. M. Grant,
J. F. Kwak and G. B. Street

IBM Research Laboratory
San Jose, California 95193, U.S.A.

Linear polyacetylene, $(CH)_x$ is one of the simplest conjugated organic polymers, and it therefore has attracted the attention of polymer chemists and physicists for some time [1]. Each carbon atom is σ bonded to one hydrogen and two neighboring carbon atoms consistent with sp^2 hybridization. The Π electrons delocalize into the bands in which carrier transport can occur. In the absence of bond alternation and Coulomb correlation the trans form of $(CH)_x$ would be a metal. In actuality the trans form of $(CH)_x$ has bond alternation and is thus a semiconductor. In a recent series of studies Shirakawa and coworkers [2,3,4] have succeeded in synthesizing polycrystalline films of $(CH)_x$ and have chemically doped these films with a variety of donors and acceptors to give n-type or p-type semiconductors. Transport and far IR transmission studies as a function of doping suggest that a semiconductor to metal transition occurs near 1 atomic % dopant concentration [3]. At the highest levels of acceptor doping, room temperature conductivities of an order a few hundred $ohm^{-1}cm^{-1}$ were obtained; a remarkably large value considering that the $(CH)_x$ films consist of tangled, randomly oriented fibrils (of average diameter 200Å) with an overall film density only one-third the theoretical density determined from X-ray diffraction studies [5].

In this short paper we will review the work done on $(CH)_x$ by our group at IBM Research San Jose. The details of most of this work were presented in March 1978 at the American Physical Society meeting in Washington, D.C., and are soon to be published in various journals. We will discuss band structure calculations [6], along with thermopower [7], electrical conductivity [7,8], Hall effect [8], optical absorption, reflectivity and photoconductivity experiments [9]. Most of the experiments have

been done on undoped $(CH)_x$ films and $(CH)_x$ films heavily doped
(~10%) with the acceptor AsF_5. We have also prepared high
conductivity $(CH)_x$ films doped with transition metals [10] (Ag
and Cu), thereby gaining new insight into the chemistry and physics
of the doping process.

We begin with our band structure results, since an
understanding of the energy bands is crucial to any interpretation
of the physical properties. The one-electron bands of $(CH)_x$ in
both the cis- and trans- conformations have been investigated by
Grant and Batra [6]. Both one-dimensional (1D) and
three-dimensional (3D) calculations have been done using the
crystal structure suggested by Baughman, et al. [5].

The major results are the following:

1)   Both cis and trans $(CH)_x$ have rather wide valence and
      conduction bands (~5eV) in the polymer chain direction.
      On the other hand, the interchain band dispersion is
      never larger than 0.3eV which implies $(CH)_x$ is a 1D-like
      material. However, this interchain interaction appears
      to be enough to suppress the Peierls instability when
      the material is doped to metallic levels of conductivity.

2)   There is a single particle band gap, of order 1.0eV, in
      cis $(CH)_x$ for uniform carbon bond lengths. There is no
      band gap in trans $(CH)_x$ for uniform bond lengths (i.e.,
      it is a metal). However, with bond alternation a gap
      appears reaching a maximum value of ~2.3eV for the extreme
      limit of pure single and double bonds.

The main conclusion of this work is that all the electronic
properties of both pristine and doped $(CH)_x$ which have been
observed to date can be interpreted within the framework of an
itinerant one-electron band structure in which Coulomb interactions
are negligible.

The optical experiments [9] give the most direct comparison
with the band theory. Our absorption measurements on undoped
$(CH)_x$ films show a steep absorption edge which can be interpreted
(as in conventional inorganic semiconductors) to give a value for
the band gap of E_g=1.2±0.1eV for cis $(CH)_x$ and E_g=0.97±0.1eV for
trans $(CH)_x$. These values agree remarkably well with the band
calculations. In addition, we have recently observed the onset
of photoconductivity at ~0.8eV in trans $(CH)_x$, which agrees with
earlier work on more poorly characterized material [11], and which
confirms the interpretation of the absorption edge as being caused
by the valence to conduction band transition. For both cis and
trans $(CH)_x$ doped with 10% AsF_5 we observe an absorption in the
near infrared which dominates the optical spectrum. We interpret
this as caused by free carrier absorption since it is also

manifested as a plasma edge in our reflectivity data. A Drude
analysis of the plasma edge gives $\varepsilon_\infty = 2.0$, $\hbar\omega_p = 2.43eV$ and
$\tau = 4.6 \times 10^{-15}$ sec.

For 10% AsF$_5$ doped films we find a room temperature dc
electrical conductivity of order 500 $\Omega^{-1}cm^{-1}$ in agreement with
the work of Chiang, et al. [3]. However, as the temperature is
lowered to 4.2°K the conductivity (σ) decreases by a factor of 5
for the Chiang, et al. [3] films and a factor of 1.5 for our films.
This distinctly <u>non-metallic</u> behavior could be caused by 1) poor
interfibril contact between metal fibers or 2) hopping via
localized states in a semiconductor gap. To distinguish these
possibilities we have performed a variety of different transport
measurements. Perhaps the most useful, thus far, has been the
thermoelectric power (S). The thermopower is a zero current
measurement, so that breaks in a current path, unless accompanied
by large breaks in the heat flow paths, do not produce sizable
effects. Therefore, in most cases, the thermopower is a direct
measurement of the intrinsic properties of a metal even when
anisotropy and crystal imperfections result in non-metallic
electrical conductivity behavior. We find, for 10% AsF$_5$ doping
in a cis-trans film, a small (~10μV/°K), positive thermopower at
300°K with a linear slope vs. temperature, extrapolating to zero
at 0°K. The sign is consistent with the acceptor doping, i.e.,
hole conduction, and the linear slope with zero intercept implies
that the fibrils are metallic. Diffusion through localized states
could give a linear thermopower but the magnitude and zero
temperature intercept would be large [12]. In addition, we find
that $\sigma(T)$ from 4.2°K down to 30mK is practically constant in value.
This is most plausibly understood as due to tunnelling between
metal particles (fibrils). Variable range hopping between
localized states predicts an exponential increase in σ as T is
lowered. The slope of the S vs. T allows us to estimate the
valence bandwidth for cis (CH)$_x$ using a simple 1D tight binding
model. With complete charge transfer to AsF$_5$ (i.e., 0.1e per CH)
we obtain a bandwidth of ~10eV in rough agreement with the better
band calculation discussed above. Our experiments [10] on silver
doping of (CH)$_x$ are strong evidence for complete charge transfer
with AsF$_5$ doping.

In an attempt to determine the number of free holes and their
mobility Seeger and Gill [8] have measured the Hall effect on 10%
AsF$_5$ doped (CH)$_x$ films. They find a positive Hall coefficient,
consistent with the thermopower, which changes very little in
magnitude between 300°K and 4.2°K (consistent with metallic
behavior). However, the magnitude of the Hall coefficient is
anomalously low and cannot be interpreted in terms of a single
band model unless one assumes a reduced value for the Hall mobility
[i.e, $R_H = (1/nec)\mu_H/\mu_D$] relative to the conductivity mobility.
Further work on this problem is in progress.

In summary then, we have briefly reviewed our theoretical and experimental work on heavily doped and undoped $(CH)_x$. Based on our results we have developed the following picture for this system. The undoped $(CH)_x$ is a semiconductor with a medium band gap; the gap being caused by bond alternation in the trans case and by the chain symmetry in the cis case. Doping with strong oxidizing agents such as AsF_5, bromine, iodine, etc. leads to complete charge transfer with mobile delocalized holes in a wide valence band of the polymer and localized charge on the dopant. The mobile holes cause metallic behavior at all temperatures. In contrast to the interpretation of Chiang, et al. [3] we believe that doping does not create localized states in the band gap (such states may exist because of disorder). Rather, the acceptor level is at an energy which falls within the valence band. The hole is localized by Coulomb interaction with the charged acceptor sites at low dopant concentrations and delocalized because of self screening at higher concentrations. These initial studies and those of others suggest there is much interesting physics and chemistry to be done on $(CH)_x$ and its modifications.

ACKNOWLEDGMENTS

We thank the ONR for partial support of this work. We also acknowledge the valuable discussions and experimental input of W. Bludau and K. Seeger during their visits to IBM San Jose and of our IBM colleagues I. Batra and T. Tani.

REFERENCES

1. For a summary and detailed references see A. A. Ovchinnikov, Soviet Physics Uspekhi 15, 575 (1973).

2. For example see To Ito, H. Shirakawa and S. Ikeda, J. Polym. Sci. Polym. Chem. Ed 13, 1943 (1975).

3. C. K. Chiang, et al., Phys. Rev. Lett. 39, 1098 (1977).

4. C. K. Chiang, et al., J. Am. Chem. Soc. 100, 1013 (1978).

5. R. H. Baughman, et al., to be published in J. Chem. Phys.

6. P. M. Grant, Bull. Am. Phys. Soc. 23, 305 (1978); P. M. Grant and I. P. Batra, submitted.

7. R. L. Greene, J. F. Kwak, T. C. Clarke and G. B. Street, Bull. Am. Phys. Soc. 23, 156 (1978) and 23, 305 (1978); also submitted to Solid State Comm.

8. K. Seeger, W. D. Gill, T. Clarke and G. B. Street, Bull. Am. Phys. Soc. 23, 305 (1978); also submitted to Solid State Comm.

9. W. Bludau, T. C. Clarke, P. M. Grant and G. B. Street, Bull.
    Am. Phys. Soc. 23, 304 (1978); to be published.

10. T. C. Clarke, R. H. Geiss, J. F. Kwak and G. B. Street,
    accepted by Chem. Comm.

11. A. Matsui and K. Nakamura, Japan J. Appl. Phys. 6, 148 (1967).

12. N. F. Mott and E. A. Davis, Electronic Processes in
    Non-Crystalline Materials (Oxford, 1971).

RECENT PROGRESS IN THE CHEMISTRY AND PHYSICS OF POLY(DIACETYLENES)

G. Wegner

Institut für Makromolekulare Chemie der Universität
- Hermann-Staudinger-Haus, Stefan-Meier-Str. 31
D-7800 Freiburg (West Germany)

I. INTRODUCTION

The search for macroscopic single crystals of polymers with the aim to study the intrinsic solid-state properties of macromolecules and their interaction in a perfect three dimensional lattice is as old as polymer science itself. Unfortunately, it seems to be an intrinsic property of high molecular weight chain molecules to aggregate under conditions of crystallization in form of highly complex morphological structures in which parts of the chains remain amorphous and form a matrix in which microscopic domains of crystalline chain segments of the same molecules are embedded. For some time it was even believed that perfect polymer single crystals do not exist at all, mainly because the phenomenon of chain folding observed in polymer crystallization is so general and structure-independent that the formation of extended chain crystals of polymers seemed to be impossible (1). The extensive research undertaken in the area of polymer crystallization leaves indead little hope for successful attempts to prepare a polymer backbone first by one or the other methods of polymer chemistry e. g. polymer charge-transfer-complexes of the type TTF-TCNQ or polyconjugated molecules or even more complex molecules like the ones proposed by Little (2) with the aim to crystallize these products afterwards into single-crystal-like or at least polycrystalline materials.

An alternative approach, namely polymerization of crystalline monomer in the solid-state so that the polymer as it grows in the parent phase retains the order of the monomer crystal did not turn out to be more successful except for the particular case of the so-called topochemical polymerization of diacetylenes as was first demonstrated by G. Wegner in 1969 (3). For the first time it became

possible not only to prepare macroscopic, nearly defect free polymer
single crystals but also to prepare macromolecules with a sequence
of conjugated carbon atoms as constitutive element of the polymer
backbone in hitherto unknown purity and stereochemical regularity.
Stimulated by the current concern in the electrical and optical
properties of fully conjugated linear polymers and organic metals
and by the growing interest in organic solid-state chemistry as a
general research subject the area of poly (diacetylenes) has since
developed into a fruitful and very active area of interdisciplinary research. The following is an attempt to critically review
the present state of knowledge on the formation and properties of
poly (diacetylenes) and to point out the direction of current research activities in this field.

II. THE BASIC FACTS

Basically the topochemical polymerization of monomers with
conjugated triple bonds ("diacetylenes") is described by

$$n \ R-C{\equiv}C-C{\equiv}C-R \longrightarrow \left[\begin{array}{c} R \\ C-C{\equiv}C-C \\ R \end{array}\right]_n \longleftrightarrow \left[\begin{array}{c} R \\ C=C=C=C \\ R \end{array}\right]_n$$

where R may be any substituent as for example

$-(CH_2)_m-CH_3$, $-CH_2-O-CO-NH-Ph$, $-CH_2-O-SO_2-(p)Tos$ etc.

The true nature of the reaction is better understood by inspection
of Fig. 1a. It shows schematically what is meant by the term topochemical polymerization: a diffusion-less solid-state transformation of a single crystal of a suitable monomer into the corresponding single crystal of the polymer. All reactivity is due to
very specific rotations of the monomers or only segments of the
monomer molecules on their lattice sites determined by the packing
properties of the molecules. Thus inevitably a crystal of extended
polymer chains is formed.

The idea that such reactions should occur was first expressed
by G. M. J. Schmidt (4) who developed a number of general rules
relating to organic solid-state chemistry (5), but the reaction of
the diacetylenes described more closely in Fig. 1b is the only
example where really polymer single crystals can be obtained.

In the parent crystal of a reactive diacetylene the molecules
are packed in a ladder-like fashion such that the ends of one triple
bond system approach the adjacent triple bond systems to a distance $\leqslant 0.4$ nm. Moreover the long axis of the triple-bond system
is inclined with regard to the translational axis to form an angle
of $\gamma \approx 45°$. Polymerization now occurs by successive tilting of
each molecule along the ladder with the minimum of movement of all

atoms from their lattice site. Thus the mode of packing and the
lattice symmetry can be retained throughout the reaction. The Polymerization is simply achieved by annealing of the colorless monomer crystals below their melting point or by high-energy or uv-ir-
radiation. Due to the formation of the conjugated backbone polymerizing crystals turn deeply colored as polymerization starts and
finally turn deep red, blue or green with typical metallic luster.
In many cases quantitative conversion is reached within a few hours
annealing time. There are also a number of cases known where polymerization stops at a limiting degree of conversion when the build-
up of lattice strain due to serious misfits between monomer packing
mode and polymer structure prevents further growth of the polymer
chains.

It should be mentioned at this point that the polymerization
of diacetylenes serves as the example for homogeneous solid-state
reactions, that is individual chains grow independent from each
other starting from points randomly distributed throughout the
lattice. Thus, a polymerizing crystal can be described as a solid
solution of extended macromolecules in the matrix of the yet unreacted parent phase. This was quite early reckognized to be the
true reason why macroscopic single crystals of the poly (diacetylenes) can be obtained (6-8) and it may be added that this is a
rather unique phenomenon in the realm of organic solid-state chemistry where most reactions are controlled by nucleation and crystal
defects rather than depend on the packing of the ideal lattice (5,9).

R. H. Baughman has combined the observed and necessary packing
geometry with the experimental observation that the polymerization
proceeds homogeneously and has tried to calculate solid-state reaction modes and relative phase reactivities of various diacetylenes
with the assumption that the lattice contribution to the activation
energy for reaction increases monotonically with the root means-
quare atomic displacements required for reaction (7,10). Although
this approach readily explains why a given modification of a particular diacetylene is reactive at all or does not polymerize
quantitatively it does not help to really predict reactivity of a
diacetylene, since polymorphism is a very common phenomenon in
the area of diacetylenes. A particularly substituted diacetylene
may have as many as five different modifications (6,11) of which
only one shows the desired reactivity. Since there is no theory
available which for a given organic molecule would predict the
possible packing modes and conditions under which they may be obtained, it is still the intuition of the experimentalist rather
than of the theoretician to find new polydiacetylenes.

The polymer chains grow via carbenes as the active intermediate as was first suggested by Takeda and Wegner (11) and was
verified by D. Bloor, M. Schwoerer and coworkers using ESR-spectroscopy as a quantitative tool (12,13). In single crystals of the

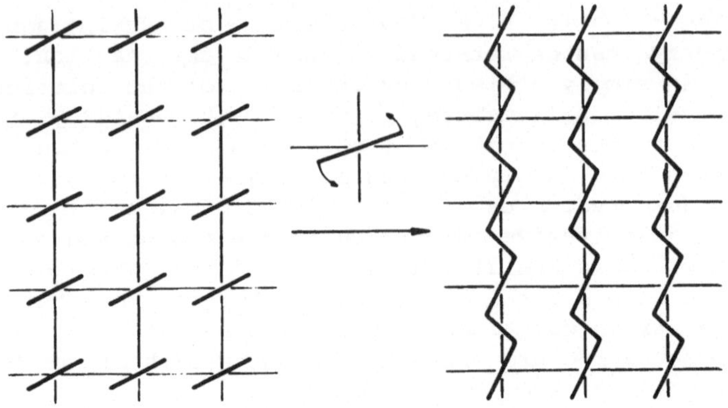

Fig. 1a. Scheme of a topochemical polymerization: transformation of a single crystal of monomers into a single crystal of polymers.

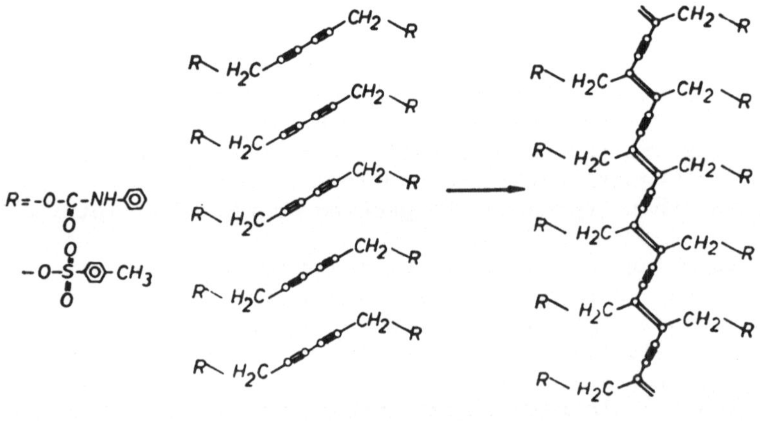

Fig. 1b. Topochemical polymerization of diacetylenes: the structure of R is explained in Tab. 3.

p-toluene sulfonate of 2,4-hexadiyne-1,6-diol (PTS) (1) which had been partially polymerized by annealing at 60°C triplet ESR-signals were observed and the zero-field splitting parameters were determined. These results were discussed in terms of Fig. 2 and it was deduced that the ESR-signal arises from a carbene localized at a nonpropagating polymer chain end. The low and positive value of the D-field splitting parameter was explained as due to a disc-shaped electron distribution with regard to the z-axis shown in Fig. 2 and consequently two mesomeric structures of the terminal group bearing the two unpaired spins were proposed.
These results readily explain the observed geometry of the Poly (diacetylene) chain as a consequence of a recombination reaction

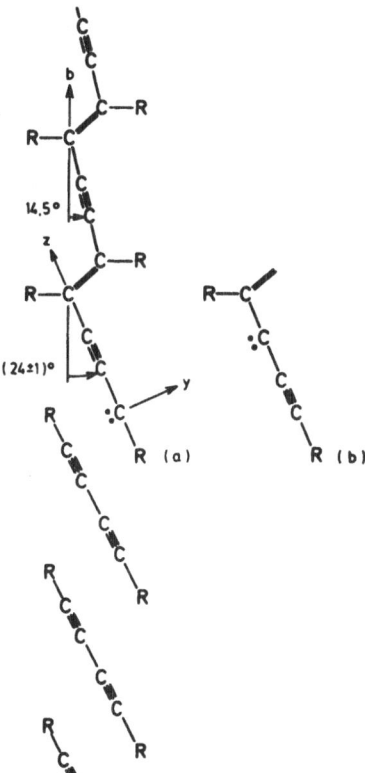

Fig. 2. Model of the active chain end species observed in PTS. y and z are the principal axes of the fine structure tensor (a) and (b) are mesomeric forms (13).

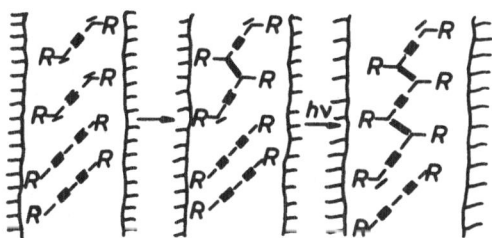

Fig. 3. Growth of the polymer chain by carbenes as active intermediates.

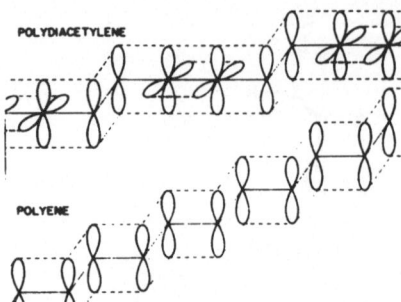

Fig. 4. Schematic description of the π-electron systems for the backbone of a polydiacetylene chain in comparison to a polyene chain.

of the thermally or photochemically excited diacetylene moiety forming a dicarbene-like structure as proposed by Wegner and depicted in Fig. 3 (5c,14). In the light of more recent results concerning the details of the polymerization kinetics this view may be an oversimplification as we shall see later.

The true nature of the bonding sequence along the polymer backbone and its quantum chemical description is a matter of debate ever since the existence of the polydiacetylenes was realized. In his first publication Wegner favoured the butatriene ($\rangle$C=C=C=C$\langle$) structure as the constitutive element of the polymer backbone (3). Soon after, the first X-ray structure analyses of poly (diacetylenes) clearly showed that in the case of the polymer hexadiyne diol-bis(phenyl urethane) (HDDU) (2) (15) and the polymer of 1 (PTS) (16) the bonding sequence is best described as a sequence of single-,triple- and double-bonds with the substituents all in trans-position with regard to the double bond (-C=C-C≡C-). Further structure analyses of other polymers showed, however, that the butatriene structure occurs at least in some derivates as for instance in the polymer of 5,7-dodecadiyne diol-1,12-bis(phenyl urethane) 3 (TCDU) and of the cyclic monomer 4 (BPG) (17,18).
Most authors, therefore, prefer to describe the polymer as a mesomeric structure with the butatriene and en-yne formulation as the limiting cases. A more concise description would be the one shown in Fig. 4 indicating the π-orbital interaction expected in the polydiacetylene chain.

The presence of a sequence of double-,single- and triple bonds in most polydiacetylenes is also favoured by the results of Raman-resonance spectroscopy. Although R. Baughman and coworkers were the first to report and analyze the Raman-spectra of a series of polydiacetylenes in terms of the en-yne structure (19,20) it was the Queen Mary College-group (21,22) who realized the importance

of the Raman-resonance effect in conjunction with the study of
the temperature dependence of the optical absorption or reflection
spectra as methods to study the fundamental and defect properties
of these crystalline materials. D. Bloor et al. used PTS which at
that time was the only polydiacetylene available in form of large
single crystals and showed (21) that a strong reflection peak
occurs at 2 eV for light polarized parallel to the polymer chain.
For light polarized perpendicularly to the chains the reflectivity
is low and featureless. Phonon side bands were also detected
above the 2 eV reflection peak due to the stretching modes of the
single, double and triple bonds of the conjugated cain. The same
chain vibrations were identified as the resonant enhanced Raman-
lines when the exciting laser line was chosen near the 2 eV absorption. It was further shown that good single crystals of fully
polymerized PTS do not show any substantial ESR signal unless the
crystals were thermally or mechanically mistreated prior to the
measurement. Only then, they show large ESR signals due to irreversible degradation. In conjunction with these observations the
question was raised how to describe the excited state of the polydiacetylene chain. Since it was already known from the work of
Schermann and Wegner (23) that PTS single crystals exhibit very
poor dark conductivity in the range of 10^{-12} $ohm^{-1}cm^{-1}$ at room temperature, Bloor considered the excited state to be an exciton, a
description which is widely accepted among the workers in the area
by now. Alternatively a MO calculation for the diacetylene chain
first undertaken by E. G. Wilson (24) treated the infinite conjugated molecule as a one-dimensional metal exhibiting van Hove
singularities. Both the vibrational side bands and the resonant
Raman-scattering could be explained in terms of this singularity.
The lack of appreciable photoconductivity in all polydiacetylenes
investigated so far is a strong argument against this description
however.

III. THE PROPERTIES OF PTS

Since the monomer PTS (1) can be readily synthesized from
commercially available starting materials in a one-step synthesis
in high purity and since large single crystals can be easily grown
by slow evaporation of a concentrated solution in acetone (25,26),
it has been widely used as a model in order to study the solid-
state polymerization as well as the physical properties of polydiacetylenes in general.
The polymer crystals, if obtained by annealing at temperatures between 50 and 75°C are highly perfect, as demonstrated by both X-ray
topography (27) and etching and cleavage studies (28,29).
PTS shows a reversible phase transition of second-order character
at temperatures below 170 K. The high-temperature form is monoclinic ($P2_1/c$), a = 14.493, b = 4.910, c = 14.936 Å, β = 118.14°,
Z = 2). The unit cell contains two symmetry related polymer chains

of different orientation extended along b, but with the planes of
the backbone parallel (16).

The transition to the low temperature phase is characterized
by torsions of the side groups attached to the polymer backbone
(30). The phenyl rings in adjacent rows of polymer chains turn in
opposite directions. This creates two different species of polymer
chains in the unit cell although the geometry of the backbone itself in terms of bond distances and angles remains unaffected.
Crystallographically the phase transition is described in essence
as a doubling of the unit cell and it does not destroy the perfection of the single crystal. The space group of the low temperature phase is $P2_1/c$ (a = 14.93, b = 4.910, c = 25.56 Å,
β = 92°, Z = 4) (30).

The same phase transition also occurs in the monomer crystal or
in partially polymerized crystals (31,32) and, therefore, is not
related to an intrensic property of the polymer chain as was previously suggested by a number of authors. Moreover, other polydiacetylenes do not show such a phase transition down to 4°K.
The occurence of two different chains in the low temperature phase
gives rise to the splitting of the optical absorption as was noted
first by D. Bloor and coworkers (21,33) and substantiated by others
(34-36). A number of different interpretations were given which -
among others - suggested distortion of the polymer chains by pola-
rons or molecular rearrangement (33,34) or chain interactions
(33,34,37). Theoretical calculations of the optical properties with
chain interaction fit the experimental results well (37) but in the
light of the crystallographic evidence described above their is no
physical reality behind these model calculations.
The details of the optical properties of the fully and partially
polymerized PTS have been studied with great care by the Queen Mary
College group including such topics as the crystal optics (38), low
temperature spectra (39) and defect induced spectra (40). In addition extensive ESR-studies were undertaken in order to detect whatever paramagnetism may be present and to correlate the findings with
theoretical predictions of the intrinsic behavior of polyconjugated chains. Fundamental models include excited conduction electrons

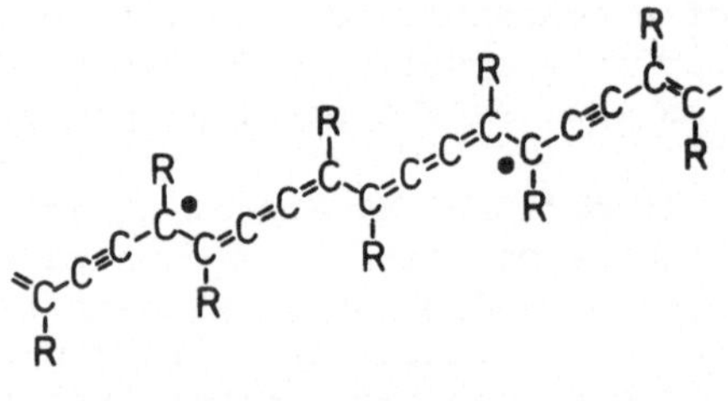

Fig. 5. Model of a hypothetical mobile defect giving rise to paramagnetic centers.

or triplet states (41,42), CT-interactions (43) and ground-state
singlet-triplet transitions induced by the thermal perturbation of
an antiferromagnetic ground state in non-bond alternating polymers
(44). Defect models include stable free radicals, bond alternation
defects of the classic Pople and Walmsley type (45) and local electron defect states in the band gap produced by structural or impurity perturbation of the polymer schain (46).
Since the poly(diacetylenes) and among them PTS unlike all the
other polyunsaturated macromolecules can be prepared in ultra-high
purity as a perfect crystal and defect states can be deliberately
induced, the significance of ESR-studies using PTS as a model to
test these theoretical assumptions is fairly obvious.
The experimental results (13,47,48) indicate a very low spin concentration in good crystals. The lack of hyperfine structure and
the almost pure lorentzian shape of the ESR line (g≐2) in single
crystals favour a mobile defect with a motionally narrowed line.
The spin concentration of typically 10^{15}-10^{16} spins/g varies somewhat with sample history and geometry. The signal intensity was
found to reversibly increase with increasing temperature in the
temperature range between 233 to 353 K with an activation energy
of 0.075 eV (13). A slightly different temperature behaviour was
found by Stevens and Bloor (48) who also measured to lower temperatures and report a curie-law dependence of the signal intensity
at temperatures below 220 K. All authors agree that the intrinsic
part of the signal is due to a defect as described by Fig. 5. Both
the length of this hypothetical defect, i. e. the distance between
the unpaired spins, and the position of the defect may fluctuate.
The proposed defect structure is similar to the one discussed by
Pople and Walmsley (45) for polyene chains (49).

The mechanical and thermomechanical properties of polymer PTS
have also received attention. The cleavage behavior and the structure of the cleavage surfaces was quite early taken as an indication
of the high   degree of perfection of these crystals and it
was infered that the concentration of chain ends should be very
small, i. e. the polymer chains extend over macroscopic distances
(29). The high perfection of such crystals was also demonstrated
recently by observation of the Brillouin-spectra of partially and
fully polymerized PTS from which the matrix elements of the elasti-
city modulus could be derived (51) (comp. also Fig. 18). A Young's
modulus of $4.38 \cdot 10^4 \text{Nmm}^{-2}$ was derived for the polymer as measured
in chain direction in good agreement with macroscopic measurements
of the stress-strain-behaviour by Batchelder and Bloor (52) and
by other authors who had measured stress-strain curves on the similar polymer HDDU (2) (53). If this value is weighted by the cross-
section per chain in the lattice a similar value is obtained as
the one known for the Young's modulus of diamond as measured in
the [110] direction (54).
Some consideration was also given to the study of the thermal expansion coefficient along the chain direction in order to gain

insight into the lattice dynamics of the polymer chains. The negative thermal expansion coefficients in chain direction for polyethylene and the polymer HDDU ($\underline{2}$) have been explained in terms of the
increasing amplitude of torsional motion of the chain with increasing temperature (55,56). Batchelder (57) has suggested that these
torsional motions are of secondary importance and he found that
the thermal expansion coefficient $\alpha_{\shortparallel}$ for PTS is small and positive
($0.9 \pm 0.2 \cdot 10^{-6} K^{-1}$) near 300 K and negative below 70 K. . From the
work of Munn (58) it becomes clear that $\alpha_{\shortparallel}$ is a complex function
of several magnitudes, among others the compressibility parallel to
the chain and the Grüneisen parameters parallel and perpendicular
to the chains. Therefore, the observed temperature dependence of
$\alpha_{\shortparallel}$ may not necessarily be indicative of torsional motions of the
chains.
Finally, it should be mentioned that the deformation and the deformation mechanisms of PTS where investigated and a number of twins
were identified (59).

IV. NEW POLY(DIACETYLENES)

The solid-state polymerization of diacetylenes has a very wide
scope and alltogether, there have been more than 200 different compounds described which undergo this reaction and - with the guidelines at hand described in the first chapter - it is easy to design
even more compounds which by principle should be reactive, if
crystallized in the suitable modification. Despite this situation
there are only a few compounds which have been investigated in
more detail and this is clearly due to the fact that many diacetylenes, although reactive in principle, to not polymerize quantitatively and/or without phase transitions which destroy the single
crystal character on polymerization. A number of compounds have
been detected, however, in the recent years which similarly to
PTS give rise to macroscopic single crystals of the polymer or at
least to stable solid solutions of the polymer in the parent phase
at the limiting conversion. In addition, polymerization of amphiphilic long chain diacetylenes in form of monolayers at the air-
water-interface or in form of multilayers build-up from the monolayers by the LB-technique has been developed as a promising new
area.

Poly(1.6-bis(N-carbazolyl)-2.4-hexadiyne) (DCH)

The most interesting among the newer systems certainly are the
1.6-bis(N-carbazolyl)-2.4-hexadiyne ($\underline{5}$) (DCH) and its polymer. DCH
was synthesized and described independently and at the same time
by Yee and Chance (60) and by V. Enkelmann and coworkers (61).
Large single crystals of the monomer are obtained by slow evaporation of a concentrated solution in DMF (62).

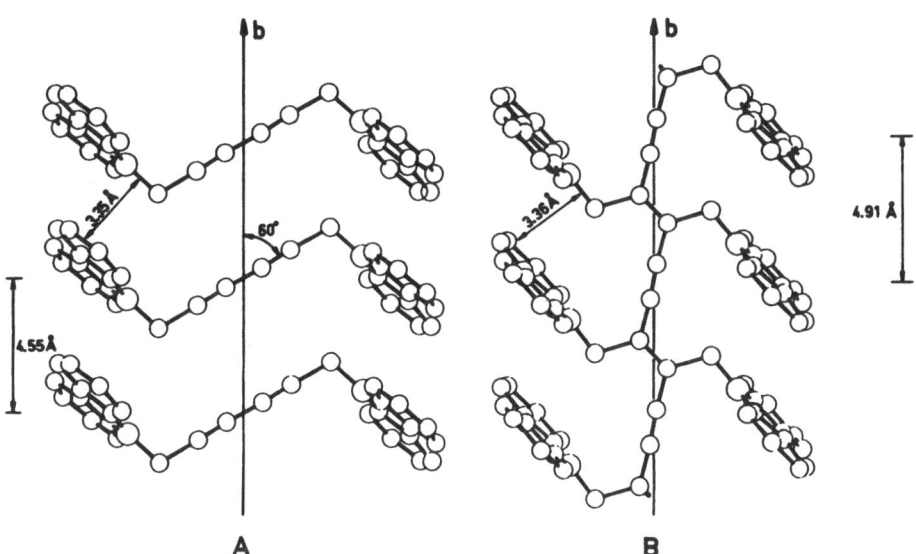

Fig. 6. Projection of the monomer and polymer structure of DCH onto the plane of the polymer backbone.

Unlike most reactive diacetylenes DCH crystals are quite stable toward the light level in the laboratory and polymerization in the X-ray beam is sufficiently slow to allow a structure determination at room temperature. The polymerization occurs by annealing at temperatures between 120 and 240°C or by irradiation with ^{60}Co-γ-rays. A dosage of about 50 M rad is sufficient to reach quantitative conversion. An autocatalytic effect is observed in both γ-ray and thermal polymerization. Thermal polymerization does not lead to single crystals of the polymer but to polycrystalline and misoriented aggregates of individual domains of the polymer. In order to obtain poly (DCH) single crystals, radiation polymerization is a necessary requirement. This as well as the "autocatalytic" effect is due to a phase change which occurs when part of the parent phase (10 percent) has been converted. The phase change proceeds homogeneously as in a martensitic transformation in radiation polymerization (at temperatures below 50°C) and heterogeneously with nucleation during annealing (61). It is worthwile to point out that the polymerizing crystals expands as much as 0.35 Å per base unit in the direction of the polymer chain without destruction of the lattice.

The monomer crystal with space group P2$_1$/c (a = 13.60, b = 4.55, c = 17.60 Å, β = 94°, Z = 2) contains 4 infinite arrays of substituted carbazolyl units stacked along the b-axis with a distance of 3.35 Å. This is about 0.6 Å more than the carbazolyl group

distance in the excimer forming sites in Poly(n-vinyl carbazole)
and similar to the excimer distance of pyrene (63). The monomer
crystals show strong phosphorescence from four traps as well as a
broad band fluorescence (61). Optically detected magnetic resonance
(ODMR) at 1.2 K showed that the four traps are localized triplet
states attributed to the carbazolyl groups (64). The interaction
between the diacetylene groups and the side groups is negligible.
An interaction of the trap states with an exciton band at 23926cm^{-1}
was proposed and   supported by temperature dependent life time
measurements (61,64).
The fluorescence and phosphorescence disappears as the polymer
chains start to grow and the polymer crystals, too, do not show
measureable emmission. It is not clear whether this is due to
self absorption, quenching of the excited state by interaction
with the polymer backbone bond system or defects created in the
course of the polymerization.
The polymer crystals with space group P2$_1$/c (a = 12.865 ,
b = 4.907, c = 17.403 Å, β = 108.3°, Z = 2) (65) show the typical
golden metallic luster of the poly(diacetylenes). The backbone
exhibits the bonding pattern of the poly(en-yne) structure (65) similar to PTS.
A projection of the crystal structure of monomer and polymer DCH
onto the plane of the polymer backbone is shown in Fig. 6. The
polymer is exceptionally stable at high temperatures and does not
decompose up to 300°C. Slow decomposition occurs at 400°C. The
polymer is sparingelly soluble in conc. sulfuric acid where it
forms a dark violet solution.

Normal incidence reflection spectra were reported by Hood
et al. (66). The lowest energy optical transition for the polymer
backbone is found at 15300 cm^{-1}, the lowest value obtained thus
far for a poly(diacetylene). The polymer chain and the carbazolyl
side groups may be treated as isolated chromophores, the former
dominating the visible portion of the spectra and the latter dominating the uv portion. The absorption peak at 15300 cm^{-1} is followed by phonon side bands which are observed as the main features
of the resonance Raman spectrum, namely the -C=C- and -C≡C-vibra-
tions at 1450 cm^{-1} and 2100 cm^{-1}, resp.

The energies for the photoconductivity onset at ca. 2.3 eV
and for the lowest energy optical transition at 1.89 eV are the
lowest reported so far for polydiacetylenes. The polymer is an electrical insulator in the dark (σ<10-15Ωcm-1) (60).
It would be interesting to study the effect of direct attachment of
the carbazolyl groups to the diacetylene linkage. Unfortunately,
this compound, 1.4-bis(N-carbazolyl)1.3-butadiyne, has proved to
be unreactive in the solid-state. The only modification of this
compound known so far shows a packing which does not allow for polymerization at all (67).

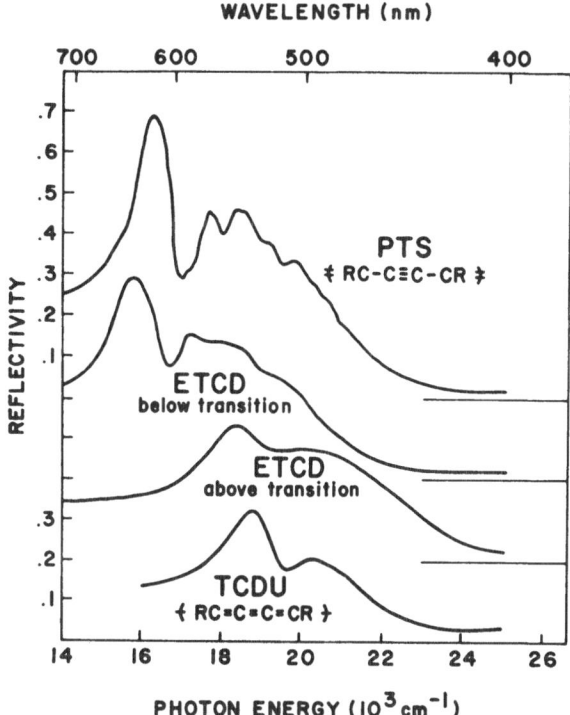

Fig. 7. Normal incidence reflection spectra for PTS, TCDU and
ETCD below and above the phase transition acc. to ref. 73
The origins are displaced as indicated by the horizontal lines.

Polydiacetylenes with cumulene structure and thermochromism

As discussed earlier, the poly(diacetylene) backbone may be represented by two mesomeric structures. Evidence for the cumulene
structure comes from two recent X-ray structure analyses of TCDU 3
and of the cyclic diacetylene 4 (BPG) (17,18). The latter is a particularly interesting case since it belongs to those compounds
which cannot be polymerized quantitatively but polymerization
stops at partial conversion without affecting the single crystal
texture although local defects may be created as indicated by a
number of features in the resonance Raman spectra (68). It is
worthwhile to note that the Raman spectra of the en-yne sequence
are not markedly different from the spectra of the butatriene sequence as can be readily seen comparing model compounds (69) and
it can, therefore, not be decided, if the cumulene or en-yne structure is present in a particular compound be merely looking to the
Raman (or IR) spectra. X-ray structure work is the only reliable

source of information on the true bond-structure till now.
Nevertheless a number of interesting observations have been made
which were suggested to be related to a thermally or pressure induced electronic transition from the en-yne to the butatriene-se-
quence and vice versa.

A number of polydiacetylenes exhibit reversible color changes
in conjunction with a temperature induced phase transition and/or
show such color changes in the course of polymerization. The latter
has been noted a number of times by Wegner et al. (5c,11,70). An
example of this color change during polymerization will be discussed below where the polymerization of diacetylenes in multilayers will be described (comp. Fig. 11 and 14).
Of the thermochromic polydiacetylenes, only ETDC (6) with
$-(CH_2)_4-OCONHC_2H_5$-side groups to the backbone has been thoroughly
investigated (69,71-73).
ETDC single crystals undergo a reversible phase transition, if heated to temperatures between $117-137^{\circ}C$ and change color from green
to red. There is considerable hysteresis observed. The lattice parameters for the low and high temperature form were reported (73),
but unfortunately there is no structure analysis available.

The important parameter, the chain repeat distance, changes from
$4.89 \overset{\circ}{A}$ in the low to $4.83 \overset{\circ}{A}$ in the high temperature form. This
was hinted as evidence for a change from en-yne to butatriene struc-
ture. Support for this idea comes also from comparison of normal incidence reflection spectra of the two modifications, if compared
with the available spectra of PTS and TCDU for which the poly(en-
yne) and butatriene structure resp. have been certified. Above the
phase transition, the ETCD spectrum is quite similar to that of
TCDU, below the transition the spectrum compares very well with
that of PTS (comp. Fig. 7) (73,74). Unfortunately, the spectrum of
poly(BPG) which is also best represented by the butatriene structure from X-ray evidence (17) shows a spectrum with maximum at
570 nm much more similar to PTS than to TCDU. TCDU itself undergoes
also a phase transition at low temperatures (<100 K) or high pressures (>1.5 kbar) accompanied by changes in the Raman and the optical spectrum similar as described for ETCD (69).
In general, it seems that whenever a irreversible or reversible
color change is seen in poly(diacetylenes) this is accompanied by
a phase transition, as is also documented for the case of long-
chain aliphatic substituted diacetylene carbonic acids and their
salts which were used to produce polymerizable mono- and multilayers (75). It is also noted that there are no gradual shifts in the
absorption spectrum going from the green or blue form to the red
form but in all cases known to the author there is a sudden shift
in the maxima of the optical spectra over about $2.5 \cdot 10^3 cm^{-1}$. In cases where a gradual shift seemed to occur as deduced from the

reflection spectra it turned out that the phase transition was controlled by nucleation and that partial conversion in microscopic
domains produced an overlap of the high and low temperature
spectra (75).

In summary the amount of information available at present does
not allow to draw decisive conclusions on the question whether the
butatriene and en-yne structures are true mesomeric forms or do
only occur in well defined crystal phase. Moreover, it is unclear,
if the observed spectral changes are due to the transition from
en-yne to cumulene structure or if they are induced by defect
states such as chain ends or large distortions of the chain. The
latter was suggested and the destorted chains were called to have
"orbital flip defects" (50) although the sharpness of the optical
transition is not in favour of such an explanation, since one would
expect a distribution or at least fluctuation of sequence length
of units between such defects giving rise to broadening of the
spectra.

Mixed crystals

Mixed crystals of organic solids are helpful in order to study the
nature of photoexcited states and the mechanism of energy transfer by deliberately introduced traps or sources for excited states.
In the field of the polydiacetylenes such crystals are even more
interesting from the additional point of view of organic solid-
state reactivity. It is expected that copolymer single crystals
can be obtained by solid-state polymerization of a single-crystal
of a solid-solution forming mixture of two different diacetylenes.

Several such systems have become available in the two recent
years from work in the authors laboratory (76,77). One of these
is described in Tab. 1. DCH forms a continuous series of mixed
crystals with the similar monomer ACH (1-(N-carbazolyl)-4-(9-anth-
racenyl) 2.4-hexadiyne). The crystals are obtained from a solution
of the monomers in DMF by slow evaporation. ACH cannot be polymerized by high-energy irradiation and only with decomposition at
high temperature near its melting point. The crystal structure of
ACH is similar to DCH but the axis in which polymerization would
be expected to proceed has an identity period of only 4.37 $\overset{\circ}{A}$ in
comparison to 4.55 $\overset{\circ}{A}$ in DCH. As can be seen from Tab. 1 this crystal parameter (b) changes smoothly with composition and so does
the solid-state reactivity. With decreasing stacking distance the
polymerization becomes more sluggish. Chemical and spectroscopical
analysis of the copolymer single crystals indicate that copolymers
are formed with the same composition as the parent phase.

Table 1: Mixed crystals (solid solutions) of DCH as one component
and solid-state polymerization behavior.

	$\overline{b}$ (Å)	yield at dose	thermal polym.

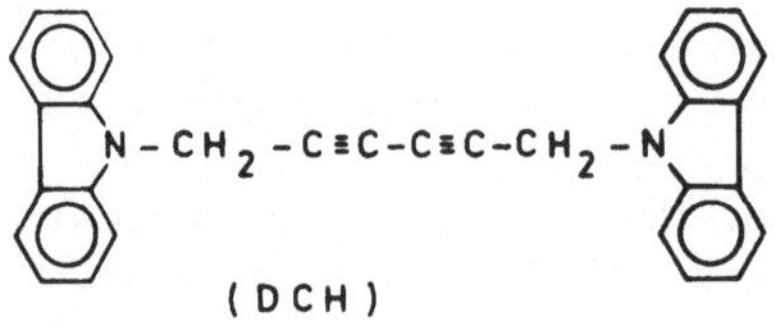

(DCH) — 4,55 — 100 / 20 Mrad — +++

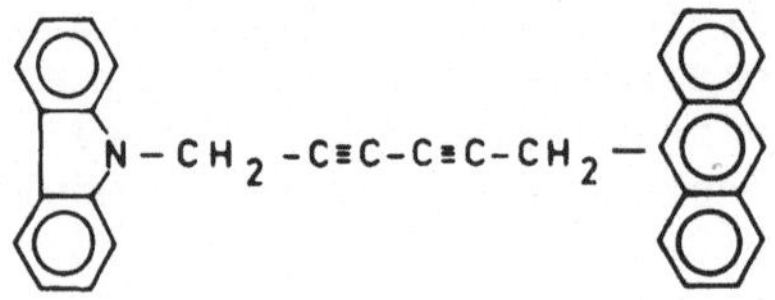

(ACH) — 4,37 — ⎯⎯ — +

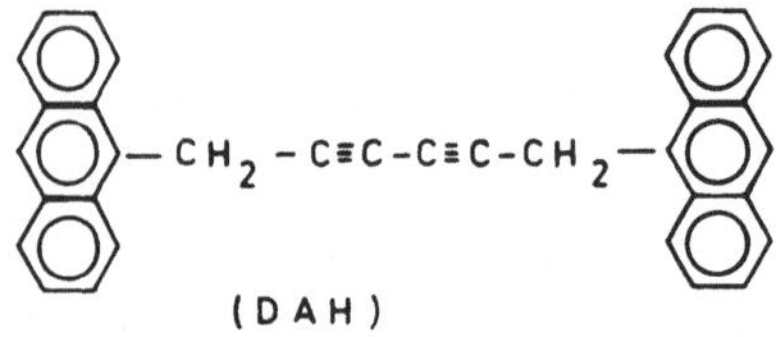

(DAH) — 4,35 — ⎯⎯ — −

DCH - ACH (25/75)	4,38	⎯⎯	
DCH - ACH (87/13)	4,49	45 / 120 Mrad	+
DCH - ACH (94/ 6)	4,53	72 / 120 Mrad	+
DCH - DAH (99/ 1)		100/ 40 Mrad	+++

Another such system is obtained, if PTS is cocrystallized
with 2.4-hexadiynylene-bis(p-chlorobenzenesulfonate) (PCS) (7) a
diacetylene, having -CH$_2$-O-SO$_2$-C$_6$H$_4$Cl side groups from a concentrated solution in acetone. PCS alone was reported to be completely
inactive and the reasons for this were given by studying the X-ray
structure (78) and it was assumed that the strong Cl-phenyl intra-
stack interaction between molecules adjacent in polymerization
direction prevents the crystal from polymerizing. V. Enkelmann (79)
showed that PCS can be obtained in a reactive modification too, if
crystallized at very high supersaturation. It is, therefore, not
too surprising that the two monomers form a series of mixed crystals considering the fact that the Van-der-Waals radii of -Cl and
-CH$_3$ are very similar.
Finally, it should be mentioned that HDDU and the diacetylene with
-CH$_2$-O-CO-NH— -side groups form also solid solutions(80).

Monolayers and Multilayers of poly(diacetylenes)

Molecularly thin films of polydiacetylenes with variable but exactly defined thickness are readily prepared by solid-state polymerization of suitable monomers in mono- or multilayers (81-85). The
basic idea is to use the ability of amphiphilic diacetylene mono-
carbonic acids to form monomolecular layers at the air-water interface of a Langmuir-trough. Acids with a total number of C-atoms
≥ 20 and a m.p. > 45°C form surface states suitable for the build-
up of multilayers from such monolayers by the LB-technique (86).
A series of suitable compounds together with a short description of
their polymerization behavior is compiled in Tab. 2. The principle
of the LB-technique is sketched in Fig. 8. By consecutive dipping
and with-drawing of a solid substrate such as a quartz or germanium
plate into and from a monolayer on the water surface consecutive
layers of molecules are transfered onto the substrate. The multilayers whose thickness is exactly given by the number of dippings
multiplied by the thickness of an individual layer are then polymerized by exposure to a UV-light source. Polymerization occurs
according to the normal mechanism without destruction of the layer
structure and with retention of the packing within the individual
layers (75,84). The poly(diacetylene)chains are all stretched out
in the plane of the substrate but do not extend over macroscopic
dimensions because the multilayers have a crystalline domain structure with the average domain diameter depending on the preparation
conditions being of the order of a few μm.
For experimental reasons related to the stability of the layers
during the transfer step it is better to work with the Cd-salts
rather than with the pure acids.
The multilayer samples are very well suited for a number of studies,
expecially if it is asked for very thin (<1000 Å) and transparent
samples. To demonstrate this, the UV-spectrum of a typical multilayer preparation in its monomer form is shown in Fig. 9. The
transmission spectrum of a similar preparation where the packing
arrangement looks like it is depicted in Fig. 10. after completion
of the photopolymerization is shown in Fig. 11.
The effect of thermochromism described earlier is very pronounced
in all multilayer polymerizations. All multilayers show, if irra-
diated, at first an absorption band with maxima at 638 and 585 nm
which by annealing or extensive photoirradiation or by treatment
with a solvent of the monomer shifts to the "red" form of the layer
with maxima at 500 and 535 nm. This is shown for a particular but
typical case in Fig. 11 where the change in the transmission spectrum with temperature on annealing of a photopolymerized multilayer
is represented. There is an isosbestic point indicating that every
molecule in the "blue" form is transformed into the "red" form
without intermediate product formation. Electron refraction data
have conclusively shown (75) that the "blue" to "red" transformation is accompanied by a phase transition which affects the packing

Table 2: Spreading and polymerization behaviour of long-chain diacetylenes with ability to form monomolecular films on water or on 10^{-3}m $CdCl_2$

| Compound $R-C{\equiv}C-C{\equiv}C-R'$ | | m.p. $[^\circ C]$ | Multilayer formation | Photopolymerization (λ <300 nm) | |
R	R'			in crystal	in multilayer[1]
$n-C_9H_{19}-$	$-(CH_2)_8-COOH$	48		++	
$n-C_{10}H_{21}-$	$-(CH_2)_8-COOH$	57	yes	++	+
$n-C_{12}H_{25}-$	$-(CH_2)_8-COOH$	61	yes	++	+
$n-C_{14}H_{29}-$	$-(CH_2)_8-COOH$	46/59[2]	yes	++	+
$n-C_{12}H_{25}-$	$-(CH_2)_3-COOH$	57	no	++,+[3]	-
$n-C_{16}H_{33}-$	$-(CH_2)_2-COOH$	93	yes	++	-
$n-C_{16}H_{33}-$	$-COOH$	75	yes	+++	+
$n-C_{12}H_{25}-$	$-COOH$	58	no	+++	-
$n-C_{10}H_{21}-$	$-(CH_2)_9-OH$	49	no	+	-

1) dark blue in less than 10 sec.: +++; dark blue in 10-30 sec.: ++; dark red after exposure >2 min.: +; no reaction; 2) phase transition at 46°C; 3) there are two modifications with different reactivity.

of the aliphatic side chains but does not destroy the layer structure.

The multilayers have been used to solve the long standing question to what the quantum yield of the photopolymerization of diacetylenes may be. This is extremely difficult to determine in an organic solid of low symmetry for reasons of crystal optics(5c). Because of their molecularly thin structure this is readily achieved using multilayers. The result of such studies is shown in Fig. 12 (88). A surprisingly high quantum yield is found which decreases with increasing conversion. The initial quantum yield amounts to about 10-14 molecules polymerized per absorbed photon; thus, the polymerization is of true chain reaction character. The results also explain the extreme sensitivity of the diacetylenes to light since a chain of ten units is already in the saturation region for the position and oscillator strength of the conjugation system; in other words, for the properties of the photoexcited state a molecule of ten base unit does not differ from an infinetely long one.

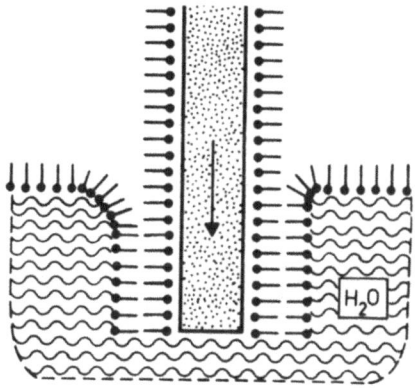

Fig. 8. Schematic representation of the transfer of monomer monolayer from the water surface to the substrate in the course of the
build-up of multilayers.

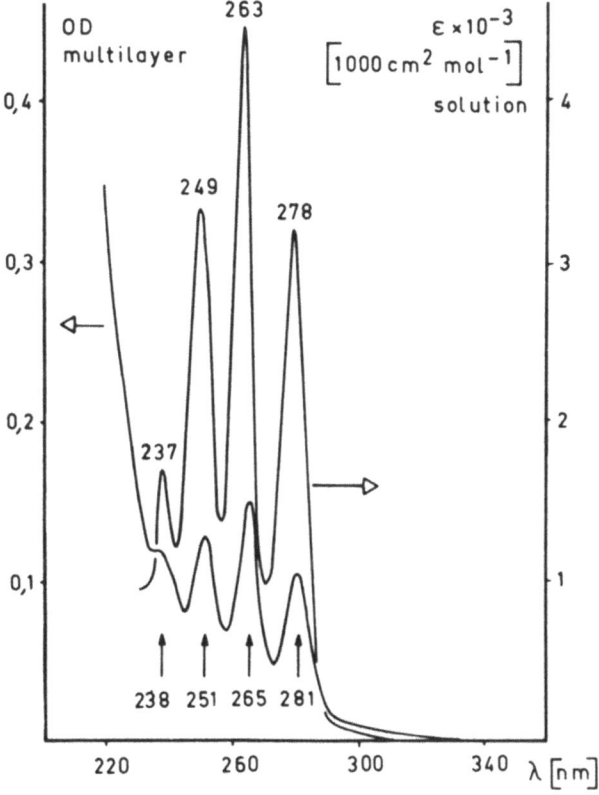

Fig. 9. UV-spectra of monomer $H_3C-(CH_2)_{15}-C\equiv C-C\equiv C-CO_2H$ (Cd-salt)
in solution (CHCl$_3$) and as multilayer (80 layers).

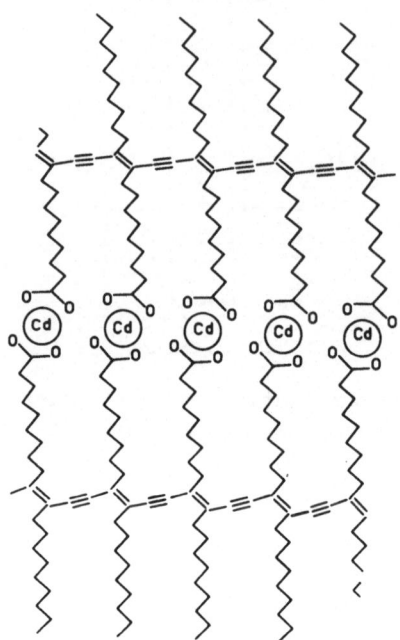

Fig. 10. Schematic representation of the poly(diacetylene)structure occuring in multilayers.

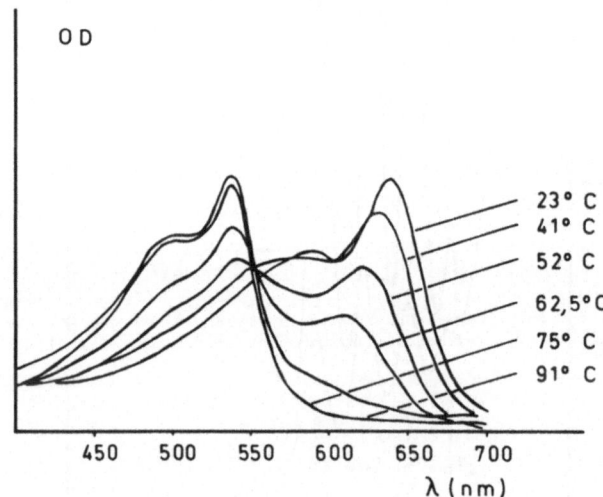

Fig. 11. Transmission spectra of multilayers of polymer of $C_{12}H_{25}-C\equiv C-C\equiv C-C_8H_{16}-CO_2H$(Cd-salt) as depending on temperature after completion of photopolymerization.

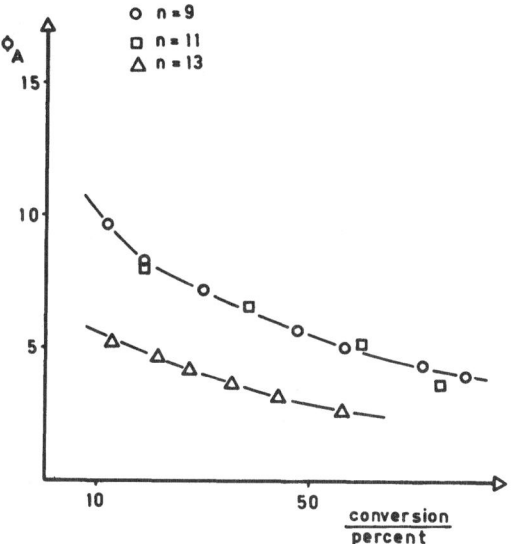

Fig. 12. Integral quantum yield ϕ_A vs. conversion for multilayers of $H_3C-(CH_2)_n-C\equiv C-C\equiv C-(CH_2)_8-CO_2H^A$(Cd-salts).

Another interesting problem is to study the photoconduction behavior of such multilayer systems. This will be discussed in the following paragraph.

V. ELECTRONIC PROPERTIES OF THE POLY(DIACETYLENES)

Because of partial π-electron delocalization along the chain direction single-crystalline polydiacetylenes are expected to show typical semiconductor properties in chain direction. In fact, a large anisotropy of the charge carrier mobility along and parallel to the chain direction has been found ($\mu_{\shortparallel}:\mu_{\perp} \approx 10^3$) (89) and the existence of a wide valence band formed by the highest π-states of the chain has been established (90-92). On the other hand, there is growing evidence that the lowest optical transition, which dominates the absorption spectrum in the visible, is an exciton rather than a band-to-band transition (93-96), the exciton being of the Frenkel type with a large charge-transfer (CT) component and not localized solely on the ene-bond or the yne-bond going along the chain. Philpott (97) has outlined this exciton theory of the lowest excited electronic states using PTS as a model. This theory considers explicetely bond excitons on the ene and yne groups, excitons arising from charge transfer between adjacent ene and yne groups, and two-exciton states to the simultaneous

Table 3: Representative examples for diacetylene monomers, polymerization behavior and physical properties of the polymer. General structure of the monomers $R-C\equiv C-C\equiv C-R'$

Formula R=R'	Common abbreviation	Representative lit.ref. Synthesis(a) and crystal struct.(b)	Polymerization	Polymer Properties
(1) $-CH_2-O-SO_2-$◯$-CH_3$	PTS	(a) 25,26,8 (b) 16,30-32	25,8,111-113	30,33,38-40 47,48,51, 99-96,103
(2) $-CH_2-O-CO-NH-PH$	HDDU	(a) 3,26 (b) 15	3,26	53
(3) $-(CH_2)_4-O-CO-NH-PH$	TCDU	(a) 71 (b) 17,18		73,74
(4) C≡C-C≡C [structure] CO-$(CH_2)_3$-CO	BPG	(a) 116 (b) 17,18		68
(5) [structure] $-CH_2-N$	DCH	(a) 60,61 (b) 61,65	60,61	60,66
(6) $-(CH_2)_4-O-CO-NHEt$	ETCD	(a) 73 (b) 73		73,74
multilayer forming compounds such as R=$-(CH_2)_{11}-CH_3$ R'=$-(CH_2)_8-CO_2H$	--	81,85	81-85,88	83,84,101

excitation of adjacent ene and yne bonds to either a singlet or a triplet state.

The model predicts the lowest singlet transition of PTS to be the most intense and to occur near 2 eV in correspondence with the experimental observations. It arrises as a collective state formed from bond excitons and CT-excitons. At $\kappa = 0$ there is no two-exciton component. It follows that molecular vibrations that perturb this mixed state, e. g., C=C-stretch and torsional vibrations, should be strongly excited by the electronic transitions as it is observed in the phonon-side bands of the visible spectra and in resonance Raman spectra. This description fits the experimental results much better than calculations based on a band to band

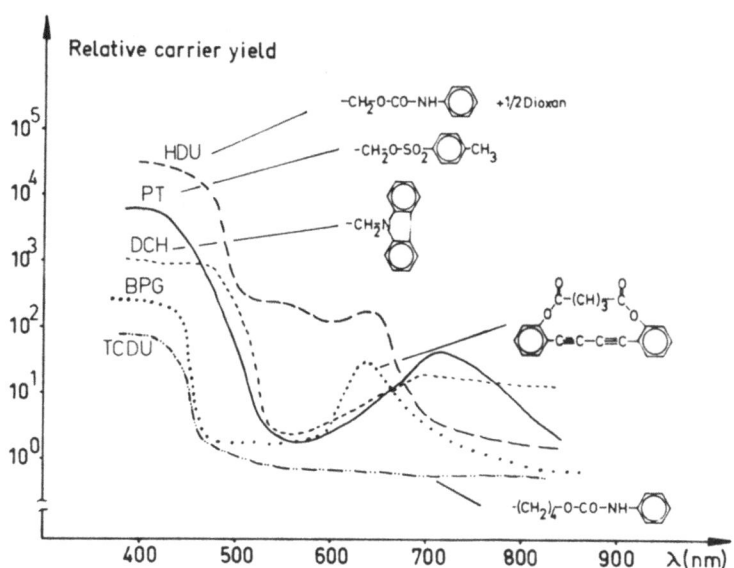

Fig. 13. Relative carrier yield of various poly(diacetylenes) measured as single crystals vs. wavelength in the photoconduction
experiment (ref. 102).

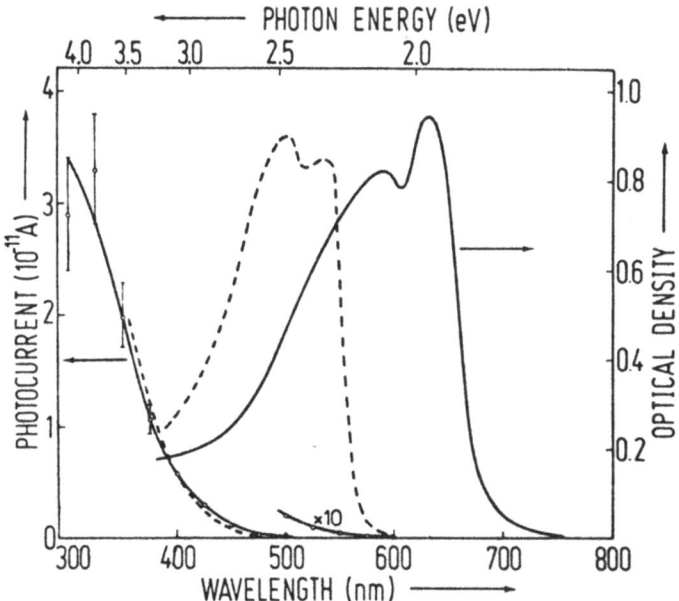

Fig. 14. Photoconduction and absorption spectra of polydiacetylene
multilayer. Full curves refer to the "blue" form, dashed curves
to the "red" form (see text). El. measurements were made with surface electrode configuration (200 V) applied over a 0.3 mm gap)
(ref. 101).

transition model taking into account possible effects arrising from
the low dimensionality of the system (24,37,98,99).

Nevertheless, poly(diacetylene) crystals are photoconductive
and the question how the charge carriers are generated is still
open. Photoconduction experiments were reported by a number of
authors (60,94,95,100,101). Fig. 13 shows the photocurrent action
spectrum of a number of single crystal poly(diacetylenes) (102)
in terms of relative carrier yields under the same experimental
conditions. It is typical for all polydiacetylenes - and all authors
agree on this - that the photocurrent action spectra display a minimum at the main peak of the crystal absorption spectrum. This
is further confirmation that the dominant crystal transition is
photoelectrically inactive and can consequently not be a valence-
to-conduction band transition. Because of the similarity between
the PTS and the TCDU photoresponse spectra it is reasonable to
conclude that the same processes are active although the two
materials have supposedly a different bonding sequence, PTS having the poly(ene-yne) and TCDU the poly(butadriene) sequence.

Analysis of the photoconduction data of polymer single crystals is complicated by the fact that in the interesting spectral
range the crystal absorption coefficient is large ($\alpha \geqslant 10^5 \text{cm}^{-1}$)
and α which is only accessible via a Kramers-Kronig-transforma-
tion of reflectivity data enters the calculations (101). This
handicap can be overcome using multilayer structures as described
in the last chapter. With assemblies containing only a few layers
the condition $\alpha \cdot d \leqslant 1$ (d being the sample thickness) can be established and the photoelectric gain must then directly reflect
the absorption spectrum of the ionizing transition. Fig. 14 shows
results obtained with such multilayers both in their "blue" and
"red" state. The absence of any correspondence between the absorption and photoconduction spectra is obvious and suggests once
more that two independent transitions are involved. Further analysis of the data interms of a model considering a valence-to-con-
duction band transition which is buried under the vibronic side
bands of the dominant exciton transition (101) yields the gap energy for the "blue" layer (absorption peak at 1.94 eV) $E_g = 2.5 \pm 0.1$
and for the "red" layer (absorption peak at 2.33 eV) $E_g = 2.6 \pm 0.1$ eV.
Similarly TCDU has a gap energy $E_g = 2.6 \pm 0.1$ eV and PTS has
$E_g = 2.1 \pm 0.1$ eV. In comparison DCH has a band gap of 2.20 eV.

The transition seems to be three-dimensional indicating finite valence and conduction band dispersion perpendicular to the
polymer chain direction.
Non-linear optical spectroscopy and determination of the non-linear
coefficients constitute additional tools to study the π-electron
behavior and π-electron delocalization in particular. An exciting
development in this area was the observation that poly(diacetylene)
crystals have unusually high third-order optical susceptibilities,

χ''' for light polarized in the chain direction (103).
Earlier work of Hermann and Ducuing (104) had shown that π-electron delocalization in the chains of polyenes results in a chain direction polarizability which is significantly dependent on the second power of the electric field. The obvious analogy between poly(enes) and poly(diacetylenes) led to the conclusion that the latter should be most interesting materials for non-linear optics experiments. Measurements on PTS and TCDU show these materials to have the highest χ'''-values known for any material, including inorganic semiconductors such as Germanium or GaAs. The theoretical explanation of the effect has been given in terms of a model assuming the optical transition near 2 eV to be a band-to-band transition (37,105, 106).Although this is not a good model as the results of the photoconduction experiments seem to indicate the high third order polarizability and optical susceptibility remains as an interesting fact. These results, coupled with the observed high damage thres - holds, could be useful in picosecond pulse applications such as the construction of parametric amplifiers, ultrafast light shutters and optical pulse sharpeners.

VI. THE KINETICS AND THE MECHANISM OF POLYMERIZATION

Considerable progress has been achieved in the elucidation of the mechanistic details of the solid-state polymerization of diacetylenes during the last two years. Most of the work has concentrated on PTS as a model substance for both thermally and photochemically induced polymerization although a few other monomers were studied, too. From earlier work which has been refered to in the first chapter, it is sufficently clear what are the general phenomena common to all diacetylenes and what are the phenomena peculiar to PTS which serve special considerations. The general phenomena are:
a) polymerization occurs by photoirradiation over the whole absorption spectrum of a given monomer,
b) thermal polymerization occurs only in the solid-state. The same product as in photopolymerization is formed,
c) polymerization proceeds homogeneously; nucleation effects are not observed although, sometimes, phase changes occur when partial conversion has been reached,
d) high-molecular weight polymers are formed in most cases,
e) the photoproduct, i. e. the newly formed polymer chains, do not work as sensitizers or quenchers during the photopolymerization,
f) triplet carbenes have been detected and were assumed to be positioned at the growing chain end.
A deeper insight became possible through further X-ray work, optical and ESR-spectroscopy as well as differential calorimetry (DSC).

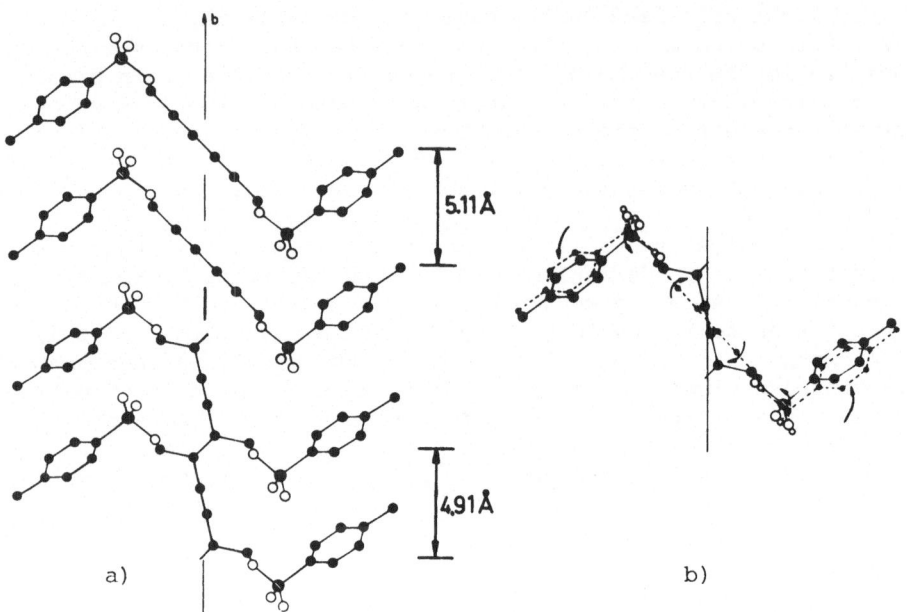

Fig. 15a). Projection of adjacent monomer (above) and polymer (below) units of PTS in their resp. crystal structures onto the plane
of the polymer backbone; b) changes in atomic positions of a single
unit, if the two structures are projected on top of each other.

X-ray results

Until recently the polymerization mechanism depicted in Fig. 2
was hypothetical in so far as the structure of a polymerizable monomer was not known due to the high reactivity of diacetylenes toward X-rays. V. Enkelmann (31,32) was the first to solve the crystal structure of a polymerizable diacetylene making use of the
observation that PTS polymerizes at 120 K so slowly that within
the time necessary to obtain a single Weissenberg photograph the
conversion does not surpass 5 percent. Thus, a structur analysis
became possible. The essence of the structural information obtained is depicted in Fig. 15. It is in agreement with the earlier
proposals in its basic features but it reveals also new details
which deserve further consideration. Note that there is no perfect
register between monomer and polymer lattice but the b-axis, i. e.
the polymerization direction, has to shrink for as much as 0.20 Å
(4 percent) going from the starting to the resulting phase. There
is also considerable movement and rotation in the side group region
as mapped in Fig. 15b.
The changes in the lattice parameters which reflect the changes in
molecular geometry occur smoothly over the whole range of

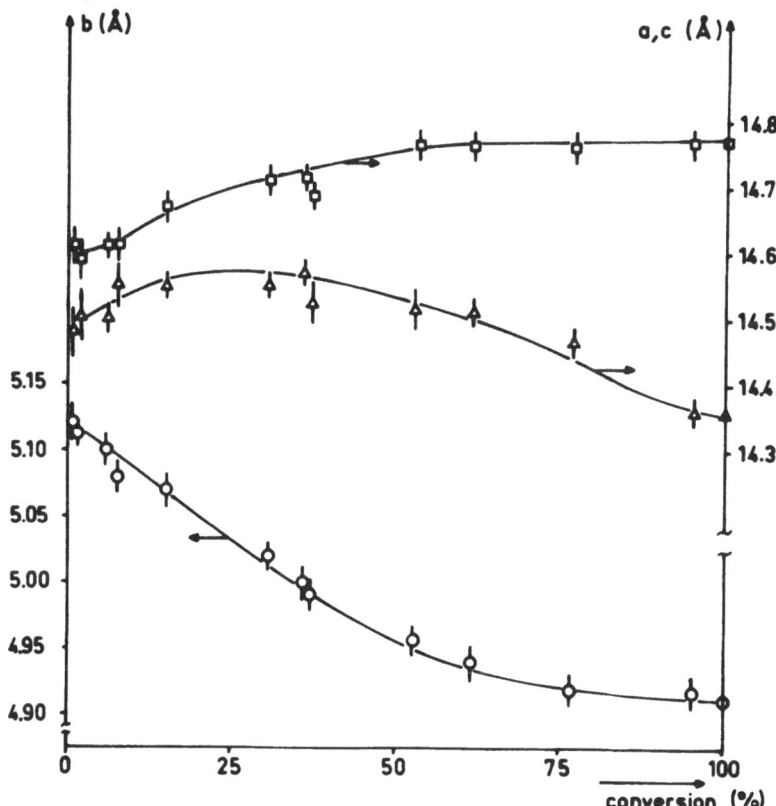

Fig. 16. Lattice parameters of PTS as depending on conversion.

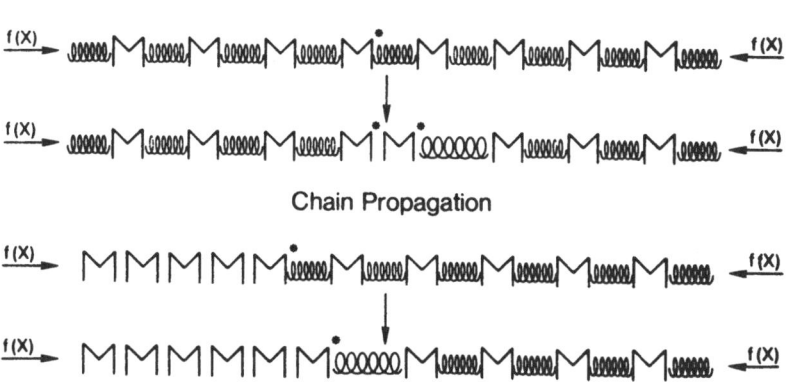

Fig. 17. Schematic representation of strain changes in going from the initial to the transition state for chain initiation and propagation according to Baughman(115). The strain is produced by forces f (X) on the monomer (M) array due to dimensional mismatch between the growing polymer chains and the parent phase.

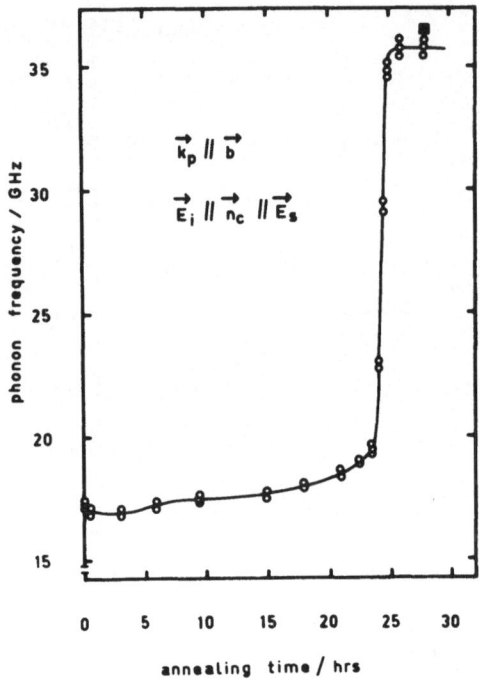

Fig. 18. Phonon frequency for
k ‖ b (chain axis) as depending on annealing time for
single crystals of PTS as
derived by Brillouin-scat-
tering.

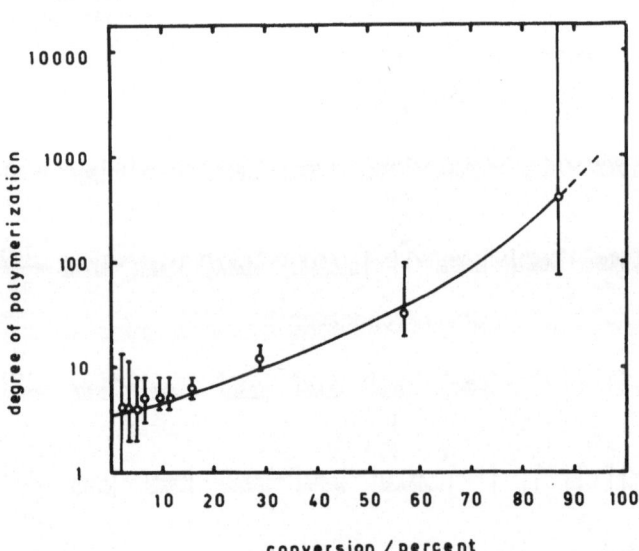

Fig. 19. Dependence of the average degree of polymerization on conversion as concluded from the data in Fig. 18 (51,32).

conversion as shown in Fig. 16. The most important parameter $\vec{b}$
which enters in all theoretical calculations concerning the kinetics
of the polymerization does not reach its final value until about
75 percent conversion has been reached.

Optical and ESR-spectroscopy

The earlier reports on triplet carbenes as the active intermediates
in chain growth (13) were substantiated and it was shown that the
same species can be obtained on photoirradiation of a PTS single
crystal at 4.2 K (107). A series of intense lines in the UV are produced during irradiation at this temperatures (108). Two main broad
absorption maxima are created independently in the red (at 665 and
715 nm) with different intensities and different kinetics. Additional weaker absorptions due to photoproducts are found at 600 nm,
500 nm and 400 nm.
When heating the crystal to 110 K all absorption lines of the low
temperature photoproducts disappear abruptly and there is marked
increase in the polymer absorption. Simultaneously with the observation of these optical absorptions, very intense triplet ESR absorptions are observed (109) from wich it is concluded, that the absorptions at $\lambda < 600$ nm are most probably due to oligomer chains
and the absorptions at $\lambda > 600$ nm due to transitions within the
triplet system of the carbene type photoproducts which are localized at distrinct carbon atoms at the oligomer chain ends.
Moreover, strong fluorescence bands were observed in partially polymerized single crystals of PTS (110). Excitation at 4.2 K yields
two emmission bands at 20486 cm^{-1} and 20802 cm^{-1} accompagnied by
vibrational side bands. These bands were also attributed to oligomer species although, alternatively, chain ends trapped at crystal
defects were made responsible as the origin of the fluorescing
states (40).
The successful determination of the quantum yield using multilayer
structures has already been mentioned (88) (comp. Fig. 12). It demonstrates clearly that the polymerization mechanism has to be described - at least in part - as a chain reaction.

Thermal polymerization

The conversion vs. time curve in the thermal polymerization of PTS
is S-shaped - a result of a dramatic autocatalytic effect observed
at about 10 % conversion (25). The nature of this effect has received much attention, although peculiar to PTS. A number of other
diacetylenes exhibit zero-order time-conversion kinetics (26). The
shape of the time-conversion curve was reinvestigated using a variety of techniques, among others DSC (111,112) and reflectance
spectroscopy (113) confirming the earlier results obtained by an
extraction technique (25).
The heat of polymerization ΔH_p was determined from the DSC results

to be -36.6 kcal mol^{-1} (-1.6 eV) (111) and the activation energy E_a for the fast and the slow part of the conversion vs. time curve was found to be the same and of the order of 22kcal mol^{-1}. Thus E_a is independent of the conversion andthe autocatalytic effect can be best understood as arising from a large increase in chain propagation length of the polymer chains. The autocatalytic effect was shown also to be present in the photo- and γ-ray polymerization (114). In thermal polymerization the ratio of the polymerization rates for the fast and the slow regime may be as high as 400 (112,114).

Baughman (115) has developed a theory based on the assumption that the autocatalytic effect has its origins in the strain energy associated with the formation of the polymer-monomer solid solution. He predicts a strain-dependent rate of chain initiation and propagation and a chain propagation lengths as a function of conversion assuming certain changes in the elastic properties of the polymerizing single crystals with increasing conversion. The basic model is sketched in Fig. 17. These changes in elastic properties have been measured quantitatively in the meantime (51) and the results are qualitatively in agreement with Baughman's predictions although there remain quantitative differences. A rather simple treatment of the data obtained from the measurement of the phonon frequency for $k_{\shortparallel}\vec{b}$ (comp. Fig. 18) in terms of the Young's modulus of the crystal in chain direction composed of stiff chains embedded in a soft matrix gives the dependence of the average degree of polymerization on conversion as shown in Fig. 19.

Although the principal ideas on the kinetics as depending on lattice strains are certainly correct further theoretical calculations are not very meaningful as long as a detailed analysis of molecular weight and molecular weight distribution is lacking. Similarly, speculations on the mechanism of the initiation reaction (113, 114) are certainly valuable and helpful for the design of further experiments but should not be taken to serious especially in the light of the yet unexplained observations of low-temperature spectroscopy (107,110).

REFERENCES

1. For an extensive discussion of the problems encountered in polymer crystallization and morphology it is refered to B. Wunderlich "Macromolecular Physics, Academic Press, New York 1976
2. W. A. Little, J. Polymer Sci. C 17, 3 (1967)
3. G. Wegner, Z. Naturforschg. 24b, 824 (1969)
4. a) F. L. Hirshfeld and G. M. J. Schmidt, J. Polymer Sci. A2, 2181 (1964), b) G. M. J. Schmidt in Reactivity of the Photo-excited Molecule, p. 227 f. Interscience (1967)

5. There are several reviews discussing the implications of organic solid-state chemistry both to preparative and structural
   chemistry, among others:
a) M. D. Cohen and B. S. Green, Chemistry in Britain, 9, 490 (1973)
b) J. M. Thomas, Phil. Trans. Roy. Soc. (London) A277, 251 (1974)
c) G. Wegner, Pure & Appl. Chem. 49, 443 (1977)
d) G. Wegner, A. Munoz-Escalona and E. W. Fischer, Makromol. Chem.
   Suppl. 1, 521 (1975)
6. J. Kaiser, G. Wegner, and E. W. Fischer, Israel J. Chem. 10,
    157 (1972)
7. R. H. Baughman, J. Polym. Sci., Polym. Phys. Ed. 12, 1511 (1974)
8. D. Bloor et al., J. Mater. Sci. 10, 1678, 1689 (1975)
9. The only other case besides the diacetylenes where formation of
   a solid solution of the growing polymer product with the parent
   monomer matrix has been reported refers to the polymerization
   of S_2N_2 to $(SN)_x$ (R. H. Baughman and R. R.Chance, J. Polym. Sci.
   Polym. Phys. Ed. 14, 2019 (1976)
10.R. H. Baughman, The Solid-State Synthesis and Properties of Pho-
   toconducting, Metallic and Superconducting Polymer Crystals in
   "Contemporary Topics in Polymer Science, Vol. 2, E. M. Pearce
   and I. R. Schaefgen, Eds. Plenum Press 1977, p. 205
11.K. Takeda and G. Wegner, Makromol. Chem. 160, 349 (1972)
12.G. C. Stevens and D. Bloor, Chem. Phys. Letters 40, 37 (1976)
13.H. Eichele, M. Schwoerer, R. Huber and D. Bloor, Chem. Phys.
   Letters, 42, 342 (1976)
14.G.Wegner, Organic Linear Polymers with Conjugated Double Bonds
   in Chemistry and Physics of One-Dimensional Metals, H. J. Keller,
   Ed., Plenum Press 1977, p. 297
15.E. Hädicke et al., Angew. Chem. 83. 253 (1971)
16.D. Kobelt and H. Paulus, Acta Cryst. B30, 232 (1974)
17.J. B. Lando, D. Day and V. Enkelmann, ACS-Polymer Preprints 18,
   1, 190 (1977)
   D. Day and J. B. Lando, J. Polymer Sci., Polym. Phys. Ed. 16,
   1009 (1978)
18.V. En kelmann and J. B. Lando, Acta Cryst. (1978) in press
19.A. J. Melveger and R. H. Baughman, J. Polymer Sci., Polym. Phys.
   Ed. 11, 603 (1973)
20. R. H. Baughman, J. D. Witt and K. C. Yee, J. Chem. Phys. 60,
    4755 (1974)
21. D. Bloor, D. J. Ando, F.H. Preston and G. C. Stevens, Chem.
    Phys. Letters 24, 407 (1974)
22.D. Bloor, F. H. Preston, D. J. Ando and D. N. Batchelder
   "Resonant Raman Scattering from Diacetylene Polymers" in:
   Structural Studies of Macromolecules by Spectroscopic Methods
   p. 91 f., K. J. Ivin, Ed. Wiley 1975
23.W. Schermann andG. Wegner, Makromol. Chem. 175, 667 (1974)
24.E. G. Wilson, J. Phys. C: Solid-State Phys. 8, 727 (1975)
25.G. Wegner, Makromol. Chem. 145, 85 (1971)
26.G. Wegner, Makromol. Chem. 154, 35 (1972)
27.J. M. Schultz, J. Mater. Sci. 12, 2258 (1976)

28.W. Schermann, J.O. Williams, J. M. Thomas and G. Wegner,
    J. Polymer Sci., Polymer Phys. Ed. 13, 753 (1975)
29.G. Wegner and W. Schermann, Colloid & Polymer Sci., 252,
    655 (1974)
30. a) V. Enkelmann and G. Wegner, Makromol. Chem. 178, 635 (1977)
    b) V. Enkelmann, Acta Cryst . B33, 2842 (1977)
31.V. Enkelmann and G. Wegner, Angew. Chem. 89, 432 (1977)
32.V. Enkelmann, R. J. Leyrer and G. Wegner, J. Amer. Chem. Soc.,
    in press
33.D. Bloor, F. H. Preston and D. J. Ando, Chem. Phys. Letters,
    38, 33 (1976)
34.B. Reimer, H. Baessler, J. Hesse and G. Weiser, phys. stat. sol.
    (b) 73, 709 (1976)
35.C. J. Eckhardt, H. Müller, J. Tylicki and R. R. Chance,
    J. Chem. Phys. 65, 4311 (1976)
36.M. Schott, F. Batallan and M. Bertault, Chem. Phys. Lett. 53,
    443 (1978)
37.a) C. Flytzanis, C. Cojan and G. P. Agrawal, Nuovo Cimento 39b,
    488 (1977)
    b) G. P. Agrawal, C. Cojan and C. Flytzanis, Phys. Rev. Letters,
    38, 711 (1977)
38.D. Bloor and F. H. Preston, phys. stat. sol. (a) 37, 427 (1976)
39.D. Bloor and F. H. Preston, ibid. 39, 607 (1977)
40.D. Bloor, D. N. Batchelder and F. H. Preston, ibid 40, 279 (1977)
41.H. A.Pohl and R. P. Chartoff, J. Polymer Sci. A2, 2787 (1964)
42.N. R. Byrel, F. K. Kleist and D. N. Stamires, J. Polymer Sci.,
    Phys. Ed. 10, 957 (1972)
43.L. S. Singer and J. Kommandeur, J. Chem. Phys. 34, 133 (1961)
44.A. A. Berlin, Vysokomol. Soed. A13, 2429 (1971)
45.J.A. Pople and S. H. Walmsley, Mol.Phys. 5, 15 (1962)
46.G. F. Kventsel and Yu. A. Kruglyak, Theor. Chem. Acta 12,1(1968)
47.G. C. Stevens and D. Bloor, J. Polymer Sci., Polym. Phys. Ed.
    13, 2411 (1975)
48.G. C. Stevens and D. Bloor, phys. stat. sol. (a) 45, 483 (1978);
    46, 141 (1978); 46, 619 (1978)
49.Some implications of such a defect and further defects derived
    from it by consecutive bond rotations ("orbital-flip-defect")
    on the spectral properties of polydiacetylenes have been dis-
    cussed (50)
50.R.H.Baughman and R. R.Chance, J. Appl. Phys. 47, 4295 (1976)
51.R.J. Leyrer, G. Wegner and W.Wettling, Ber. Bunsenges. Phys.
    Chem. 82, 697 (1978)
52.D. N. Batchelder and D. Bloor "Strain Dependence of the Vibra-
    tional Modes of a Diacetylene Crystal", Preprint
53.R. H. Baughman, H. Gleiter and N. Sendfeld, J. Polymer Sci.,
    Polym. Phys. Ed. 13, 1871 (1975)
54.F. C. Frank, Proc. R. Soc. London, Ser. A 319, 127 (1970)
55.Y. Kobayashi and A. Keller, Polymer 11, 114 (1970)
56.R. H. Baughman and E. A. Turi, J. Polymer Sci., Polym. Phys.
    Ed. 11, 2453 (1973)

57. D. N. Batchelder, J. Polymer Sci., Polymer Phys. Ed. 14, 1235 (1976)
58. R.W. Munn, J. Phys. C. 5, 535 (1972)
59. R. J. Young, D. Bloor, D. N. Batchelder and C. L. Hubble J. Mater. Sci. 13, 62 (1978) and references therin
60. K. C. Yee and R. R. Chance, J. Polym. Sci., Polym. Phys. Ed. 16, 431 (1978)
61. V. Enkelmann, G. Schleyer, G. Wegner, H. Eichele and M. Schwoerer Chem. Phys. Letters 52, 314 (1977)
62. The monomer undergoes a photochemical isomerization if solutions are exposed even to day light for some time. Crystallization should be carried out in the dark.
63. A. Warshel and E. Huler, Chem. Phys. 6, 463 (1974)
64. H. Eichele, M. Schworer and J. K. von Schütz, Chem. Phys. Letters, 56, 208 (1978)
65. P. A. Apgar and K. C. Yee, Acta Cryst. B34, 957 (1978)
66. R. J. Hood, H. Müller, C. J. Eckhardt, R. R. Chance and K.C.Yee, Chem. Phys. Letters 54, 295 (1978)
67. J. J. Mayerle and M. A. Flandera, Acta Crst. B34, 1374, (1978)
68. D. Bloor, W. Hersel and D. N. Batchelder, Chem. Phys. Letters 45, 411 (1977)
69. Z.Igbal, R. R.Chance and R. H. Baughman, J. Chem. Phys. 66, 5520 (1977)
70. G. Wegner, G. Arndt, H. J. Graf and M. Steinbach in "Reactivity of Solids" J. Wood et al. Eds., Plenum Press 1977, p. 487 f.
71. G. J. Exharos, W. M. Risen and R. H. Baughman, J. Amer. Chem. Soc. 98, 481 (1976)
72. R. R. Chance, R. H. Baughman, A. F. Preziosi and E. A. Turi, Bull. Amer. Phys. Soc. 22, 408 (1977)
73. R. R. Chance, R. H. Baughman, H. Müller and C. J. Eckhardt, J. Chem. Phys. 67, 3616 (1977)
74. R. H. Baughman and R. R.Chance "Fully conjugated polymer crystals: Sòlid-state synthesis and properties of the polydiacetylenes" Ann. New York Acad. Sci., in press
75. B. Tieke, Thesis, University of Freiburg, 1978
76. V. Enkelmann and G. Schleier, unpublished
77. H. Eichele, Thesis, University of Bayreuth, 1978, p. 124 f.
78. J. J. Mayerle and T. C. Clarke, Acta Cryst. B34, 143 (1978)
79. V. Enkelmann, private communication
80. C. Kröhnke, private communication
81. B. Tieke, G. Wegner, D. Naegele and H. Ringsdorf, Angew. Chem. Int. Ed. 15, 764 (1976)
82. B. Tieke, H.-J. Graf, G. Wegner et al., Colloid & Polymer Sci., 255, 521 (1977)
83. B. Tieke and G. Wegner, in "Topics in Surface Chemistry", Plenum Press 1978 p. 121
84. B. Tieke, G. Lieser and G. Wegner, J. Polym. Sci., Polym. Chem. Ed., in press
85. D. Day and H. Ringsdorf, J. Polym. Sci. Polym. Letters Ed. 16, 205 (1978)

86. K. B. Blodgett,J. Amer. Chem. Soc. 57, 1007 (1935); for a complete description of the technique and its application see ref. (87)
87. H. Kuhn and D. Möbius, Angew. Chem. Int. Ed. 10, 620 (1971)
88. B. Tieke and G. Wegner, Makromol. Chem. 179, 1639 (1978)
89. K. Lochner, B. Reimer and H. Bässler, Chem. Phys. Letters 41, 388 (1976)
90. J. Knecht, B. Reimer and H. Bässler, Chem. Phys. Letters 49, 327 (1977)
91. G.C. Stevens, D. Bloor and P. M. Williams, Chem. Phys. 28, 399 (1978)
92. J. Knecht and H. Bässler, Chem. Phys. in press
93. B. Reimer and H. Bässler, phys. stat. sol. (a) 32, 435 (1975)
94. R. R. Chance and R. H. Baughman, J. Chem. Phys. 64, 3889 (1976)
95. K. Lochner, B. Rei mer and H. Bässler, phys. stat. sol. (b) 76, 533 (1976)
96. R. Jankowiak, J. Kalinowski, B. Reimer and H. Bässler, Chem. Phys. Letters 54, 483 (1978)
97. M. R. Philpott, Chem. Phys. Letters 50, 18 (1977)
98. D. S. Boudreaux and R. R. Chance, Chem. Phys. Letters 51, 273 (1977)
99. M. Kerteśz, J. Koller and A. Azman, Chem. Phys. 27, 273 (1978)
100. H. Müller, C. J. Eckhardt, R. R. Chance and R. Baughman, Chem. Phys. Letters 50, 22 (1977)
101. K. Lochner, H. Bässler, B. Tieke and G. Wegner, phys. stat. sol. (b) 1978 in press
102. V. Enkelmann, personal communication
103. C. C. Sauteret et al. Phys. Rev. Letters 36, 956 (1976)
104. J. P. Hermann and J. Ducuing, J. Appl. Phys. 45, 5100 (1974)
105. C. Cojan, G. Agrawal and C. Flytzanis, Phys. Rev. B2, 909 (1977)
106. G. P. Agrawal, C. Cojan and C. Flytzanis, Preprint
107. R. Huber, M. Schworer, C. Bubeck and H. Sixl, Preprint "The Sign of D in the Triple Carbene of a Polydiacetylene Single Crystal) (PTS)
108. H. Sixl, W. Hersel and H. C. Wolf, Chem. Phys. Letters 53, 39 (1978)
109. C. Bubeck, H. C. Wolf and H. Sixl, Preprint "Photopolymerization of Diacetylene Single-Crystals II. ESR-Spectroscopy
110. H. Eichele and M. Schwoerer, phys. stat. sol. (a) in press
111. R.R. Chance, G. N. Patel, E. A. Turi and Y. P. Khanna, J. Amer. Chem. Soc. 100, 1307 (1978)
112. A.R.McGhie, P.S. Kalyanaraman and A. F. Garito, J. Polymer Sci., Polym. Phys. Ed. in press
113. R. R. Chance and J. M. Sowa, J. Amer. Chem. Soc. 99, 6703 (1977)
114. R. R. Chance and G. N. Patel, J. Polym. Sci., Polym. Phys. Ed. 16, 859 (1978)
115. R.H. Bauhgman, J. Chem. Phys. 68, 3110 (1978)
116. R. H. Baughman and K. C. Yee, J. Polymer Sci., Pol. Chem. Ed. 12, 2467 (1974)

STUDIES OF THE ADSORPTION OF HALOGENS BY POLYDIACETYLENES

D. Bloor, C.L. Hubble and D.J. Ando

Department of Physics
Queen Mary College
Mile End Road, London, E1 4NS, UK

The adsorption of electron-attracting species such as halogens or AsF_5 by fibrous crystalline polyacetylene, $(CH)_n$, films yields materials with high room temperature electrical conductivity [1] . Solid state polymerization of diacetylenes, $R-C\equiv C-C\equiv C-R$, is capable of producing conjugated polymers, $(=CR-C\equiv C-CR=)_n$, in a highly ordered form [2]. We have undertaken studies of the adsorption of halogens by polydiacetylenes to see if effects comparable with those observed for polyacetylene occur.

The large highly perfect single crystals obtained from monomers such as the bis(p-toluene sulphonate) diacetylene (TS), $R = CH_2-O-SO_2-\emptyset-CH_3$, and bis-phenylurethane, $R = CH_2-O-CO-NH-\emptyset$, are unsuitable for such work on two counts. First, they have a small surface-volume ratio and since diffusion into the crystal is likely to be slow only a small fraction of the sample will be affected by the adsorbed species. Secondly, reaction with the large sidegroups, which occupy most of the crystal lattice, is likely, such reaction is found for TS crystals. We have, therefore, worked with three polydiacetylenes with simple sidegroups which were prepared in the form of thin fibres or films. These are the polymers obtained from 2,4-hexadiyne-1-ol (1H), 3,5-octa diyne-1,8-diol (OD) and pentacosa-10,12-diynoic acid (PCD).

The 1H monomer was prepared by the method of Armitage et al. [3] and was purified by crystallization and sublimation. The monomers stack with adjacent OH groups forming a strong hydrogen bonded chain parallel to the polymer chain direction [4]. The monomer molecules are distorted, hindering polymerization which proceeds slowly for thermal and uv initiation. The polymer residue obtained by dissolving the monomer matrix at low conversion is a fibrous material with a dull green metallic lustre. However, because of the slow polymerization rate, typical samples have weights of only a few milligrams.

A polymer residue, degassed in a vacuum of better than 10^{-4} torr and exposed to iodine vapour at room temperature under vacuum, took on a black lustrous appearance. The iodine content of this sample was 1.2 atoms of iodine per monomer unit, this fell to a stable value of 0.9 atoms of iodine per monomer unit after desorption in a high vacuum. This sample was found to have a radical spin concentration of 1.8×10^{17} spins/gram. The build up of the radical signal during exposure to iodine vapour for a second sample is shown in figure 1.

Adsorption of I_2 from hexane solution was also studied. Figure 2 shows two typical sets of data. The open circles are for a 2.1 mg sample immersed in solution under inert gas and the full circles are for a 1.3 mg sample immersed in solution under a normal atmosphere. In both cases a very rapid initial rise in radical signal is followed by a slow decay of the spin concentration.

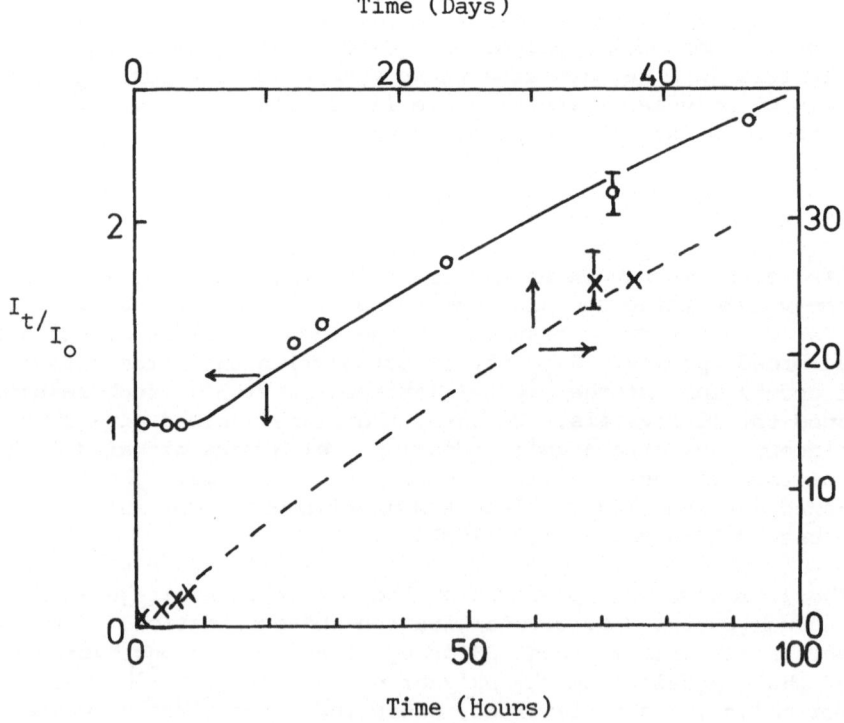

Figure 1

Conductivity measurements were not possible on the small samples available.

The OD monomer was prepared by the method of Bowden et al.[5]. The monomer units pack in hydrogen bonded sheets [6] within which the polymer chains form. The monomer is moderately reactive and lustrous yellow sheet-like polymer residues are obtained by dissolving the monomer matrix at low conversion. These residues do not change colour on exposure to iodine under vacuum or in hexane and there is no detectable change in weight. The EPR radical signal is small and remains so on exposure to iodine. Conductivity measurements indicate an increase on exposure to iodine, but this must be treated with some caution in view of the other negative results.

OD polymer residues exposed to bromine vapour decomposed completely in a period of a few days.

UV polymerized multilayers of PCD on glass substrates were supplied by Prof. G. Wegner [7]. Electrodag electrodes were used to attach leads to the films, typical electrodes were a few mm in length separated by less than 1 mm. The initial resistance was high, $>10^{13}\Omega$, but fell on exposure to iodine vapour. Results for a well behaved film as a function of iodine vapour pressure

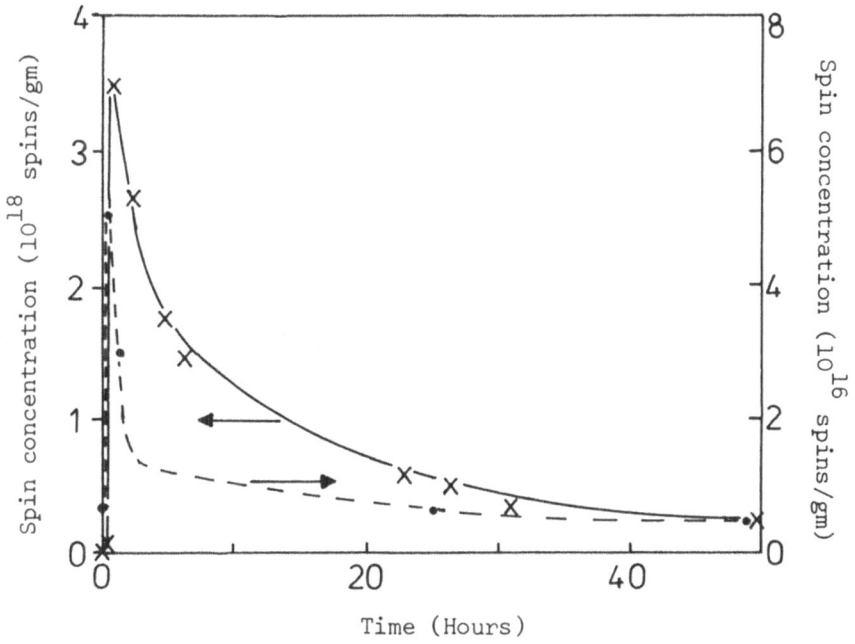

Figure 2

are shown in figure 3.   In the absence of iodine the conductivity
fell rapidly to about 30 times its initial value.   The increase in
conductivity is comparable with that reported for polyacetylene
but the sample remains essentially an insulator as the initial
resistance of the film was high.

During exposure the film changes colour from blue to pink, a
similar colour change occurs when the film is disordered on
heating about 80°C.   In some cases the blue—red colour change was
accompanied by a more rapid but less regular increase in
conductivity.

The results discussed above show some evidence for the
occurrence of charge transfer on iodine adsorption.   However, the
impermanent nature of the effects, together with the drastic
decomposition observed for bromine exposure, suggest that
subsequent reaction occurs, probably by iodine attack at the backbone triple bond, which is more reactive than the double bond in
polyacetylene.

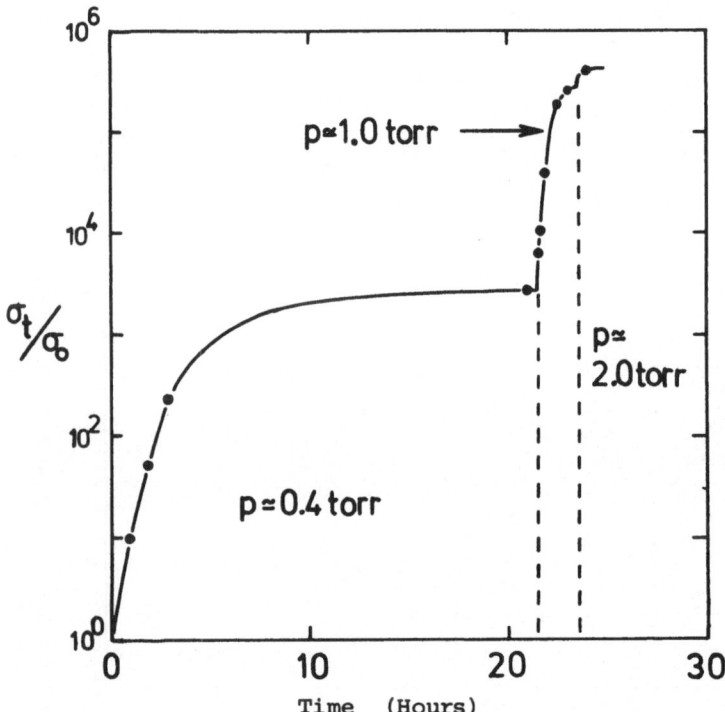

Figure 3.   Variation in conductivity of a polydiacetylene
multilayer exposed to iodine vapour at the approximate
partial pressures shown.

Raman spectra failed to reveal any lines characteristic of iodine ions but fluorescence characteristic of disorder was observed from PCD multilayers. This is compatible with reaction, after an initial charge transfer, leading to some disruption of the polymer chains.

In view of the ease with which 1H residues adsorb iodine, the lack of adsorption by OD is surprising, possibly the molecular packing in the polymerized hydrogen bonded sheets is too dense to allow iodine penetration.

Although iodination of polydiacetylenes has some effect, these are only transitory and do not lead to samples with useful electrical properties.

REFERENCES

1. C.K. Chiang et al. J. Am. Chem. Soc. 100, 1013 (1978)
2. G. Wegner, Pure and Appl. Chem., 49, 443 (1977)
3. J.B. Armitage, E.R.H. Jones and M.C. Whiting, J. Chem. Soc., 1993 (1952)
4. D.A. Fisher, D.N. Batchelder and M.B. Hursthouse, Acta Cryst. (in press)
5. K. Bowden, I. Heilbron, E.R.H. Jones and K.H. Sargeant, J.Chem. Soc., 1579 (1947)
6. D.A. Fisher, D.J. Ando, D.N. Batchelder and M.B. Hursthouse, Acta Cryst. (in press)
7. B. Tieke, W. Lieser and G. Wegner, J. Polym. Sci. Chem. Ed. (in press)

SOLID STATE REACTIVITY OF SOME BIS(AROMATIC. SULPHONATE)

DIACETYLENES

D. Bloor, D.J. Ando, D.A. Fisher and C.L. Hubble

Department of Physics
Queen Mary College
Mile End Road, London, El 4NS, UK

The topochemical nature of the solid state polymerization of diacetylenes is well known [1]. The influence of the packing of monomer units on reactivity [2] and that of elastic strain in partially polymerized crystals on the reaction kinetics have been discussed [3]. In order to investigate these factors further we have studied the reactivity of a series of closely related monomers. The general formula of these monomers is:

$$R-C\equiv C-C\equiv C-R, \qquad R = -CH_2-O-SO_2-C_6H_4-X$$

where X is CH_3, F, Cl, Br, H and OCH_3. In addition the monomer with the benzene structure replaced by a naphthyl unit was prepared.

The bis(p-toluene sulphonate), X = CH_3, (abbreviated TS) has been extensively studied and provided the basis for the discussion of reaction kinetics [3]. The naphthyl sulphonate and the benzene sulphonates with X = F, Cl and Br are all unreactive. The X-ray structure of the chloro-benzene sulphonate [4] shows that the side group interactions favour a more parallel packing of the diacetylene units than in TS, which is unfavourable for solid state polymerization. All these unreactive monomers have similar crystal morphologies suggesting that such unfavourable packing occurs in them all.

Reactive forms have been found for the benzene sulphonate and the methoxy-benzene sulphonate (MBS). In this paper we discuss the polymerization behavior of MBS and compare it with that of TS.

MBS monomer was prepared by the reaction of p-methoxy-benzene sulphonyl chloride with 2,4-hexadiyne-1,6-diol, analogous to the

preparation of TS [5]. Samples were purified by recrystallization
from a range of solvents. Two crystal modifications were obtained,
first large plate-like crystals of an unreactive modification and
secondly small dendritic crystals of a reactive modification.
Crystallization behaviour is unusual in that both forms can be
obtained from the same solution for a number of different solvents.
The proportion of the two forms appears to be controlled by
crystallization kinetics rather than purity.

The unreactive modification is formed from monomer units with
a different molecular conformation than that of TS. The side
groups are rotated at the -O- unit to form S shaped, rather than
nearly linear, molecules which pack with too large a separation of
the diacetylene units for reaction to be possible.

Crystals of the reactive modification are small and twinned so
that a full structure determination has not been possible. However,
spectroscopic studies suggest that the crystal structure must be
similar to that of TS. In particular the Raman spectrum, shown in
figure 1, is similar, but not identical, to that of TS. The
frequency shifts observed during polymerization are larger than
those for TS [6]. The data for the double and triple bend stretching
vibrations are shown in figure 2, the former shows the resonant
interaction previously observed in TS [6,7]. Comparison with the

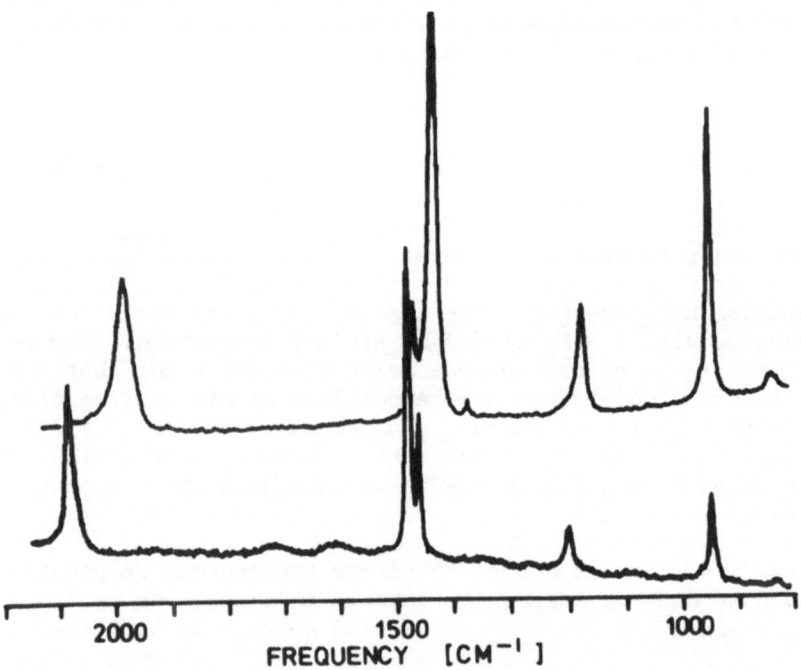

Fig.1. Raman spectrum of MBS partially polymerized crystals; upper
spectrum 1% conversion to polymer, lower spectrum 96% conversion .

data for a strained TS polymer fibre [7] shows that the initially formed polymer is subjected to the equivalent of a 5% tensile strain.

Comparable results are obtained from the optical spectra. Diffuse reflection at low conversion, see figure 3, gives an absorption energy of 2.19 eV while specular reflection from polymer crystals gives a peak absorption at 1.98 eV, these values are respectively slightly higher and lower than those found for TS. Using the data on the optical absorption shift of a strained TS polymer fibre [8] indicates an initial polymer strain of just over 5% for MBS.

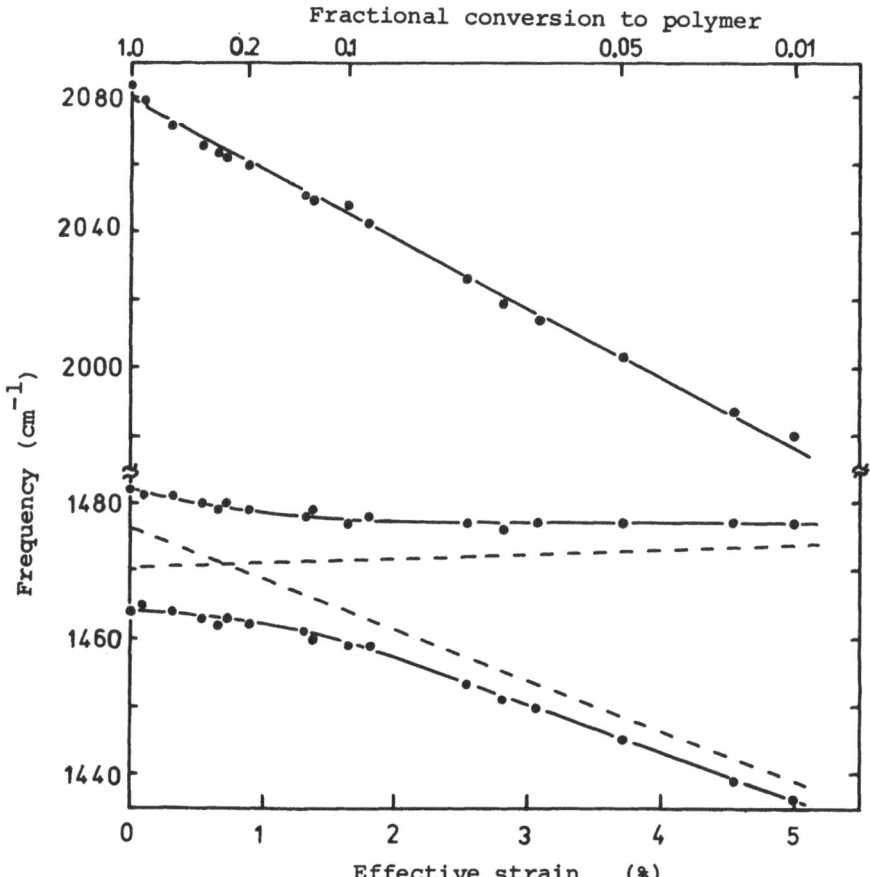

Figure 2. Variation of the C=C and C≡C stretching vibrational frequencies during the polymerization of MBS. The calculated unperturbed C=C and sidegroup mode frequencies are shown by the dashed lines.

The degree of conversion to insoluble polymer at 60°C for MBS and TS is shown in figure 4. The dashed curves are calculated from [3] for a 5% lattice concentration for TS and zero lattice contraction for MBS.

Thus, the Raman and optical spectra indicate that the lattice contraction for MBS is somewhat larger than that of TS since the observed frequency shifts are larger. The kinetic data, however, fit a model with no dimensional changes in the case of MBS.

It is apparent in figure 2 that most of the polymer strain in MBS is removed at low levels of conversion to polymer. Hence, after 2 hours at 60°C less than 1% strain remains and the remainder of the polymerization occurs near to zero lattice contraction. Since the Raman frequencies of MBS and TS polymer are similar, they must have similar elastic constants. Thus, to explain the observed behaviour either MBS monomer must have a significantly lower elastic constant than TS, which seems unlikely, or the higher initial polymerization rate reflects a longer initial chain length for MBS than TS. The assumptions of [3], which render the calculated reaction kinetics insensitive to initial chain length, appear not to be valid for MBS. Further kinetic studies are in hand in order to test this hypothesis.

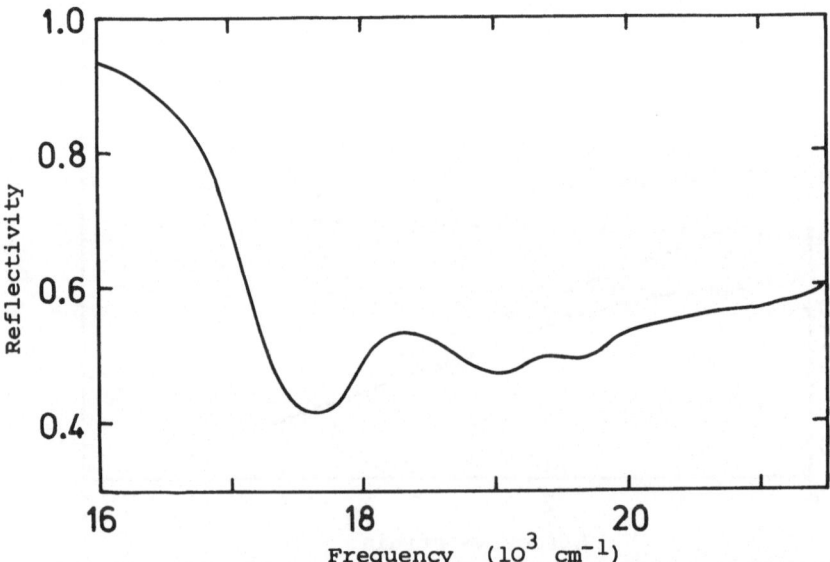

Figure 3.   Diffuse reflection spectrum of microcrystalline
MBS at about 1% polymer content.

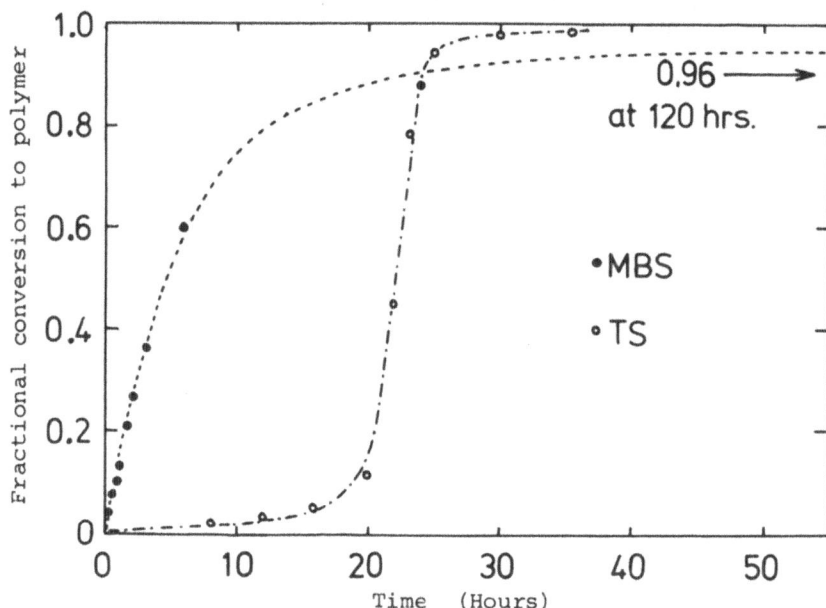

Figure 4.   Polymerization curves of MBS and TS at 60° C.

REFERENCES

1. G. Wegner, Pure and Appl. Chem. 49, 443 (1977)
2. R.H. Baughman, J.Polym.Sci.Phys.Ed. 12, 1511 (1974)
3. R.H. Baughman, J.Chem.Phys. 68, 3110 (1978)
4. J.J. Mayerle and T.C. Clarke, Acta Cryst. B34, 143 (1978)
5. G. Wegner, Makromol. Chemie 145, 85 (1971)
6. D. Bloor, R.J. Kennedy and D.N. Batchelder, to be published
7. D.N. Batchelder and D. Bloor, J.Polym.Sci.Phys.Ed. in press
8. D.N. Batchelder and D. Bloor, J.Phys.C.Sol.St.Phys. 11,
    L629 (1978)

KINETICS OF SOLID STATE POLYMERIZATION OF 2,4-HEXADIYNE-1,6-DIOL-

BIS (P-TOLUENE SULFONATE)[†]

A.F. Garito*, A.R. McGhie and P.S. Kalyanaraman

Department of Physics and L.R.S.M.,
University of Pennsylvania
Philadelphia, Pennsylvania 19104

Diacetylene molecular crystals have long been known to react in the solid state to form high molecule weight polymers. The first polymer structure proposed by Seher[1], and by Bohlman and Inhoffen[2] was a linear, ladder-type polymer. The 1,4-addition reaction in Figure 1 leading to a fully conjugated linear chain structure was demonstrated by Wegner[3] using general arguments from other solid state reactions and related crystallographic data.

The general polymerization mechanism for diacetylenes is shown in Fig. 1. Polymerization is believed to proceed via either a) a carbene species or b) a radical species which then reacts via 1,4 addition to give a polymer with either a butatriene[4] or acetylenic[5] backbone. Both mesomeric forms have been identified by x-ray structural analysis.

Many diacetylene monomers with different R groups have been synthesized and their polymerization behaviour investigated[3,6]. Polymerization can be induced thermally, photochemically, using ionizing radiation and mechanically. Reactivity varies from 0 to 100% depending on R and the polymorphic form of the diacetylene crystal grown. Few systems have been found which go to 100% conversion but of those that do 2,4 hexadiyne-1,6-diol bis (p-toluene-sulfonate (PTS) has been the most extensively studied[3,7,8,9,10,11].

Solid state polymerization of PTS can be polymerized thermally in times ranging from two months at room temperature to two hours near its melting point, ~90°C. The unique feature of the polymerization behavior is the occurence of an induction period during which polymerization proceeds very slowly followed by an 'autocalalytic' reaction during which the polymerization rate dramatically increases by a factor of up to 200[7],[8]. The usual explanation for such behaviour involves an incubation time necessary for the nucleation of a polymer crystal phase followed by rapid growth[12]. In the case of PTS, however, this does not apply as the polymer is completely soluble in the monomer crystal. In fact the only change in structure after polymerization is a 5% contraction along the crystallographic 'c' direction[9], i.e., the chain axis. This is believed to be responsible for the 'autocatalytic' effects via the lattice strain generated by the 5% mismatch, and Baughman[13] has developed a mechanical model based on this concept which is consistent with the observed phenomena. In the induction period polymer chains are initiated but their length is determined by the crystal strain field. Estimated chain lengths in this region are 10-20 units. At approximately 10% conversion, independent of the temperature of polymerization, crystal strain is relieved as conversion to polymer structure takes place and the final polymer crystal is strain free. Experimentally it has been found that polymerization during the induction period and during the fast polymerization regime have the same activation energy E_A = 22.3 ± 1.0 kcal/mole as determined by a variety of techniques; extraction measurements[9],[14], spectroscopically[14],[15] and by thermal analysis on both single crystal[7],[8] and polycrystalline samples[8],[11]. It was concluded that the rate determining step was that of chain initiation and that the actual rate of polymerization was determined by the number of monomer units which polymerize during the chain propagation step.

We have recently reported on the kinetics of isothermal polymerization of PTS single crystals using a differential scanning calorimeter to follow the heat of reaction with time[7]. Similar studies have also been carried out by Patel et al.[11] on polycrystalline PTS with qualitatively similar results. Fig. 2 shows a typical series of isothermal polymerization runs, at different temperatures, for single crystals of PTS all taken from the same batch. These were analysed as an induction period followed by a fast polymerization regime obeying first order kinetics over the range 20-95% polymerized. This yielded a rate constant, k and an induction period t_0 defined as the time at which the fast polymerization appeared to start. Arrhenius plots of log k vs $1/T$ and log $(1/t_0)$ vs $1/T$ are shown in Fig. 3 and both processes have the same activation energy. It has also been shown that different batches of single crystals and polycrystals all yield the same activation energy, E_A = 22.3 ± 0.9 kcal/mole. However, it was observed that the induction period for all samples, at constant temperature, was the same to within ±15% whereas the first order rate constants varied by up to 300%. We have interpreted these results as supporting the strain controlled polymerization theory[9],[13] in which the induction period is controlled by lattice strain and is only slightly

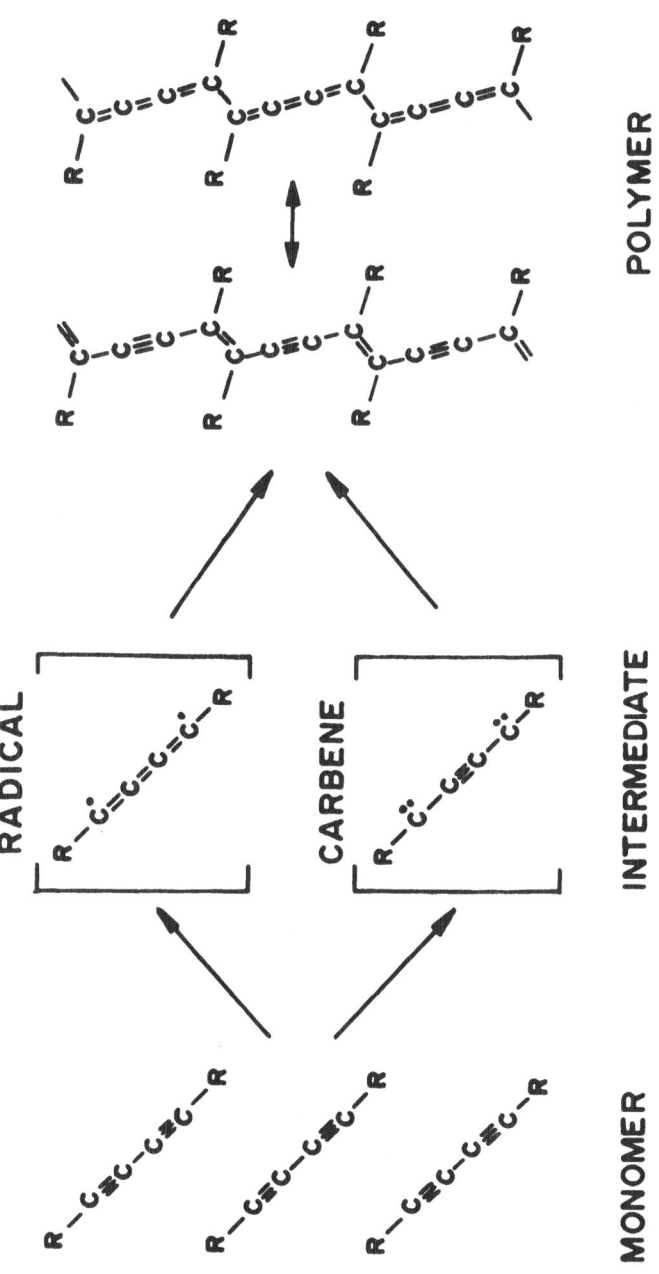

Figure 1 Mechanism for solid state polymerization of 1,3 diacetylenes. Proposed mechanisms include either a diradical or carbene intermediate. Final polymer can exist in either acetylenic or butatriene mesomeric forms.

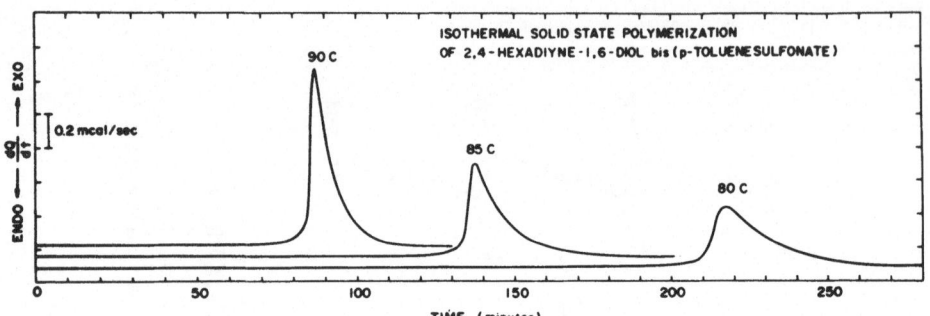

Fig. 2.   DSC curves for isothermal solid state polymerization of
          PTS single crystals at 90°C, 85°C and 80°C. (from Ref. 7).

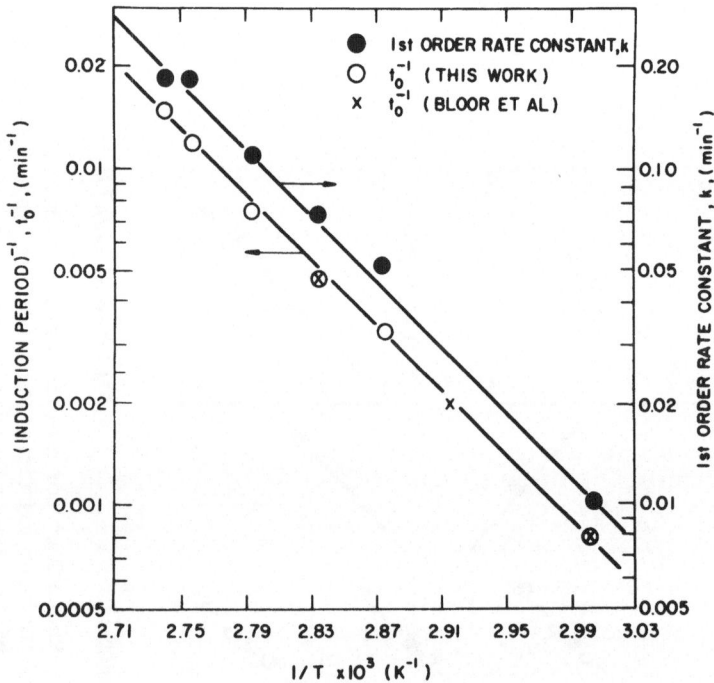

Fig. 3.   Arrhenius plots of first order rate constant, k for the
          fast polymerization regime, vs temperature, and the
          reciprocal induction period, t_0^{-1}, vs temperature (from
          Ref. 7).   Also included is the induction period of Bloor
          et al. (x) from Ref. 9.

dependent on purity whereas fast polymerization is no longer strain dependent and chain propagation can occur until a terminating site is reached. The latter process is therefore expected to be more dependent on the chemical impurity concentration. In our samples greater purity gave larger values of k but this has not been studied systematically to date. Evidence that lattice strain prevented chain growth without causing chain termination was given by the fact that samples of the same batch of crystals which were allowed to anneal at room temperature for up to 50 days prior to isothermal anneal at 90°C gave maximum polymerization rates which were 30% larger than those for fresh crystals from the same batch.

Significant differences in the kinetics were observed when polycrystalline samples of small crystallite size, $\approx$10-100μ, were isothermally annealed. A large fraction of polymer, ~40%, formed before the maximum rate of polymerization was reached compared to single crystal or large polycrystalline sample values of ~20%. These results imply that in fine particles lattice strain is more easily relieved and consequently chains can grow larger during the induction period. This predicts that these samples should have the shortest induction periods but this was not observed.

The other important quantity which can be obtained by DSC is the heat of polymerization. For two batches of single crystals we obtained an averaged value of ΔH_p = -31.2 $\pm$ 0.9 kcal/mole. This value compares favorably with the heat of reaction derived from Dewar bond length/bond energy calculations[11] of ΔH_p = -31.5 kcal/mole for PTS. For polycrsytalline samples, we have obtained a larger value of ΔH_p = -38.8 kcal/mole which compares favorably with the polycrystalline value -36.5 kcal/mole obtained by Patel et al.[11] The reason for the difference in ΔH_p for single crystal and polycrystalline samples may involve increased thermal decomposition, cross linking of polymer chains, or other surface induced effects. DSC studies on other thermally polymerizable diacetylenes have also yielded values of ΔH_p in the range 30-40 kcal/mole suggesting that this value may be intrinsic to diacetylenes. The relatively low value of ΔH_p has been used[11] as supported for the diradical mechanism of chain initiation despite spectroscopic evidence that a triplet carbene species has been detected during the fast polymerization regime[16,17].

In summary, our DSC kinetics measurements of the solid state polymerization of PTS distinctly show slow polymerization takes place during an induction period followed by a rapid transition to a fast polymerization regime which appears to obey first order kinetics over ~20-95% polymer formation. Rate-controlling chain initiations in both regimes proceed with an activation energy E_A = 22.3 $\pm$ 0.9 kcal/mole. The ratio of polymerization rates in the fast polymerization regime to that in the induction period is k/k_o $\approx$ 150, corresponding to final chain lengths of order 1-3 x 10^3

monomer units. The heat of polymerization ΔHp measured for single crystals is -31.2 ± 0.9 kcal/mole, and consistently higher values have been observed for polycrystalline (-38.8 ± 1.5) and aged single crystal (-38.0 ± 1.5) samples which may be associated with extra contributions coming from thermal decompositon, cross-linking, or surface induced effects.

†This work was supported by the Advanced Research Projects Agency and in part by NSF-MRL and NATO.

REFERENCES

1. A. Seher, Ann. 589, 222 (1954).
2. F. Bohlman and E. Inhoffen Ber., 89, 1276 (1956).
3. G. Wegner, Makromol Chem. 145, 85 (1971), G. Wegner, Z. Naturforsch. 24b, 824 (1969).
4. D. Day & J.B. Lando, J. Polym. Sci. Polym. Phys. Ed., (in press).
5. D. Kobelt & E.F. Paulus, Acta. Cryst. B30, 232 (1973).
6. E. Baughman & K.C. Yee, J. Polym. Sci. Polym. Chem. Ed. 12, 2467 (1974), R. Baughman, J.D. Witt & K.C. Yee, J. Chem. Phys. 60, 4755 (1974).
7. A.R. McGhie, P.S. Kalyanaraman, A.F. Garito, J. Poly. Sci. Polym. Letters,
8. A.R. McGhie, P.S. Kalyanaraman, A.F. Garito, 5th International Symposium on the Organic Solid State, Brandeis University, June 1978. Proceeding to be published in Mol. Cryst. & Liquid Cryst.
9. D. Bloor, L. Koski, G.C. Stevens, F.H. Preston & J.D. Ando, J. Mater. Sci. 10, 1678 (1975).
10. R.R. Chance & G.N. Patel, J. Polym. Sci. Polym. Phys. Ed. 16, 859 (1978).
11. G.N. Patel, R.R. Chance, E.A. Turi & Y.P. Khanna, J. Amer. Chem. Soc. (In Press).
12. B. Wunderlich, Makromolecular Physics Vol. 2, Academic Press, NY 1971.
13. R.H. Baughman, J. Chem. Phys. 68, 3110 (1978).
14. R.R. Chance & G.N. Patel, J. Polym. Sci. Polym. Phys. Ed 16, 859 (1978).
15. R.R. Chance & J.M. Sowa, J. Amer. Chem. Soc. 99, 6703 (1977).
16. G.C. Stevens & D. Bloor, Chem. Phys. Letters 40, 37 (1976).
17. H. Eichele, M. Schwoerer, R. Huber & D. Bloor, Chem. Phys. Letters 42, 342 (1976).

INTERCALATION COMPOUNDS OF GRAPHITE

F. Lincoln Vogel

Department of Electrical Engineering and Science and
Laboratory for Research on the Structure of Matter
University of Pennsylvania, Philadelphia, Pa.   19104

Intercalation compounds of graphite are of interest, both
scienfically and technologically because of unusual properties that
derive mainly from high degree of crystal anisotropy. This structure, composed of atoms tightly bonded in loosely stacked planes,
produces a two dimensionality that yields a high in-plane strength
and elastic modulus, high electrical conductivity, selective catalytic and chemical reactions and other interesting phenomena. Since
a wide range of variables can control these effects, the materials
scientist is afforded an opportunity for the design of synthetic
materials rather than relying upon the limited properties of natural
materials. Thus in this conference on molecular metals the exposure
of unconventional properties is to be expected. The science of
materials that produces this design information lies at the juncture
of chemistry and physics, employing the former for the intelligent
synthesis of materials and the latter for rationalization of properties. This paper will present intercalation compounds of graphite
from the point of view of a quest for a synthetic material of high
electrical conductivity.

A number of reviews have appeared on graphite intercalation
compounds over the years starting in 1959 (1) with Hennig and with
Rüdorff, then Ubbelohde and Lewis in 1960 (2). An extensive review
was given in 1965 by Herold, Platzer and Setton (3) in Les Carbones. In 1976 (4) is the review by Ebert, notable for its list
of chemical species that have been intercalated in graphite and
Gamble and Geballe (4) whose review is part of a treatment of all
inclusion compounds. In May of 1977 an international conference
on graphite intercalation compounds was held in La Napoule, France
which was attended by the majority of active researchers in the free
world. The proceedings of the conference (5) constitute an

extensive and complete overview of the field at that time. This
paper is based largely on the research reported in that conference.

This review will begin with relevant considerations of the
parent material, graphite, then proceed to the donor intercalation
compounds of graphite since they are the simplest, being formed
primarily from alkali metal atoms, then finally to the acceptor
compounds which are formed from acid molecules and which display
the most interesting characteristics.

GRAPHITE – STRUCTURE AND PROPERTIES

In graphite the two important bondings are associated with the
carbon 2s and 2p electrons. The sp^2 hybridized orbitals constituting
the σ bonds are formed into a planar hexagonal network. The remaining p_z orbital, the π bond, loosely holds the planes together.
The in-plane closest approach of the carbon atoms is 1.35 Å. The
planes are stacked in an ABAB ... sequence as shown in Figure 1
with half of the atoms in one plane directly over the one below at
a spacing of 3.35 Å. The three atom unit cell that is shown in
Figure 1 has dimensions a_0 = 2.46 Å and c_0 = 6.70 Å. The anisotropy of properties results from this difference in bond strength
parallel and normal to the hexagonal planes.

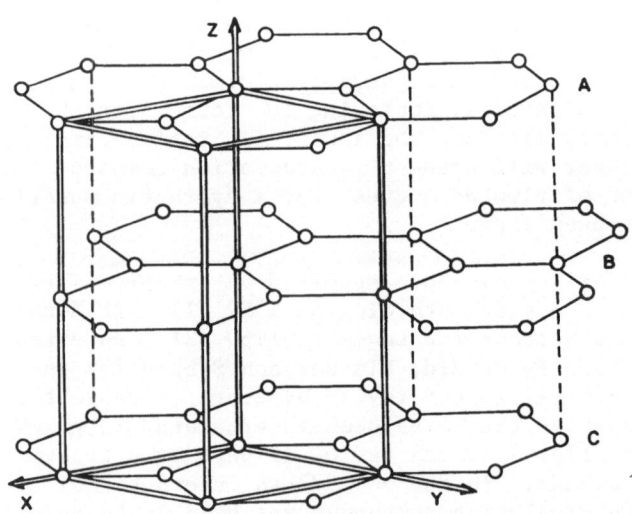

Figure 1. Graphite Crystal Structure.

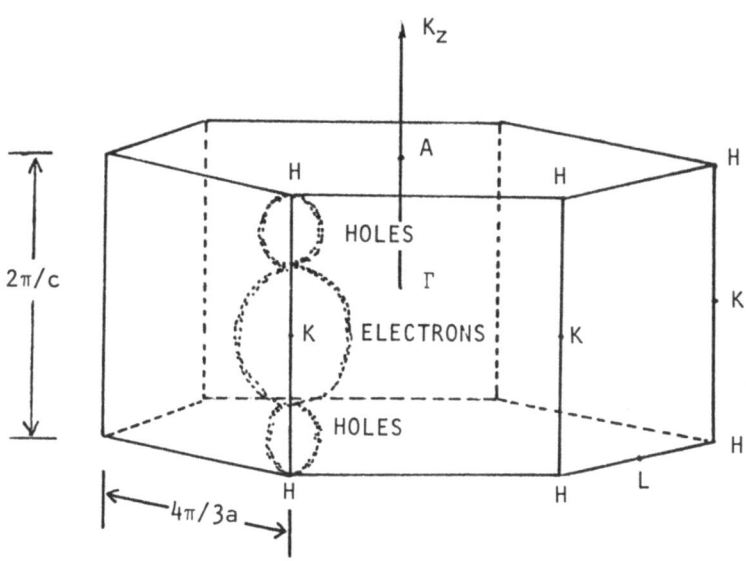

Figure 2.   The Brillouin Zone and the Fermi Surface of Graphite (11).

Fundamental physical studies of graphite are, in most cases, best carried out in single crystals. This placed a severe restriction on graphite research for many years since natural crystals of appropriate size and purity were obtained by exploration and extraction of material from such diverse places as Ceylon, Madagascar and Ticonderoga, New York. In the period around 1960 to 1964 Dr. Arthur Moore, working in Prof. Ubbelohde's laboratory at Imperial College developed a method (6) for fabricating synthetic graphite by decomposing a hydrocarbon gas and hot pressing the resulting graphitic material. The product is called "highly oriented pyrolytic graphite" (HOPG) and is characterized crystallographically by a c axis spread of a few tenths of a degree and a random orientation of the a axis. Since most properties of interest in graphite and its crystal compounds are isotropic within the hexagonal plane, HOPG is a very adequate substitute for single crystals. The ready availability of these crystals has been instrumental in the rapid development of the field in the last few years. In particular, our program at Penn has benefited from the generosity of Dr. Moore, now at Union Carbide Research Laboratory, who has provided us with this material.

The electrical conductivity of pure graphite crystal, which has been studied by Spain (7)(8) using a d.c. four point method and more recently in our laboratory (9) using an r.f. induction

Table 1. Transport Properties of Synthetic Graphite
and Natural Graphite

	Synthetic (HOPG)	Natural (Ticonderoga)
Room Temperature:		
σ_a $(\Omega^{-1}cm^{-1})$	[*]2.3×10^4	[**]2.6×10^4
σ_a $(\Omega^{-1}cm^{-1})$	[*]5.9	[*]~ 10
σ_a/σ_c	$4,000$	$2,700$
μ_a $(cm^2 volt\text{-}sec)$	[*]$12,400$	[**]$11,000$
n $(No./cm^3)$	[*]1×10^{19}	[**]1.5×10^{19}
Liquid Helium Temperature:		
$\sigma_a (\Omega^{-1}cm^{-1})$	4.1×10^5	[**]1.2×10^6

[*]Measured with direct current (8)
[**]Measured at r.f. (10)

method clearly reflects the high crystal anisotropy. Table 1 shows
a summary of values obtained. The room temperature a axis con-
ductivity of 2.6×10^4 $\Omega^{-1}cm^{-1}$ and c axis conductivity of 10 $\Omega^{-1}cm$
are similar for both HOPG and natural graphite. At liquid helium
temperature, however, a large difference exists in a axis conductivity. With the best HOPG crystals $\sigma_{LHe}/\sigma_{RT} = 18$ but with the
best natural crystals $\sigma_{LHe}/\sigma_{RT} = 46$. This difference is attributable to phonon scattering from defects. The relatively high mobility value (copper – a highly conducting metal – has an electron
mobility of ~50 cm^2/volt-sec) of graphite is probably a consequence
of the rigid, resonantly bonded hexagonal structure of the graphite
layer planes.

The Fermi surface of pure graphite consists of six sets of ellipsoids, one at each corner of the hexagonal Brillouin zone, as is
shown in Figure 2. Each set has an electron segment in the middle
with a hole ellipsoid above and below. A good discussion of energy
bands in pure graphite by Spain (11) appears in the proceedings of
the La Napoule conference.

DONOR INTERCALATION COMPOUNDS

The loose coupling of the hexagonal layers of graphite allow
for the formation of intercalation compounds by insertion of chemical species between these layers. When the inserted specie is a
metal, such as from Group I in the Periodic Table, the graphite
lattice acts as a giant anion attracting electrons from the alkali

Table II.   Electrical Conductivity of Alkali Metal Intercalated
            Graphite at Room Temperature

Compound and Stage	Conductivity σ_a in $\Omega^{-1}cm^{-1}$
K-Cg-1	1.1×10^5
2	1.7×10^5
3	2.1×10^5
Rb-Cg-1	1.0×10^5
2	1.5×10^5
Cs-Cg-1	1.0×10^5
3	1.2×10^5
Li-Cg-1	1×10^5

metal. The composition is described by stage, the definition of
which is portrayed in Figure 3 for the potassium compound. Each
layer is filled to a fixed planar density and a shift occurs in
the graphite stacking sequence so that the AB relationship between adjacent layers is changed to AIA on either side of the intercalated layer, where I indicates the interposed intercalant
layer. Although little change occurs in the in-plane carbon interatomic distance a large expansion accompanies the interpolation
of the intercalate layer and so the stage of the compound can be

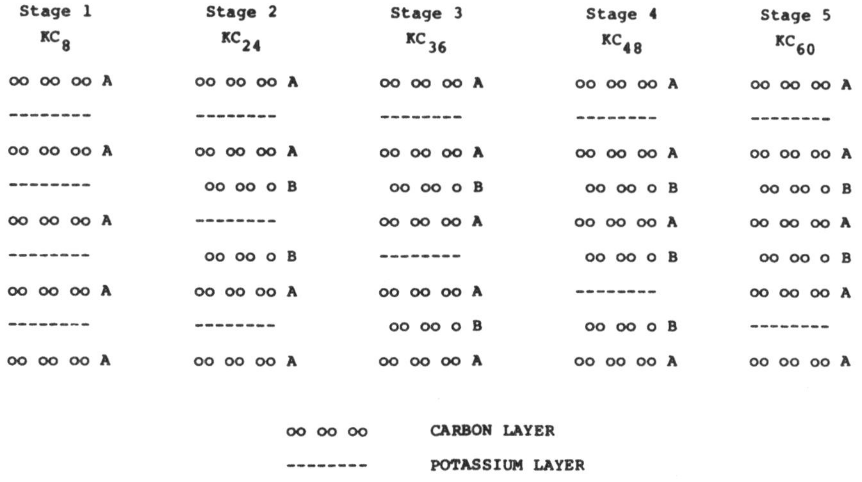

Figure 3.   The Stage Structure for Potassium Graphite (14).

determined from the lattice repeat distance by the following relationship:

$$I_c = d_i + (n - 1) \ 3.35 \ \text{Å}$$

Since the repeat distance is uniform for each stage, staging is
considered as a form of interlayer ordering.

Intralayer ordering also occurs in the donor intercalation
compounds of graphite with the exact form depending on the atom
size and the stage itself. For the larger atoms such as K, Rb and
Cs the unit cell contains nine atoms in the arrangement shown in
Figure 4 with the formula MC_8. In this type of structure the alkali metal atoms form a $2a_0 \times 2a_0$ superlattice in the plane of the
carbon hexagons. Since there are four possible positions for the
intercalate atom in this plane, designated as α, β, γ and δ in
Figure 4, interlayer ordering occurs in this type of compound by
the intercalate taking alternate positions in successive layers
such as $A\alpha A\beta A\gamma A\delta A \ldots$. The closest approach of potassium atoms in
the b.c.c. metal crystal is 4.624 Å and so fitting them into a tri-
angular planar lattice 4.92 Å on a side represents good economy of
packing. Rubidium atoms are 4.88 Å in diameter – about the same
size as potassium and therefore form the same totally ordered first
stage structure at room temperature. However, above room temperature, the order disappears (12) in steps as shown in Figure 5.

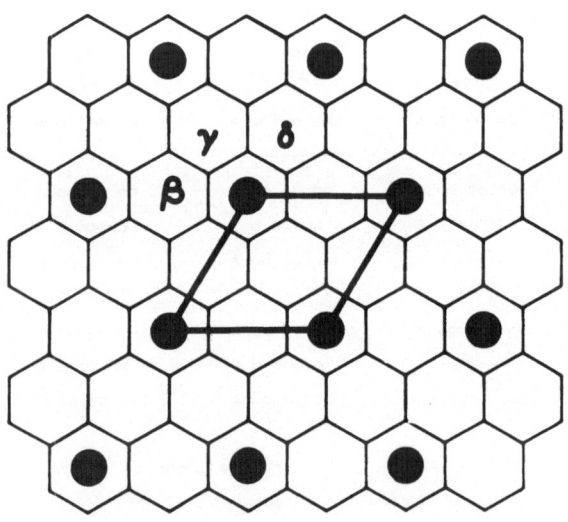

Figure 4.   The KC_8 Structure.

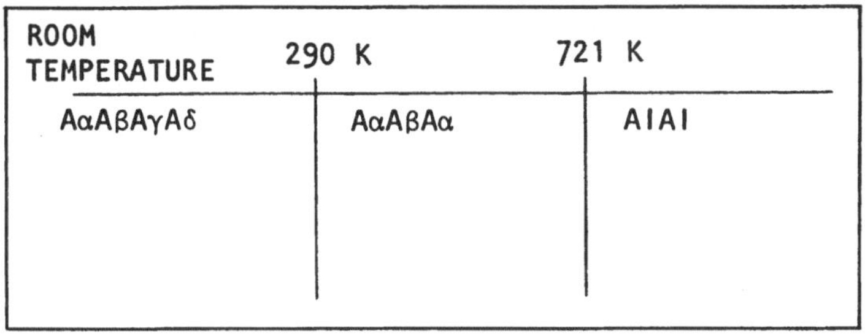

ROOM TEMPERATURE	290 K	721 K
AαAβAγAδ	AαAβAα	AIAI

Figure 5.   Transformations in RbC_8 (12).

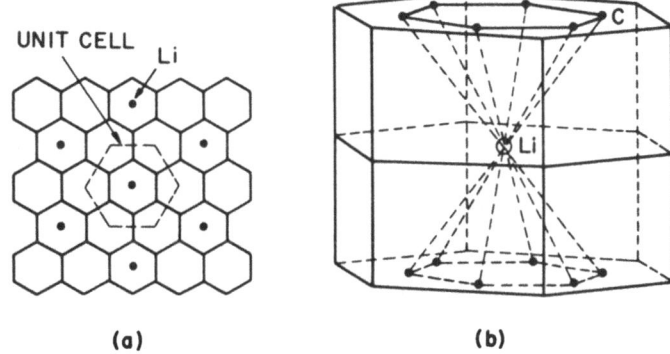

(a)   (b)

Figure 6.   The LiC_6 Structure.

·Lithium with an atomic diameter equal to 3.038 Å is somewhat smaller than the other alkali metals first discussed and so it can be accommodated by a higher density arrangement shown in Figure 6. In this unit cell there are seven atoms with a stage 1 chemical formula of LiC_6 and the Li atoms occupy positions 4.26 Å apart.   The small size of the Li atom allows use of the interstitial space immediately above and below a carbon layer and the stage 1 room temperature sequence becomes AαAαAα ... .

Many types of ternary compounds can be formed with alkali metals and graphite and these display a fascinating variety of combinations and structures.   The simplest case is provided by the heavy alkali metals where it is known that the action of two of them leads to the ternary compound $M_xM'_{1-x}C_8$ ($0 < x < 1$) which may be considered as a solid solution of MC_8 and $M'C_8$ (13).   Sodium is particularly of interest since it is generally believed that it will not intercalate alone, although there is some disagreement on this point (14).   Sodium will however form a ternary with Cs or K but when combined with lithium only the binary LiC_6 is obtained.

Hydrogen can be made to form a ternary with KC_8 that has the formula $KC_8H_{2/3}$. However, this ternary is really a two stage compound having the structure shown in Figure 7a, which was determined by x-ray diffraction. Proof of this structure was obtained in a clever experiment by Guerard (15) in which the stage 2 ternary was reintercalated with potassium, resulting in the structure of Figure 7b. It follows that this new layer can not be hydrogenated. Such compounds as this have interest for the storage of hydrogen because the density of that element is greater in the ternary compound than in the pure liquified gas.

Aromatic molecules such as benzene or toluene have been shown to form ternary alloys with the alkali metal binary compounds (16) (17). Usually only the stage 2 or higher binaries react to form a ternary of lower stage and the metal and aromatic molecule occupy the interstitial space together. The intimate physical and electronic contact under these conditions suggests possibilities for catalytic action.

The alkali metal compounds can be synthesized by contacting a good grade of graphite with the metal in either the molten or vapor state. Vapor state intercalation is preferred, and has led to the general adoption of the "tube à deux boules" method shown in Figure 8 that was first used by Fredenhagen (18) in 1926 and extensively applied by Herold in the 1950's (19). The vapor pressure of the reactant is controlled by temperature, T_r with T_g held somewhat higher. The stage obtained is determined by the temperature difference $T_g - T_r$. This method is used for many other intercalants that are solid or liquid at T_r.

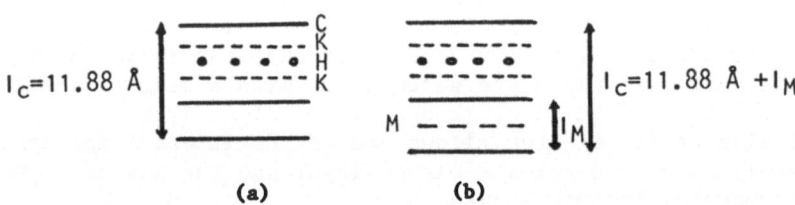

(a)   (b)

Figure 7.   The Stage Structure of Potassium Hydrogen Graphite Compounds.

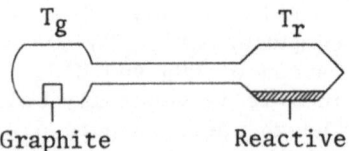

Figure 8.   Méthode de Tube à Deux Boules (19).

Table III.   Selected Conductivity Values of
             Acceptor Intercalated Compounds of Graphite

Compound and Stage	σ_a $\Omega^{-1}cm^{-1}$	σ_c $\Omega^{-1}cm^{-1}$
HNO_3 – 1	1.7×10^5	2
HNO_3 – 2	3.3×10^5	–
HNO_3 – 3	2.9×10^5	–
HNO_3 – 4	2.4×10^5	–
AsF_5 – 1	5.0×10^5	0.23
AsF_5 – 2	6.3×10^5	0.24
AsF_5 – 3	5.8×10^5	0.26
SbF_5 – 1	3.5×10^5	–
SbF_5 – 2	4.0×10^5	–
SbF_5 – 3	1.0×10^6	–
SbF_5 – 6	5.8×10^5	–
$FeCl_3$ – 1	1.1×10^5	10
$FeCl_3$ – 2	2.5×10^5	–

The insertion of alkali metal atoms into the graphite layer planes alters the physical properties markedly.   There is a transfer of electrons to the graphite with a consequent increase in c axis and a axis conductivity. The a axis conductivities are given in Table II for a number of donor elements. The intercalated conductivities are all about the same – 1×10^5 $\Omega^{-1}cm^{-1}$, regardless of electronegativity, polarizability or whatever (20). This value is roughly five times that of pristine graphite. The c axis increases with ultimate values in the 10^3 to 10^4 $\Omega^{-1}cm^{-1}$ range indicating ratios of 10 to 100. Complete analysis of the transport properties in donor compounds is complicated by the fact that apparently there are two carriers present (21). However, based on estimates derived from magnetoresistance and Hall measurements on a series of stage 1, 2, and 3 compounds Foley (22) has estimated that the mobility at room temperature is degraded about 2 orders of magnitude compared to graphite, i.e. $\mu_{graphite} = 10^4 cm/volt-sec$ compared to $\mu_{MC_x} = 10^2$.

The anisotropy is also evident in reflectivity measurements on stage 1 Cs and K compounds done by Zanini (23). With the plane of polarization normal to the c direction, strong metallic behavior is observed compared to the case where the plane of polarization is parallel to the c axis. See Figure 9.

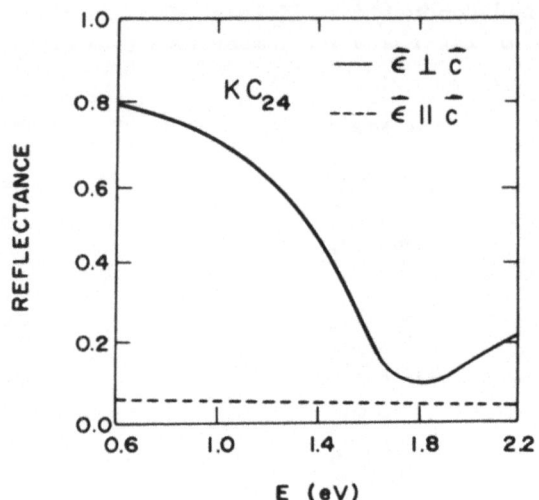

Figure 9. Reflectance of KC_{24} (23).

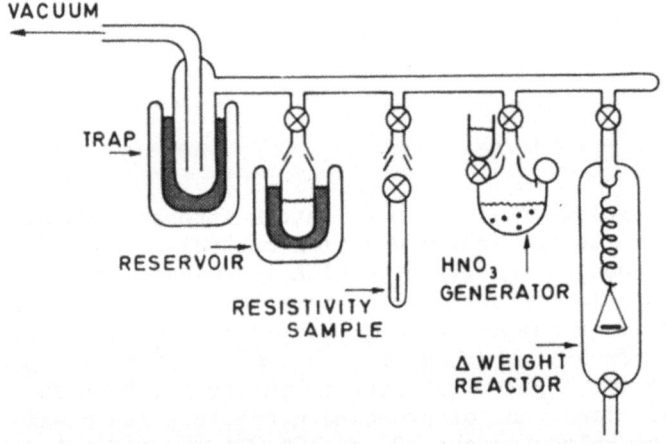

Figure 10. Vacuum line apparatus for synthesis of graphite
            compounds (24).

ACCEPTOR INTERCALATION COMPOUNDS

Nonmetal molecular species that intercalate into graphite in
general produce acceptor compounds. Examples of these are bromine,
sulfuric acid, nitric acid and many halides and oxides. Ebert's
review (4) lists well over sixty chemical species that form acceptor compounds with graphite and it seems fairly safe that his list
is not exhaustive.

The chemical treatment of graphite crystals with molecules of an acceptor, as with the alkali metals, proceeds best by exposure to the vapor. A prototype (24) vacuum line for accomplishing this in a controlled manner is shown in Figure 10. Provision is made to atmosphere control, distillation of intercalant, weight change determination and in situ measurements either connected to the system or by sealing off. Also, a controlled partial pressure of gaseous intercalant can be obtained by volumetric calibration of the system. Data obtained this way on the graphite-AsF_5 system are shown in Figure 11. Intercalation was observed in progress by monitoring the conductivity by rf induction and the thickness by cathetometry while the graphite crystal was exposed to 0.3 atmospheres of AsF_5-gas. The prominent feature of this curve is the existence of staging whereby the sample progresses from a very dilute compound to stage 4, then 3, then 2 and finally stage 1.

Interesting perspectives on the mechanisms by which intercalation occurs have been obtained. In one set of experiments J.G. Hooley (25) observed the progress of intercalation on a cylinder of graphite long in the c direction. The side was engraved with five scale markings at equal intervals from the end as shown in Figure 12. When the sample was exposed to bromine vapor while the distance between markings was recorded it was discovered that intercalation proceeds down the sample starting from the c face ends. It is known from the "ash tray" effect, that is, the depression

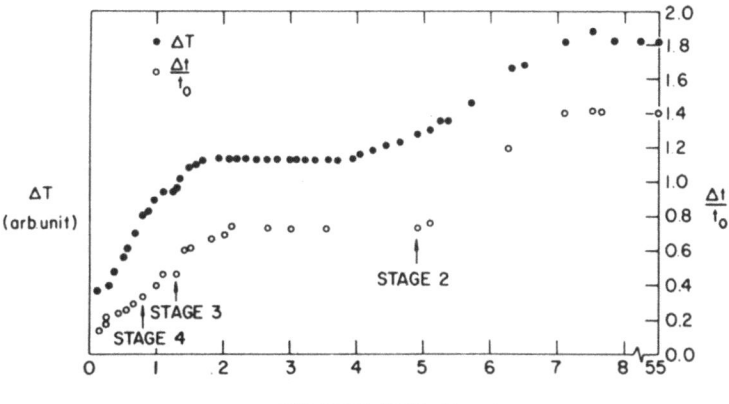

REACTION TIME (hours)

Figure 11. Increase of Thickness and In-Plane Conductivity of Graphite AsF_5 Compounds. ΔT is incremental period of rf circuit indicating conductivity, $\Delta t/t_0$ is fractional increase in thickness over original (33).

that occurs in the center of the c face of a flat intercalating sample, that diffusion of intercalant takes place parallel to the hexagonal layers. Hooley's experiment makes it clear that the diffusion does not take place uniformly but starts at the c faces at the end. Hooley obtained additional proof that intercalation is initiated at the ends by sealing off the c face ends with acetate cement. Then, on exposure to bromine no reaction occurred despite ready access of the side faces where the intercalant enters into the vapor. This experiment was replicated by the author using HNO_3 as the intercalant and gold films for the stop.

The question of how the intercalant is transported to the interior of the crystal and ionized was addressed by Forsman (26). It had long been observed that a brown gas, probably NO_2, accompanies the intercalation of graphite by nitric acid. This observation led Forsman to propose the following reaction steps for the process:

$$2HNO_3 \rightarrow 2H_2O + NO_3^- \text{ (ads)} + NO_2^+ \text{ (ads)} \tag{1}$$

$$Cg + NO_3^- \text{ (ads)} + NO_2^+ \text{ (ads)} \rightarrow Cg^+ + NO_3^- + NO_3^- \text{ (ads)} + NO_2 \text{ (ads)} \tag{2}$$

$$Cg + NO_3^- \text{ (ads)} + NO_2 \text{ (ads)} \rightarrow Cg^+ NO_3^- + NO_2 \tag{3}$$

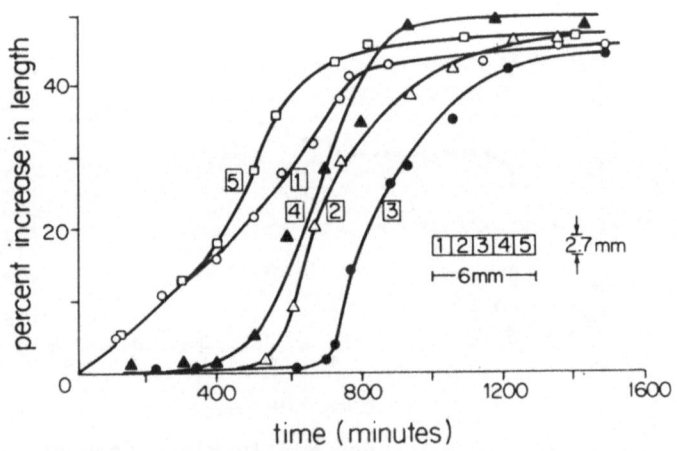

Figure 12. The Hooley Experiment (25) indicating intercalation of sample from c face ends. Sample shown in insert.

The brown gas is evolved when the neutral NO_2 is desorbed. Intercalation of acceptors then can be viewed as a two step process.
The first step is oxidation of the carbon free radical at the edge
of the planes in the graphite crystal which converts it to the
carbonium ion. This positive charge delocalizes into the lattice
taking the NO_3^- anion with it for the second step of the reaction.
The quantitative presence of NO_2 (27)(28) has now been used to confirm this reaction mechanism for nitrate intercalation.

Applying the Forsman mechanism to the intercalation of graphite
with SbF_5 resulted in the following set of chemical equations:

$$2SbF_5 \rightarrow SbF_6^- \text{ (ads)} + SbF_4^+ \text{ (ads)}$$

$$Cg + SbF_6 \text{ (ads)} + SbF_4^+ \text{ (ads)} \rightarrow Cg^+ + SbF_6^- \text{ (ads)} + SbF_4$$

$$Cg^+ + SbF_6^- \text{ (ads)} \rightarrow Cg^+ SbF_6^-$$

$$2SbF_4 \rightarrow SbF_5 + SbF_3$$

Two predictions of interest are made by this set of reactions.
The first is that SbF_3 is a product of the reaction and this has
been found to be the case (29). The second is that the electrically
active specie is SbF_6^- which may be more difficult to prove in view
of the likelihood that the antimony fluoride is probably present in
the intercalated layer as a macromolecule.

In view of the clear indications in the mechanisms of intercalation that the passage of atoms, ions and molecules is parallel to
the hexagonal network in forming the graphite compounds, the staging illustrated in Figure 11 is difficult to rationalize. All of
the atom motions in the mechanism are in the $\perp c$ direction but the
change from one stage to the next requires mass movement in the
$||c$ direction. The model of Daumas and Herold (30) shown in Figure
13 has been proposed as the most likely way of filling and emptying
inserted layers. The shift from an A position to a B position that
must accompany the insertion of a layer implies the movement of a
partial dislocation in the plane translated. The Daumas and Herold
model provides for domains of intercalation which have partial dis-
locations in the graphite planes at the boundary. Stage changes
then are made by relatively small $\perp c$ motions between domains of
different stage.

The electrical conductivity properties of the acceptor intercalation compounds have received considerable attention since the
work of Ubbelohde (24) indicated that high in-plane conductivities
are possible. However, the earlier work was unable to completely
overcome the three major difficulties in performing electrical
transport measurements:

Table IV. Transport Properties of Selected Acceptor Compounds in Graphite

Compound and Stage	a-Axis Resistivity ρ_a in $\mu\Omega$ cm	Anisotropy ρ_c/ρ_a	Mobility μ-cm^2/volt sec	Carrier Density n x 10^{20}/cm^3	Intercalate Density N_i x 10^{21}/cm^3	Fractional Ionization
AsF$_5$ - 1	2.5	9×10^5	1980	12.6	5.9	.21
AsF$_5$ - 2	1.6	3×10^5	3200	12.2	4.2	.29
(d_i = 8.10 Å)						
HNO$_3$ - 2	3.5	1.4×10^5	1750	10.2	2.9	.18
(d_i = 7.80 Å)						
FeCl$_3$ - 1	9.5	1×10^4	920	7.1	5.4	.13
FeCl$_3$ - 2	4.3	-	1630	8.9	4.8	.18
(d_i = 9.46 Å)						

$$\sigma = Ne\mu$$

$$\mu = \sqrt{\frac{\Delta\rho}{\rho \cdot H^2}} \cdot 10^8$$

1. Electrical Contacts and highly corrosive acids are incom-
    patible.

2. Expansion of as much as 150% can occur during intercala-
    tion.

3. Extremely high c to a axis anisotropy of conductivity makes
    it impossible to achieve uniform current throughout the
    cross section of a sample.

The work of Zeller, Denenstein and Foley of our laboratories made it
possible to obtain valid electrical transport information on these
systems. It was found (31) that during intercalation of graphite
with such acceptors as AsF_5 the anisotropy increases from 10^4 to
10^6 which make the usual four point measurement invalid. They
developed an r.f. induction apparatus (9) that completely avoids
the difficulties mentioned previously. The a axis conductivity
values listed in Table III were obtained by this method (32)(33)
(34)(35) with the exception of the SbF_5-stage 3 which was obtained
by the author on cored wires (36). The c axis measurements were
obtained at d.c. Based on the data in Table III several generalizations can be made:

1. Intercalation of graphite with acceptor molecules can re-
    sult in very high in-plane electrical conductivities – in
    some cases greater than copper and silver. (Electrical
    resistivity of copper is 5.9×10^5 $\Omega^{-1}cm^{-1}$ and silver
    6.3×10^5 $\Omega^{-1}cm^{-1}$.)

2. The highest conductivity does not occur at stage 1 but
    rather in the vicinity of stage 2 or 3.

3. There appears to be an inverse relationship between
    σ_a and σ_c such that the acceptors that produce the highest
    in-plane conductivities produce the lowest c axis conduc-
    tivities. High anisotropy appears to be a requirement for
    high in-plane conductivity.

Further details of the electrical transport in acceptor compounds have been obtained from magnetoresistance measurements and
Table IV displays some of this unpublished data obtained by Zeller,
Pendrys, Perrachon and the author. Two major conclusions can be
drawn from these data:

1. On intercalation, the carrier mobility of the acceptor com-
    pounds is reduced from its value of over ten thousand in
    graphite to several thousand. Contrary to some expecta-
    tions, the lower the resistivity the higher the mobility.

2. Between 10% and 20% of the molecules intercalated are
    ionized and contributing carriers – in this case holes to
    the conduction scheme.

The reflectivity of the AsF$_5$ compounds have been studied in some detail by Hanlon and Fischer (37) and Figure 14 shows the reflectance spectrum for stages 1 through 4 superimposed on the reflectance spectrum of graphite. By stage 4 the plasma edge is already well developed and the position of the minimum progresses in energy according to stage as would be expected. Using a suitable set of assumptions Hanlon was able to deduce optical conductivities for the stage 1 and 2 compounds which agree with the d.c. conductivity.

The changes that take place in the Fermi surface on intercalation with donor or acceptors is described by Fischer (38) and shown in Figure 15. With donors the center (electron) ellipsoid expands into the end surfaces, becoming one large ellipsoid. In the case of acceptors the end (hole) pieces enlarge at the expense of the center ellipsoid, becoming cylindrical.

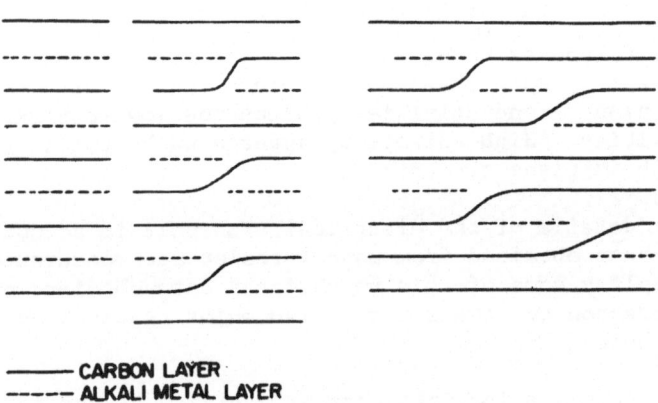

CARBON LAYER
ALKALI METAL LAYER

Figure 13.   The Stage Domain Model of Daumas and Herold (30).

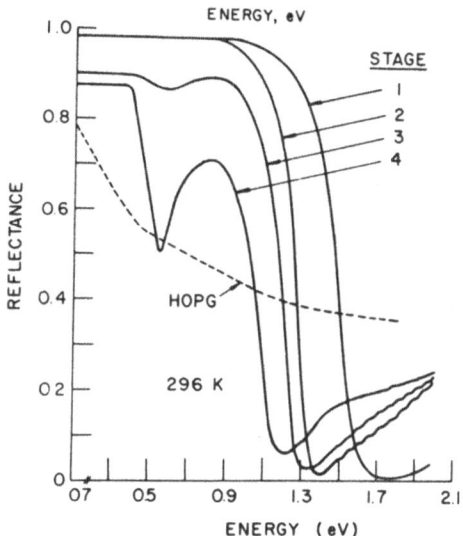

Figure 14. Reflectance of Graphite-AsF$_5$ Compounds by Stage Compared to Graphite (37).

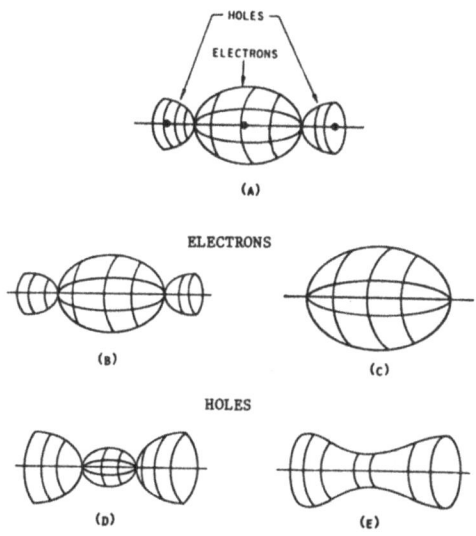

Figure 15. Progression of Fermi Surfaces from Graphite (A) above to (B) and (C) in middle with donors or (D) and (E) bottom with acceptors.

REFERENCES

1. G.R. Hennig. Prog. Inorg. Chem. 1, 125 (1959).
   W. Rüdorff, Adv. Inorg. Chem. 1, 223 (1959).
2. A.R. Ubbelohde and L.A. Lewis, "Graphite and its Crystal Compounds," Oxford, 1960.
3. A. Herold, N. Platzer and R. Setton, Les Carbones, t. 2, 462.
4. L.B. Ebert, Ann. Rev. Mat. Sci. 6, 181 (1976).
   F.R. Gamble and T.H. Geballe, Treatise on Solid State Chemistry, Vol. III, p. 89, Plenum, 1976.
5. Mat. Sci. and Eng. 31, 1977.
6. A.W. Moore, A.R. Ubbelohde and D.A. Young, Proc. Roy. Soc. A280, 153 (1964).
7. I.L. Spain, A.R. Ubbelohde and D.A. Young, Proc. Roy. Soc. A262, 345 (1967).
8. I.L. Spain, Chem. and Phys. Carbon 8, 1 (1973).
9. C. Zeller, A. Denenstein and G.M.T. Foley, submitted to Rev. Sci. Instr.
10. C. Zeller and L.A. Pendrys, unpublished.
11. I.L. Spain, Mat. Sci. Eng. 31, 183 (1977).
12. W.D. Ellenson, D. Semmingson, D. Guerard, D.G. Onn and J.E. Fischer, Mat. Sci. Eng. 31, 137 (1977).
13. A. Herold, D. Billaud, D. Guerard and P. Lagrande, Mat. Sci. Eng. 31, 25 (1977).
14. W. Rüdorff, Chimie 19, 489 (1965).
15. D. Guerard, P. Lagrange and A. Herold, Mat. Sci. Eng. 31, 29 (1977).
16. G. Merle, I. Rashkov, C. Mai and J. Gole, Mat. Sci. Eng. 31, 39 (1977).
17. L. Bonnetain, Ph. Touzain and A. Hamwi, Mat. Sci. Eng. 31, 45, (1977).
18. K. Fredenhagen and G. Cadenbach, Z. Anorg. Chem. 158, 249 (1926).
19. A. Herold, Bull. Soc. Chim. Fr., 999 (1955).
20. A.R. Ubbelohde, Proc. Fifth Carbon Conf., p.1 (1961).
21. J.E. Fischer, Mat. Sci. Eng. 31, 211 (1977).
22. G.M.T. Foley, unpublished.
23. M. Zanini and J.E. Fischer, Mat. Eng. 31, 169 (1977).
24. A.R. Ubbelohde, Proc. Roy. Soc. A304, 25 (1968).
25. J.G. Hooley, W.P. Garby and J. Valentin, Carbon 3, 7 (1965).
26. W.C. Forsman, Proc. Thirteenth Carbon Conf., p. 153 (1977).
27. S. Laughin, R. Grayeski and J.E. Fischer, J. Chem. Phys. (October 1978).
28. W.C. Forsman, F.L. Vogel, D.E. Carl and Jeffery Hoffman. To be published in Carbon.
29. W.C. Forsman and D.E. Carl. Submitted to Carbon.
30. N. Daumas and A. Herold, C.R. Serie C268, 373 (1969).
31. G.M.T. Foley, C. Zeller, E.R. Falardeau and F.L. Vogel, Sol. State Comm. 24, 371 (1977).
32. F.L. Vogel, H. Fuzellier, C. Zeller and E.H. McRae, to be published in Carbon.

33. F.L. Vogel, G.M.T. Foley, C. Zeller, E.R. Falardeau and J. Gan, Mat. Sci. Eng. 31, 261 (1977).

34. F.L. Vogel, J. Gan and T.C. Wu, Proc. Fifth London International Conf. on Carbon and Graphite (1978).

35. Jean Bernard Perrachon, Masters Thesis, University of Pennsylvania, 1978 (unpublished).

36. F. Lincoln Vogel, J. Mat. Sci. 12, 982 (1977).

37. L.R. Hanlon, E.R. Falardeau and J.E. Fischer, Solid State Comm. 24, 377 (1977).

38. J.E. Fischer, "Electronic Properties of Graphite Intercalation Compounds," chapter in "Intercalated Layer Materials," F. Lévy, ed. (D. Reidel, Holland) (in press).

ANISOTROPY OF GRAPHITE INTERCALATION COMPOUNDS

J. E. Fischer

Department of Electrical Engineering and Science and
Laboratory for Research on the Structure of Matter
University of Pennsylvania, Philadelphia, Pa. 19104

INTRODUCTION

Graphite intercalation compounds consist of an alternating sequence of one or more hexagonal carbon layers separated by monolayers of intercalated atoms or molecules (1). The resulting compounds invariably exhibit metallic behavior parallel to the layers,
with σ_a of order $10^5 \Omega^{-1} cm^{-1}$ at the saturated limit of one intercalant layer per carbon layer. This behavior is essentially independent of intercalant species, although metallic intercalants such
as Li, K, Rb, etc. give n-type conductivity whereas electron acceptor intercalants (Br_2, HNO_3, AsF_5, etc.) lead to p-type conduction.
A much greater variability is observed in properties perpendicular
to the layers. Metallic intercalants lead to large σ_c, suggesting
relatively mild overall anisotropy, while acceptor intercalation
causes a strong reduction in σ_c and anisotropies (measured by σ_a/σ_c)
as large as 10^6 at 300 K. This unusual and highly variable anisotropy is manifested not only in the crystallographic, electronic
and mechanical properties, but also more subtly in the variation of
critical phenomena among different compounds (2). The large degree
of variability in interlayer coupling makes this system attractive
for systematic studies of physical phenomena predicted to occur in
two dimensions, since one can vary the strength of c-axis interactions over a range of 10^5 or more while keeping the a-axis properties essentially constant. Furthermore, the slight variation in σ_a
values with different intercalants appears to correlate with anisotropy; the higher σ_a/σ_c, the higher is σ_a. Thus, understanding the
anisotropy may also provide a partial solution to the problem of
optimizing σ_a.

In this paper I discuss the various factors which influence

anisotropy. Dilation of the graphite interlayer spacing, necessary
to accommodate the intercalant, would decrease σ_c (increase the
anisotropy), while orbital mixing between carbon π and intercalant
valence electrons would have the opposite effect. The 2D symmetry
of the intercalant layer plays a role in controlling the extent of
s-π mixing in alkali metal compounds. In dilute compounds, screening of intercalant layers by free carriers (3) may have the effect
of enhancing the anisotropy even further.

DONORS VS. ACCEPTORS

It has long been presumed that the charge transfer process in
graphite intercalation compounds leads to ionic bonding between
intercalant and graphite layers, with the graphite layers acting as
amphoteric macro-ions with respect to donor or acceptor intercalants. (4) In the simplest possible case, such a bonding scheme
would lead to insulating behavior along the $\bar{c}$-axis. This model
actually works quite well for acceptor compounds but not for donor
compounds, as can be shown by comparing the actual layer spacings
with those predicted from the ionic radii of various intercalants.
Such a procedure strongly suggests that partial covalent bonding,
which is due to s-p hybridization, rules out the ionic model for
donor compounds.

The influence of hybridization on crystal structure is best
displayed by comparing the actual thickness of a carbon-interca-
lant-carbon sandwich (denoted by I_c) with the value predicted by a
purely ionic model which assumes no change with intercalation of
the orbital configuration of the carbon atoms. This model assumes
that the valence electron (or hole) of the intercalant becomes delocalized in the resonant π-bond system of graphite monolayers,
hence the ionic bonding is between intercalant ions and two-dimensional sheets of charge on the graphite layers. In this model I_c
is given by

$$I_c^2 = (d_{ion} + d_C)^2 - a^2$$

where d_{ion} is the usual Goldschmidt diameter of the singly-charged
ion, a is twice the in-plane C-C bond length and d_C is the effective diameter of $sp^2 p_z$ carbon covalent bonding in graphite. The
definition of d_C is somewhat arbitrary; in what follows, d_C is
taken as the average of the two interlayer C-C distances corresponding to staggered and eclipsed configurations along the $\bar{c}$-axis.
It is important to note that d_C is a valid concept only if the orbital configuration of carbon atoms is unchanged by intercalation.

The results of the analysis are shown in Table 1. The first
column gives the observed layer spacings for several compounds.

Table 1

	I_c (measured)	I_c (ionic model)	ratio	σ_a (300 K)	σ_c (300 K)
graphite	3.35	–	–	2.5×10^4	10
LiC_6	3.70	4.43	0.83	2.6×10^5	1.7×10^4
BaC_6	5.26	6.36	0.83	$> 10^5$	30
KC_8	5.40	5.92	0.91	1.0×10^5	3×10^3
RbC_8	5.65	6.23	0.91	1.0×10^5	–
CsC_8	5.94	6.66	0.89	1.0×10^5	–
C_6HNO_3	7.78	8.03	0.97	3.0×10^5	2.0
C_8AsF_5	8.11	8.12	1.00	4.0×10^5	0.25

These are equivalent to the $\bar{c}$-axis repeat distances in stage 1 compounds if screw axes on the metal sublattice are neglected. The second column is the distance that would obtain with spherical ions nested between carbon hexagons in the flanking graphite layers, assuming the carbon p_z orbitals are unchanged from their configuration in graphite. The third column gives the ratio of actual to predicted values. The last two columns give room temperature values of electrical conductivity parallel and normal to the layers (σ_a and σ_c respectively). The latter should be a rough indication of the net effects of $\bar{c}$-axis dilation (which would give low σ_c) and hybridization (tending to increase σ_c).

Our first observation is that the model is reasonably good for molecular compounds but not for metal compounds. Consider first the molecular compounds. We conclude that the dominant effect here is $\bar{c}$-axis dilation, and that any orbital mixing is too weak to seriously perturb the carbon p_z orbitals. This is confirmed by the small σ_c values for C_6HNO_3 and especially C_8AsF_5; examination of a large number of molecular compounds suggests that the correlation between large layer spacing and small σ_c is semiquantitative. This in turn implies that stage 1 acceptor compounds can be treated as a collection of noninteracting graphite monolayers with E_F reduced to account for charge transfer. (5) The fact that the ionic model works so well for acceptors is somewhat fortuitous and cannot be construed to mean that all intercalant molecules carry unit charge. The preponderance of experimental evidence is in the opposite direction, i.e. only a fraction of molecules are ionized (or each molecule carries small fractional charge). (6)

Turning to the metallic compounds, we note that the ionic model seriously underestimates the spacing between carbon layers flanking a metal layer and that the discrepancy is worse for MC_6 than for MC_8 compounds. A variety of results indicate that metal intercalants are indeed ionized, so d_{ion} should be a good estimate for the effective intercalant diameter. The inadequacy of the ionic model in predicting I_c for donor compounds must therefore be attributed to problems with d_c. The most natural explanation is that the orbital configuration of carbon is changed due to hybridization with metal valence s orbitals, resulting in partial covalent bond formation which pulls the carbon layers in relative to the ionic model result. Table 1 further implies that hybridization below E_F is stronger for the MC_6 compounds (Ba, Li) than for MC_8 (K, Rb, Cs), an observation which is borne out by band theory and charge density, as discussed below.

The tabulated σ_c values for donor compounds are much larger than for acceptor compounds. Large values of σ_c are clearly incompatible with a 2-D ionic model. The highest σ_c value among donor compounds is for LiC_6 (7), in which the room-temperature anisotropy is only ~15. This can be attributed entirely to structural considerations rather than s-p mixing whereas the high value of σ_c in KC_8 is due to metal-carbon hybridization at the Fermi surface. As discussed below, this curious result derives from the different Brillouin zone symmetries associated with the MC_6 and MC_8 in-plane metal lattices. The compound BaC_6 is potentially quite interesting; preliminary σ_a measurements (8) on very poor samples give values of $10^5 \Omega^{-1} cm^{-1}$, and Montelbano (9) obtained $\sigma_c = 30 \ \Omega^{-1} cm^{-1}$, so that anisotropy is quite high, at least 3000. The reason is twofold: the large layer spacing relative to LiC_6 decouples the π orbitals on adjacent carbon planes, and the MC_6 lattice does not lead to s-p mixing near E_F.

Several other observations point up the differences in anisotropy between donor and acceptor compounds. During intercalation reactions, the acceptor compounds are much more susceptible to exfoliation or self-cleaving, and are quite easy to cleave with tape (as is graphite) whereas LiC_6 is much harder to tape-cleave. Polarized reflectance data on MC_8 compounds again indicate strong metallic behavior in both directions. (10) The c-axis sound velocity of MC_8 compounds is substantially higher than in pure graphite. (11) Finally, in donor compounds the conduction electron spin resonance linewidth increases with intercalant mass and concentration, suggesting spin-orbit interaction as the dominant relaxation mechanism. (12) This in turn implies that the conduction electron wave function has substantial amplitude at the metal nucleus, i.e. the charge is not strongly localized along c. The situation in acceptors is reversed (13); linewidths increase as the intercalant concentration decreases.

The σ_c magnitudes in Table 1 suggest a band mechanism for

donors but hopping-type c-axis transport for acceptors. The available temperature-dependence data is not inconsistent with this suggestion. For pure graphite σ_c is on the borderline of band theory,
with $\lambda \sim 2$ layer spacings, and is still controversial.

STRUCTURAL EFFECTS

In the previous section we showed how donor and acceptor compounds may be differentiated by considering the chemical interactions between carbon and intercalant layers. Here we deal with the
further division of donor compounds by considering different epi-
taxial arrangements of metal layers on the graphite substrate. The
relatively small atomic radii of Li and Ba favor a hexagonal close-
packing where the M-M nearest neighbor distance is 4.24 A; the larger radii of heavy alkalis force a more dilute 2D structure with
M-M distance 4.90 A. As shown in Figure 1, this leads to different
relationships between the basal plane unit cells of the original
graphite (light lines) and the two classes of compound (heavy lines),
which in turn leads to different Brillouin zone symmetries hence
different band structures.

A zero-order estimate of the <u>differences</u> between MC_8 and MC_6
band structures can be made by considering the folding of the 2-D
graphite bands (14) into the respective zones dictated by the two
in-plane structures of Figure 1. For a single layer of pure graphite, shown in Figure 2a, the π and π^* bands are degenerate at K
and the Fermi energy is located at the degeneracy. In Figs. 2b
and 2c we show composite 2D band structures of graphite plus metal
before hybridization, for MC_8 and MC_6, respectively. Mapping of
graphite states at Γ is neglected, since these lie well above or
below the metal s-band (shown as a dashed curve). The separation
between the s-band minimum and E_F is roughly the difference between
metal ionization potential and graphite work function, of order
1-2 eV.

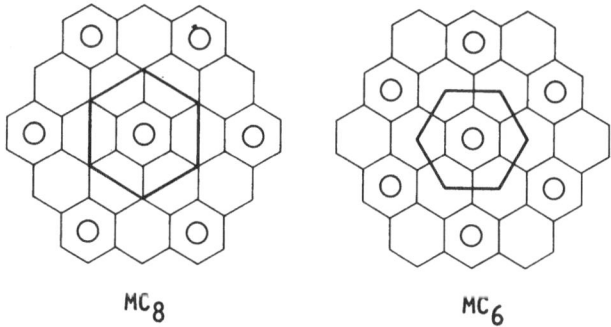

MC_8 MC_6

Fig. 1. Basal plane crystal structures of stage 1 donor compounds.

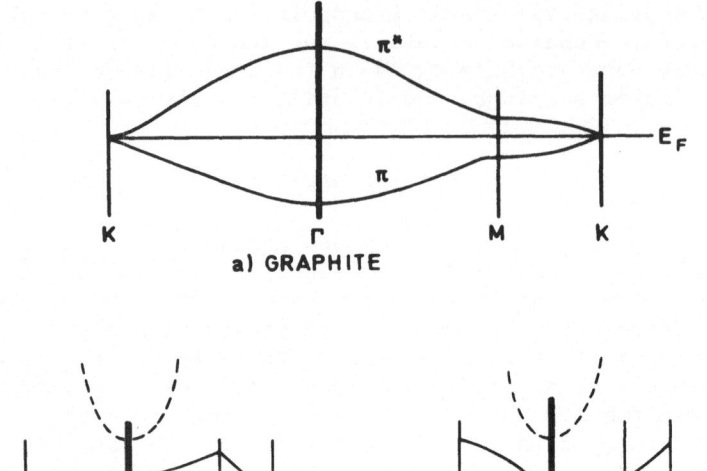

Fig. 2. Schematic band structures displaying the effects of basal
plane zone folding. a) pure graphite, for which the unit
cell is the light hexagon in Fig. 1. b) Folded into the
MC_8 zone, in which the M point appears at Γ. c) Folded
into the MC_6 zone; now the K point maps onto Γ. In b) and
c), the metal s-band is shown dashed, before turning on
s-π mixing.

In the MC_8 structure, Γ contains the original Γ and M points,
whereas in MC_6 Γ contains the K point. As a result, the energy
denominators which control the amount of hybridization will be
smaller for MC_8 than MC_6. So, all other things being equal, we
expect more hybridization in the former. Turning on the interaction
leads to two effects: a distortion of the E(k) curves where
E(s-band) and E(graphite) are close together, and an increase in E_F
to account for the extra electrons in the system. This simple construction demonstrates that the Fermi surface of MC_8 is likely to
contain low-anisotropy contributions from the metal s-band, whereas
MC_6 will be essentially graphite-like. This explains the high σ_c
value for KC_8 relative to BaC_6. In the LiC_6 case, the relatively
small increase in I_c (3.35 A to 3.70 A) indicates that π-overlap
from adjacent carbon layers is still important, and is responsible
for the high σ_c value in this material.

Detailed calculations of KC_8 (15) and LiC_6 (16) confirm the above ideas. The Fermi surface of KC_8 contains cylindrical and spherical pockets which are partially connected. The cylinders arise from newly-occupied 2-D graphite π-bands, since the 5.40 Å layer spacing is enough to decouple the π-orbitals on adjacent layers; and the spherical pocket is attributable to s-p mixing. The latter is responsible for the high σ_c value. In LiC_6 the spherical pocket is absent but the cylinders are warped and fluted by π-overlap, giving rise to large σ_c for a different reason. Although BaC_6 has not been studied theoretically, one expects neither spherical pockets, due to the MC_6 symmetry; nor warping of the cylinders, because the layer spacing is large.

Detailed comparison of the KC_8 and MC_6 calculations also explains why the ionic model is worse for the latter than the former (Table 1). The valence band of KC_8 is identical to the folded 2-D graphite result, indicating no s-p mixing below E_F. In LiC_6, on the other hand, the presence of Li removes many of the degeneracies produced by the folding, leading to gaps as large as 2 eV within the valence band. This can be taken as evidence for s-p mixing, and detailed charge density calculations give clear evidence for partial covalent bonding between Li and its 12 nearest neighbor carbons. The anisotropy in mechanical properties is thus expected to be lower in MC_6 than in MC_8.

PHASE TRANSITIONS

Two kinds of phase transition have been observed in graphite compounds. (2) The lower temperature transition mainly involves a change in c-axis stacking of intercalant layers, while the upper transition corresponds to the onset of in-plane disorder, sometimes identified as 2D melting. The systematics of the critical temperatures have not been studied in detail, nor have the structural rearrangements been completely worked out. However, a provocative degree of correlation already exists between phase transition behavior and anisotropy. For example, in the highly anisotropic HNO_3 compounds the disorder temperature is 250 K independent of stage, that is, the onset of disorder in one HNO_3 layer is unaffected by other layers. This is not unexpected. In potassium compounds, on the other hand, the upper transition (which presumably corresponds to in-plane disorder) moves to higher T as the K layer separation increases, implying a coupling between K layers separated by 1 or 2 C layers. The most surprising result is that the lower transition in K compounds, presumably the c-axis stacking rearrangement, occurs at 94 $\pm$2 K for stage 1, 2 and 3. Even though c-axis interactions are involved in defining the structures above and below the transition, the K layer separation does not affect the critical temperature. Finally, in AsF_5 compounds the onset of disorder occurs gradually over a broad temperature range, as might

be expected from its extreme anisotropy.   There is clearly much
fertile ground here for detailed studies, both structural and electronic.

CONCLUSION

The variable anisotropy of graphite intercalation compounds
can be understood as a combination of chemical and structural
effects.   Their relative contributions in a particular compound
can be exploited to provide prototype systems with essentially
constant in-plane properties while the c-axis properties vary by
five orders of magnitude.   Such systems should prove ideal for the
systematic exploration of 2-D phenomena.

REFERENCES

1.   J.E. Fischer and T.E. Thompson, Physics Today, July 1978,
     p. 36.
2.   J.E. Fischer, "Electronic Properties of Graphite Intercalation
     Compounds," chapter in "Intercalated Layer Materials," F. Lévy,
     ed. (D. Reidel, Holland) (in press).
3.   L. Peitronero, S. Strässler, H.R. Zeller and M.J. Rice,
     Phys. Rev. Lett. (to be published).
4.   A.R. Ubbelohde, Proc. Roy. Soc. A304, 25 (1968).
5.   F. Batallan, J. Bok, I. Rosenman and J. Melin, Phys. Rev.
     Lett. 41, 330.
6.   G. Dresselhaus and M.S. Dresselhaus, Mat. Sci. Eng. 31, 235
     (1977).
7.   C. Fuerst, preliminary data.
8.   M.J. Moran, unpublished.
9.   L.A. Girifalco and T. Montelbano, J. Mat. Sci. 11, 1036 (1976).
10. M. Zanini and J.E. Fischer, Mat. Sci. Eng. 31, 169 (1977).
11. W.D. Ellenson, D. Semmingson and J.E. Fischer, Mat. Sci. Eng.
     31, 137 (1977).
12. P. Delhaes, Mat. Sci. Eng. 31, 225 (1977).
13. R. Setton, Thesis, Orléans, France, 1971 (unpublished);
     S.K. Khanna, E.R. Falardeau, A.J. Heeger and J.E. Fischer,
     Solid State Comm. 25, 1059 (1978).
14. G.S. Painter and D.E. Ellis, Phys. Rev. B1, 4747 (1970).
15. T. Inoshita, K. Nakao and H. Kamimura, J. Phys. Soc. Japan,
     43, 1237 (1977).
16. N.A.W. Holzwarth, S. Rabii and L.A. Girifalco, Phys. Rev. B
     (in press).

HIGH ELECTRICAL CONDUCTIVITY GRAPHITE INTERCALATION COMPOUNDS

F. Lincoln Vogel and Claude Zeller

Department of Electrical Engineering and Science and
Laboratory for Research on the Structure of Matter
University of Pennsylvania, Philadelphia, Pa. 19104

Intercalation compounds of graphite of the acceptor type have
been demonstrated to have electrical conductivities superior to
those of the best known natural metals copper and silver. This
accomplishment is of considerable scientific and technological interest because: 1) these materials are highly anisotropic and provide a highly two dimensional medium for solid state experimentation and 2) one can forsee the development of conductors of electricity that are more efficient than those currently in use. Further, there is a promise of creating a material which has an electrical conductivity still higher than those experienced to date —
perhaps as high as ten times that of copper. This paper will suggest a path that leads in that direction.

The intrinsic electrical conductivity of graphite at room temperature whether measured on good natural crystals or highly ori-
ented pyrolytic graphite (HOPG) is $\sigma_a = 2.6 \times 10^4 \Omega^{-1}\text{cm}^{-1}$ parallel
to the hexagonal network and $\sigma_c = 0.1$ to $1.0\ \Omega^{-1}\text{cm}^{-1}$ normal to the
hexagonal layers.[1][2] The range in reported values of the latter
figure is probably due to a variation of planar defect concentrations in the samples measured.

The conductivity can be expressed as:

$$\sigma = ne\mu$$

where n = carrier density in No./cm^3
  e = electronic charge in coulombs
  μ = mobility in cm^2/volt-sec

From Hall and magnetoresistance measurements these parameters have been determined to be, for pure graphite:

$$n = 2 \times 10^{19}/cm^3$$

$$\mu = 10^4 cm^2/volt\text{-}sec$$

For comparison with this in-plane conductivity of graphite, the electrical conductivity of copper is 23 times greater at $\sigma_{Cu} = 5.9 \times 10^5 \Omega^{-1}cm^{-1}$, copper having about one electron per atom and an electron mobility of 50 $cm^2/volt\text{-}sec$.

With respect to the electronic structure of graphite intercalation compounds, two types of species have been studied. The alkali metals, when inserted in graphite, behave as donors and increase the in-plane conductivity by a factor of six to eight[3]. On the other hand, a large number of mineral acids and Lewis acid halides act as acceptors when inserted in graphite and increase the conductivity by a factor of fifteen or more[4].

The highest values of electrical conductivity reported in the literature so far have resulted from the intercalation of graphite with SbF_5[5] and AsF_5[6]. In one set of experiments graphite powder intercalated with SbF_5 at a concentration to give a stage 3 compound was loaded into a copper ampoule 6.4 mm in O.D. and swaged down to a 1 mm O.D. wire. By subtracting out the conductivity of the copper sheath, the SbF_5-graphite was determined to have a conductivity of $1 \times 10^6 \Omega^{-1}cm^{-1}$. Also, a single measurement on an HOPG crystal intercalated to stage 6 exhibited a conductivity of $5.9 \times 10^5 \Omega^{-1}cm^{-1}$ — exactly the same as copper[7]. Another series experiments with AsF_5-intercalated HOPG yielded a maximum conductivity at stage 2 of $6.3 \times 10^5 \Omega^{-1}cm^{-1}$ — slightly higher than the conductivity of silver. While some problem with reproducibility exists there seems little doubt that experiments are accurate in depicting a higher electrical conductivity in these materials than is available from the best naturally occurring metals silver and copper.

As stated earlier in this paper, graphite exhibits a strong anisotropy in its intrinsic a and c axis conductivity — a factor of about 10^4. The relatively low values of conductivity and strength in the c direction are attributable to the weak p_z orbital overlap between the layers. When acid acceptor molecules are inserted between the graphite layers, because of their lack of a strong s character in the bonding orbitals, s-p hybridization does not occur as it does with donor intercalation. So, the graphite p_z orbitals are simply separated further, causing a reduction of strength and conductivity in the c direction. Consequently, the anisotropy of conductivity of the AsF_5 and SbF_5 intercalation compounds is increased to the enormous value of 10^6. An anisotropy of this magnitude makes impossible the use of conventional four point bridge samples of normal dimensions for either Hall or conductivity measurements since the basic assumption of uniform current flow between voltage probes cannot be satisfied. Thus, it is necessary to rely on a contactless induction method for determination of conductivity (which can be done most satisfactorily) and a contactless magnetoresistance method for determination of carrier mobility (which is less satisfactory). From magnetoresistance measurements performed in this way it has been determined that the mobility is reduced by a factor of five or six over the intrinsic graphite value and the carrier concentration is increased by a factor of about 45. Since the concentration of intercalant in the compound is in the mid $10^{21}/cm^3$ range, the ionization factor, f, is between 10 and 20%. To summarize then, the intercalation of graphite with strong acid fluorides which results in an electrical conductivity equal to or slightly greater than that of copper and silver is achieved in materials having an intercalant ionization factor of 15% and a carrier mobility of 2000 cm^2/volt-sec.

How can both of these variables be increased in view of their interdependence? The mobility can be increased by attaining high chemical and physical perfection in the host graphite to minimize defect scattering, and by maintaining stoichiometric composition with intercalants that do not produce additional scattering. Carrier density can be increased by intercalation with stronger acids to the point where scattering is adversely affected.

ACKNOWLEDGEMENT

The authors wish to thank L.A. Pendrys and T.C. Wu for providing some of the experimental data used, and Prof. Ian Spain for very helpful discussions.

REFERENCES

(1) I.L. Spain, Chem. and Phys. Carbon 8, 1 (1973).
(2) C. Zeller, G.M.T. Foley and F.L. Vogel, J. Mat. Sci. 13, 1114 (1978).
(3) A.R. Ubbelohde, L.C.F. Blackman and J.F. Mathews, Nature 183, 454 (1959).
(4) A.R. Ubbelohde, Proc. Roy. Soc. A309, 297 (1969).
(5) F.L. Vogel, J. Mat. Sci. 12, 982 (1977).
(6) G.M.T. Foley, C. Zeller, E.R. Falardeau and F.L. Vogel, Sol. State Comm. 24, 371 (1977).
(7) F.L. Vogel, J. Gan and T.C. Wu, Fifth London International Conference on Carbon and Graphite. September 1978.

SALTS OF AROMATIC CATIONS AND RELATED GRAPHITE SALTS

N. Bartlett, R. N. Biagioni, G. McCarron, B. McQuillan, and F. Tanzella

Department of Chemistry, University of California
Materials and Molecular Research Division
Lawrence Berkeley Laboratory
Berkeley, California 94720

INTRODUCTION

In 1974 Bartlett & Richardson[1] used salts of the O_2^+ ion to prepare the first salt of the hexafluorobenzene cation:

$$C_6F_6 + O_2^+AsF_6 \longrightarrow C_6F_6^+AsF_6^- + O_2$$

The salt $C_6F_6^+AsF_6^-$ is a simple paramagnet and the crystal structure is related to that of cesium chloride with each discrete ion rhombohedrally coordinated by eight ions of the other kind.

This synthesis opened up the possibility of using the O_2^+ salts (& other powerful oxidizers) in the preparation of salts containing other delocalized-electron cations. We have extended our studies to (1) other perfluoroaromatics (2) hydrocarbon aromatics (3) graphite and (4) boron nitride (layer form).

DISCUSSION

Aromatic Cation Salts

The synthesis[2] of salts derived from perfluorotoluene, perfluoropyridine & perfluorotriazine, followed approximately the same procedure as for $C_6F_6^+$. Like that salt, all of these salts contain discrete cations. Moreover all are unstable at ordinary temperatures.

We were particularly interested in the preparation of salts of

benzene itself. The interaction of benzene with either $O_2^+AsF_6^-$
or $C_6F_6^+AsF_6^-$ in SO_2ClF as solvent, produces a blue-black solid
which is admixed with a colorless crystalline solid. This same
colorless solid is produced in high yield by the interaction of
arsenic pentafluoride with benzene in hydrogen fluoride or SO_2ClF.
It has the composition $C_6H_5AsF_4$ and may[3] be the salt $(C_6H_5)_2AsF_2^+$-
AsF_6^-:

$$C_6H_6 + AsF_5 \longrightarrow C_6H_5AsF_4 + HF$$

The blue-black diamagnetic solid produced in the O_2^+ and $C_6F_6^+$
oxidations of benzene is evidently polymeric, being of low solubility and low volatility. The solid appears to be comparable to
powdered graphite in its conduction of electricity. It seems
unlikely that this solid could be the sought after salt $C_6H_6^+AsF_6$
and is more likely to be a polyphenyl or a salt of such a polymer.
It is noteworthy that the benzene cation is a poorly described
species & is evidently difficult to preserve. This is not the case
for the higher molecular weight aromatics.[4]

Extension of the synthetic approach used for the synthesis of
$C_6F_6^+$ salts to the higher molecular weight aromatics was of considerable interest because it appeared possible, at least for the
monopositive large-ring-system cations, that some overlap of the
ring systems might occur. Such overlap could result in extension
of the electron delocalization to the crystal as a whole. The
highly stable anions such as AsF_6^-, PtF_6^-, etc. have an effective
thickness of ~ 4.7 Å , and since aromatic molecules have an effective thickness comparable with graphite (of a little less than
3.4 Å) ring overlap appears to be feasible.

There are practical problems in the syntheses. The higher
molecular weight aromatics are usually not very soluble in solvents which are compatible with our oxidizers, & if the aromatic is
not entirely in solution, unreacted starting material becomes coated
by the salt formed in the oxidation. Moreover the dioxygenyl salts
are of low solubility in the useable solvents. It is usually necessary for the O_2^+ salts to be present in stoichiometric quantity
since they cannot be easily separated from the desired product. A
strategy for removing an excess of the dioxygenyl salt, which can
occasionally be employed, is to convert it to the $C_6F_6^+$ salt and
then decompose the latter at room temperature (all of the decomposition products are volatile).

$$2 C_6F_6^+AsF_6^- \longrightarrow C_6F_8 + C_6F_6 + 2 AsF_5$$

Happily we have found that for the hydrocarbon aromatics and
the higher molecular weight perfluoroaromatics the lower ionization
potentials permit the $C_6F_6^+$ salts to be used as the oxidizers. Not
only does the low thermal stability mean that excess reagent is

readily removed, but these salts are more soluble than their O_2^+ analogues. The perfluoronaphthalene salt[2] can be prepared in this way

$$C_6F_6^+AsF_6^- + C_{10}F_8 \longrightarrow C_{10}F_8^+AsF_6^- + C_6F_6$$

The perfluoronaphthalene salts, like the $C_6F_6^+$ salts are simple paramagnets and the crystallographic & e.s.r. data demonstrate that the cations are discrete.

Oxidation of the large hydrocarbon aromatics has yielded products which are more likely to involve overlapping cations. Thus both anthracene and coronene interact with $O_2^+AsF_6^-$ to yield blue-black solids which conduct electricity. Structural information is however presently lacking.

Graphite and Boron-Nitride Salts

Most of our work has involved oxidation of graphite and the synthetic and structural work have had, as their main focus, the settling of the question of the validity of the salt formulation $C_x^+X^-$ for the strong-acid-anion intercalates. Thus while certain of the intercalates of graphite with AsF_5 have recently been described[5] as having electrical conductivity superior to that of graphite, information on the nature of the intercalated AsF_5 has remained meager. It has even been recently suggested that the intercalated species were largely molecular AsF_5. This has been shown to be contrary to the behavior one anticipates for the strong oxidizer AsF_5.[7] Moreover, our work with the oxidizer SO_3F, in which both $C_{12}SO_3F$ and $(BN)_4SO_3F$ had been achieved, persuaded us that salt formulations were appropriate for the fluorosulfates (i.e., $C_{12}^+SO_3F^-$ and $(BN)_4^+SO_3F^-$) and by extension would also be appropriate for intercalates of the higher ionization potential anions AsF_6^-, PtF_6^-, IrF_6^- and OsF_6^-. It seemed probable that the AsF_5 oxidation of graphite produced at least some AsF_6^-.

The synthesis of the hexafluorometallates ($MF_6^- = Os$, Ir, Pt) was undertaken for two reasons. The first was to obtain a series of salts containing nearly isodimensional intercalated species, MF_6, of graded electron affinity (PtF_6 having the highest known molecular electron affinity and IrF_6 and OsF_6 having electron affinities successively ~ 1 eV less). The second was to use magnetic susceptibility measurements, as an adjunct to x-ray diffraction studies, to settle the constitution of the salts.

Treatment of pyrolytic graphite with arsenic pentafluoride yielded material of approximate composition C_8AsF_5. This is a blue solid and corresponds to the material first described by Selig and his co-workers.[8]

Single crystals of graphite (previously characterized by x-ray precession photography) were similarly treated with AsF_5 and over-pressure of two or three atmospheres of the gas was maintained over the crystal by enclosing the crystal and gas in a quartz capillary. By slow intercalation much order can be maintained and there is a good possibility that adequate intensity data can eventually be collected to show the disposition of the arsenic fluoride species with respect to the graphite lattice. It appears that the unit cell of the first stage graphite/AsF_5 material is pseudohexagonal (possibly monoclinic) with "a_0" = 4.92, "c_0" = 16.20 Å, V = 340.03 Å. The observed 00ℓ reflections have $\ell = 2n$ and it seems probable that the gallery height in the intercalate (i.e., the spacing between adjacent carbon sheets) is 8.10 Å. This is appro-priate spacing for an octahedral AsF_6^- species fitted between graphite sheets of thickness $\sim$ 3.4 Å.

A parallel study of graphite oxidized by $O_2^+AsF_6^-$ was also made. The oxidation:

$$graphite + O_2^+AsF_6^- \longrightarrow graphite^+AsF_6^- + O_2 \uparrow$$

was carried out, at $\sim$ -20°, using SO_2ClF as a dispersive agent and solvent. Gravimetric work, using pyrolytic graphite, has esta-blished that a first stage material of limiting composition C_8AsF_6 is generated when an excess of O_2AsF_6 is used. Single crys-tals of graphite were similarly converted to a first stage mater-ial and proved to be hexagonal with a_0 = 4.92 and c_0 = 8.10, V = 170 Å^3.

The a_0 value (4.92 Å) for the unit cell of the $C_8^+AsF_6^-$ salt is twice that of graphite and this indicates that the carbon net-work has essentially the same dimensions as in the unoxidized car-bon. It is reasonable to assume that the AsF_6^- species are centered, with respect to the graphite network. If, as is pre-sently indicated, c_0 = 8.10 Å and not a simple multiple of it, then this implies an eclipsing of the graphite sheets.

The unit cell of C_8AsF_6 also implies close packing of the AsF_6^- species. The unit cell volume of 170 Å^3 can be divided into a car-bon network component and an AsF_6^- component. Since the unit cell contains 8 carbon atoms and since, from the graphite unit cell, the volume occupied by 4 carbon atoms of graphite is 35 Å^3, we can con-clude that the volume available for the AsF_6^- is $\sim$ 100 Å^3. But this is the effective packing volume for hexafluoride molecules. The C_8AsF_6 stoichiometry probably therefore represents a packing limit rather than a graphite oxidation limit.

The similarity of the C_8AsF_6 and C_8AsF_5 x-ray data suggested that the gallery dimensions were being determined by a common species: AsF_6^-. X-ray K-shell absorption-edge spectra for the two materials when compared with like spectra of AsF_6^- salts and a spectrum of AsF_3 confirm this view and show that the AsF_5 must oxidize the graphite according to the equation:

$$3 \; AsF_5 + 2 \; e^- \longrightarrow 2 \; AsF_6^- + AsF_3 \quad .$$

The electron affinities of the hexafluorides of osmium, iridium and platinum are estimated to be: OsF_6, 165; IrF_6, 190 and PtF_6, 215 kcal/mole[-1], these being based, for the last value on calorimetry and a lattice energy estimate for $O_2^+PtF_6^-$ and, for the others, on interpolation between the PtF_6 value and the more certain value for $E(WF_6) = 104$ kcal/mole[-1] from the work of Cooper et al.[9]

Graphite incorporates each of OsF_6 and IrF_6 at room temperature to form first stage materials (blue) of composition C_8MF_6. Both are stable under vacuum to at least 250°. At $\sim$ 500° C_8IrF_6 ignites spontaneously: $C_8IrF_6 \rightarrow (CF)_n + Ir_{(m)} + CF_4 + C_2F_6$, etc.

Single crystal work has shown that C_8OsF_6 can be made in a highly ordered state and the unit cell is very like that given for C_8AsF_6. Again the symmetry appears to be hexagonal with $a_0 = 4.92$ Å and $c_0 = 8.1$ Å. So far, no single crystal data have been obtained for the iridium and platinum cases and attempts to produce them by introducing the vapors into single crystals of graphite have resulted in complete loss of _ab_ plane order.

The magnetic data for the osmium and iridium salts have been obtained and show that the metals are quinquevalent. Thus C_8OsF_6 magnetic susceptibility obeys a Curie-Weiss relationship and has a magnetic moment similar to that of the simple cubic salt $SF_3^+OsF_6^-$. Similarly, C_8IrF_6 shows a temperature independent paramagnetism ($\chi_M \approx 7.5 \times 10^{-4}$ c.g.s units above 17 K) akin to that of the $A^+IrF_6^-$ (A = K, Cs, NO) salts.

These magnetic results, combined with the x-ray diffraction findings are fully in accord with the formulation $C_8^+MF_6^-$ for the Os and Ir cases. Our preliminary studies for the graphite/PtF_6 indicate that the intercalated species is PtF_6^{2-} and the final first stage material seems to be $C_{12}PtF_6$. The double negative PtF_6^{2-}, in $C_{12}PtF_6$, results in a higher electron withdrawal from the graphite than in $C_8^+MF_6^-$ and the incomplete filling of the galleries is presumably a consequence of the high carbon charge as well as the greater repulsive interaction of the guests.

Recently we have carried out a heat capacity (C) study[10] of the first-stage graphite fluorosulfate[7] (of composition $C_{8.2}SO_3F$), over the temperature range $0.3 \rightarrow 1.5$°K. The plot of C/T versus T^2 is nearly linear and indicates metallic behavior. The intercept, at 0°K, has a value of 15.0 $\mu J/_g$-K^2. On the reasonable basis that

the carriers are associated with the carbon network this gives
γ = 30.1 μJ/g carbon-K^2. Natural Madagascar graphite has a value
γ = 1.15 μJ/$_g$-K^2. Thus the number of carriers in the fluorosulfate
appears to be an order of magnitude greater than in graphite itself. This is not inconsistent with the formulation $C_{8.2}^+SO_3F^-$.
Similar studies on $(BN)_4SO_3F$ are in progress.

EXPERIMENTAL

Syntheses

All reactants & products are moisture sensitive, hence all
preparations & manipulations were carried out with strict exclusion of water. The O_2^+ salts and $(SO_3F)_2$, used in the oxidations,
were prepared as previously described.[11,12] The preparations were
carried out in Teflon FEP or quartz vessels on a vacuum line and
solids were manipulated in a Vacuum Atmospheres Corporation DRILAB.

X-Ray Diffraction

Samples for x-ray diffraction studies were loaded into quartz
capillaries in a DRILAB. Single crystals of graphite were obtained
from a graphite containing marble by dissolving the latter in conc.
HCl. Selected single crystals were subjected to precession photography before and after intercalation. To allow for the expansion,
accompanying intercalation, each crystals was suspended in quartz
wool within its capillary.

Heat Capacity Measurement for $C_{8.2}SO_3F$

Graphite (7.67 g), contained in a Pyrex bulb, was allowed to
interact with $(SO_3F)_2$ (8.2 g) at room temperature for $\sim$ 30 min.
Unconsumed $(SO_3F)_2$ was removed under vacuum until the solid attained
constant weight at room temperature. The net weight increase of
7.71 g signifies a stoichiometry $C_{8.20}SO_3F$ for the solid & this
was conformed by carbon analysis (Found: 49.94%. $C_{8.2}SO_3F$ requires
49.87%). An x-ray powder photograph conformed the C axis spacing
of 7.66 Å appropriate for the first stage compound.[7]

The solid (9.84 g) was mixed with dry Kel-F oil (5.24 g) and
was loaded into a copper can provided with a tight fitting lid.
This sample was used for the heat capacity measurements[10] over the
temperature range 0.3 to 1.5°K.

ACKNOWLEDGMENTS

This work was supported by the Division of Chemical Sciences, Office of Basic Energy Sciences, U.S. Department of Energy and an I.B.M. fellowship to R. N. Biagioni.

REFERENCES

1. T. J. Richardson and N. Bartlett, J. Chem. Soc., Chem. Comm., 427, 1974.

2. N. Bartlett et. al. Submitted to Angew. Chem.

3. The compound $C_6H_5AsF_4$ is reported to be liquid at room temperature, see: W. C. Smith, J. Amer. Chem. Soc., 82, 6176, 1960.

4. For example: I. C. Lewis and L. S. Singer, J. Chem. Physics, 43(8), 2712, 1965.

5. E. R. Falardeau, G. M. T. Foley, C. Zeller and F. L. Vogel, J. C. S. Chem. Communs., 389, 1977.

6. J. E. Fisher, Electronic Properties of Graphite Intercalation Compounds, Physics and Chemistry of Materials with Layered Structures, F. Levy, D. Reidel, Eds., Dordrecht Holland 1977 (in press). L. B. Ebert and H. Selig in Abstracts Franco-American Conference on Intercalation Compounds of Graphite, May 23-27, La Napoule, France.

7. N. Bartlett, R. N. Biagioni, B. McQuillan, A. S. Robertson and A. C. Thompson, J. C. S. Chem. Communs., 200, 1978.

8. J. Binenboym, H. Selig, S. Sarig, J. Inorg. Nucl. Chem., 38, 2313, 1976.

9. C. D. Cooper, R. N. Compton and P. W. Reinhardt, Abstracts of Papers for the IXth International Conference on the Physics of Electronic and Atomic Collisions, J. S. Risley, and R. Geballe, Eds., Vol. 2, p. 922, Seattle, Wash., 24-30 July 1975.

10. R. Biagioni, J. Ho, J. Boyer, N. Phillips, and N. Bartlett, to be published.

11. J. Shamir and J. Binenboym, Inorg. Chim. Acta, 2, 37, 1968; D. E. McKee and N. Bartlett, Inorg. Chem., 12, 2738, 1973.

12. M. Wechsberg, P. A. Bulliner, F. O. Sladkey, R. Mews and N. Bartlett, Inorg. Chem., 11, 3063, 1972.

THE CHEMISTRY AND PHYSICS OF POLYTHIAZYL,

$(SN)_x$, AND THE POLYTHIAZYL HALIDES

G. B. Street and W. D. Gill

IBM Research Laboratory

San Jose, California 95193

ABSTRACT: The chemistry and physics of $(SN)_x$ and its halogen derivatives are reviewed and a model consistent with present electronic and structural data is presented for $(SNBr_{0.4})_x$ involving polythiazyl cationic chains with Br_3^- ions and bromine inserted between them.

INTRODUCTION

In this paper we will summarize the results which have been reported for pristine $(SN)_x$ since the publication of previous reviews[1-4] but the bulk of the paper will be devoted to a discussion of work on the polythiazyl halides. At the time of the discovery of the interesting electrical properties of $(SN)_x$ extensive efforts were made to prepare derivatives and analogs of this unusual polymer.[3] All efforts to prepare analogs by replacing sulfur with selenium have so far been unsuccessful.[5] However in 1976 Bernard et al.[6] reported the formation of an intercalation compound of $(SN)_x$ and bromine. In a general survey of the interaction of $(SN)_x$ with donors and acceptors we had observed that halogens, in particular bromine, were effective in increasing the electrical conductivity of $(SN)_x$. The influence of halogens on the conductivity of $(SN)_x$ was first mentioned by Yoffe[2] though no details were reported. Since 1976 the polythiazyl halides have been studied extensively by several groups and some consensus on the interpretation of the data is beginning to emerge.

SUMMARY OF RECENT WORK ON $(SN)_x$

A large number of experiments lead to the conclusion that polymeric $(SN)_x$ is an anisotropic three-dimensional semi-metal which exhibits bulk type II superconductivity caused by the usual BCS mechanism.[4] The transition temperature ~0.26°K is strongly dependent on the polymer chain perfection and other unknown types of disorder. The nature of the structural disorder in $(SN)_x$ is still under investigation.[7,8] $(SN)_x$ and its bromine derivative remain unique examples of superconducting polymers. As we shall discuss in more detail later in this section a partial Meissner effect has now been observed for $(SN)_x$ clarifying the intrinsic nature of the superconductivity.[9] Thus the hydrogen impurity in the $(SN)_x$ crystals would not appear to be important in determining the superconductivity. Although they were not able to identify the source of this hydrogen specifically, Smith et al.[10] have shown that the impurity has a molecular weight of 220±12 and is formed by reaction with water either during or after the polymerization of S_2N_2.

The structure of $(SN)_x$ has been redetermined using neutron diffraction.[11] The results are consistent with the earlier X-ray structure[12] except for some measure of disagreement on the population of defect sites. Love et al.[13] have reported two new forms of $(SN)_x$ distinguished partly by their modified physical properties, and partly by the color of their two unidentified precursors isolated from the products of vapor phase decomposition of S_4N_4. We have observed similar reddish-brown crystals which polymerize in part to give $(SN)_x$ if the pressure of nitrogen in the pyrolysis apparatus is allowed to rise.[14] In our experience the color of the crystals was shown by mass spectrometric and melting point data to be due to contamination with S_4N_2. Love et al.[15] have prepared crystals of $(SN)_x$ by the low temperature photopolymerization of solution grown S_2N_2. Despite the low growth temperature their electrical properties are similar to those grown at room temperature. Smith[16] has reexamined the vapor species observed on pyrolysis of S_4N_4 vapor over silver and quartz wool and has confirmed that silver wool is a more effective catalyst if S_2N_2 is the desired polymerization product.

Several new band structure calculations have been published for $(SN)_x$,[17-20] for S_2N_2,[17,20] and for $(SCH)_x$.[18] Most calculations have been in good agreement with XPS or UPS data and with experimental values of the density of states at E_f and the effective mass. Batra[21] has calculated the XPS of $(SN)_x$ including the photoionization cross-section and finds much improved agreement with experiment. Angular resolved photoemission experiments have been reported by Ley et al.[22] and by Koch and Grobman.[23] The dispersion of the bands parallel to b measured by Ley et al. was slightly less than calculated using the OPW method.[24]

A calculation of the plasmon excitations in $(SN)_x$ has been made by Ruvalds et al.[25] They have shown that for an electron gas with $m_\perp/m_\parallel = 1.9$ very good agreement is obtained with the data of Chen et al.[26] Stolz et al.[27] measured the plasmon dispersion in an electron loss experiment for wavevector k parallel and perpendicular to b. The reflectivity was calculated and is in good agreement with the optical reflectivity. A discontinuity in the dispersion at $k = 0.4 A^{-1}$ was attributed to possible interference with free-electron-like single particle excitations.

Wendel[28] has used a simplified 4 parameter valence force model to describe the features of the phonon data. This model shows that interchain sulfur-nitrogen forces are important, and that internal strains play an essential role. The linear compressibilities of $(SN)_x$ have been measured by Clarke[29] using X-ray diffraction techniques for hydrostatic pressure up to 20 kbar. He finds $k_a \approx k_c \approx 2.5 k_b$ with the magnitude of k_b comparable to ionic solids at 1 bar and to covalent solids at 20 kbar. All lattice spacings contract under pressure and the intralayer compressibility $k_{[201]}$ is only slightly less than k_a and k_c.

The temperature dependence of the reflectivity of $(SN)_x$ has been reported by Kaneto et al.[30] On cooling a slight blue shift of the plasma edge was observed which is attributed to a change of ω_p. This is opposite to the results of Grant et al.[31] who report a red shift on cooling which they attribute to volume effects. Grant and Welber[32] have also reported a weak red shift of the Drude edge under hydrostatic pressure, consistent with attributing the temperature shift to a volume effect.

The pressure dependence of normal conductivity was investigated in detail for pressures up to 40 kbar by Friend et al.[33] The data is in good agreement with earlier results (to 10 kbar) of Gill et al.[34] The T^2 dependence of resistivity[35] was followed with pressure. Between 3 and 8 kbar two regimes of T^2 dependence were observed both saturating at room temperature. At higher pressure electron-phonon scattering dominates the resistivity. The effects of uniaxial stress on conductivity were reported by Chiang et al.[36] who found anisotropy of the measured stress coefficients of ~100.

Conductivity and susceptibility measurements were reported by Kaneto et al.[37] for $(SN)_x$ crystals before and after γ-irradiation. A conductivity minimum in the γ-irradiated samples was attributed to the presence of magnetic impurities (Kondo effect). On irradiation the susceptibility changed from temperature independent Pauli paramagnetism to Curie-Weiss behavior indicating the presence of ~2.5×10^{20} cm^{-3} paramagnetic centers. Transverse magnetoresistance (i//b) measurements between 1.8 and 296K have been investigated by Kaneto et al.[38] Negative magnetoresistance observed at low magnetic field and low temperature (T<77K) is attributed to spin-dependent scattering at chain breaks or defects with localized spins. The magnetoresistance anisotropy in the a-c plane was measured and found to be in reasonable agreement with the predictions from band structure anisotropy by Beyer et al.[39]

The most notable new result related to the superconducting properties of (SN)$_x$ was the observation of a partial Meissner effect by Dee et al.[9] demonstrating that superconductivity is without question a bulk property. This result indicates that (SN)$_x$ is a coupled filamentary type II superconductor. The observed transition was broad and considerable flux trapping was present which contributed to failure of an earlier attempt by Dee et al.[40] to observe the effect.

The pressure dependence of T_c up to 17 kbar has been measured by Müller et al.[41] with the interesting observation that above 8 kbar T_c suddenly decreases to ~100mK and then decreases with increasing pressure. Before 8 kbar, T_c increased linearly with pressure in reasonable agreement with previous observations of Gill et al.[34] A sudden decrease in T_c near 9 kbar has also been reported by Bickford et al.[42] Müller et al. attributed this sudden change in T_c to presence of a new metallic phase. Bickford et al. also studied the pressure dependence of the critical magnetic field. Anisotropy of the perpendicular critical field appears to come from intrinsic band structure anisotropy. The superconducting properties are concluded to be those of a collection of weakly coupled fibers.

Tunnel junctions on (SN)$_x$ crystals have been investigated by Chaikin et al.[43] Although good tunnel junctions were formed no superconducting gap in the (SN)$_x$ could be detected down to 60mK. Observation of a broad zero-bias anomaly suggested that the surface of the (SN)$_x$ crystal consists of small non-superconducting particles similar to (SN)$_x$ films.

The electrical properties of (SN)$_x$ films have been extensively studied. Films deposited at low substrate temperature (T_s=100K) reported by Yoshino et al.[44] have a conducting activation energy of 0.1eV. Conductivity, thermoelectric power, Hall effect and magnetoresistance have been measured by Beyer et al.[45,46] for films deposited on substrates at temperatures T_s between -10C and 50C (see Fig. 1). With increasing T_s there is a dramatic increase in conductivity with nearly metallic behavior achieved at the highest T_s. No superconductivity was observed in even the most metallic of these films. A model of a heterogeneous

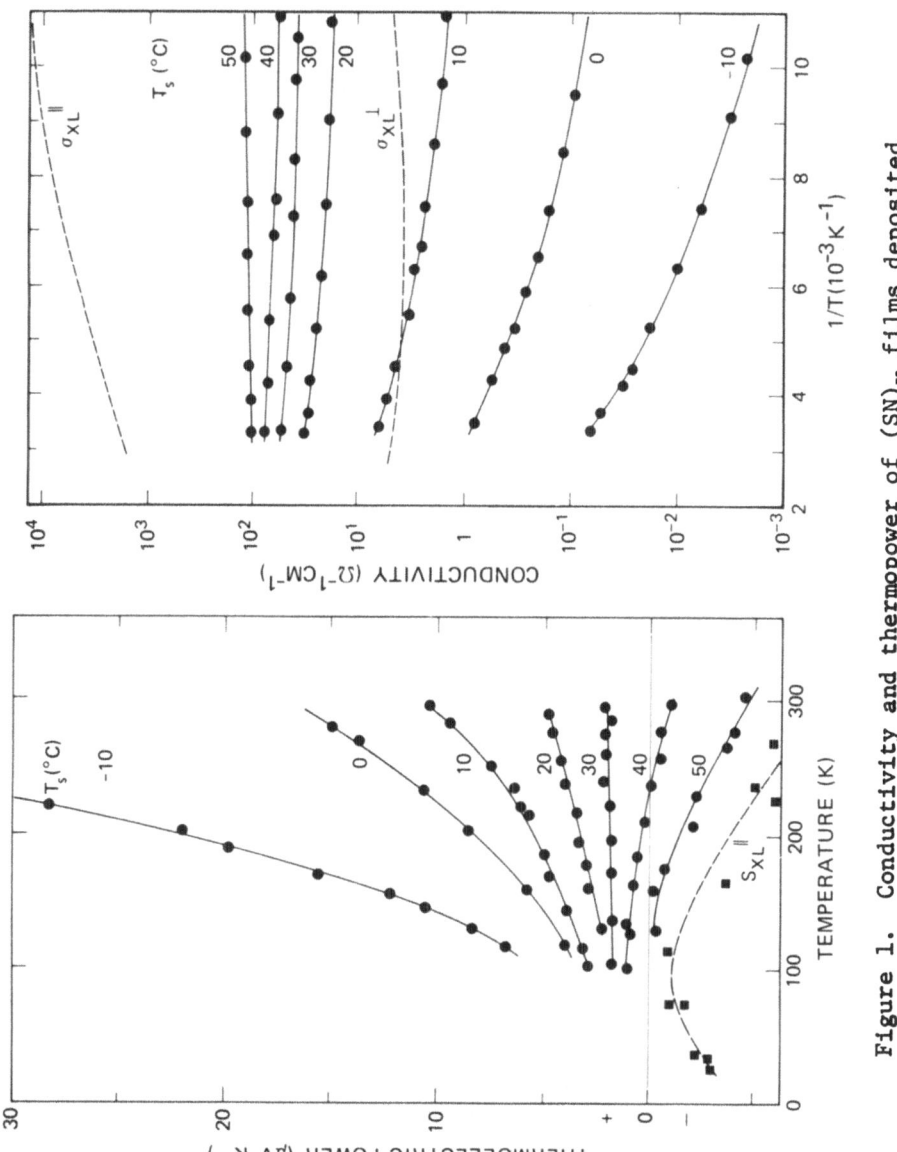

Figure 1. Conductivity and thermopower of $(SN)_x$ films deposited at various substrate temperatures T_s.

conductor explains these results but suggests the existence of
a gap in the density of states correlated with $(SN)_x$ chain length.
Such a gap might also explain the absence of superconductivity
in $(SN)_x$ films. $(SN)_x$ films have interesting properties for
electrode materials. Scranton et al.[47,48] have shown that $(SN)_x$
is more electronegative than gold and can form Schottky barriers
which are significantly larger than those produced by gold on
n-type semiconductors. Cohen and Harris[49] have examined solar
cells consisting of films of $(SN)_x$ on GaAs.

POLYTHIAZYL HALIDES

The $(SN)_x$ chain structure and crystal structure[12] are shown
in Fig. 2. The unit cell of $(SN)_x$ contains two, almost flat,
translationally inequivalent, centrosymmetrically related $(SN)_x$
chains. These two inequivalent chains (type 1 and type 2)
alternate along the c-axis whereas chains of the same type are
adjacent along the "a" axis. The chains all lie in the $10\overline{2}$ plane
where type 1 and type 2 alternate. The ready cleavage which
takes place at these $10\overline{2}$ planes suggests the possibility of
intercalating various molecules between them. The well known
adducts of S_2N_2[50] and S_4N_4[51] with Lewis acids such as $SbCl_5$ and
BCl_3 have led to unsuccessful attempts to intercalate these
species into $(SN)_x$. However during a survey of a large number
intercalates, Street and Gill noted that the halogens Br_2, I_2,
ICl and IBr all increased the conductivity of films or crystals
of $(SN)_x$. Bromine is approximately five times more effective
than the other halogens, it increases the conductivity of $(SN)_x$
by a factor of ten.[52] Accordingly the bromine derivatives have
been the most extensively studied by ourselves as well as
others.[53-55]

FORMATION OF POLYTHIAZYL HALIDES

When crystals of $(SN)_x$ are exposed to bromine vapor the
crystals change color from gold to blue-black and expand in
directions perpendicular to the chain axis. In order to avoid
side products it is important to dry the bromine over $CaCl_2$ and
remove the HBr by treating with anhydrous copper sulfate before
use. Exposure to the vapor pressure of bromine at room
temperature leads to a composition $(SNBr_{0.55})_x$ which on pumping
in vacuum (10^{-5} Torr) at room temperature for one hour gives a
final composition of $(SNBr_{0.4})_x$. At this composition the crystal
has expanded by ~50% in volume perpendicular to the chain axis.
This expansion is accompanied by some exfoliation and microscopic
cracking but the external habit and fibrous nature of the crystals
are retained. The flotation density increases from $2.32 g/cm^3$ to
$2.65 g/cm^3$. Though this composition is not unique it does appear

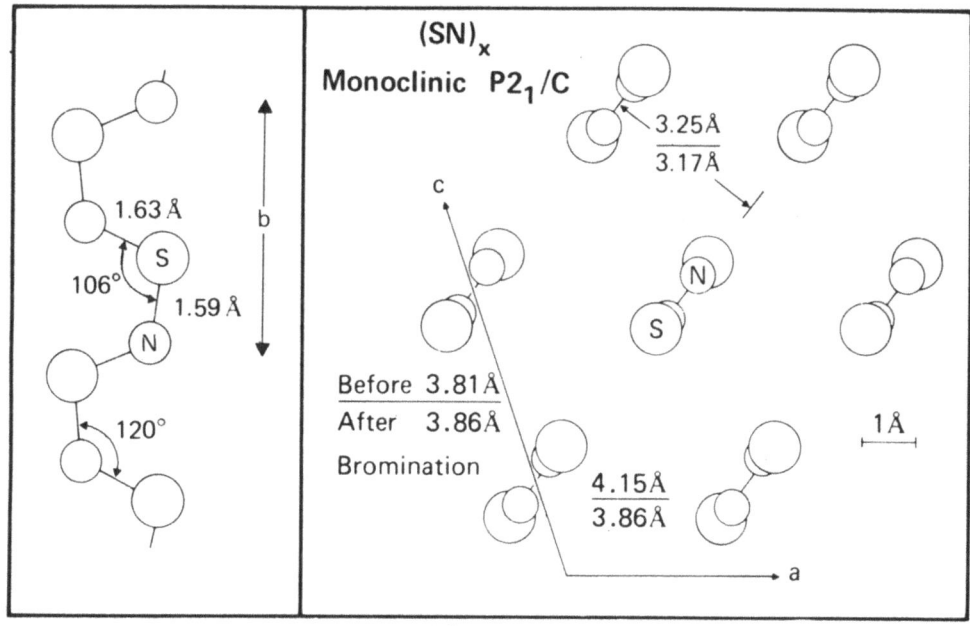

a = 4.153Å
b = 4.439Å   β = 109.7°
c = 7.637Å

Figure 2. Structure of $(SN)_x$ looking down the chain axis.

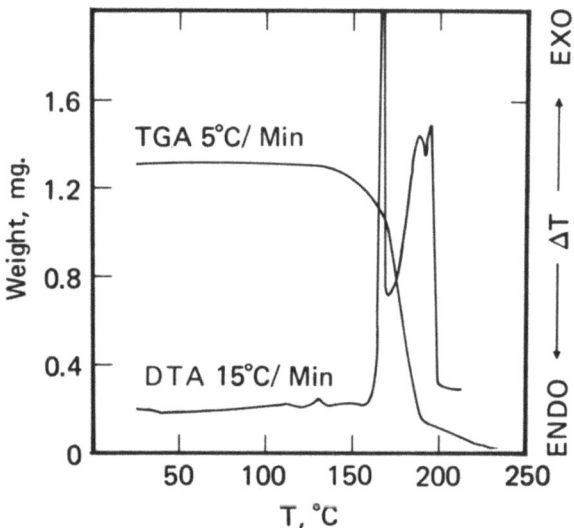

Figure 3. Thermal analysis data for $(SNBr_{0.4})_x$.

to have the optimum electrical properties and it has been the
most extensively studied. As the thermal analysis[56] data in
Fig. 3 make clear, virtually any composition of lower bromine
content can be obtained by heating a sample of $(SNBr_{0.4})_x$ in
vacuum for various periods of time. It is important to brominate
with bromine vapor because $(SN)_x$ reacts with liquid bromine to
give significant amounts of a side product $S_4N_3Br_3$.[57] Brominated
$(SN)_x$ reacts much more rapidly on exposure to air than pristine
$(SN)_x$ giving rise to NH_4Br and ammonium sulfur-oxygen compounds.[58]

Brominated $(SN)_x$ can also be prepared by the direct
bromination of S_4N_4.[59-61] Bromine vapor causes the S_4N_4 crystals
to turn black and expand rapidly destroying their external habit.
After removal of free bromine, the resulting, electrically
conducting, black powder has a composition of $(SNBr_{0.4})_x$. IR,[56]
Raman,[62,65,66] thermal analysis[56] and mass spectrometry[58] all
suggest that this material is the same as that obtained by direct
bromination of $(SN)_x$. X-ray powder data[59,61] show a large broad
peak corresponding closely to the $10\overline{2}$ interchain spacing of $(SN)_x$.
The poorly crystalline nature of brominated S_4N_4 is not unexpected
since the crystal structure of S_4N_4 does not provide as favorable
a route for solid state polymerization as that of S_2N_2. At lower
bromine contents e.g., $(SNBr_{0.25})_x$ the X-ray powder patterns are
of higher quality showing several more lines,[61] however our own
experience confirms the latter authors doubts, about the
authenticity of the sharp X-ray lines they report for $(SNBr_{0.4})_x$
and $(SNBr_{0.25})_x$. In the case of S_4N_4 it is even more important
than for $(SN)_x$ to brominate with bromine vapor in order to avoid
$S_4N_3Br_3$ formation.[57] ICl and IBr vapor also react readily with
S_4N_4 to form conducting solids.[59,61] Because of the more facile
dissociation of IBr into I_2 and Br_2 at room temperature[63] we have
not investigated this reaction further. However, $SN(ICl)_{0.36}$
(prepared from S_4N_4) has a pellet conductivity[59] of 50 $\Omega^{-1}cm^{-1}$,
similar to pellets of $(SNBr_{0.4})_x$ or $(SN)_x$. At 125° Iodine can
be made to react with S_4N_4 to give a conducting solid, but this
material has not been examined further. Mass spectroscopically
$SN(ICl)_{0.36}$ heated to 45°C was found to give off I_2, ICl, HI,
HCl, S_2Cl_2 and NSCl.[58] No Cl_2 and less surprisingly no sulfur
iodine species such as NSI were observed. Though we were able
to prepare $SN(ICl)_{0.36}$ from S_4N_4 most attempts to react S_4N_4 with
carefully purified ICl lead to polythiazyl compounds in which
the iodine, chlorine ratio is less than unity.[57] Thus, though
we believe S_4N_4 reacts with ICl to give halogenated $(SN)_x$, the
end product, in our experience, is not phase pure and contains
the halides of the $S_4N_3^+$ ion.[57] However Akhtar et al.[61] report
reactions of S_4N_4 with both ICl and IBr to give compositions
$SN(ICl)_{0.4}$ and $SN(IBr)_{0.4}$. Further studies will be required
before we fully understand the reaction of S_4N_4 with ICl and IBr.
Exposure of $(SN)_x$ crystals to ICl vapor for short periods of time

leads to surface reaction only, whereas longer exposure tends to destroy the $(SN)_x$.

MOLECULAR NATURE OF THE INTERCALATED HALOGEN

Extensive efforts have been directed to elucidating the nature of the bromine after intercalation into $(SN)_x$. Bromine in graphite is known to be only weakly ionized Br_2 molecules[63] but Raman spectroscopy[64-66] shows that the situation in brominated $(SN)_x$ is more complex. Raman data reveal two strong fundamentals. The first at ~150 cm^{-1} has been assigned without ambiguity to the symmetric stretch of the tribromide ion, Br_3^-. The second band at ~230 cm^{-1} has been assigned as the asymmetric stretch of the same Br_3^- and also as the stretching frequency of a Br_2 molecule, down shifted from the 325 cm^{-1} observed for gaseous bromine by its strong association with the $(SN)_x$ lattice. In order to distinguish between these two possibilities Macklin et al.[68] examined the IR of brominated $(SN)_x$ and found no evidence for the 230 cm^{-1} peak which would be expected to be IR active if it were the asymmetric stretch of a linear Br_3^-. On this basis we conclude that both Br_2 and Br_3^- are present in brominated $(SN)_x$. From the similarity in the Raman spectra of $(SNBr_{0.4})_x$ prepared from S_4N_4 we assume the bromine is present as Br_2 and Br_3^- in this case also.[62] According to the polarization of the Raman peaks observed for brominated $(SN)_x$ the Br_2 and Br_3^- appear to be oriented with their axes parallel to the $(SN)_x$ chains. EXAFS studies[69] are also consistent with this orientation and they also show that the Br_2 and Br_3^- are present in similar concentrations. Magnetic susceptibility measurements demonstrate the absence of such paramagnetic species as Br_2^-.[70]

Although they have not been investigated in such detail, from Raman[65] and electron diffraction studies,[71] crystals of $(SN)_x$ treated with ICℓ vapor, appear to contain no trihalide species but simply ICℓ molecules adsorbed on the surface of the $(SN)_x$.

MASS SPECTROSCOPIC STUDIES

The original experiments of Bernard et al.[6] pointed out that the bromine could be removed from the $(SN)_x$ by heating in vacuum. This is shown quantitatively in Fig. 3. Smith and Street[58] and Allen et al.[72] have examined the species eminating from the surface of $(SNBr_{0.4})_x$ as a function of temperature. Both studies agree that initially there are several bromine species including Br_2, NSBr, HBr and S_2Br_2. The concentrations of these various bromine species as a function of time of heating at 85°C are shown in Fig. 4. The S_xN_y species are also shown. The most striking result is the rapid fall off of the intensity of the

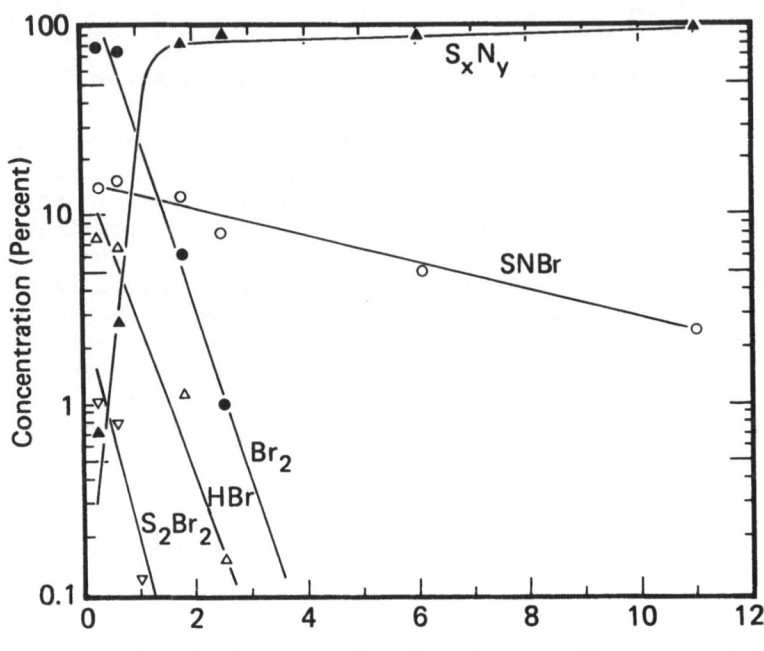

Figure 4.   Concentration as a function of time for the various
gas-phase species volatilized from $(SNBr_{0.4})_x$ upon heating at
85°C.

Br_2 peak such that after 3–4 hours the only significant bromine
species is NSBr.   These results can be rationalized in terms of
the two forms of bromine, indicated spectroscopically, Br_2 and
Br_3^-.   The molecular Br_2 would be expected to come off more readily
explaining its early dominance in the first two hours of heating.
Dissociation of the tribromide ion would also lead to free bromine
and Br^-.   This Br^- could react with the sulfur in the $(SN)_x$ chain
and subsequently volatilize as NSBr.   The appearance of NSBr as
an intermediate in the reactions of bromine with sulfur–nitrogen
compounds[57] has been reported previously.   These results would
imply that brominated $(SN)_x$ samples, prepared by debromination
of $(SNBr_{0.4})_x$, would not be identical to the same compositions
prepared by direct bromination of $(SN)_x$.

It is interesting to note that both of the mass spectroscopic studies reveal that the concentration of hydrogen in brominated $(SN)_x$ is at least a factor of ten less than in pristine $(SN)_x$. Apparently the hydrogen is removed as HBr.

X-RAY AND ELECTRON DIFFRACTION STUDIES OF $(SNBr_{0.4})_x$

Though bromination of $(SN)_x$ crystals leaves their fibrous nature and external habit unchanged, it raises havoc within the crystal causing such an increase in disorder and twinning that a complete single crystal X-ray structure determination is not possible.[54,58] Thus we are left with indirect methods of determining the mode of incorporation of Br_2 and Br_3^- in the crystal. The currently available data has been obtained from electron diffraction[54] and X-ray powder diffraction[6,54,65b] techniques as well as EXAFS[69] and Raman[64-66] spectroscopy. In their original experiments Bernard et al.[6] concluded from X-ray powder data that only the repeat unit in the chain direction remained unchanged on bromination. We have confirmed[52,54] that the chain axis repeat unit does not change from X-ray precession photographs. Baughman[65b] showed that as the bromine content increased, the separation between chains of the same type, d(100), decreased and the separation between chains of the opposite type, d(002) increased, until their separation was essentially the same, 3.62Å. Simultaneously the separation between the planes of chains d(10$\bar{2}$) decreased. The magnitude of these changes in the lattice are shown in Fig. 2 and they clearly demonstrate that some bromine is intercalating into the $(SN)_x$ lattice.

Electron diffraction studies[54] of $(SNBr_{0.4})_x$ show diffraction patterns typical of $(SN)_x$ but with considerable streaking perpendicular to b^* resulting from the significantly increased degree of disorder and (100) twinning (see Figs. 5a,b). From the width of the diffraction spots it is estimated that the fibers are approximately 30Å in diameter. Similar results are obtained from dark field electron micrographs shown in Fig. 5c. The fiber diameter of these crystals before bromination was ~70Å. Superimposed on this $(SN)_x$-like diffraction pattern are diffuse lines perpendicular to b^* at (0,1/2k,0) corresponding to an essentially one-dimensional commensurate superlattice oriented along the chain axis with a repeat unit twice that of pristine $(SN)_x$. In addition to this superlattice, diffuse lines appear near 2b/3, 2b/5 and 2b/7 (Fig. 4). The presence of all these superlattice lines has been confirmed by R. Comes[73] using diffuse X-ray scattering. On heating a sample to ~100°C in the electron microscope these latter lines become irreversibly less intense. On the other hand on cooling to ~150°K the 2b superlattice disappears but reappears on heating.[74] This reversible transition appears quite sharp occurring over ~5° temperature interval.

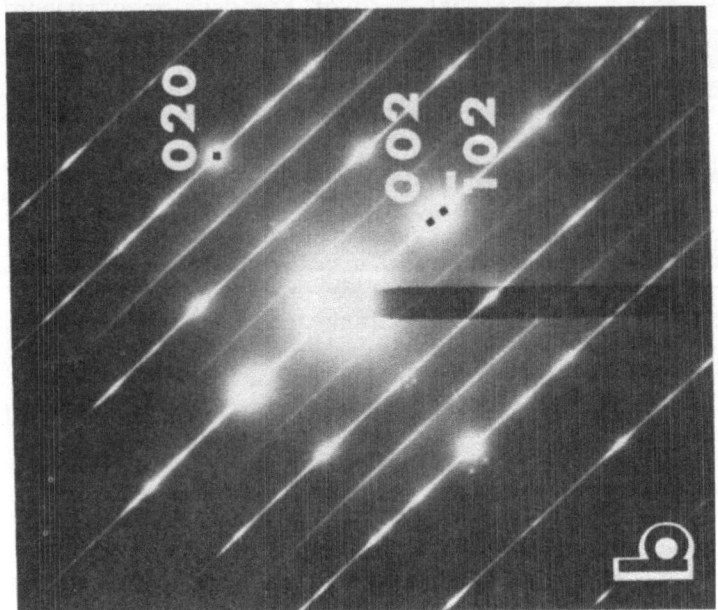

Electron diffraction pattern from $(SNBr_{0.4})_x$ fibers oriented similarly to (a). The simultaneous occurrence of (002) and $(10\bar{2})$ reflections and the larger streaks are attributed to extensive twinning.

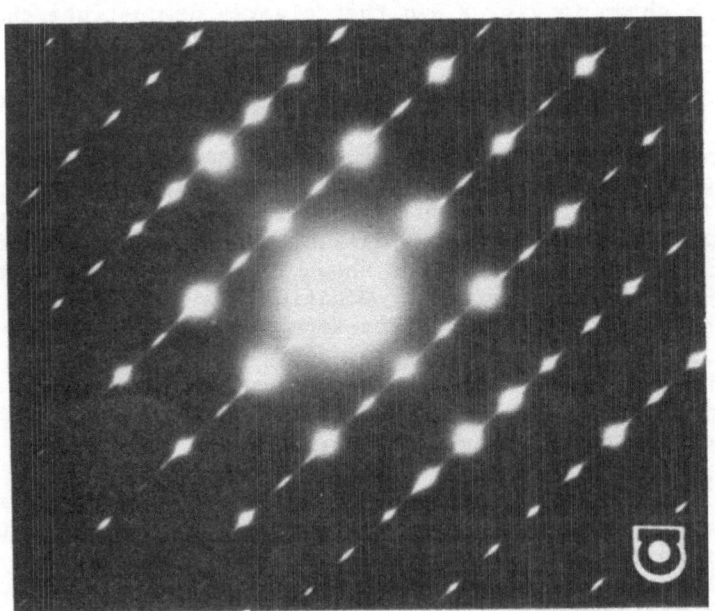

Figure 5.   Electron diffraction pattern from $(SN)_x$ fibers showing the b*c* reciprocal lattice net.

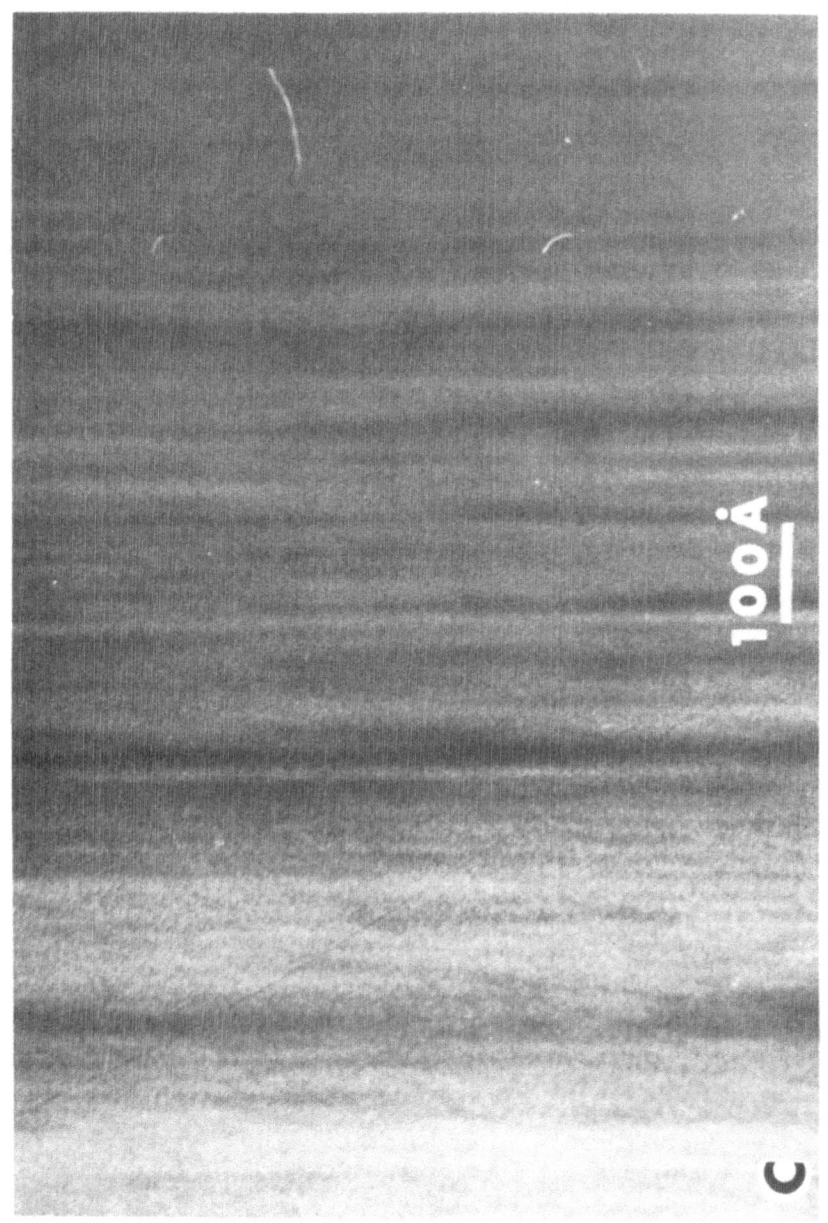

Figure 5. Electron micrograph of $(SNBr_{0.4})_x$ fibers with $\sim 30 \text{Å}$ diameters.

Though we do not understand the origin of the lines near 2b/3, 2b/5 and 2b/7 their irreversible disappearance at 100°C suggests they may be associated with the adsorbed Br_2 species. We believe that the 2b superlattice is associated with the ordering of the linear Br_3^- ion along the $(SN)_x$ chains. The reason for the disappearance of this 2b superlattice at low temperatures is not yet clear.

Extended X-ray fine structure absorption, EXAFS, measurements have been carried out on $(SNBr_{0.4})_x$ crystals by Morawitz et al.[69] The EXAFS structure using the bromine K_α edge at 13.5 KeV was investigated as a function of crystal orientation and temperature (300K to 4K) to determine the orientation of the bromine relative to $(SN)_x$, the bromine coordination and the nature of the transition at 150K causing disappearance of the 2b superlattice. Analysis of these data indicates that the bromine molecules are aligned along the $(SN)_x$ b-axis, and that a mixture of Br_3^- and Br_2 is present with the fraction of Br_3^- increasing at lower temperature. This is consistent with infrared studies.[67]

ELECTRONIC PROPERTIES

The conductivity of films and crystals of $(SN)_x$ exposed to bromine to form $(SNBr_{0.4})_x$ are observed to increase by about an order of magnitude[52-55] to a value of $\sigma_\| \approx 2-4 \times 10^4$ $\Omega^{-1}cm^{-1}$ at 300K. On cooling to 4.2K Gill et al.[54] report a conductivity increase of ~10 (Fig. 6) while Chiang et al.[55] find an increase of ~100. Both authors[35,54] also find the temperature dependence weaker than the T^2 behavior of $(SN)_x$. Thermopower measurements by Gill et al.[54] compared with $(SN)_x$ data[75] are shown in Fig. 6. On bromination the thermopower changes sign from negative to small positive values consistent with the bromine acting as an acceptor. The simple linear temperature dependence is typical of a metal. The conductivity perpendicular to the $(SNBr_{0.4})_x$ fibers, unlike that of pristine $(SN)_x$ (Fig. 6), is metallic down to 3°K below which it is temperature independent.[54] This behavior suggests that $(SNBr_{0.4})_x$ is more three-dimensional than $(SN)_x$ itself. Gill et al.[75] have recently reported on the pressure dependence of normal conductivity. In contrast to $(SN)_x$ where σ increases dramatically under pressure,[37,38] the conductivity of $(SNBr_{0.4})_x$ only increases about 10 to 20% under 10 kbar.

Polarized reflectance measurements[54,55] show a strong red-shift of the plasma edge out of the visible. Figure 7 shows reflectance of a crystal before and after bromination.[54] However Drude-Lorenz analysis suggests that within experimental error there is little or no change in the plasma frequency or the optical scattering time on bromination. The shift of the plasma edge is primarily due to an increase in the background dielectric

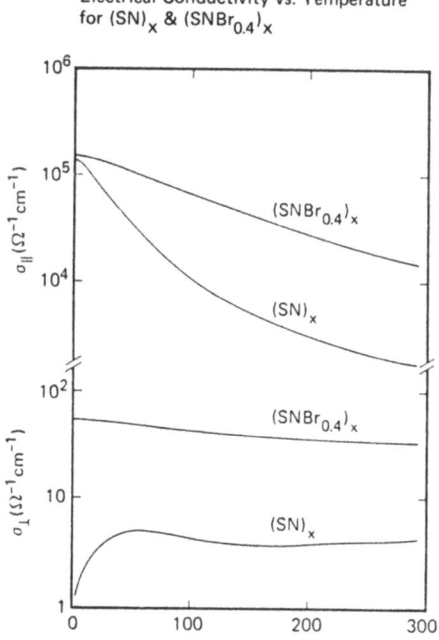

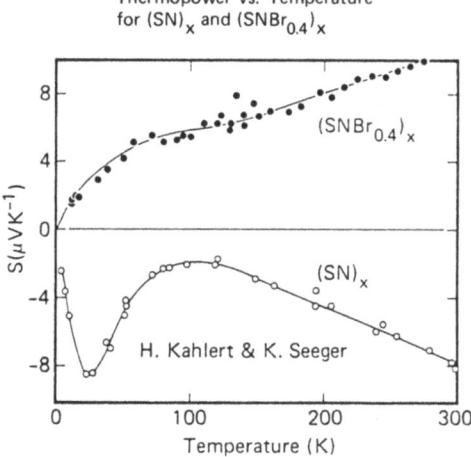

Figure 6. Comparison of conductivity and thermopower for $(SN)_x$ and $(SNBr_{0.4})_x$ crystals. Thermopower of $(SN)_x$ taken from data of Kahlert and Seeger.[75]

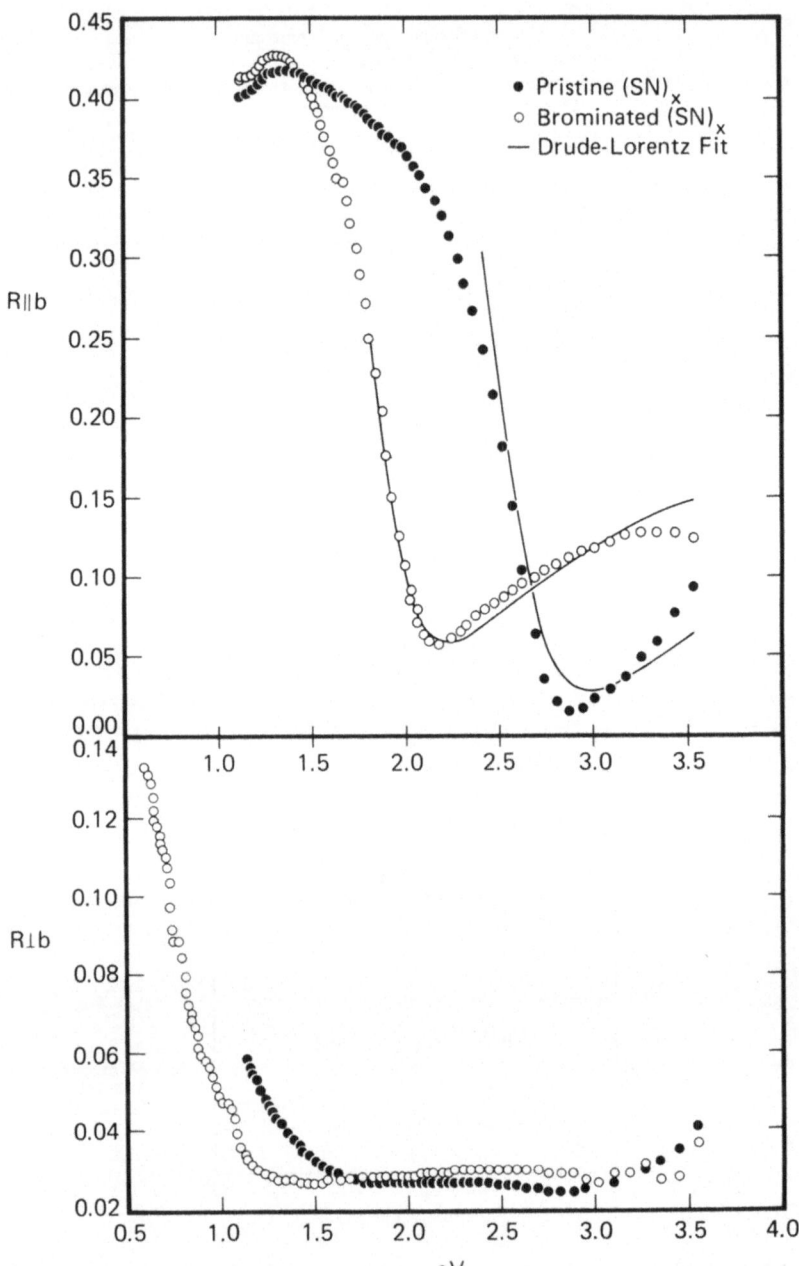

Figure 7. Polarized reflectance of the same (SN)$_x$ crystal before and after bromination.[54]

constant on bromination. From the lack of an edge in their
perpendicular reflectivity Chiang et al.[55] conclude that
interchain coupling is weakened by bromination. This conclusion
is contrary to that of Dee et al.[9] and Kwak et al.[76] based on
superconductivity data discussed below and the normal state
conductivity perpendicular to the fibers as measured by Gill
et al.[54]

Initial measurements[54] of T_c on $(SNBr_{0.4})_x$ crystals showed
negligible change from $(SN)_x$ values, however more recent
measurements[76] show that T_c increases about 20% on bromination.
The superconducting transition in $(SNBr_{0.4})_x$ is much sharper than
in $(SN)_x$ as shown in Fig. 8. The width of this transition in
$(SN)_x$ was in part ascribed to disorder[4] however $(SNBr_{0.4})_x$ is
much more disordered. This suggests that the width of the
transition is due to superconducting fluctuations[4] and their
absence in $(SNBr_{0.4})_x$ suggests that it is more three-dimensional
than $(SN)_x$. The pressure dependence of T_c is also qualitatively
different from $(SN)_x$.[77] Instead of increasing as in $(SN)_x$,[34,41]
T_c of the brominated material decreases monotonically to about
150mK under 10 kbar (see Fig. 9) as would be expected for a
three-dimensional s-p superconductor. Dee et al.[9] have observed
a complete Meissner effect in $(SNBr_{0.4})_x$ crystals. Their data
show a much sharper transition than in pristine $(SN)_x$ indicative
of more tightly coupled superconducting fibers in the $(SNBr_{0.4})_x$.
Critical field studies by Kwak et al.[76] show that in contrast to
$(SN)_x$, $H_\perp$ is well behaved showing no anomalous curvature when
plotted against T_c. These data are interpreted by these authors
as indicative of increased interfiber coupling in the
superconducting sense i.e., $(SNBr_{0.4})_x$ is more three-dimensional
than $(SN)_x$.

The magnetic susceptibility of brominated $(SN)_x$ has been
measured as a function of temperature by Scott et al.[70] For
$(SNBr_{0.4})_x$ a small Curie-like contribution is observed below 30K.
From this contribution an upper limit on the concentration of
S=1/2 species, such as Br_2^-, of 2×10^{-4} molar is obtained. At
higher temperature the susceptibility increases linearly with
temperature. The excess paramagnetism increases with increasing
bromine content.

Conductivity and magnetoresistance measurements of $(SN)_x$
exposed to bromine, iodine and ICl are reported by Philipp and
Seeger.[78] Their results indicate that only bromine significantly
improves the conductivity. After halogen treatment, a negative
magnetoresistance component increases in magnitude which the
authors attribute to increased $(SN)_x$ chain interruptions.

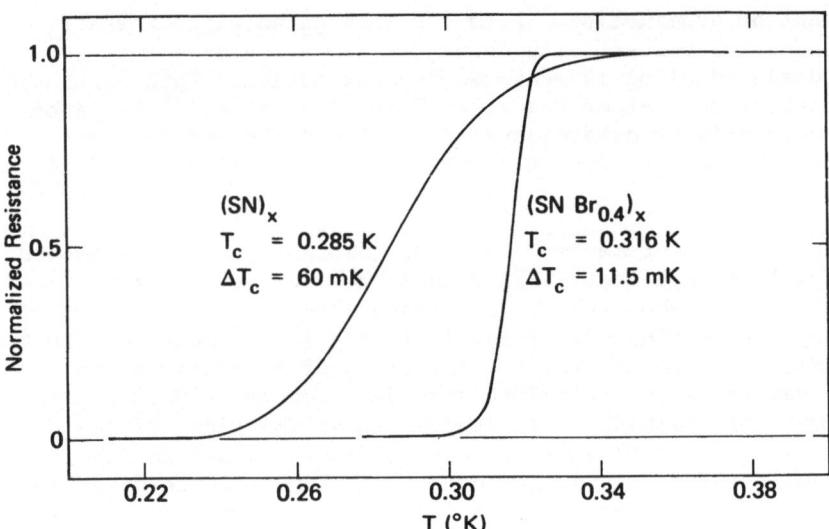

Figure 8.   The super conducting transition in $(SN)_x$ and $(SNBr_{0.4})_x$.

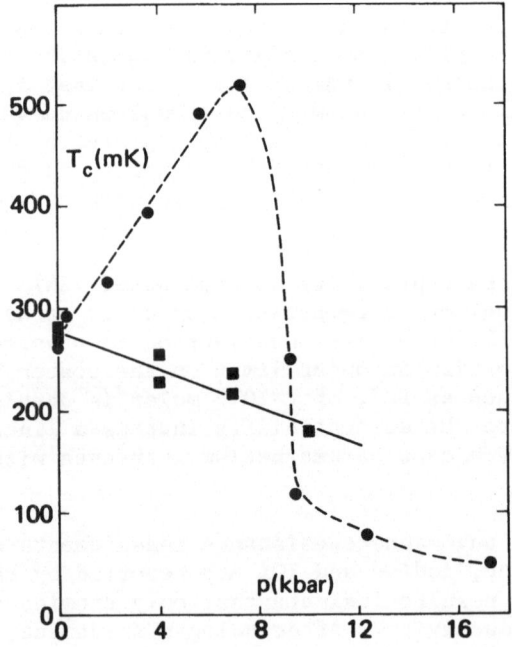

Figure 9.   Pressure dependence of T_c in $(SN)_x$ and $(SNBr_{0.4})_x$. The $(SN)_x$ data is taken from Müller et al.[41]

MODEL FOR $(SNBr_{0.4})_x$

From the spectroscopic data already discussed it appears that the intercalated bromine exist as Br_2 and Br_3^-. Both species are aligned with their axis parallel to the $(SN)_x$ chain axis. The changes in the lattice constants perpendicular to b, particularly the decrease in $d(10\bar{2})$ strongly suggest that the bromine is not intercalating between the $(10\bar{2})$ planes but instead resides in the $(10\bar{2})$ plane, occupying chain sites. The lack of three-dimensional order in $(SNBr_{0.4})_x$ implies that the bromine occupies these sites in a random fashion in the "a" and "c" directions. On the other hand along the chain axis the bromine species must occupy specific sites in a periodic fashion in order to give rise to the observed "2b" superlattice. We believe that the Br_3^- ions are responsible for this "2b" superlattice. Geometrically this is consistent with the size of Br_3^-.[56] From the observation that bromine prefers to bond to sulfur rather than nitrogen, both bromine species would be expected to align themselves along the chain axis so as to maximize sulfur-bromine interactions and minimize nitrogen-bromine interactions. A possible packing of the Br_3^- ion between $(SN)_x$ chains is given in Fig. 10.

In a fibrous material such as $(SN)_x$ the bromine will also be attached to the outside of the fibers. As the fiber diameter becomes smaller, the amount of bromine which can be accommodated in this form becomes significant relative to that which is intercalated between the chain. In the case of $(SNBr_{0.4})_x$ with its ~30Å diameter fibers it is possible to accommodate all of the bromine in the monolayer on the outside of the fibers[56] (Fig. 11). Iqbal et al.[65b] have observed that the small angle X-ray scattering present in pristine $(SN)_x$ disappears in $(SNBr_{0.4})_x$. These results indicate the absence of density fluctuations perpendicular to the chains. In order to account for this absence of density fluctuations Iqbal et al.[65b] estimated that 10% of the bromine has entered the $(SN)_x$ lattice while some 90% resides in the interfibrillar regions.

The large increase in conductivity and the approximately constant plasma frequency suggest that the main effect of bromination on the electronic properties of $(SN)_x$ is to increase the dc scattering lifetime.[54] Increased free carrier density due to charge transfer from bromine cannot by itself account for the tenfold increase in conductivity. The large conductivity change, small change in T_c and small pressure dependence of σ have suggested a model for the electronic properties[77,79] which we outline here. In $(SN)_x$ the resistivity is well described by a T^2 dependence and is therefore believed to be dominated by electron-hole scattering processes.[35] Efficient scattering can be expected on the basis of the geometry and separation of the

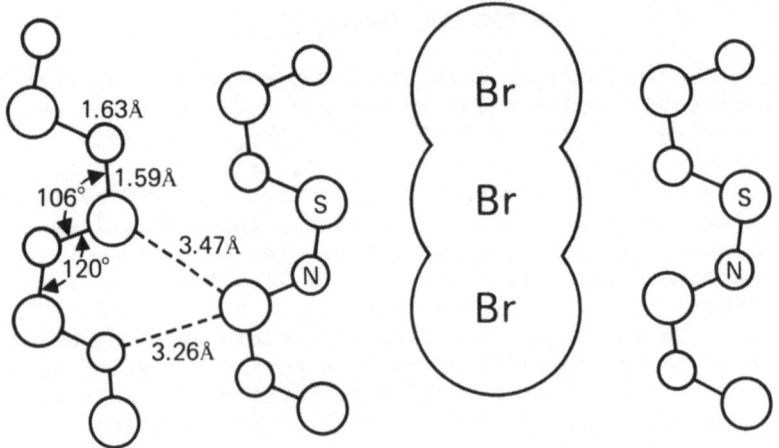

Figure 10.   Structure of brominated (SN)$_x$ showing a possible packing scheme for Br$_3^-$ chosen to maximize S–Br interactions.

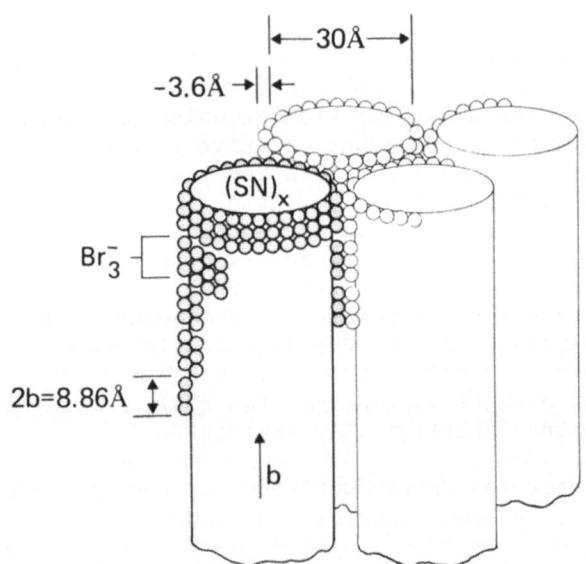

Figure 11.   Bromine cladding the outside of (SN)$_x$ fibers.

hole and electron pockets of the Fermi surface.[80] Suppression
of the electron-hole scattering process in brominated $(SN)_x$ can
be reasonably expected from Fermi surface considerations.
Assuming that charge transfer from the bromine removes
approximately 0.1 electrons/SN unit from the conduction bands,
the Fermi level is lowered by about leV. Lowering E_F drastically
effects the shape of the Fermi surface, expanding the volume of
the hole pockets and shrinking or possibly even eliminating the
electron pockets. Consideration of the effects of hydrostatic
pressure on the band structure of pristine $(SN)_x$ suggests that
the electron pocket may be shifted from the zone center under
high pressure resulting in reduction of the electron-hole
scattering probability. Suppression of electron-hole scattering
processes would result in increased conductivity by increasing
the scattering lifetime. This would be signalled by a change
from a T^2 dependence of ρ to a more linear temperature dependence
typical of electron-phonon scattering. Although not linear, the
trend toward smaller temperature exponents is observed both in
brominated $(SN)_x$[55] and in $(SN)_x$ under high pressure.[37] In
addition one expects a much reduced pressure dependence of σ in
brominated $(SN)_x$ since electron-phonon scattering processes will
not be as sensitive to the details of the Fermi surface. In fact
the pressure dependence of σ for brominated crystals[77] can be
adequately accounted for by lattice stiffening effects usually
dominant in metals. Thus, this model invoking Fermi surface
changes due to charge transfer on bromination and subsequent
suppression of electron-hole scattering, explains the major
features of the electronic properties in brominated $(SN)_x$.

ACKNOWLEDGMENTS

We thank P. M. Grant and R. L. Greene for critically reviewing
this manuscript and the ONR for partial financial support.

REFERENCES

1. H. P. Geserich and L. Pintschovious, Advances in Solid State
    Physics, 16, 65 (1976).

2. A. D. Yoffe, Chemical Society Reviews 5, 51 (1976).

3. G. B. Street and R. L. Greene, IBM J. of Res. and Develop.
    99, 21 (1977).

4. R. L. Greene and G. B. Street, "Chemistry and Physics of
    One-Dimensional Metals," H. J. Keller (ed.) Plenum Press,
    New York, 1977 p. 167.

5. G. Wolmershäuser, C. R. Brulet and G. B. Street, to appear in Inorg. Chem.

6. C. Bernard, A. Herold, M. Lelaurain and G. Robert, Compr. Rend. (c) 283, 625 (1976).

7. J. Petermann and J. M. Schultz, submitted to J. of Materials Sci.

8. J. M. Schultz and J. Petermann, submitted to Phil. Mag.

9. R. H. Dee, D. H. Dollard, B. G. Turrell and J. F. Carolan, Solid State Commun. 24, 469 (1977).

10. R. D. Smith, J. R. Wyatt, D. Weber, J. J. DeCorpo and F. E. Saalfeld, Inorg. Chem. 17, 1639 (1978).

11. G. Heger, S. Klein, L. Pintschovious and H. Kahlert, J. Solid State Chem. 23, 341-347 (1978).

12. M. J. Cohen, S. F. Garito, A. J. Heeger, A. G. MacDiarmid, C. M. Mikulski, M. S. Saran and J. Kleppinger, J. Amer. Chem. Soc. 98, 3844 (1976).

13. P. Love, G. Myer, H. I. Kao, M. M. Labes, W. R. Junken and C. Elbaum, J.C.S. Chem. Commun.

14. G. B. Street, unpublished data.

15. P. Love, H. I. Kao, G. H. Myer and M. Labes, J.C.S. Chem. Comm., 1978, to appear Ann. New York Acad. Sci.

16. R. D. Smith, to be published.

17. W. Y. Ching, J. G. Harrison and C. C. Lin, Phys. Rev. B 15, 5975 (1977).

18. T. Yamabe, K. Taneka, A. Imamura, H. Kato and K. Fukui, Bull. of Chem. Soc. Japan 50, 798 (1977).

19. S. Suhai and J. Ladik, Solid State Commun. 22, 227 (1977).

20. I. P. Batra, S. Ciraci and W. E. Rudge, Phys. Rev. B 15, 5858 (1977).

21. I. P. Batra, Phys. Rev. B 17, 4114 (1978).

22. L. Ley, H. J. Stolz, T. Grandke and J. Azoulay, Proc. 5th UVU Conf., Montpellier, Sept. 1977.

23. E. E. Koch and W. D. Grobman, Solid State Commun. <u>23</u>, 49 (1977).

24. W. E. Rudge and P. M. Grant, Phys. Rev. Lett. <u>35</u>, 1799 (1975).

25. J. Ruvalds, F. Brosens, L. F. Lemmens and J. T. Devresse, Solid State Commun. <u>23</u>, 243 (1977).

26. C. H. Chen, J. Silcox, A. F. Garito, A. J. Heeger and A. G. MacDiarmid, Phys. Rev. Lett. <u>36</u>, 525 (1976).

27. H. J. Stolz, E. Petri and A. Otto, Phys. Stat. Sol. (b) <u>80</u>, 289 (1977).

28. H. Wendel, J. Physics C: Solid State Phys. <u>10</u>, L1 (1977).

29. R. Clarke, Solid State Commun. <u>25</u>, 333 (1978).

30. K. Kaneto, K. Yoshimo and Y. Inuishi, J. Phys. Soc. Japan <u>43</u>, 1013 (1977).

31. P. M. Grant, P. Mengel, E. M. Engler, G. Castro and G. B. Street, Bull. Amer. Phys. Soc. <u>21</u>, 254 (1976).

32. P. M. Grant and B. Welber, Bull. Amer. Phys. Soc. <u>22</u>, 372 (1977).

33. R. H. Friend, D. Jérome, S. Rehmatullah and A. D. Yoffe, J. Physics C: Solid State Phys. <u>10</u>, 1001 (1977).

34. W. D. Gill, R. L. Greene, G. B. Street and W. A. Little, Phys. Rev. Lett. <u>35</u>, 1732 (1975).

35. C. K. Chiang, M. J. Cohen, A. J. Heeger, C. M. Mikulski and A. G. MacDiarmid, Solid State Commun. <u>18</u>, 1451 (1976).

36. C. K. Chiang, A. J. Heeger and A. G. MacDiarmid, Phys. Lett. <u>60A</u>, 375 (1977).

37. K. Kaneto, K. Tanimura, K. Yoshino and Y. Inuishi, Solid State Commun. <u>22</u>, 383 (1977).

38. K. Kaneto, M. Yamamoto, K. Yoshino and Y. Inuishi, Solid State Commun. <u>26</u>, 311 (1978).

39. W. Beyer, W. D. Gill and G. B. Street, Solid State Commun. <u>23</u>, 577 (1977).

40. R. H. Dee, A. J. Berlinsky, J. F. Carolan, E. Klein,
    N. J. Stone, B. G. Turrell and G. B. Street, Solid State Commun.
    23, 303 (1977).

41. W. H. G. Müller, F. Baumann, G. Dammer and L. Pintschovius,
    Solid State Commun. 25, 119 (1978).

42. L. R. Bickford, R. L. Greene and W. D. Gill, Phys. Rev. B 17,
    3525 (1978).

43. P. M. Chaikin, P. K. Hansma and R. L. Greene, Phys. Rev. B 17,
    179 (1978).

44. K. Yoshino, Y. Makio, K. Kaneto and Y. Inuishi, Jap. J. Appl.
    Phys. 16, 1047 (1977).

45. W. Beyer, W. D. Gill and G. B. Street, Solid State Commun.
    27, 343 (1978).

46. W. Beyer, H. Mell and W. D. Gill, Solid State Commun. 27,
    185 (1978).

47. R. A. Scranton, J. Appl. Phys. 48, 3838 (1977).

48. R. A. Scranton, J. S. Best and J. O. McCaldin, J. Vac. Sci.
    Technol. 14, 930 (1977).

49. M. J. Cohen and J. S. Harris, submitted to Appl. Phys. Lett.

50. K. J. Wynne and W. L. Jolly, Inorg. Chem. 6, 107 (1967).

51. R. L. Patton and W. L. Jolly, Inorg. Chem. 8, 1392 (1969).

52. G. B. Street, W. D. Gill, R. H. Geiss, R. L. Greene and
    J. J. Mayerle, J.C.S. Chem. Commun., 407 (1977).

53. M. Akhtar, J. Kleppinger, A. G. MacDiarmid, J. Milliken,
    N. J. Cohen, A. J. Heeger and D. L. Pebbles, J.C.S. Chem.
    Commun. 473 (1977).

54. W. D. Gill, W. Bludau, R. H. Geiss, P. M. Grant, R. L. Greene,
    J. J. Mayerle and G. B. Street, Phys. Rev. Lett 38, 1305
    (1977).

55. C. K. Chiang, M. J. Cohen, D. L. Peebles, A. J. Heeger,
    M. Akhtar, J. Kleppinger, A. G. MacDiarmid, J. Milliken and
    M. J. Moran, Solid State Commun. 23, 607 (1977).

56. G. B. Street, S. Etemad, R. H. Geiss, W. D. Gill, R. L. Greene
    and J. Kuyper, Anal. New York Acad. Sci., 313, 737 (1978).

57. G. Wolmershäuser and G. B. Street, Inorg. Chem. 17, 2685
    (1978); J. J. Mayerle, G. Wolmershäuser and G. B. Street, to
    be published.

58. R. D. Smith and G. B. Street, Inorg, Chem. 17, 941 (1978).

59. G. B. Street, R. L. Bingham, J. I. Crowley and J. Kuyper,
    J.C.S. Chem. Commun. 464 (1977).

60. M. Akhtar, C. K. Chiang, A. J. Heeger and A. G. MacDiarmid,
    J. Chem. Soc. Chem. Commun. 846 (1977).

61. M. Akhtar, C. K. Chiang, A. J. Heeger, J. Milliken and
    A. G. MacDiarmid, Inorg. Chem. 17, 1539 (1978).

62. H. Temkin and G. B. Street, unpublished results.

63. D. A. Platt, D. L. Chung and M. S. Dresselhaus, to be
    published.

64. H. Temkin, D. B. Fitchen, W. D. Gill and G. B. Street, Anal.
    New York Acad. Sci., 1978.

65. (a) Z. Iqbal, R. H. Baughman, J. Klepping and A. G. MacDiarmid;
    (b) Solid State Commun. 25, 409 (1978); Anal. New York Acad.
    Sci., 1978.

66. H. Temkin and G. B. Street, Solid State Commun. 25, 455
    (1978).

67. D. M. Yost, JACS 55, 552 (1933).

68. J. Macklin, W. D. Gill and G. B. Street, to be published.

69. H. Morawitz, W. D. Gill, G. B. Street and D. Sayers, Bull.
    Amer. Phys. Soc. 23, 304 (1978); H. Morawitz, W. D. Gill,
    P. M. Grant, G. B. Street and D. Sayers, to be published.

70. J. C. Scott, J. D. Kulick and G. B. Street, Bull. Amer. Phys.
    Soc. 23, 304 (1978).

71. J. Thomas, R. H. Geiss and G. B. Street, to be published.

72. W. N. Allen, J. J. DeCorpo, F. E. Saalfeld and J. R. Wyatt,
    Chem. Phys. Lett., 1978.

73. R. Comes, APS Bull. 23, 424 (1978).

74. R. H. Geiss and J. Thomas, unpublished result.

75. H. Kahlert and K. Seeger, <u>Physics of Semiconductors</u>, Proc.
    13th Intern. Conf. Rome, F. G. Fumi (ed.), Tipografia Marves,
    Rome, 1976.

76. J. F. Kwak, R. L. Greene and G. B. Street, Bull. Amer. Phys.
    Soc. <u>23</u>, 384 (1978).

77. W. D. Gill, J. F. Kwak, R. L. Greene, K. Seeger and
    G. B. Street, Bull. Amer. Phys. Soc. <u>23</u>, 382 (1978).

78. A. Philipp and K. Seeger, to be published.

79. P. M. Grant, R. L. Greene, W. D. Gill and J. F. Kwak, to be
    published.

80. P. M. Grant, W. E. Rudge and I. B. Ortenburger, <u>Lecture Notes
    in Physics</u>, Vol. <u>65</u>, <u>Organic Conductors and Semiconductors</u>,
    Springer-Verlag, Berlin, 1977.

ELEMENTAL SULFUR - NEW QUESTIONS TO AN OLD PROBLEM

Max Schmidt
Universität Würzburg
Am Hubland, D-8700 Würzburg

What actually is sulfur? Is that a real question? Certainly not if we are content with the answer, trivial today, that it has the simplest stoichiometric composition conceivable - merely a series of atoms of mass number 16. This answer, however, clarifies only the centuries-long arguments about the composition of sulfur, a matter that was connected very intimately with the development of chemistry from ancient times through alchemy to the present day. Nevertheless, this answer is of no further help if we want to know more about the properties of sulfur. Actually, this element presents an extremely complicated chemical system which is still far from understood. According to the external conditions, that system is comprised of sulfur molecules, each containing from two to several hundred thousand atoms mixed together in great variety: S_x ($x = 2$ to $\approx 10^6$!).

We are dealing here with true dynamic equilibrium mixtures. Sulfur atoms have a surprisingly well developed tendency and ability to combine with one another to form chain or ring molecules. In this respect only carbon is superior to sulfur.

Sulfur atoms have six outer electrons, with the two unpaired 3p electrons covalently bonded to neighboring atoms. Therefore the resulting chains cannot be collinear but must be zig-zag. Such zig-zag

chains, of course, are not planar, because each sulfur atom still
has two further electron pairs which prevent free rotation around the
sulfur-sulfur bond, producing a favored dihedral angle.

According to whether we proceed upwards or downwards when passing from the third to the fourth atom in the middle plane, we have
the beginning of a right- or a left-handed helix. Sulfur chains, and,
of course also all the many sulfanes and sulfane-derivatives, i.e.,
compounds containing chains of sulfur atoms, thus must have helical
conformations. A helix with zero translation leads to a non-planar
ring.

Thus, the main factors determining the conformation of compounds
with isolated or cumulated sulfur-sulfur bonds are the bond distance
d, bond angle α, and dihedral angle γ.

In disulfane, H_2S_2, a rotation barrier of between 8 and 29 kJ/
mol has been calculated by different authors. Because there is practically no steric hindrance between the two hydrogen atoms, in this
simplest molecule the sulfur-sulfur bond distance is 2055 $\overset{o}{A}$ - a nearly ideal bond - and the dihedral angles are 91.3^o and 90.6^o, respectively.

If we pass from the simple case of isolated to cumulated sulfur-
sulfur bonds, then interactions between next nearest neighbors will
influence the exact geometry of the molecules significantly. The
simplest chemical compound containing cumulated sulfur-sulfur bonds
is elemental sulfur itself.

In spite of the complexity of the system elemental sulfur and
its incurable, confused nomenclature there is one very important fact
in this system. It is the fact that under normal conditions only one
compound of sulfur atoms is thermodynamically stable, namely the eight-
membered cyclooctasulfur, S_8. For reasons mentioned before, this ring
naturally is not planar. The crown-shaped molecules crystallize in an
orthorhombic structure up to 95.4^oC where they pass into monoclinic
crystals. These eight-membered rings decompose partially at $\sim119^oC$,

forming other molecules which depress the solidification temperature; sulfur has no melting point in the correct sense of the word. The sulfur-sulfur distance in this ring is 2.06 $\overset{o}{A}$, the bond angle is 108°, and the dihedral angle is 98°. The difference of these latter two angles from the ideal of 90°, as in disulfane with its isolated sulfur-sulfur bond in the system with eight cumulated bonds, is caused mainly by interactions between next nearest neighbors. If there are at least two cumulated sulfur-sulfur bonds with two other ligands R or two more sulfur atoms (at least five sulfur atoms altogether), then there are three different possible conformations, namely cis, d-trans, and l-trans.

Cyclohexasulfur, S_6, also has an all-cis arrangement. It was the only other definitely known sulfur molecule besides S_8 at the beginning of our synthetic work. The properties of S_6 concincingly demonstrate the fact that in symmetrical ring molecules of sulfur atoms, bond angles and dihedral angles are not independent but are interconnected. At a given bond angle α, ring closure is possible only at a certain dihedral angle γ; the latter one is determined by the ring size and the number of cis, d-trans, and/or l-trans arrangements. Therefore sometimes the dihedral angle is not optimum, for example, at the minimum of the torsion potential, as with S_6. As in S_8, the bond distance is ~ 2.06 $\overset{o}{A}$. The bond angle is 101°, somewhat smaller than that for S_8 (108°), but γ is only 75° while for the stable cyclooctasulfur, it is 98°. This results in a rather high ring strain in the chair formed six-membered ring. The very light-sensitive cyclohexasulfur is decomposed by visible light as well as by moderate warming up to $\sim 50^{o}$C. Chemically it reacts with both nucleophiles and electrophiles in redox reactions that are up to 10,000 times faster than those of S_8.

We asked ourselves whether it might be possible to build new thermodynamically unstable molecules containing atoms of only one kind, i.e., new modifications of an element, by kinetically controlled syntheses. This was indeed possible, There new preparative methods enabled us to synthesize the thermodynamically unstable ring molecules $S_5(?)$, S_6, S_7, S_9, S_{10}, S_{11}, S_{12}, S_{18} (two allotropic forms), S_{20}, and $S_{24}(??)$. Two review articles describe the synthetic routes as well as some properties of the new sulfur allotropes (1,2).

A rather big surprise was the first synthesis of cyclododeca-sulfur. In 1949, Pauling published a very important paper for sulfur chemistry (3).On the basis of his considerations and calculations, Pauling predicted for the still-hypothetical 12-membered sulfur ring an even higher ring strain and thus greater instability than in the extremely sensitive and labile cyclohexasulfur. This prediction proved to be wrong. Cyclododecasulfur, S_{12}, crystallizes in pale yellow needles. It does form regular solutions in CS_2, but its solubility is surprisingly low - less than 1 % of that of S_8. The biggest surprise, however, was the thermal stability as well as the stability towards light. This allotrope of elemental sulfur may be stored in light at room temperature for many years without decom-position. Furthermore, whereas the most stable form of sulfur, S_8, partly decomposes with ring opening and thus seemingly melts at $\sim 119^oC$, the new sulfur modification may be heated up to 148^oC. At this temperature it melts and decomposes, forming the usual and very complex mixture of molten sulfur. This unexpected stability and solu-bility made it possible to separate the new sulfur form from a large excess of S_8. Today we know that S_{12} is formed in the course of dif-ferent reactions besides our original synthetic redox reaction, for instance during the thermal decomposition of cyclooctasulfur-mon-oxide, S_8Om (4) or also by the photo-induced decomposition of S_6. In fact, we could isolate S_{12} from ordinary cooled sulfur melts after preheating S_8 to $\sim 120^o-130^oC$.

The x-ray determination revealed a highly symmetrical 12-mem-bered ring. In this ring, however, the atoms are not arranged in two planes as is the case in Pauling's hypothesis, but in three planes; six atoms form a regular hexagon as the middle plane, three atoms are in a second plane above, and the other three in the third

plane below the middle plane. With this conformation bond parameters
are optimized and are similar to those found in the stable S_8.

A short time after the synthesis and structure determination
of the 12-membered ring and without knowing of this synthesis,
Tuinstra, as a consequence of very interesting and rather sophisticated considerations and calculations, predicted a sufficient stability at room temperature and the absolutely correct and precise
conformation for the still-hypothetical S_{12} ring (5).

With respect to its three predecessors, any atom in a chain can
make a choice from two different positions (as already mentioned
earlier), namely the fourth atom initiates with its three predecessors (1) a right-handed or (2) a left-handed screw. If n is the
number of atoms forming a molecule, the number of possible different
molecular conformations is now $2^{(n-3)}$ $(n > 2)$. With respect to only
crystalline modifications, we can exclude all configurations in
which the successive choices (1) and (2) are distributed at random;
that means that only those configurations will be considered in
which a distinct sequence of choices (1) and (2) is indefinitely
repeated (that is, of course, also all ring molecules of that type).

The repeating unit is a distinct sequence of choices from (1)
and (2); in the molecule this sequence corresponds to a row of
Successive atoms, the relative positions of which are fixed if bond
and dihedral angles are kept constant. Such a repeating sequence of
atoms is called a motif (or module). In general, a motif will coincide with its successor in the molecular string after a translation
combined with a rotation in space. This implies that in general these
molecular conformations are helical. In this argument, as mentioned
earlier, ring molecules may be considered as special helical molecules with pitch zero.

The hitherto known sulfur rings will be discussed under this
point of view. A speculative consideration follows from these discussions: one can think of a large number of rather stable neutral

sulfur rings composed of parallel helices linked end on. With normal
bond distances, bond angles, and torsion angles they have the com-
position S_{6+n6} (S_{12}, S_{18}, S_{24}, S_{30}, S_{36}...). These theoretical molecules might well be major constituents of sulfur melts and of inso-
luble sulfur.

The very complicated sulfur melt can no longer be explained by
the equilibria:

$$nS_8 \rightleftharpoons n \cdot S_8 \cdot \rightleftharpoons S_{8n}$$

There are convincing facts that in the melt at any given temperature
there are also molecules other than S_8 and multiples of S_8. They will
be described.

At the present time, the scientific basis of practical applications of elemental sulfur unfortunately is not very strong in many
respects. The situation has been improved, however, in the years, but
much more work is needed so that we can influence the properties of
elemental sulfur to meet the requirements of different practical
applications.

Literature
1. Schmidt, M., Angew.Chem. (1973) 85, 474.
2. Meyer, B., Chem.Rev. (1976) 76, 367.
3. Pauling, L., Proc.Natl.Acad.Sci. U.S. (1949) 35, 495.
4. Steudel, R., Angew.Chem. (1972) 84, 344.
5. Tuinstra, F., "Structural Aspects of the Allotropy of Sulfur and
   the Orther Divalent Elements", Uitgeverij Waltmann, Delft, 1967.

SURFACE STUDIES OF ELECTRICALLY-CONDUCTING "CsSn$_2$I$_5$"

Steven L. Suib, John E. Chalgren and Galen D. Stucky[1]
Richard J. Blattner[2] and Paul F. Weller[3]

[1]School of Chemical Sciences and Materials Research
 Laboratory, University of Illinois, Urbana, IL 61801
[2]Materials Research Laboratory, Urbana, IL 61801
[3]Western Illinois University, Macomb, IL 61455

INTRODUCTION

The gel method of crystal growth is a rather simple technique for the controlled crystallization of certain compounds.[1] Single crystals of CsSn$_2$I$_5$ have been grown in sodium metasilicate gels and identified by X-ray powder and single crystal diffraction methods and suitable chemical analyses. This report deals with the synthesis, chemical characterization and surface properties of CsSn$_2$I$_5$.

SYNTHESIS

A 1.72 M stock solution of sodium metasilicate was used in all gel preparations. A mixture of 6 ml stock solution, 6 ml distilled water and 6 ml cesium iodide was initially prepared. To this solution, 15 ml of 2 M glacial acetic acid was added. This mixture was then poured into a 5 cm diameter test tube and allowed to set. After hardening of the gel, 6 ml of a saturated solution of SnI$_2$ dissolved in concentrated HI was added to the top of the set gel. This last step was carried out in a glove bag under nitrogen. The test tube was then capped with a rubber stopper and placed in a 35°C water bath. In general a 1 M CsI solution was used and crystallization started after one hour. The crystals were separated from the gel using spatulas after crystal size reached a maximum.

CHARACTERIZATION

After 5 days both CsSnI$_3$ and CsSn$_2$I$_5$ single crystals are observed in the gel matrix. The CsSnI$_3$ is yellow with a needle-like habit. The $CsSn_2I_5$ crystals are octahedral in shape but vary in color depending on whether the test tube has been exposed or kept absent from ultraviolet or visible light or air. When grown in the absence of light the crystals are orange and insulators. The same $CsSn_2I_5$ crystals when grown in the light are black and are good semiconductors. On exposure to air or light the orange crystals develop a black coating. The black crystals grown in the presence of light also have an orange inner matrix.

Atomic absorption analyses (Cs, Sn) and Volhard and Mohr titrations (I) as well as ion selective electrode analysis (I) have not yielded any detectable difference between the Cs/Sn/I ratios of the orange and black forms of the crystals.

The orange $CsSn_2I_5$ form crystallizes in the tetragonal system in the not uniquely determinable space group P4cc (#103) or P4/mcc (#124) with lattice constants a = 6.40 Å and c = 16.05 Å. A single crystal X-ray structure determination of this system is presently underway.

A series of 4-probe single crystal and pressed powder conductivity measurements were made on the black $CsSn_2I_5$ crystals. The conductivity of these single crystals is 100 ohm^{-1} cm^{-1}.

It is noted here that SnO is black and soluble in aqua regia whereas SnO_2 is white to translucent and insoluble in aqua regia. The black $CsSn_2I_5$ crystals when put into aqua regia dissolve except for a thin transparent film which remains in solution.[2]

SURFACE STUDIES

An Auger electron spectroscopy study was undertaken to determine the surface composition of the $CsSn_2I_5$ system. Figure 1a shows the Auger spectrum of the black single crystal form of $CsSn_2I_5$. The spectrum shows a doublet at 426 eV and 430 eV indicative of tin and a peak at 510 eV due to oxygen. It is noted that peaks for Cs and I are not present. The intensity ratio of the 430 eV/426 eV peaks for Sn is characteristic of chemical bonding.[3] In the case of the crystal surface this ratio is greater than unity. This intensity ratio was also observed on bulk tin oxide powder. In the case of the crushed crystal in Figure 1b this ratio is less than unity. In the crushed crystal spectrum new peaks at 520 eV due to iodine and a doublet at 563 eV and 575 eV due to cesium are now apparent. The crushed crystal was prepared by crushing a single crystal into indium foil and immediately placing this into the Auger spectrometer under nitrogen. The orange color due to the crushing could still be seen after the spectrometer was pumped down to 10^{-10} torr.

$$CsSn_2I_5$$

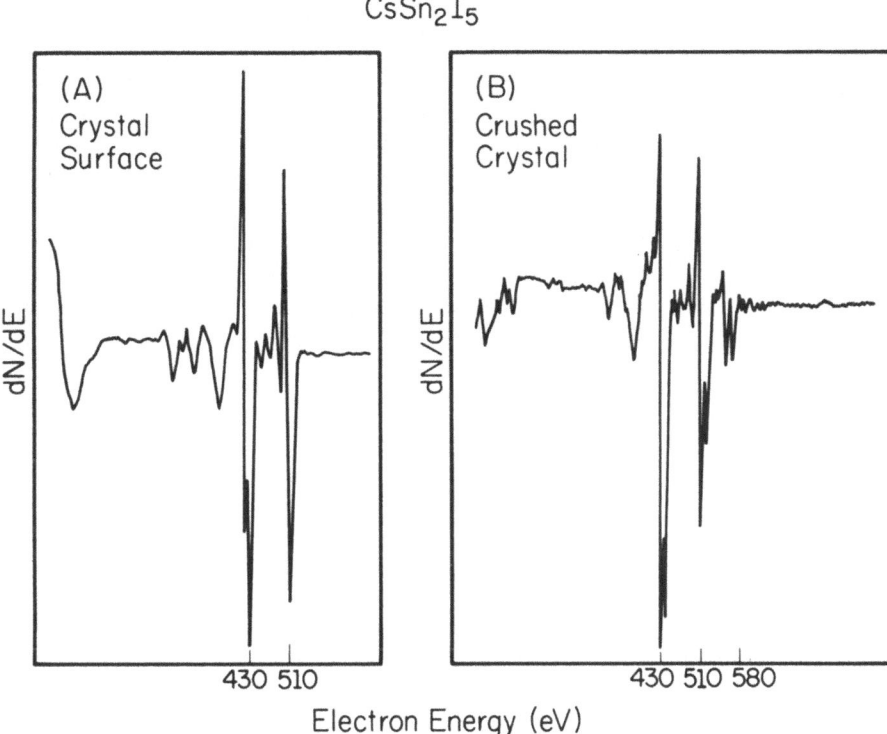

Electron Energy (eV)

Figure 1. Auger spectra of single crystal $CsSn_2I_5$: (a) crystal surface; (b) crushed crystal.

Figure 2a shows the spectrum observed after profiling through a significant portion of the oxide layer. At this depth Cs and tin oxide are present. Iodine does not appear to be present at this point. The Auger spectrum for the crushed crystal is shown in Figure 2b for comparison.

CONCLUSIONS

These data suggest the following in depth structure for these crystals:

1. An orange core consisting of $CsSn_2I_5$;

2. A transition region containing SnO (black) and possibly cesium in which the iodine level may be below the Auger detection limit;

3. A surface layer of tin oxide SnO_2 (transparent) in which both cesium and iodine levels are below the Auger detection limit.

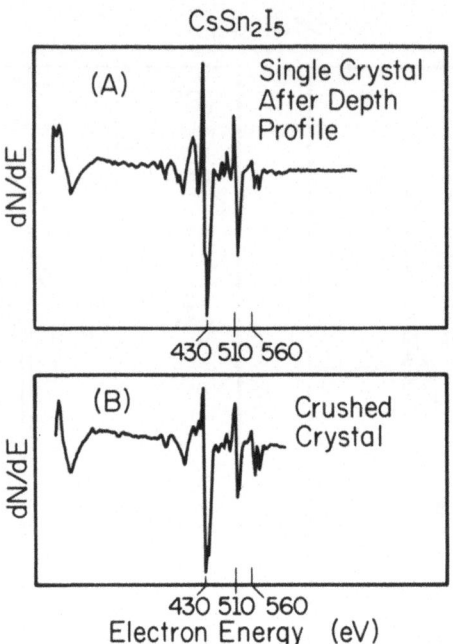

Figure 2. Auger spectra of single crystal $CsSn_2I_5$: (a) after depth profile; (b) crushed crystal.

These conclusions account for the color and the electrical conductivity of this system since SnO_2 is known to be a good conductor even in the form of a thin film. We are presently studying the tin Mössbauer and ESCA of the black and orange forms to determine whether interstitial tin metal is involved in the black $CsSn_2I_5$ conductor. It is concluded that the electrical conductivity is due to conduction on the surface rather than through the crystal.

REFERENCES

1. H. K. Henisch, Crystal Growth in Gels (Pennsylvania State University Press, Ltd., University Park, 1973).

2. D. L. Zellmer, Ph.D. Thesis, Analysis and Electrochemistry of Semiconducting Tin Oxide, Urbana, Illinois, 1969.

3. T. A. Carlson, Photoelectron and Auger Spectroscopy (Plenum Press, New York, 1975).

ONE-DIMENSIONAL PARTIALLY OXIDIZED TETRACYANOPLATINATE METALS:

NEW RESULTS AND SUMMARY*

Jack M. Williams and Arthur J. Schultz

Chemistry Division, Argonne National Laboratory

9700 S. Cass Avenue, Argonne, Illinois, 60439, USA

INTRODUCTION

In this section we have been requested to review the status of research dealing with the physical and chemical properties of partially oxidized tetracyanoplatinate (POTCP) complexes and their analogues of which the prototype is $K_2[Pt(CN)_4]Br_{0.3} \cdot 3H_2O$, "KCP(X)" where X = Br^- or Cl^-.[1] If such a request for an up-to-date review had been proferred less than 5 years ago, there would have been very few partially oxidized tetracyanoplatinate POTCP analogues, other than KCP(Cl) and KCP(Br), to discuss. Fortunately, in the past two years a large number of new POTCP complexes with unique physical properties have been synthesized. The new salts display among other things, vastly *different* materials properties, differing degrees of partial oxidation (DPO) of the Pt atoms, unusual and variable crystal structures, and in some cases electrical conductivities (greatly increased!) over that of prototypic KCP(Br,Cl). Our search to characterize POTCP materials is based on the goal of determining what determines their unique properties, as expressed in 1974 by Bloch and Weisman:[2] *"It appears likely that the partially oxidized Krogmann salts are chemically unique systems in which, by accident, a set of rather exacting energetic and structural criteria are simultaneously satisfied. If this is so, then the synthetic search for analogues is likely to be frustrating."*

---

*Research at Argonne National Laboratory is performed under the auspices of the Division of Basic Energy Sciences of the U. S. Department of Energy. J.M.W. wishes to acknowledge N.A.T.O. (grant no. 1276).

It is our interest in this section to first review the salient research results derived for KCP(X) to date, which will include a brief description regarding the experimental conditions required in order to synthesize these systems, and then to enlarge on this information and discuss previously unknown POTCP one-dimensional metals. We have presented a limited review of POTCP materials recently.[3] To that end we will not deal extensively with the physics or theory of these materials and refer the reader to other recent reviews.[4] The other main group of partially oxidized Pt salts are the oxalates and they have been reviewed by Miller and Epstein.[4]

I.   METALLIC CONDUCTIVITY IN ONE-DIMENSIONAL TRANSITION METAL $\underline{d}^8$ SYSTEMS

The outstanding features associated with POTCP salts are anisotropic physical properties, the most noteworthy of which is their 1-dimensional (1-D) metallic conductivity along the metal chain axis. This takes place via electron delocalization along the overlapped d_z2 orbitals as depicted in Fig. 1. The genesis of

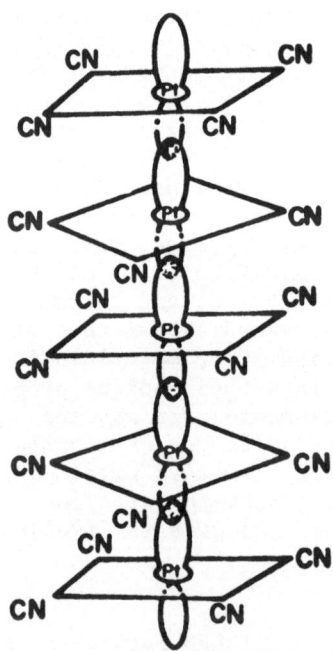

Fig. 1.   Diagram of overlapped d_z2 orbitals of platinum which result in metal "chain" formation.

such systems, as pointed out by Krogmann[5] and Underhill[6], generally requires that two important conditions must be met:

(1) The coordinated metals must "stack" in a columnar fashion, as shown in Fig. 1, in such a manner as to allow d_z2 orbital overlap. It is generally observed that square-planar metal coordination promotes such stacking interactions which in turn may result in metal-metal separations only ∿0.01-0.3 Å longer than in the parent metal. Coordinating ligands which promote these interactions are limited mainly to Cl^-, CN^-, CO, and $(C_2O_4)^{-2}$.

(2) The coordinated metal must be capable of forming nonintegral formal oxidation states via chemical reactions causing only partial oxidation. The partial oxidation of the Pt metal results in a partially filled band which is required for high electrical conductivity. In the case of POTCP salts (vide infra) the DPO may be varied from ∿2.2-2.4. As the DPO is increased, the Pt-Pt intrachain separation *decreases* dramatically and, in general, the metallic conductivity along the chain axis rises markedly.

Krogmann[5] and Underhill[6] have both pointed out that the metals given in Table 1 have accessible $\underline{d}^8$ electron configurations and therefore a great tendency to form square-planar coordination complexes which form columnar structures. To date, and in decreasing order of occurrence, Pt, Ir, and Rh have been observed to form 1-D metal complexes. It does appear that Pd^{+2} and Ni^{+2} form weakly interacting metal chains in the *bis*-(diphenylglyoximato)tri-iodide derivatives of Pd and Ni.[7] In the Ni derivative, for example, the M—M distance (3.27 Å), is much longer than in Ni metal (2.49 Å). The long M—M distance appears to be due in part to the nonplanar configuration of the ligands about the Ni atom. Since it is our intent to review, as best we can, the POTCP series of 1-D conductors, the ensuing discussion will focus particularly on the relationship between crystal and molecular structure and the understandable ways in which they vary from one POTCP complex to another.

TABLE 1. METALS WITH ACCESSIBLE $\underline{d}^8$ ELECTRON CONFIGURATIONS

Configuration	Metal (oxidation state)			
$3d^8$	Fe(0)	Co(+1)	Ni(+2)	Cu(+3)
$4d^8$	Ru(0)	Rh(+1)	Pd(+2)	Ag(+3)
$5d^8$	Os(0)	Ir(+1)	Pt(+2)	Au(+3)

The Partial Oxidation of Pt^{+2} Salts and Methods for the Rapid
Synthesis of New POTCP Complexes

Numerous square-planar TCP complexes of Pt^{+2} exist such as
$K_2[Pt(CN)_4] \cdot 3H_2O$ [Pt-Pt = 3.48 Å], which form columnar stacks containing infinite Pt-Pt chains. However, since the Pt^{+2} salts are
not partially oxidized their metal-metal repeat separations are
typically 0.3–0.8 Å longer than in Pt metal (2.78 Å).[8] In addition to being nonmetallic in appearance, they are electrical insulators due to the filled d_z2 band [conductivity $\sim$5 x 10^{-7} Ω^{-1}
cm^{-1} for $K_2[Pt(CN)_4] \cdot 3H_2O$]. Upon partial oxidation of most *mono*-
valent cation TCP salts of Pt^{+2}, the formation of the $Pt^{2.2-2.4}$
salt and a partially filled d_z2 conduction band produces some very
dramatic changes: (a) the conductivity along the Pt chain increases by a factor of $\sim$$10^6$ or greater as the Pt-spacings diminish
to <2.95 Å, and (b) crystals with metallic lusters ranging from
gold to bronze to copper are formed. However, the appearance of
a metallic luster does not guarantee 1-D metallic conductivity.
This may be checked using polarized specular reflectance spectroscopy, as pointed out by Krogmann and Geserich[9] (see Fig. 2), and
we have used this technique to demonstrate that $M_{1.75}[Pt(CN)_4] \cdot xH_2O$,
$M=K^+$, Rb^+, and Cs^+, and 1-D metals.[10] As shown in Fig. 2,

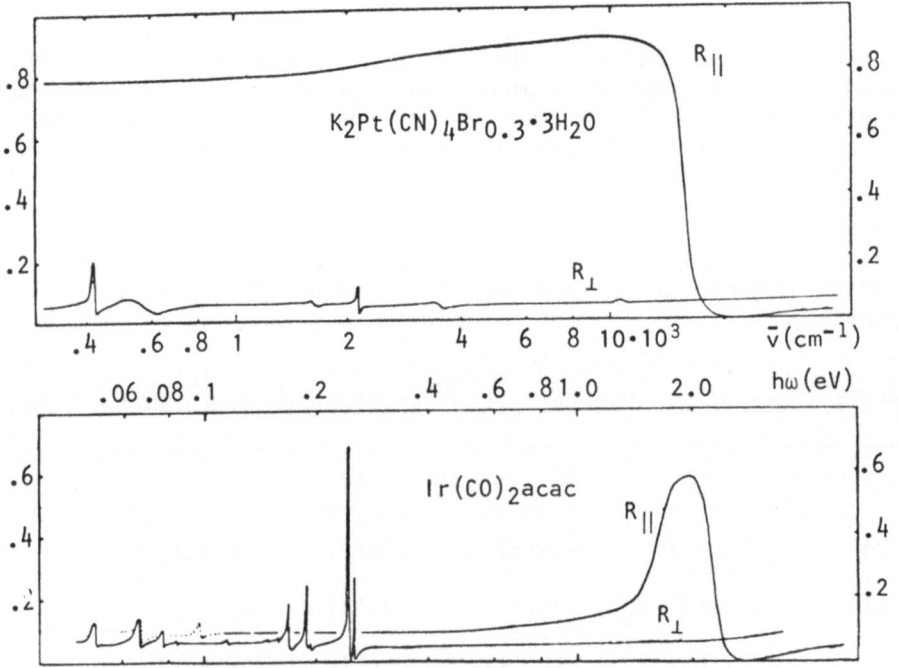

Fig. 2.   Polarized specular reflectance spectra of KCP(Br)(1-D
metal) and $Ir(CO)_2$ acac (1-D semiconductor) [after
Krogmann and Geserich (9)].

$K_2[Pt(CN)_4]Br_{0.3} \cdot 3H_2O$ strongly reflects polarized light parallel to the chain direction $[R_{||}]$ even in the far infrared, while the $Ir^{+1.0}$ complex, with its completely filled valence shell, shows a maximum and rapid drop-off in $R_{||}$. Thus, even though both complexes appear to be metallic, the Ir complex is certainly not a 1-D metal. The sharp rise in $R_{||}$ simply means that a plasma edge occurs, i.e., that the collective modes of the electrons in the crystal are not strictly localized and can be excited. The fact that the $R_{||}$ does not rise to 100% also means that an *energy gap* is created. The physical property changes which accompany partial oxidation may be explained in terms of a molecular-orbital model, or in terms of band theory.[5] In the latter, partial oxidation (for example by Cl^- in the case of $K_2[Pt(CN)_4] \cdot 3H_2O$) results in electron removal from the highest occupied band, thereby forming a partially filled band with the expected metallic properties.

We would now like to briefly mention the traditional methods for synthesis of anion and cation deficient POTCP salts which are indicated in 1 and 2 below:

(1)   Anion deficient KCP(Br):

$$\frac{5}{6}K_2[Pt^{+2.0}(CN)_4] \cdot 3H_2O + \frac{1}{6}K_2[Pt^{+4.0}(CN)_4Br_2] \overset{H_2O}{\rightarrow}$$

$$K_2[Pt^{+2.3}(CN)_4]Br_{0.3} \cdot 3H_2O.$$

(2)   Cation deficient K(def)TCP:

$$K_2[Pt^{+2.0}(CN)_4] \cdot 3H_2O + H_2O_2 \overset{H^+}{\rightarrow} K_{1.75}[Pt^{+2.25}(CN)_4] \cdot 1.5H_2O.$$

In formula (1) above, KCP(Br) is prepared by mixing the unoxidized and oxidized starting materials $[\frac{5}{6}Pt^{2.0} + \frac{1}{6}Pt^{4.0} \rightarrow Pt^{\sim 2.3}]$; in formula (2), K(def)TCP is prepared by peroxide oxidation of the $Pt^{2.0}$ salt. It is important to note that in these somewhat special cases it now appears that the stoichiometries of the products are *not* dependent on the ratios of the starting materials, but only on the final oxidation state of Pt. However, as we will show later, in the case of the $(F-H-F)^-$ systems the reactant concentrations are of extreme importance and allow production of *different* POTCP salts using the *same* cation and anions. Indeed, for the $(F-H-F)^-$ systems, method (1) above is not feasible because the appropriate $(FHF)^-$ derivative of the Pt^{+4} salt apparently cannot be prepared.

What we wish to stress is that the long and tedious methods given above (only final synthetic steps were given) have been largely replaced by electrochemical techniques. The technique generally involves oxidation of Pt^{+2} salts, with a 0.5-1.5 V DC source using Pt electrodes, which usually results in the production of the desired POTCP salt in *minutes (or hours)* rather than weeks.

These techniques have been discussed in detail elsewhere.[11] It
should be stressed that although numerous Na$^+$ salts have been discussed in the literature, in all likelihood it appears that none
have been prepared to date.

In the next section we will discuss the crystal and molecular
structures of selected POTCP salts in order to illustrate the
special relationships which are coming to light which relate
structure and physical properties such as electrical conductivity.
The interplay of these relationships serves as a guide to the
synthesis of new materials, often with enhanced conductivity in
considerable excess of that of KCP(Br).

II.   GENERAL COMMENTS REGARDING THE CRYSTAL AND MOLECULAR STRUCTURES OF 1-D POTCP COMPLEXES

Before discussing the molecular and crystal structures of
analogues of KCP(Br), the prototype complex for the POTCP series
of 1-D conductors, certain conspicuous observations can be drawn,
i.e., *the basic molecular geometries formed by POTCP complexes
are, by and large, somewhat predictable because they are guided
by (i) the basic integrity of the Pt-Pt chains once partial oxidation has occurred, (ii) by the columnar stacking of the square-
planar TCP moieties, and (iii) by the limited range of allowed
Coulombic and hydrogen bonding crystal binding patterns.* We have
pointed out elsewhere[3] that the structures of the starting materials
themselves [Pt^{2+} and Pt^{4+} salts], and the M$^+$...$^-$N≡C interactions
which influence the Pt-Pt overlap, contain structural geometries
similar to many known POTCP complexes. However, only one item of
obvious importance need be mentioned here in as much as it pertains
to the synthesis of POTCP salts in general. As pointed out in
equations (1) and (2) above regarding the synthesis of KCP(Br), the
Pt^{2+} salt is *hydrated* and the Pt^{4+} salt is *not*. Our subsequent
structural studies have revealed that in the Pt^{4+} salt, K$_2$[Pt(CN)$_4$-
Br$_2$], two Br$^-$ ions lie *within* the 8-fold coordination sphere of
the K$^+$ ion. However, additional studies have shown that if the
Pt^{4+} salt is hydrated, e.g., Ba[Pt(CN)$_4$Br$_2$]·4.5H$_2$O and Na$_2$[Pt(CN)$_4$-
Br$_2$]·2H$_2$O, and *no* X$^-$ ions reside within the metal-ion coordination
sphere, then it does not seem to be possible to prepare the POTCP
salt in an aqueous based system. (For a detailed discussion, see
ref. 12). The use of other solvent systems for the synthesis of
some POTCP salts may be indicated. Obviously the knowledge of the
structural geometries of Pt^{+2} and Pt^{+4} starting materials may play
a role in guiding future syntheses of POTCP salts.

Anion Deficient $K_2[Pt(CN)_4]Br_{0.3} \cdot 3.0H_2O$, KCP(Br)

Understanding the crystal structure of KCP(Br) is of fundamental importance in order to explain the unique electrical conductivity associated with the Pt-Pt chains in this system. The
basic crystal structure first reported by Krogmann and Hausen[13]
showed the unusually short [Pt-Pt ≃ 2.89 Å] and linear Pt-atom
chains although the water molecule and K^+ distribution required
revision, as illustrated in Fig. 3. The Pt atoms occupy the positions (00$\underline{z}$) and are *equally spaced*, though this is not required in
space group $\underline{P4mm}$. The $Pt(CN)_4^{-1.70}$ groups are not planar, as
shown in Fig. 4, and because of Coulombic $K^+ \ldots {}^-N{\equiv}C$ interactions
the Pt chain is alternately compressed or expanded in a pairwise
fashion (Fig. 5).[14]

Our group was the first to locate a "second site"[14] of then
unknown importance with respect to the electrical conductivity of
KCP(Br) and labeled as Br* in Fig. 3 and H_2O* in Fig. 4. With due
caution we stated[15] that "We are aware, of course, that the second
site could represent an oxygen atom (of an OH⁻ or H_2O molecule),"

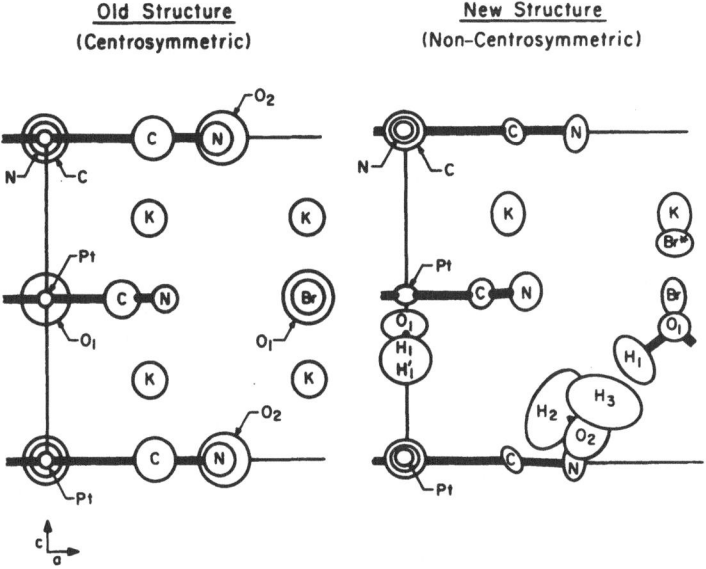

Fig. 3.   The old (x-ray)[13] and revised (neutron)[14] structure of
         $K_2[Pt(CN)_4]Br_{0.3} \cdot 3H_2O$. Note that in the revised structure
         the atom labeled as Br* is a highly disordered H_2O molecule.

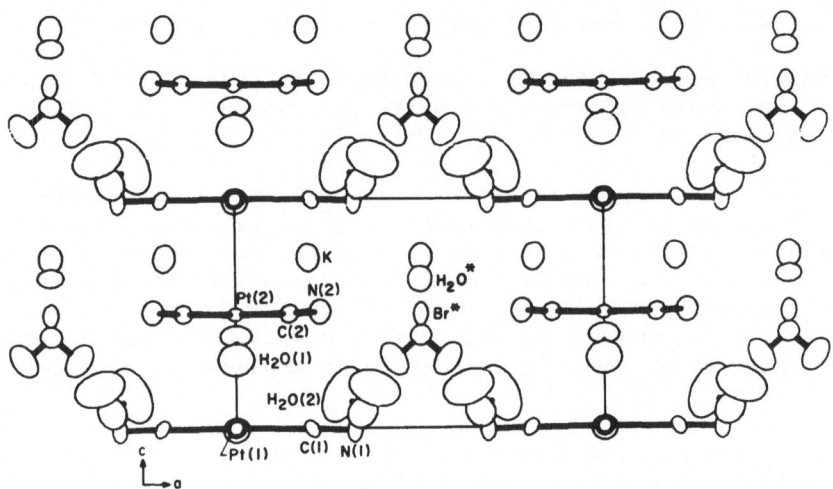

Fig. 4.   A *b*-axis half-cell projection of the structure of K_2Pt-$(CN)_4Br_{0.3}\cdot3H_2O$.[14] Only independent atoms in the asymmetric unit are identified. The Pt atom at the origin is overlapped by C and N atoms. The disordered Br^- and H_2O sites are labelled as Br* and H_2O*. Note the asymmetric placement of the K^+ ions and H_2O molecules in the unit cell.

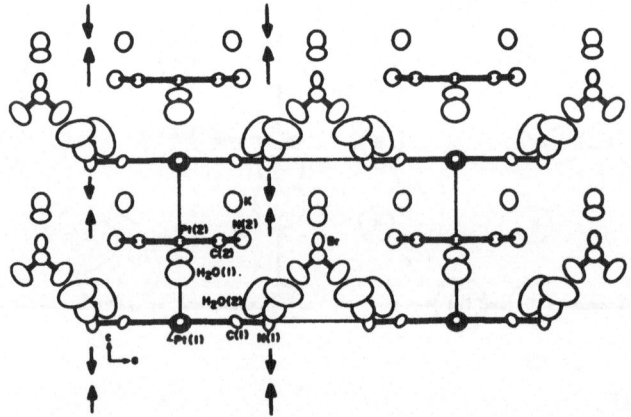

Fig. 5.   Diagram showing the allowed longitudinal distortions of the Pt chain in $K_2[Pt(CN)_4]Br_{0.3}\cdot3H_2O$ as caused by the $K^+\ldots^-N\equiv C$ interactions. The required $\overline{4mm}$ symmetry and alternating layers of K^+ ions dictate $\overline{the}$ directions in which the K^+ ions alternately compress and expand the Pt chains.

but with room temperature data we could not resolve this fine point. Numerous subsequent crystal structure studies at liquid He temperature, on both hydrogenated and deuterated materials indicate that the "second site" is partially occupied by a water molecule with a highly disordered orientation of the hydrogen atoms.[16] Thus, the *central portion* of the unit cell in KCP(Br) is alternatively occupied by either a Br^- ion at $(1/2,1/2,1/2)$(labelled Br* in Fig. 4, 60% occupancy) or a water molecule at $(1/2,1/2,0.67)$(H_2O* in Fig. 4, 40% occupancy). These arguments lead to a hydration value of 3.2 H_2O, which is consistent with the thermogravimetric analyses reported by Peters and Eagen.[16] However, a 3.0 hydrate is indicated from the plot in Fig. 6, which has been confirmed by other investigators.[17] Furthermore, there appear to be inconsistencies concerning the disordered orientation of the water molecule as determined by NMR *vs* diffraction methods.[16] The rather drastic effects of water content on the electrical conductivity in KCP(Br) are summarized by Drosdziok and Engbrodt.[18]

Of further importance is that KCP(Br) undergoes a gradual transition from a room temperature 1-D conductor (σ = 5–800 Ω^{-1} cm^{-1}) to a low temperature insulating state.[4] The observation of

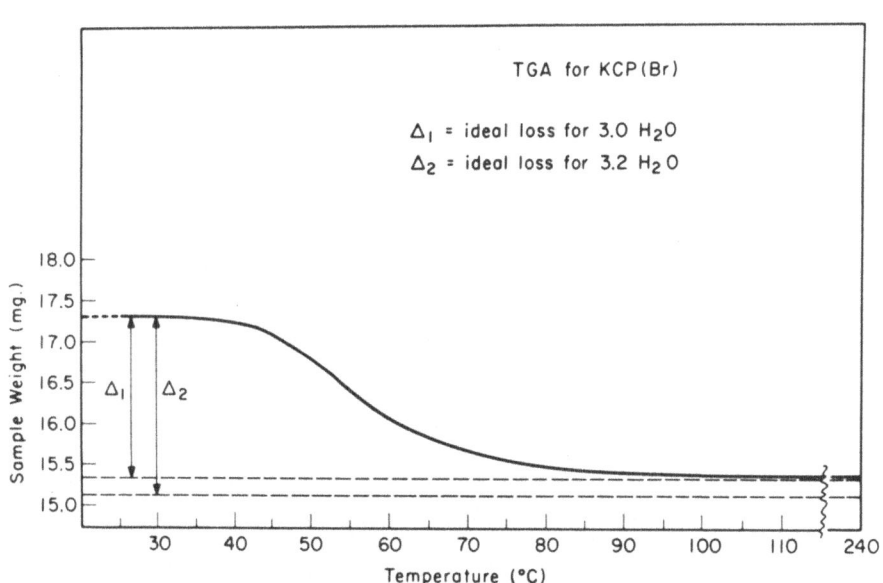

Fig. 6.   Thermogravimetric analysis of $K_2[Pt(CN)_4]Br_{0.3}\cdot 3.0H_2O$, KCP(Br), using a heating rate of 4°C/min. Note that KCP(Br) is a 3.0 hydrate if prepared by the method of Saillant et al.[48] and is not a 3.2 hydrate.

a giant Kohn anomaly[19,20] and a low temperature three-dimensional superlattice[19,21,22] from various x-ray diffuse scattering and inelastic neutron scattering studies have been proposed as evidence for a Peierls transition in KCP(Br). At room temperature, the scattering experiments indicate dynamic, sinusoidal displacements of the platinum atoms along each chain, with no coherence between neighboring chains. The repeat period of 6.67 $\underline{c}'$, where $\underline{c}'$ is the Pt-Pt distance, is exactly that expected for a Peierls distortion. At 77°K and below, a 3-D ordering occurs and the distortion becomes static. Neutron scattering studies by Lynn et al.[23] indicate that the sinusoidal Pt chain distortion involves the entire $Pt(CN)_4^{-1.7}$ unit which responds to periodic fluctuations in electronic charge due to a charge density wave (CDW) formed by the Pt d_{z^2} electrons. In structural terms, the $Pt(CN)_4^{-1.7}$ groups "ride" with the CDW and at liquid He temperatures the CDW displacement of the Pt atoms is only 0.025 Å from their perfectly equal spacing derived by classical structure analysis. Therefore, the *average* crystallographic structure of KCP(Br) is invariant with temperature but is modulated by a CDW, to which the $Pt(CN)_4$ groups respond.

Anion Deficient $Rb_2[Pt(CN)_4]Cl_{0.3} \cdot 3H_2O$, RbCP(Cl)

In a fashion analogous to KCP(Br), RbCP(Cl) has been characterized by single crystal neutron diffraction, x-ray diffuse scattering, and thermogravimetric analysis.[24] It appears that RbCP(Cl) is isostructural (space group $\underline{P4mm}$), with KCP(Br), except that (at room temperature) only one halide site can be identified, i.e., there does not seem to be any way to prove directly the presence of a "defect" water molecule in RbCP(Cl) (vide infra). That RbCP(Cl) is a 3.0 hydrate is indicated in Fig. 7. The crystal structure is given in Fig. 8 which definitely indicated that the Pt-atom chain in RbCP(Cl) is "dimerized", i.e., the Pt-Pt distances occur in discrete pairs within the perfectly linear chains. An x-ray diffuse scattering study[24] has established the Pt oxidation state as +2.31 (2), which is consistent with the stoichiometry obtained from elemental analyses. Because of the isostructural nature of RbCP(Cl) and KCP(Cl), it would appear that the replacement of the smaller K^+ ion by the larger Rb^+ ion results in lattice expansion along $\underline{c}$, which produces larger Pt-Pt intrachain separations. Thus, the average intrachain Pt-Pt separation (2.90 Å) is slightly larger than in KCP(Br)(2.88 Å) and KCP(Cl)(2.87 Å). It then appears reasonable that the unequal and slightly longer Pt-Pt separations in RbCP(Cl), compared to those in KCP(X) apparently result in *increased* electron localization along the Pt-atom chain and the order of magnitude decrease in electrical conductivity of RbCP(Cl) compared to KCP-(Cl).[25] The alternate expansion (Pt-Pt = 2.924(8) Å) and contraction (Pt-Pt = 2.877(8) Å) of the Pt chain arising from $Rb^+...\overline{N}\equiv C$ interactions, is illustrated by arrows in Fig. 9. The degree of

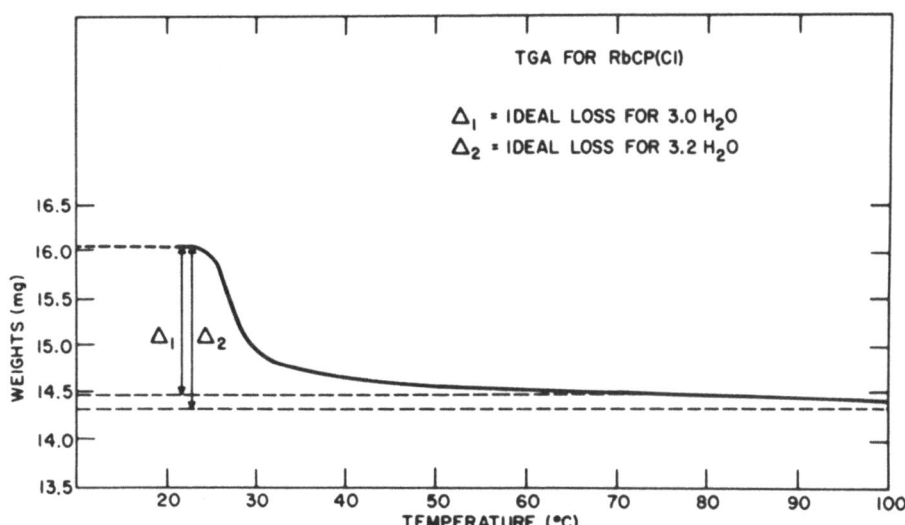

Fig. 7.   Thermogravimetric analysis of $Rb_2[Pt(CN)_4]Cl_{0.3} \cdot 3.0H_2O$, RbCP(Cl)(heating rate 4°C/min).

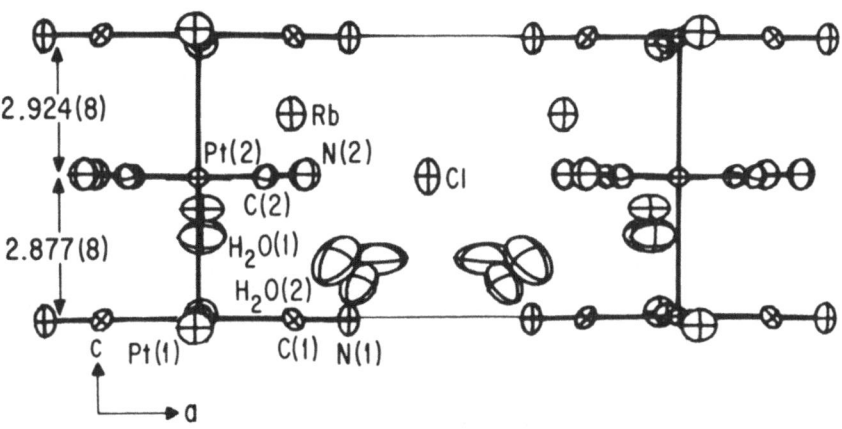

Fig. 8.   Drawing of the unit cell (50% probability ellipsoids) of $Rb_2[Pt(CN)_4]Cl_{0.3} \cdot 3.0H_2O$ showing the linear Pt atom chain, which contains *un*equal Pt-Pt separations, and the asymmetric location of the Rb^+ ion and the H_2O molecules. The average Pt-Pt distance in RbCP(Cl) is ~0.03 Å longer than in KCP(Cl) or KCP(Br). Note that only one site is indicated for Cl^- (see text) for this room temperature structure.

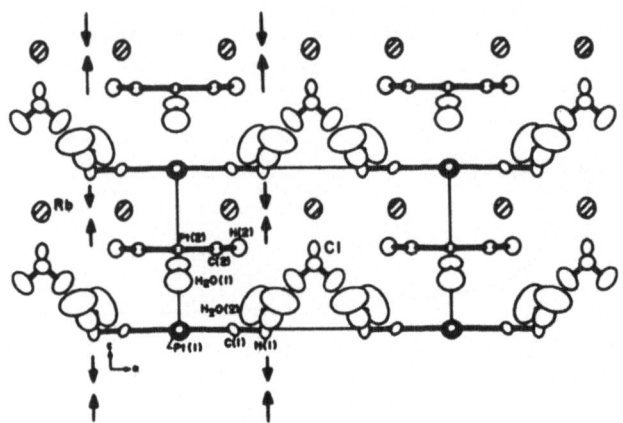

Fig. 9.   Diagram showing the $CN^- \ldots Rb^+$ interactions (indicated by arrows) which result in canting of the cyanide groups toward the cation and dimerization of the Pt atom chain in $Rb_2[Pt(CN)_4]Cl_{0.3} \cdot 3.0H_2O$

chain dimerization decreases at 110°K, i.e., the Pt-Pt distances become 2.862(6) and 2.885(6) Å with a net decrease in the average Pt-Pt spacing of ∿0.02 Å to 2.87 Å (see Fig. 10). Since the conductivity also decreases at low temperatures, it appears that the Peierls transition, and not the degree of dimerization, is the major factor in determining the temperature dependence of the conductivity.

Although we were unable to completely verify the presence of a "second site" or "defect water" site in RbCP(Cl), we present the final difference Fourier nuclear density map in Fig. 11, which indicates the existence of a small peak *oppositely* directed along z (0.5,0.5,∿0.3) when compared to KCP(X). We could not verify the site as H_2O or Cl^-, using least-squares refinement techniques, and because for crystals I and II the data were obtained at room temperature it was impossible to determine the exact nature of what appears to be a "second site" analogous to what we first observed in KCP(X). In an attempt to better understand this question of additional sites in the crystal structure we have collected very high precision single crystal neutron diffraction data[26] at 110°K and our preliminary results indicate that there are in fact *two* additional sites, at (0.5,0.5,∿0.35) and at (0.5,0.5,∿0.67)(see Fig. 11). Although we were unable to prove, using least-squares

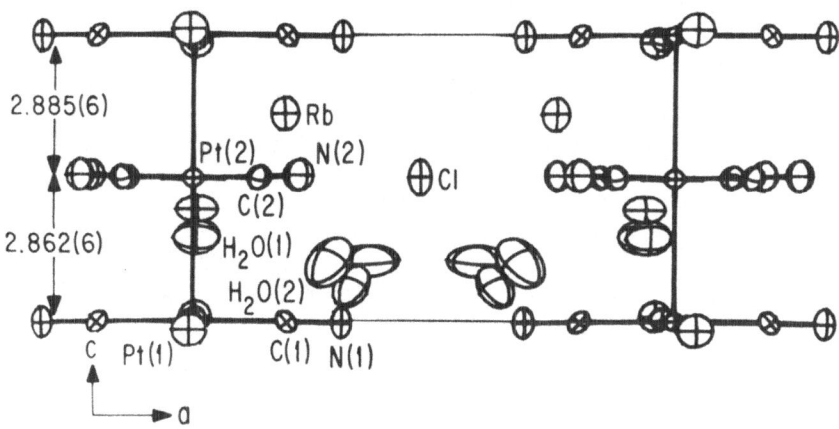

Fig. 10.   Drawing of the unit cell of $Rb_2[Pt(CN)_4]Cl_{0.3} \cdot 3H_2O$ using
           neutron data collected at 110°K. Note that the degree of
           chain dimerization *decreases*, compared to the room temper-
           ature structure (Fig. 8) and that the average Pt-Pt
           spacing decreases by ~0.02 Å at 110°K.

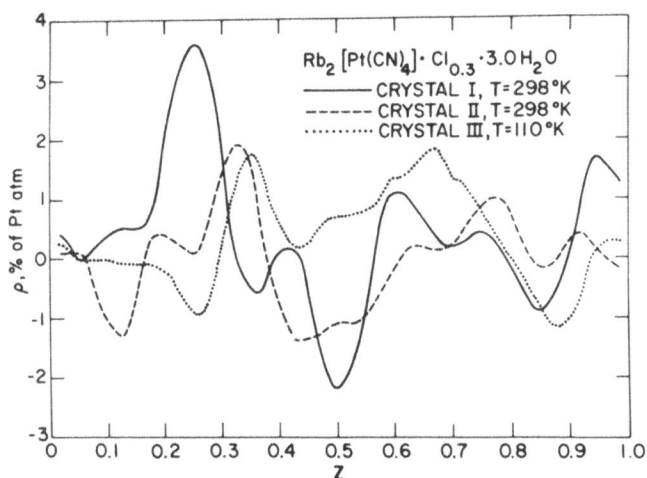

Fig. 11.   Difference Fourier density line $(0.5, 0.5, \underline{z})$ for $Rb_2[Pt-(CN)_4]Cl_{0.3} \cdot 3.0H_2O$ from the neutron data. Crystals I, II,
           and III weighed 1.4 mg, 14.9 mg, and 10.0 mg, respectively.
           A possible "second site" at $\underline{z} \simeq 0.3$ could not be confirmed
           from least-squares analysis of room temperature data but
           was apparent in data collected at 110°K.[26]

refinement techniques, that these extra sites are disordered H_2O
sites we may be reasonably certain that this is the case.

Anion Deficient $(NH_4)_2[Pt(CN)_4]Cl_{0.3} \cdot 3H_2O$, $NH_4CP(Cl)$

Because the ionic radii of Rb^+ and NH_4^+ cations are very nearly
equal (1.48 Å), the molecular structures of the POTCP salts are, as
expected, very similar (space group P4mm) with some noticeable exceptions. Both the NH_4^+ and Rb^+ salts contain dimerized Pt atom
chains. However, the pattern of long and short separations are reversed (see Fig. 12). In the case of the NH_4^+ salt both x-ray and
neutron diffraction data were collected and the Pt–Pt chain repeat
distances are, as in the Rb^+ salt, *different* at 2.914(26) and 2.947-
(26) Å (neutron), or, more precisely, 2.910(5) and 2.930(5) (x-
rays).[27] The average Pt–Pt separation (2.92 Å) in $NH_4CP(Cl)$ is
~0.04 Å longer than that observed in KCP(Br). The combined effects
of the increased, and now *unequal*, intrachain Pt–Pt separations in
$NH_4CP(Cl)$ appear to be the main cause of its diminished conductivity
compared to KCP(Br). A factor which supports this hypothesis is that
in RbCP(Cl) the Pt–Pt distances are also of unequal length (2.924(8)
and 2.877(8) Å) and RbCP(Cl) also has a lower room temperature conductivity than KCP(Br).

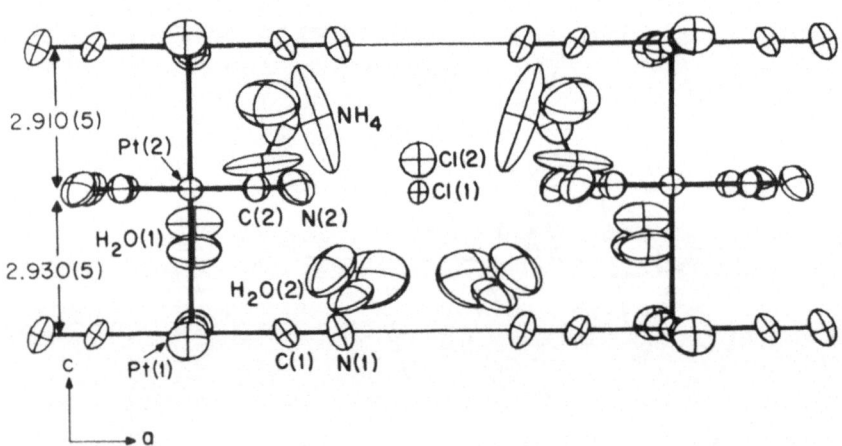

Fig. 12.   Drawing of a half-unit cell (50% probability ellipsoids)
           of $(NH_4)_2[Pt(CN)_4]Cl_{0.3} \cdot 3H_2O$ showing the linear Pt atom
           chains, the *unequal* Pt–Pt separations, and the asymmetric
           placement of H_2O molecules and NH_4^+ ions in different
           halves of the unit cell. The drawing of the thermal ellip-
           soids is based on the neutron diffraction data while the
           Pt–Pt spacing is from the x-ray diffraction experiment (see
           text and ref. 27). The sites labelled Cl(1) and Cl(2) may
           also contain H_2O or even H_3O^+.

Numerous synthetic problems have been encountered elsewhere
with regard to the exact Cl⁻ content of NH₄CP(Cl) and, not surprisingly, the neutron and x-ray data yielded different results regarding placement of the Cl⁻ ion.[28] While a "two-site" model for
the Cl⁻ ion (and water?) was obtained from the neutron data [Cl⁻ at
(0.5,0.5,0.4838) and (0.5,0.5,0.6055)], only one site at (0.5,0.5,
0.5) was derived from the x-ray data. (However, the Cl⁻ x-ray determined occupancy factor could not be refined by least-squares
methods to a reasonable value.)[27]

Even though NH_4^+ and Rb^+ have very nearly equal cationic radii
and their POTCP salts are isomorphous, their low-temperature electrical conductivities [NH₄CP(Cl) << RbCP(Cl)] are vastly different.[25]
As illustrated in Figs. 13 and 14, the H-bonding schemes involving
water molecule protons in RbCP(Cl), KCP(Br), and NH₄CP(Cl) are essentially identical. However, because the NH_4^+ ion forms H-bonding
interactions in NH₄CP(Cl), the salt contains three N—H···⁻N≡C and
one N—H···O bond. Therefore, it appears that the conductivity
variations observed in these salts may also result from subtle
crystal binding changes due to different cations and anions.

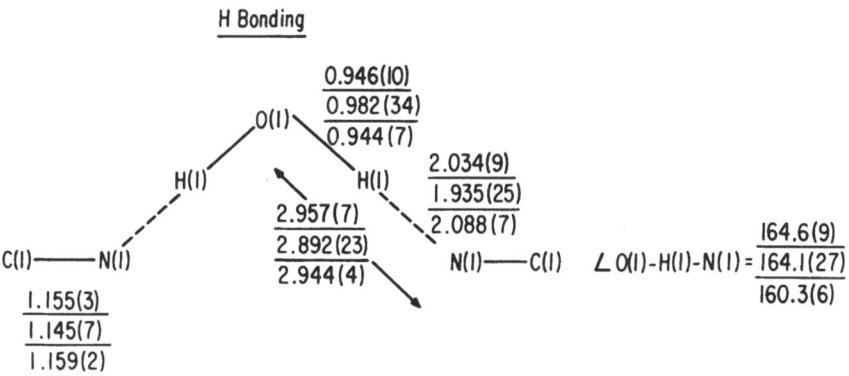

Fig. 13. Composite drawing showing the room temperature H₂O(1)
molecule hydrogen bonding interactions in Rb₂[Pt(CN)₄]-
Cl₀.₃·3.0H₂O (top) in (NH₄)₂[Pt(CN)₄]Cl₀.₃·3H₂O (middle),
and in K₂[Pt(CN)₄]Br₀.₃·3.0H₂O (bottom). The isostructural nature of all three compounds is evident from the
nearly equal hydrogen bonding patterns (see also Fig. 14).

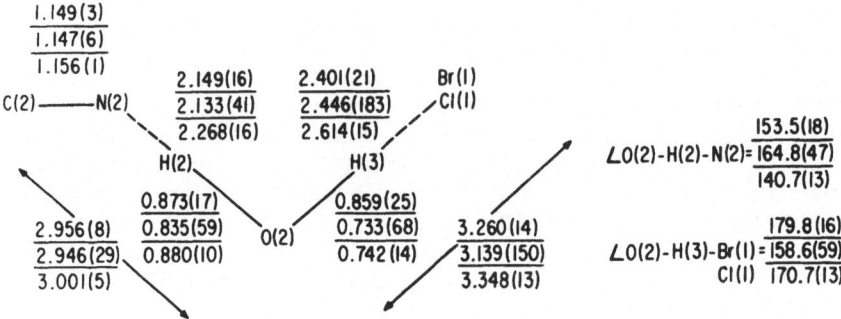

Fig. 14.   Composite drawing indicating the room temperature $H_2O(2)$
           molecule hydrogen bonding interactions in RbCP(Cl) (top),
           NH_4CPCl (middle), and KCP(Br) (bottom.

Anion Deficient (and Anhydrous) $Cs_2[Pt(CN)_4]Cl_{0.30}$, CsCP(Cl)

The Cs^+ derivative of KCP(X) represents an interesting depar-
ture in composition and structure from its hydrated K^+ and Rb^+ analogues, which both occur in space group P4mm. However, CsCP(Cl)
occurs in space group I4/mcm and is apparently anhydrous (a possible
trace of H_2O was detected in a TGA analysis). An x-ray analysis[29]
has shown that the Pt-Pt chain is perfectly linear (Fig. 15) and the
intrachain repeat is constrained to (c/2) = 2.859(2) Å. This Pt-Pt
chain repeat separation is slightly less than that in KCP(Br) (2.88
Å) and therefore, as expected, their electrical conductivities are
very similar.[30]   No trace of residual scattering density attributable
to extra halide or H_2O was evident in the final difference Fourier
electron density maps.   With the exception of the placement of a
single Cl^- anion along the (0,1/2,z) line, CsCP(Cl) is isomorphous
and isostructural with anhydrous $Rb_2[Pt(CN)_4](FHF)_{0.40}$ and $Cs_2[Pt-
(CN)_4](FHF)_{0.39}$, which are members of the first series of *homologous*
POTCP salts, which are discussed in the next section.

It is relevant at this time to point out that ions such as Na^+,
K^+, and Mg^{2+}, which have a high (z^2/r) value are generally hydrated

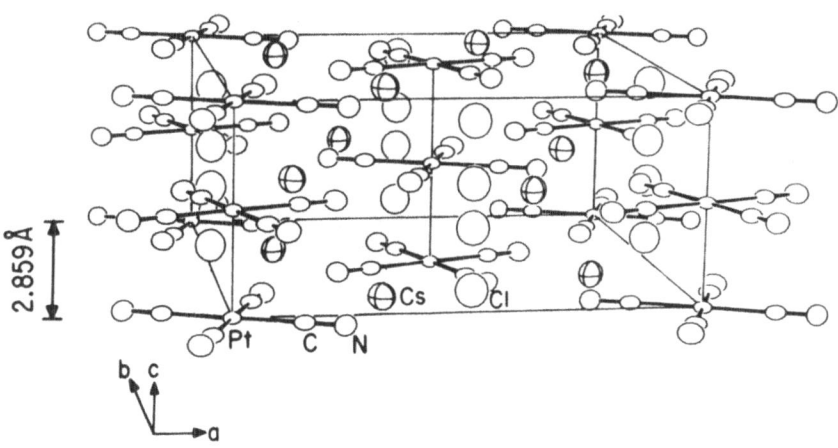

Fig. 15. A perspective view of the unit cell of $Cs_2[Pt(CN)_4]Cl_{0.3}$.[29]
Note the lack of H_2O and the existence of only one Cl^-
site. All thermal ellipsoids are shown at the 50% probability level.

to a much larger extent than ions such as Rb^+ or Cs^+. Thus, the
MCP(X) salts, where M = K, Rb, and X = Cl, Br, all of which are isolated as hydrates, exemplify this point. However, it is generally
agreed that the tendency toward hydration is not solely dependent
on either the cation or anion alone but rather on the particular
combination of cation and anion.[31] Therefore, one might expect
salts having a large disparity in their cation versus anion size to
be hydrated, and salts containing ions of equal size to be anhydrous.
This may explain why RbCP(Cl), with r_{Rb^+}/r_{Cl^-} = 0.81, is hydrated
whereas CsCP(Cl), with r_{Cs^+}/r_{Cl^-} = 0.93, is anhydrous.

Cation Deficient $K_{1.75}[Pt(CN)_4] \cdot 1.5H_2O$, K(def)TCP

When $M_2[Pt(CN)_4] \cdot 3H_2O$, M = K^+, Rb^+, or Cs^+, is partially oxidized (using H_2O_2 or electrolytically) in the *absence* of halide ion
a new type of POTCP salt is formed, *viz* $M_{1.75}[Pt(CN)_4] \cdot 1.5H_2O$. For
example, the new coppery-colored K^+ conducting salt has a room temperature conductivity of $\sim$70–100 Ω^{-1} cm^{-1}, which also decreases with
decreasing temperature.[32] Only one structure of this type has been
characterized and that is the $K_{1.75}[Pt(CN)_4] \cdot 1.5H_2O$ shown in Fig. 16.

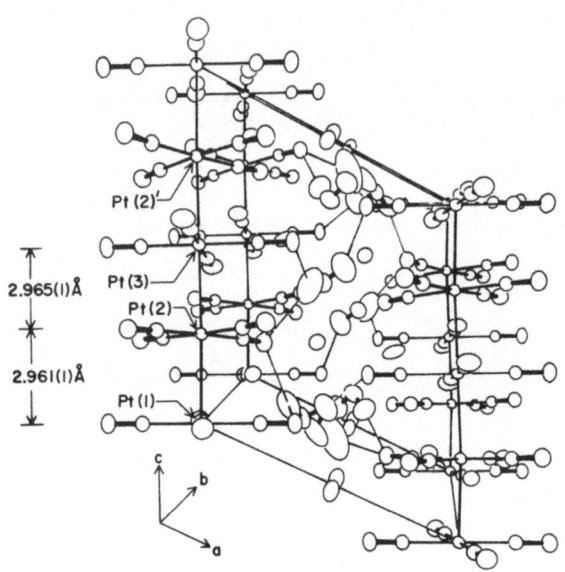

Fig. 16.   The unit cell drawing[33] of $K_{1.75}[Pt(CN)_4]\cdot 1.5H_2O$ showing
            the nonlinear Pt(1)-Pt(2)-Pt(3) chain which extends along
            c and has equal Pt-Pt separations [bond angle 173.25(3)°].
            Thermal ellipsoids are scaled to enclose 50% probability.
            Inversion centers occur at Pt(1) and Pt(3) only.   The
            Pt(2) atom is displaced 0.170 Å perpendicular to the c-
            axis.   Hydrogen bonds from H_2O to cyanide group nitrogen
            atoms (N···H < 2.6 Å) are indicated by faint lines and K^+
            ions are shown without bonding interactions.

The most surprising finding from both x-ray and neutron diffraction[33]
studies is that the salt contains a novel "zig-zag" Pt-atom chain
(see Fig. 17) with two independent but equal Pt-Pt repeat separations which are ∿0.08 Å longer than in KCP(Br) (Pt-Pt = 2.88 Å).
In KCP(Br) the Pt-chains are longitudinally distorted ∿0.01 Å but
this distortion is very small compared to the large transverse distortion of 0.170 Å perpendicular to c in K(def)TCP, as shown in
Fig. 17.   The origin of the Pt-chain distortion in K(def)TCP arises
from the asymmetric K^+ placement about Pt(2), resulting in a charge
imbalance of $K^+\cdots{}^-N{\equiv}C$ Coulombic interactions with concomitant Pt-
chain distortion.   Polarized specular reflectance data have shown
that the K, Rb, Cs(def)TCP salts are all 1-D metals (see Fig. 18).[10]
Infrared analyses of the cation-deficient salts, which demonstrate
that some of the IR active vibrations are sensitive to the oxidation
state of Pt, have also been reported.[34]

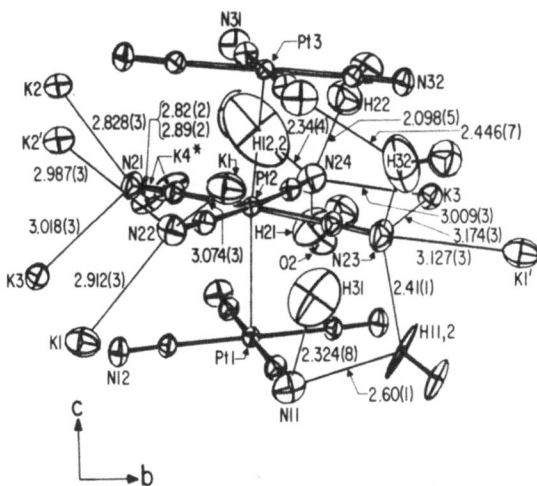

Fig. 17. Diagram showing atom labels and bond distances for the
coordination spheres of Pt(2) and its cyanide nitrogens.
The transverse Pt chain distortion is in the -b direction
as would be expected due to the unbalanced $K^+ \cdots {}^-N \equiv C$ Coulombic interactions (a total of six in the -b direction
and only two in +b). Only interactions to Pt(2) nitrogens
are indicated since symmetry inversion centers at Pt(1)
and Pt(3) prohibit any transverse displacement of these
atoms.

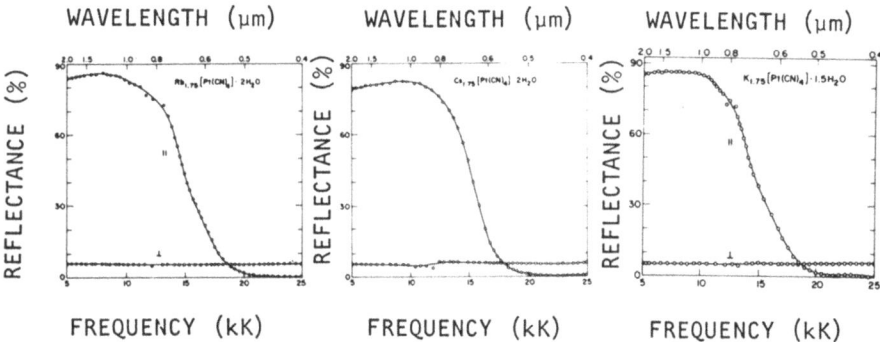

Fig. 18. Polarized specular reflectance[10] from $K_{1.75}[Pt(CN)_4] \cdot 1.5$-
H_2O, $Rb_{1.75}[Pt(CN)_4] \cdot 1.5H_2O$, and $Cs_{1.75}[Pt(CN)_4] \cdot 1.5H_2O$.
Reflectance parallel and perpendicular to the Pt chains
are labelled as $\parallel$ and $\perp$, respectively. The Pt chain in
$K_{1.75}[Pt(CN)_4] \cdot 1.5H_2O$ is bent [Pt(1)-Pt(2)-Pt(3) bond
angle 173.2°].

From a room temperature x-ray diffuse scattering study,[35]
which exhibits sheets of diffuse scattering perpendicular to the Pt
chain direction similar to KCP(Br), it is apparent that the associated superlattice is closely *commensurate* with the c-axis and corresponds to the $2k_F$ wavevector (where k_F is the Fermi wavevector)
predicted from the chemical stoichiometry. The superlattice repeat
is therefore 8 times the average Pt-Pt distance. Neutron scattering
studies of Carneiro et al.[36] show the presence of a Kohn anomaly and
it is concluded that a half-filled conduction band is present. The
commensurate nature of the K(def)TCP superlattice may account for
its lower conductivity compared to KCP(Br), for which the 6.67 c'
superlattice is incommensurate. We have also obtained similar x-
ray diffuse scattering results for the Rb^+ and Cs^+ deficient salts.

III. THE FIRST *HOMOLOGOUS* SERIES OF 1-D POTCP METALS CONTAINING THE (FHF)⁻ AND F⁻ ANIONS

A new series of POTCP salts containing the triatomic bifluoride
anion $(FHF)^-$ have been prepared.[37,38] These results demonstrate the
existence of non-identical POTCP complexes which contain identical
cations and anions, i.e., they constitute the first homologous
series of POTCP complexes.

As indicated in Table II, synthesis of the POTCP complexes may
be accomplished electrolytically or chemically (H_2O_2) from a solution containing $M_2[Pt(CN)_4]$, MF and HF $(M = K^+, Rb^+, or Cs^+)$. In

TABLE II. PREPARATION OF PARTIALLY OXIDIZED TETRACYANOPLATINATE
COMPOUNDS CONTAINING (FHF)⁻ OR F⁻ ANIONS

Compound	[MTCP]	[MF]	[HF]	Oxidation
$K_2[Pt(CN)_4](FHF)_{0.30} \cdot 3H_2O$	0.4	2.9	9.6	1.5 V or H_2O_2
$Rb_2[Pt(CN)_4](FHF)_{0.40}$	0.2	1.9	8.7	1.5 V
$Rb_2[Pt(CN)_4](FHF)_{0.26} \cdot 1.7H_2O$	0.2	1.9	8.7	H_2O_2
$Cs_2[Pt(CN)_4](FHF)_{0.39}$	0.2	2.6	8.7	1.5 V or H_2O_2
$Cs_2[Pt(CN)_4](FHF)_{0.23}$	0.056	1.0	0.076	H_2O_2
$[C(NH_2)_3]_2[Pt(CN)_4](FHF)_{0.26} \cdot xH_2O$	0.8	0	28.9	1.5 V
$Cs_2[Pt(CN)_4]F_{0.19}$	0.3	1.6	a	1.5 V

aIn this preparation, the pH was maintained at 9.0 ± 0.1 during
electrolysis by the addition of CsOH.

the case of $KCP(FHF)_{0.30} \cdot 3H_2O$ and $CsCP(FHF)_{0.39}$, either synthetic method yields identical products given the same initial starting material concentrations. However, with the Rb^+ analogues, the two oxidation procedures yield POTCP complexes that differ in their degree of partial oxidation (DPO), Pt-Pt repeat distance, space group, and degree of hydration (see Table III).

The most remarkable finding involves the anhydrous and iso-structural homologues $CsCP(FHF)_{0.39}$ and $CsCP(FHF)_{0.23}$.[37,38] The only difference between the two complexes is the anion content, which alters the degree of partial oxidation (DPO) (i.e., the Pt oxidation state) and the Pt-Pt intrachain separation (see Table III). This is accomplished by varying the initial concentration of the reactants, as presented in Table II. Since the highest possible degree of "loading" of $(FHF)^-$ into the lattice would yield $CsCP(FHF)_{0.50}$ (see the following section), this suggests a type of solid solution formation and that numerous other homologues may exist. In the case of the $KCP(X)_{0.30}$ ($X = Cl^-$, Br^-) salts, only a

TABLE III.   CONDUCTIVITY AND CRYSTAL DATA OF $(FHF)^-$ CONTAINING POTCP SALTS

Complex	Space Group	Pt-Pt (Å)	$\sigma_\parallel$ (ohm^{-1}cm^{-1})[a]
$K_2[Pt(CN)_4](FHF)_{0.30} \cdot 3H_2O$	P4mm	$2.923(5)$[b]	c
$Rb_2[Pt(CN)_4](FHF)_{0.40}$	I4/mcm	$2.798(1)$	1600
$Rb_2[Pt(CN)_4](FHF)_{0.26} \cdot 1.7H_2O$	C2/c	2.89	c
$Cs_2[Pt(CN)_4](FHF)_{0.39}$	I4/mcm	$2.833(1)$	2000
$Cs_2[Pt(CN)_4](FHF)_{0.23}$	I4/mcm	$2.872(2)$	250-350
$Cs_2[Pt(CN)_4]F_{0.19}$	Immm	$2.886(1)$	c
$[C(NH_2)_3]_2[Pt(CN)_4](FHF)_{0.26} \cdot xH_2O$	c	2.90	c
$K_2[Pt(CN)_4]Br_{0.30} \cdot 3H_2O$[d]	P4mm	$2.890(2)$	700-1050
$K_{1.75}[Pt(CN)_4] \cdot 1.5H_2O$[d]	P$\bar{1}$	$2.963(2)$	115-125

[a]These are the maximum values of four probe d.c. conductivities obtained at room temperature on single crystals. Electrodes were gold wire (0.001 in.) and contacts were made with Aquadag.
[b]This is the average of the two independent Pt-Pt separations of 2.918(1) and 2.928(1) Å.
[c]Under study
[d]These compounds are included for comparative purposes.

single stoichiometry is obtained which is totally independent of the concentrations or relative concentrations of the starting materials.

In addition, we have synthesized the first known POTCP complex containing the monofluoride, F^-, anion. If $Cs_2[Pt(CN)_4]$ is electrolyzed in a basic solution containing CsOH and CsF, crystals with the apparent stoichiometry $Cs_2[Pt(CN)_4]F_{0.19}$ are obtained. This is the first 1-D POTCP complex with body-centered orthorhombic symmetry, which is analogous to the body-centered tetragonal $(FHF)^-$ complexes, except that the a and b axes are no longer equivalent.

The single crystal room temperature conductivities of some of the $(FHF)^-$ salts have been measured (Table III). As one would predict, those complexes with the shorter Pt-Pt separations and correspondingly higher DPO also tend to have higher conductivities. In fact, the room temperature conductivities of $Rb,CsCP(FHF)_{0.4}$ are approximately double that of KCP(Br).

The Crystal Structure of $Rb_2[Pt(CN)_4](FHF)_{0.40}$, $RbCP(FHF)_{0.40}$

The crystal structure of $RbCP(FHF)_{0.40}$ is shown in Fig. 19 and consists of columnar chains of square-planar $Pt(CN)_4^{-1.60}$ groups which stack parallel to $\underline{c}$.[39] The derived Pt-Pt distance (exactly $c/2$) of 2.798 Å is only ~ 0.02 Å longer than in Pt metal itself and is the shortest yet obtained for any POTCP salt. The shortened Pt-Pt spacings are consistent with the high DPO of 0.40. The Pt-Pt distances are all equivalent, the TCP moiety is constrained to perfect planarity and surprisingly, the metallic gold-colored salts are completely anhydrous. The $(FHF)^-$ ions, whose presence was verified by IR spectroscopy,[40] are presumably linear and are also aligned in a conspicuous parallel pattern co-linear with $\underline{c}$ (the Pt-Pt chain direction). The Cs^+ analogue (DPO $\cong 0.39$) is isostructural and also contains a very short Pt-Pt chain repeat distance of 2.833(1) Å.[41]

The absence of interchain hydrogen bonding as a means for crystal stabilization indicates that the binding forces must be due to stronger $[Rb^+, Cs^+ \cdots X^-]$ interactions alone. This accounts for the short *inter*chain Pt-Pt separation in RbCP(FHF) of 8.98 Å, which is nearly 1 Å shorter than in hydrated KCP(Br) (9.91 Å). It is not clear at this time whether the interchain coupling is increased, due to the short interchain distance and strong Coulombic interactions, or is decreased (with higher anisotropy) due to the lack of interchain hydrogen bonding. However, it is apparent from Fig. 20 that the temperature dependence of the conductivity of the $(FHF)^-$ salts is different from that of KCP(Br). Analysis of these data by Wood and Underhill[42] reveals that the three-dimensional ordering of the Peierls distortions occurs at $\sim 100°K$ for KCP(Br) and $\sim 73°K$ for

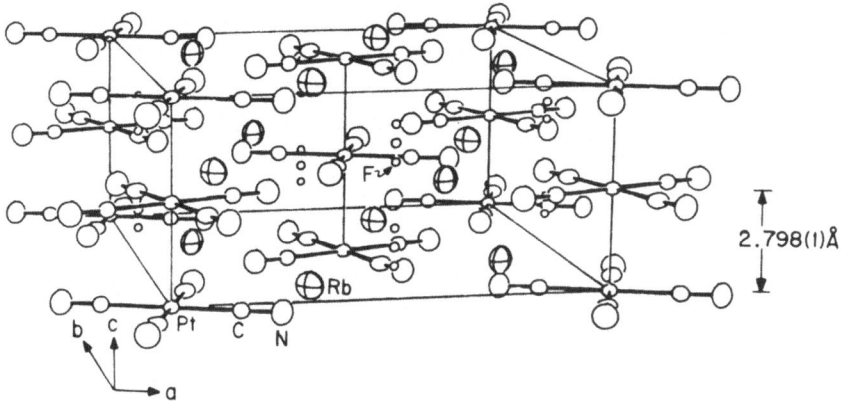

Fig. 19. Perspective view of the unit cell of $Rb_2[Pt(CN)_4]$-
          $(FHF)_{0.40}$. The small circles are the partially occupied
          fluorine positions. The thermal ellipsoids of the other
          atoms are drawn with a scale of 50% probability. The
          Pt-Pt spacing is the shortest known of any POTCP salt.

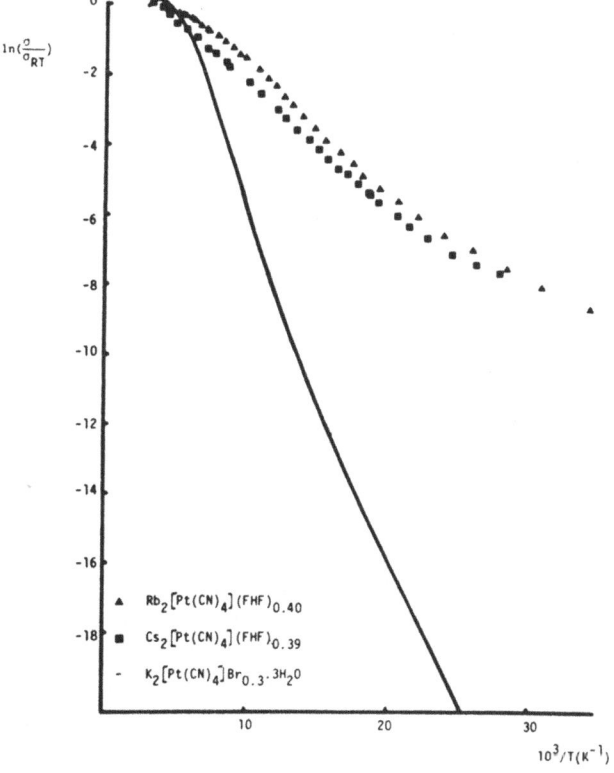

Fig. 20. The ratio of the conductivity to the room temperature
          conductivity vs inverse temperature.

CsCP(FHF) and RbCP(FHF).   It has been postulated[43,44] that the
Peierls transition from a metallic to a low-temperature insulating
state in 1-D conductors can be suppressed by *increasing* the interchain coupling.   Regardless of the underlying cause, because of
their higher room temperature conductivity and lower 3-D ordering
temperature, RbCP(FHF) and CsCP(FHF) are clearly "better metals"
than KCP(Br,Cl).

The Crystal Structure of $K_2[Pt(CN)_4](FHF)_{0.30} \cdot 3H_2O$, KCP(FHF)

KCP(FHF)$_{0.3}$ is essentially isostructural[45] with KCP(Br), both
of which possess a DPO = 0.30.   Although the Pt-Pt spacings are
equal (2.88 Å) in KCP(Br), they are different in KCP(FHF)$_{0.3}$ [Pt-
Pt = 2.928(1) and 2.918(1) Å as illustrated in Fig. 21.   The most
striking finding is that the average Pt-Pt separation in the (FHF)$^-$
salt is $\sim$0.04 Å longer than in the Br$^-$ derivative.   Since the degrees of partial oxidation (DPO) of the Pt atoms in KCP(Br) and
KCP(FHF) are apparently equal at 0.30, the increased Pt-Pt separations in the latter do not appear to be due to a decreased DPO.
Therefore, although identical cations are involved in these two isostructural materials which both contain disordered anion arrays

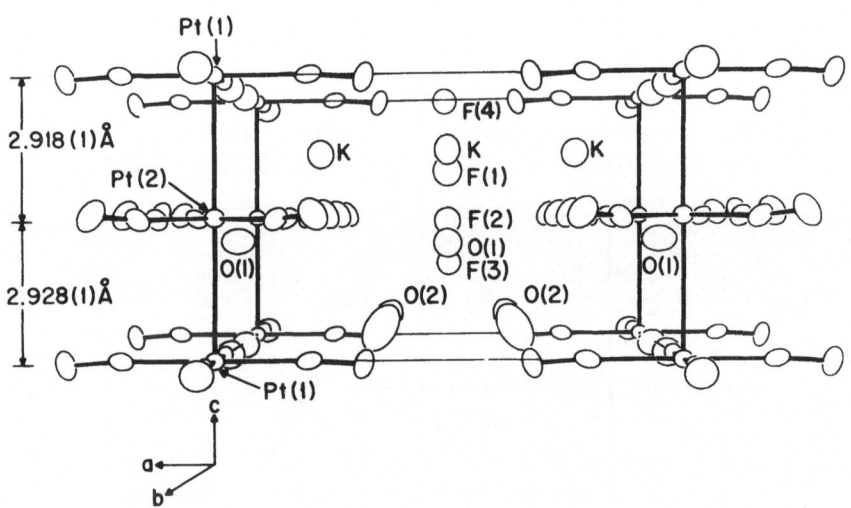

Fig. 21.   Drawing of the unit cell of $K_2[Pt(CN)_4](FHF)_{0.3} \cdot 3H_2O$
showing the Pt atom chain and the *un*equal Pt-Pt distances.
Hydrogen atoms are not shown.

aligned parallel to $\underline{c}$, it appears that the presence of the triatomic (FHF)$^-$ ion is the driving force behind the increased Pt-Pt chain distance. Finally, although no H atoms were located in this study, we can only speculate that the close isomorphism of KCP(Br) and KCP(FHF) results in the same pattern of hydrogen bonding.

IV. THE STRUCTURE OF $Cs_2[Pt(CN)_4](N_3)_{0.25} \cdot 0.5H_2O$, $CsCP(N_3)$

The azide anion $(N_3)^-$ and the bifluoride anion $(F-H-F)^-$ are both linear and are both ~ 2.28 Å in length. However, attempts to electrolytically synthesize $(N_3)^-$ salts of $M_2[Pt(CN)_4] \cdot xH_2O$, $(M = K, Rb, and Cs)$ were successful for the Cs^+ derivative only. The 1-D complex is tetragonal (space group $\overline{P4b2}$)[46] and the crystal structure contains perfectly linear Pt-atom chains (Pt-Pt = $\underline{c}/2$ = 2.877(1) Å) as shown in Fig. 22. A point of some interest is that the crystal structure of $CsCP(N_3)$ is nearly isostructural with the POTCP complexes Rb, $CsCP(FHF)_{0.4}$ and $CsCP(Cl)$. The similarities are readily observable upon direct comparison of Figures 15, 19 and 22.

The DPO of 0.25(2) was determined from x-ray diffuse scattering techniques and is in complete agreement with the elemental analyses.[46] While our final x-ray difference Fourier maps were virtually feature-less, with no indication of independent sites for water molecules in the crystal lattice, the chemical analyses consistently suggest the presence of H_2O. Therefore, we believe that in all likelihood the partially occupied azide sites are alternatively occupied by water molecules. Additional justification for this model is the anoma-lously high atomic multiplier obtained in the least-squares refine-ments for one of the two independent azide nitrogen atoms.

V. THE FIRST POLYATOMIC-TRIANIONIC POTCP COMPLEX, $Rb_3[Pt(CN)_4](O_3SO \cdot H \cdot OSO_3)_{0.49} \cdot H_2O$, RbCP(DSH)

An unusual POTCP complex containing the disulfatohydrogen ion, $Rb_3[Pt(CN)_4](O_3SO \cdot H \cdot OSO_3)_{0.49} \cdot H_2O$, "RbCP(DSH)", has been prepared electrolytically. The complex is unique in that it contains a poly-atomic trianion and a ratio of three alkali metal ions per Pt atom, instead of the usual two.[47]

The structure is triclinic $(P\overline{1})$ and comprises square-planar $Pt(CN)_4$ groups, which stack along $\underline{a}$, and are linked primarily by electrostatic $Rb^+ \cdots {}^-N \equiv C$ interactions (along $\underline{b}$) as shown in Fig. 23. The central portion of the cell is occupied by two $(SO_4)^{2-}$ groups related by an inversion center at (1/2,1/2,1/2), with one short $O \cdots O$ distance of 2.52(2) Å. This suggests the existence of a strong H-bond; hence, the likelihood that the complex anion is a novel $(O_3SO \cdot H \cdot OSO_3)^{3-}$ anion. The Pt-atom chain is precisely linear

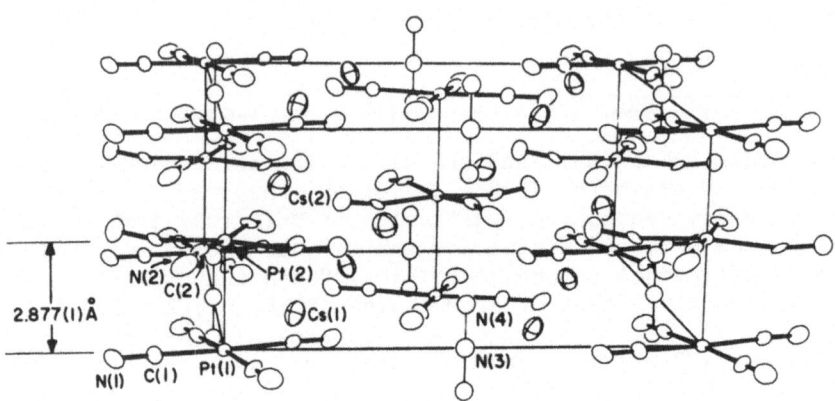

Fig. 22.   Perspective view of the unit cell of $Cs_2[Pt(CN)_4](N_3)_{0.25}$ ·
            $0.5H_2O$. The Pt-Pt chains are perfectly linear ($\underline{c}/2$) and
            the linear azide anions are co-linear with the Pt chains.
            Thermal ellipsoids are drawn at 50% probability.

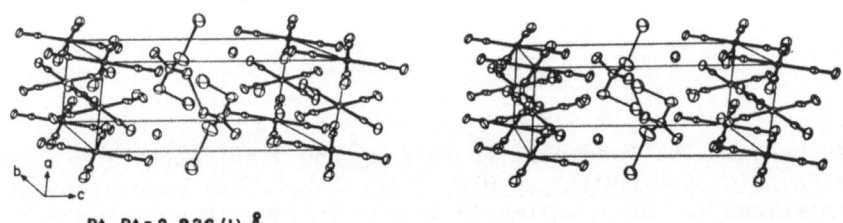

Fig. 23.   A stereoscopic view of the unit cell of $Rb_3[Pt(CN)_4]$-
            $(O_3SO \cdot H \cdot OSO_3)_{0.49} \cdot H_2O$ with thermal ellipsoids scaled to
            50% probability. Hydrogen bonding interactions are indi-
            cated by unshaded bonds and the O-H-O separation in the
            disulfatohydrogen ion is 2.52(2) Å. The Rb atoms are
            shown without any bonds drawn to them.

and the Pt-Pt separations ($a/2$) are required crystallographically to be equal (2.826(1) Å). These very short Pt-Pt distances are ∿0.05 Å shorter than in KCP(X) and quite comparable to those in $Cs_2[Pt(CN)_4](FHF)_{0.39}$.

The formulation of 0.49 ± 0.01 disulfatohydrogen ions per Pt atom is based on least-squares refinement of the sulfate ion occupancy parameters in the x-ray structure. The derived $(SO_4)^{2-}$ site occupancy factors are essentially unity indicating that RbCP(DSH) may be the first POTCP salt which is *stoichiometric* with respect to both cation and anion. The least-squares derived degree of partial oxidation (DPO) for Pt would therefore be +0.49. From the stoichiometry derived from the elemental analyses, the DPO is ∿0.47.

A more reasonable DPO value of ∿0.40, roughly predictable from a Pt-Pt distance of 2.83 Å, may be accommodated if the structure contains ∿0.1 additional H^+ ions per Pt. This is not entirely unexpected since the highly acidic medium used in the preparation might result in HSO_4^- or H_3O^+ ion formation. An x-ray diffuse scattering experiment is presently in progress to ascertain the DPO.

Our room temperature 4-probe electrical conductivity measurements on single crystals of RbCP(DSH) have yielded values of σ_{max} = 2000–2100 Ω^{-1} cm^{-1}. These values may be compared with σ_{max} = 850 Ω^{-1} cm^{-1}, which we obtained for KCP(Br).

VI. SUMMARY AND CONCLUSIONS

It is clear that our newly gained knowledge regarding the syntheses, structures, and electrical conductivities of 1-D POTCP compounds has greatly increased in the last two years. We have learned a great deal regarding the subtle interplay between these properties and this knowledge will serve to guide us in future research efforts.

The following generalizations concerning the hydration state of anion deficient compounds appear to be valid:

(1) Hydrated POTCP complexes form primitive tetragonal lattices (P4mm) with the M^+ cations in one half of the unit cell and water molecules in the other half. Furthermore, the cations are located between the planes of the $Pt(CN)_4$ groups. Compounds in this class include KCP(Cl), KCP(Br), RbCP(Cl), $(NH_4)CP(Cl)$ and KCP(FHF).

(2) Anhydrous POTCP complexes are body-centered tetragonal (I4/mcm) and the cations occupy sites in the same plane as the Pt-$(CN)_4$ groups, which may tend to maximize $M^+ \cdots ^-N\equiv C$ interactions. Compounds in this class are CsCP(Cl), $RbCP(FHF)_{0.40}$, $CsCP(FHF)_{0.39}$ and $CsCP(FHF)_{0.23}$. As previously discussed, $RbCP(N_3)_{0.25}$ should also be included in this class.

The intrachain Pt-Pt separations in POTCP complexes are dependent on three factors:

(1) The Pt-Pt separation is inversely related to the DPO.

(2) Among isostructural compounds, those with the smaller alkali metal cations have a shorter Pt-Pt distance. For example, RbCP(FHF) < CsCP(FHF), and KCP(Cl,Br) < RbCP(Cl). However, smaller cations also tend to form hydrated salts, which increases the Pt-Pt separations with respect to anhydrous salts with larger cations.

(3) Intermolecular steric interactions between cyano ligands within a chain are not the limiting factor in the reduction of the Pt-Pt repeat distance. In RbCP(FHF)$_{0.40}$, which has the shortest known Pt-Pt separation (2.798(1) Å), the rotation angle between neighboring TCP groups is only 40.4(6)°, whereas the minimum intramolecular steric interaction would be obtained with an angle of 45°.

Finally, it is now apparent that the more metallic complexes (higher conductivity, lower 3-D ordering temperature) are anhydrous, with short intrachain Pt-Pt separations *and* short interchain distances. The smaller temperature dependence of the conductivity of the anhydrous salts is probably due to the absence of interchain hydrogen bonding interactions and the presence of strong, highly temperature independent, Coulombic interactions. Thus, the synthesis of POTCP complexes, with a Pt-Pt separation less than the internuclear distance in Pt metal (2.78 Å), would require the preparation of an anhydrous Na$^+$, K$^+$, or even, Li$^+$ salt. Thus, a technique must be devised to synthesize POTCP complexes with small alkali metal cations in an anhydrous medium.

ACKNOWLEDGEMENTS

The authors wish to gratefully acknowledge the co-workers who have contributed so heavily to the synthesis and physical characterization of the materials reported here: J. A. Abys, R. Besinger, D. Blair, R. W. Broach, R. K. Brown, C. C. Coffey, E. Corless, T. Cornish, K. Cornett, M. Davis, N. P. Enright, J. R. Ferraro, J. Francisco, E. Gebert, H. M. Gerdes, D. Gerrity, L. Graziano, H. J. Guggenheim (Bell Laboratories), T. Hall, M. Iwata, P. L. Johnson, K. D. Keefer, J. Kincaid, T. Koch, G. C. Lee, K. A. Leslie, S. C. Lin, T. Lynch, J. S. Miller (Rockwell International), G. Needham, D. McManus, R. "Zap" Maffly, J. L. Petersen, S. W. Peterson, T. Pragovich, A. Reis, F. K. Ross (Virginia Polytechnic Institute and State University), K. Stearley, L. Stecher, A. E. Underhill, D. Vidusek, D. M. Washecheck, C. Werth, C. White, J. Willett and P. Young.

REFERENCES

1. Please note that in this review we refer to KCP(Br) and KCP(Cl) jointly as KCP(Br,Cl) or KCP(X). However, there is a mixed Br^--Cl^- halide complex KCP(Br-Cl) which has been reported: J. S. Miller and R. J. Weagley, Inorg. Chem. <u>16</u>, 2965 (1977).

2. A. N. Bloch and R. B. Weisman, in *Extended Interactions Between Metal Ions in Transition Metal Complexes*, ed. L. V. Interrante, Am. Chem. Soc. Sym. Ser. <u>5</u>, 369 (1974).

3. G. D. Stucky, A. J. Schultz, and J. M. Williams, Ann. Rev. Mater. Sci. <u>7</u>, 301 (1977).

4. J. S. Miller and A. J. Epstein, Prog. Inorg. Chem. <u>20</u>, 1 (1976); I. F. Shehegolev, Phys. Status Solidi, <u>12</u>, 9 (1972); H. G. Schuster, ed., *One-Dimensional Conductors*, Heidelberg: Springer (1975); H. J. Keller, ed., *Low-Dimensional Coopera- tive Phenomena*, New York: Plenum (1975); H. R. Zeller, Fest- körperprobleme, <u>13</u>, 31 (1973); H. J. Keller, ed., *Chemistry and Physics of One-Dimensional Metals*, New York: Plenum (1976).

5. K. Krogmann, Angew. Chem. Int. Ed. Engl. <u>8</u>, 35 (1969).

6. A. E. Underhill, in *Low-Dimensional Cooperative Phenomena*, ed. H. J. Keller, New York: Plenum, p. 287 (1975).

7. A. Gleizes, T. J. Marks, and J. A. Ibers, J. Am. Chem. Soc. <u>97</u>, 3545 (1975) and J. S. Miller and A. Epstein, Prog. Inorg. Chem. <u>20</u>, see p. 46 (1976).

8. *Handbook of Chemistry and Physics*, 53rd Ed., 1972-1973. Cleveland:Chem. Rubb. Co. p. F-80.

9. K. Krogmann and H. P. Geserich, in *Extended Interactions Between Metal Ions in Transition Metal Complexes*, ed., L. V. Interrante, Am. Chem. Soc. Sym. Ser. <u>5</u>, 350 (1974).

10. R. L. Musselman and J. M. Williams, J. Chem. Soc., Chem. Comm. 186 (1977).

11. See for example: H. Terrey, J. Chem. Soc., 202 (1928); J. S. Miller, Science <u>194</u>, 189 (1976); J. M. Williams, D. P. Gerrity, and A. J. Schultz, J. Am. Chem. Soc. <u>99</u>, 1668 (1977), and J. M. Williams *et al.*, *Inorg. Synth.* <u>19</u> (1978).

12. T. R. Koch, P. L. Johnson, D. M. Washecheck, T. L. Cornish, and J. M. Williams, Acta Cryst. <u>B33</u>, 3249 (1977).

13. K. Krogmann and H. D. Hausen, Z. Anorg. Allg. Chem. 358, 67
    (1968). A simultaneous study of KCP(Br) was carried out by
    A. Piccinin and J. Toussaint, Bull. Soc. Roy. Sci. Liege 36,
    122 (1967).

14. J. M. Williams, J. L. Petersen, H. M. Gerdes and S. W.
    Peterson, Phys. Rev. Lett. 33, 1079 (1974).

15. J. M. Williams, M. Iwata, S. W. Peterson, K. A. Leslie, and
    H. J. Guggenheim, Phys. Rev. Lett. 34, 1653 (1975).

16. See the following articles and references therein: G. Heger,
    H. J. Deiseroth, and H. Schultz, Acta Cryst. B34, 725 (1978);
    C. Peters and C. F. Eagen, Inorg. Chem. 15, 782 (1976).

17. (a) Private communication with J. R. Miller, Rockwell Inter-
    national, CA. (b) See footnote 15 of T. Takahashi, H. Akagawa,
    H. Doi, and H. Nagasawa, Solid State Commun. 23, 809 (1977).
    (c) S. Drosdziok and M. Engbrodt, Phys. Status Solidi B, 72,
    739 (1975). (d) Private communication with H. R. Zeller,
    Brown Boveri Research Center, Baden Switzerland.

18. S. Drosdziok and M. Engbrodt, Phys. Status Solidi B, 72, 739
    (1975) and references therein.

19. For reviews and background on X-ray and neutron scattering
    studies see (a) R. Comes, One-Dimensional Conductors, ed.
    H. G. Schuster, Springer-Verlag, p. 32 (1975); (b) B. Renker,
    L. Pintschovious, W. Gläser, H. Rietschel and R. Comes, ibid.,
    p. 53; (c) B. Renker and R. Comes, Low-Dimensional Cooperative
    Phenomena, ed. H. J. Keller, New York: Plenum Press, p. 235
    (1975).

20. B. Renker, H. Rietschel, L. Pintschovius, W. Gläser, P. Brüesch,
    D. Kuse and M. J. Rice, Phys. Rev. Lett. 30, 1144 (1973).

21. R. Comes, M. Lambert and H. R. Zeller, Phys. Status Solidi B,
    58, 587 (1973).

22. B. Renker, L. Pintschovius, W. Gläser, H. Rietschel, R. Comes,
    L. Liebert and W. Drexel, Phys. Rev. Lett. 32, 836 (1974).

23. J. W. Lynn, M. Iizumi, G. Shirane, S. A. Werner and R. B.
    Saillant, Phys. Rev. B12, 1154 (1975).

24. J. M. Williams, P. L. Johnson, A. J. Schultz and C. C. Coffey,
    Inorg. Chem. 17, 834 (1978).

25. A. E. Underhill, D. M. Watkins and D. J. Wood, J. Chem. Soc.,
    Chem. Commun., 805 (1976).

26. R. K. Brown and J. M. Williams, Inorg. Chem. (submitted for publication).

27. P. L. Johnson, A. J. Schultz, A. E. Underhill, D. M. Watkins, D. J. Wood and J. M. Williams, Inorg. Chem. 17, 839 (1978).

28. A. E. Underhill, D. M. Watkins and D. J. Wood, J. Chem. Soc., Chem. Commun., 805 (1976); A. E. Underhill, D. M. Watkins and D. J. Wood, J. Chem. Soc., Chem. Commun., 392 (1977); and ref. 15 of ref. 27.

29. R. K. Brown and J. M. Williams, Inorg. Chem., in press (1978).

30. A. E. Underhill, D. J. Wood and J. M. Williams (work in progress).

31. S. G. Phillips and R. J. P. Williams, *Inorganic Chemistry*, Oxford, p. 76 (1966).

32. K. Carneiro, K. Jacobsen and J. M. Williams (work in progress).

33. K. D. Keefer, D. M. Washecheck, N. P. Enright and J. M. Williams, J. Am. Chem. Soc. 98, 233 (1976); J. M. Williams, K. D. Keefer, D. M. Washecheck and N. P. Enright, Inorg. Chem. 15, 2446 (1976); A. H. Reis, S. W. Peterson, D. M. Washecheck and J. S. Miller, J. Am. Chem. Soc. 98, 234 (1976).

34. J. R. Ferraro, L. J. Basile and J. M. Williams, J. Chem. Phys. 67, 742 (1977).

35. A. J. Schultz, G. D. Stucky, J. M. Williams, T. R. Koch and R. L. Maffly, Sol. State Commun. 21, 197 (1977).

36. K. Carneiro, J. Eckert, G. Shirane and J. M. Williams, Sol. State Commun. 20, 333 (1976).

37. J. M. Williams and A. J. Schultz, Annals of the New York Acad. of Sci. 00, 0000 (1978).

38. J. M. Williams, A. J. Schultz, K. B. Cornett and R. E. Besinger, J. Am. Chem. Soc. 00, 0000 (1978).

39. A. J. Schultz, C. C. Coffey, G. C. Lee and J. M. Williams, Inorg. Chem. 16, 2129 (1977).

40. L. J. Basile, J. R. Ferraro and J. M. Williams, J. Chem. Phys. 66, 4941 (1977).

41. A. J. Schultz, D. P. Gerrity and J. M. Williams, Acta Cryst. B34, 1673 (1978).

42. D. J. Wood, A. E. Underhill, A. J. Schultz and J. M. Williams, (submitted to Sol. State Commun.)

43. M. Thielemans, R. Deltour, D. Jerome and J. R. Cooper, Sol. State Commun. 19, 21 (1976).

44. B. Horovitz and A. Birnboim, Sol. State Commun. 19, 91 (1976).

45. R. K. Brown, P. L. Johnson, T. J. Lynch and J. M. Williams, Acta Cryst. B34, 1965 (1978).

46. R. K. Brown, D. A. Vidusek and J. M. Williams, Inorg. Chem. (submitted for publication).

47. A. J. Schultz, J. M. Williams, R. K. Brown and R. E. Besinger, Inorg. Chem. (submitted for publication).

48. R. B. Saillant, R. C. Jaklevic and C. D. Bedford, Mat. Res. Bull. 9, 389 (1974).

UNIVERSAL BEHAVIOUR OF PLATINUM CHAIN COMPOUNDS, DEDUCED FROM ELECTRICAL CONDUCTION RESULTS

Kim Carneiro

Physics Laboratory I, University of Copenhagen

Universitetsparken 5, DK 2100 Copenhagen, Denmark

INTRODUCTION

During the last few years the substantial interest in quasi one-dimensional (1D) conductors has led to an enormous amount of both theoretical and experimental evidence. However, the detailed descriptions of the behaviour of individual compounds, often leaves little to the benefit of a simple and unified picture of the field. It seems therefore worthwhile, in the scope of this workshop, to point out where such "universal behaviour" exists, and it is the purpose of this paper to present the platinum chain compounds as a well-defined class of simple 1D systems; and it will be shown that the real compounds exhaust most of the interesting parameter space that theorists have outlined. For simplicity, we shall confine our description to the electrical conductivity, with minor degression to what can be learned from diffuse X-ray scattering.

Since the occurrence of what is now considered the prototype quasi 1D conductor $K_2Pt(CN)_4Br_{0.3}\cdot 3H_2O$, a variety of conducting Pt-compounds has been prepared. The planar ligands, surrounding the Pt-ions may be cyano groups $(CN)_4$, or oxalate groups $(C_2O_4)_2$, leading to "CP's" or "OP's" respectively. The partial oxidation of the platinum has been accomplished by oxidizing halides, "(H)", or by making cation deficient salts, "C(def)". And finally the cation may be a metal, an "A" standing for the ammonium ion NH_4^+ or a "G" for guanidinium $C(NH_2)_3^+$. Using these code names the compounds to be discussed here are $KCP(Br)$,[1-2] $ACP(Cl)$,[3] $GCP(Br)$,[4] $K(def)CP$,[5] $Rb(def)CP$,[5] $Mg(def)OP$,[6] and $Co(def)OP$,[6] all compounds of which the electrical conductivity has been reported.

THEORY

Considering electron-phonon interaction only, mean field theories give the following expression for the transition temperature, owing to the Peierls instability:[7-8]

$$T_P = \alpha T_F \exp\{-1/\lambda\} \tag{1}$$

where $T_F = \varepsilon_F/k_B$ is the Fermi temperature and λ is a dimensionless electron-phonon coupling constant. The prefactor α depends on the model for the electron band in the metallic state, above T_P. Below T_P a gap $\Delta(T)$ exists, and at zero temperature, it has been found that:

$$\Delta(0) = 4 \, T_F \exp\{-1/\lambda\} \tag{2}$$

However, the ordinary concept of a phase transition at the critical temperature, T_P, breaks down in 1D. Critical fluctuations, not considered in a mean field theory, prevent phase transitions from taking place at finite temperatures, and leaves T_P as a scale temperature, below which an apparent or fluctuating gap $\Delta(T)$ is present. What re-establishes the phase transition in real compounds is the interchain electron-electron coupling, η. In quasi 1D systems, a real phase transition takes place at a temperature T_{3D}, where:

$$T_{3D}/T_P = f(\eta) \quad , \quad f < 1 \quad . \tag{3}$$

$f(\eta)$ has been calculated in the small[9] and intermediate[10] η-range. Hence the relevant phase diagram, shown in Fig.1, for such a quasi 1D conductor, reveals three regions: A 1D metallic phase above T_P, a 1D intermediate phase between T_P and T_{3D}, and a 3D semiconducting phase below T_{3D}. Fig.1 also shows the characteristics of the lattice distortion, as originally observed by Comes et al. in KCP(Br).[11]

It is therefore natural to describe the electrical conduction by a gap $\Delta(T)$ satisfying (2), and varying rapidly around T_{3D}. In this paper we use the following expression:

$$\Delta(T) = \Delta(0)/(\exp\{-\beta(T-T_{3D})\}+1) \quad , \tag{4}$$

where β determines the slope of $\Delta(T)$ at T_{3D}. The conductivity then becomes:

$$\sigma(T) = \sigma_0(T) \exp\{-\Delta(T)/k_B T\} \tag{5}$$

where $\sigma_0(T)$ is determined by the conduction mechanism. However,

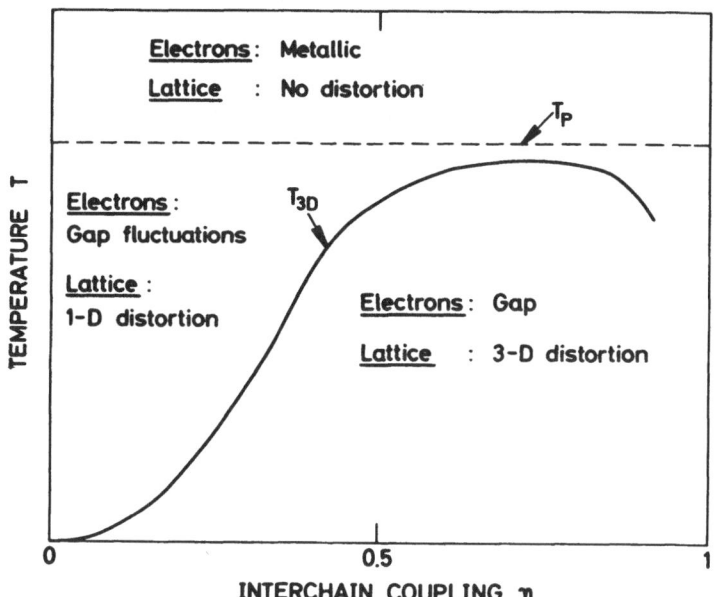

Figure 1. Phase diagram of a quasi one-dimentional conductor with
a Peierls instability. T_P is the mean field transition
temperature, and T_{3D} is the three dimensional ordering
temperature.

as pointed out earlier,[12] experiments do not yield $\Delta(T)$ directly,
but rather the following function:

$$D(T) = - \partial \, \ell n \, \sigma(T)/\partial(1/T)$$

$$(6)$$

$$= -T^2 \partial \, \ell n \, \sigma_0(T)/\partial T + \Delta(T) - T \, \partial\Delta(T)/\partial T$$

Thieleman et al.[2] in their study of KCP(Br) under pressure, used
the diffusion approximation for $\sigma_0(T)$ and found the expression:

$$D(T) = -\tfrac{1}{2}T + \Delta(T) - T \, \partial\Delta(T)/\partial T \qquad (7)$$

to be in good agreement with their experiment. We find it more
convenient at first to neglect the variation of $\sigma_0(T)$ and use the
expression:

$$D(T) = \Delta(T) - T \, \partial\Delta(T)/\partial T \quad . \qquad (8)$$

Fig. 2 shows the three functions $\Delta(T)$, $\sigma(T)$, and $D(T)$, the latter
two computed both according to (7) and to (8). It seems appropriate

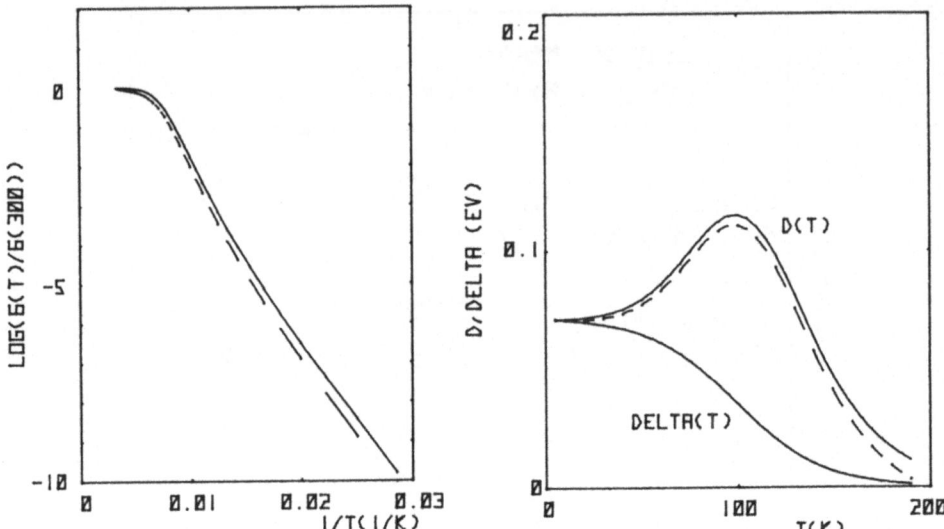

Figure 2.   Characteristic behaviour of the conductivity of a quasi
            one-dimensional conductor. The left side shows
            logσ(T) vs. 1/T from (5) and to the right is shown
            the gap Δ(T) from (4) and the experimentally available
            function D(T) . Full lines are from (8) and dashed
            lines from (7). The parameters chosen are appropriate
            for KCP(Br).

at this point to mention that although no physical significance is
to be attributed to the functional form of our Δ(T) , it agrees
rather well with the shape of the corresponding temperature dependence of the lattice distortion d(T) as determined by diffuse
neutron scattering.[13] And d(T) and Δ(T) should be proportional.

RESULTS AND DISCUSSION

     In Fig.3 we summarize the comparison between the experimental-
ly deduced D(T) from (6) for the eight compounds listed above,
and the calculated D(T) from (8). Also shown are the calculated
Δ(T) and the specific values of D(O) and T_{3D} . The overall
agreement indicates the relevance of the expression for Δ(T) ,
and the specific compounds are discussed separately as follows.

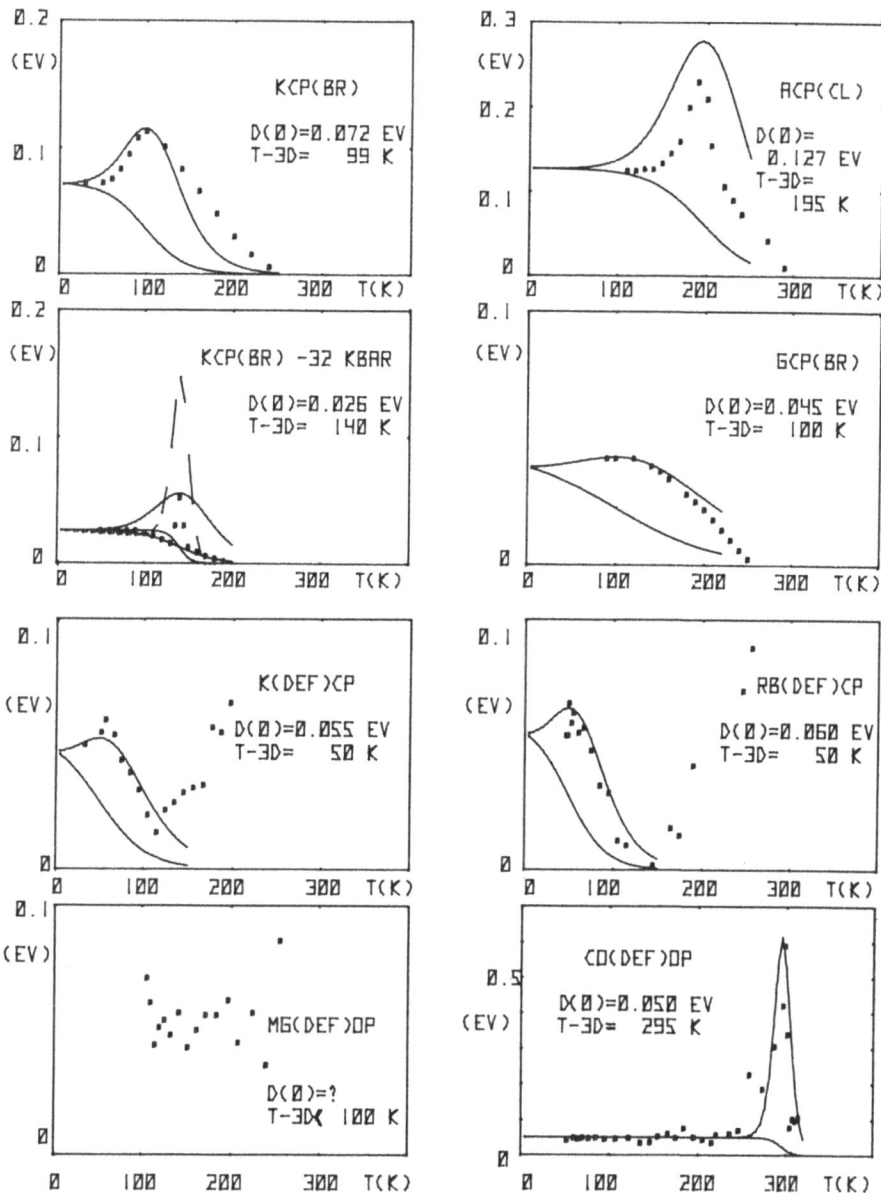

Figure 3. Comparison between experimental results (■) , and calculations of Δ(T) and D(T) .

KCP(Br) at ambient[1] and high pressures[2] and ACP(CL) are
isostructural, and from these three compounds one clearly identi-
fies the three regions of Fig.2. At low temperatures, a well-
defined gap $D(0) = \Delta(0)$ appears, the characteristic structure
around T_{3D} leads into the intermediate region, and finally the
high temperature region where the measured $D(T) \lesssim 0$. It should
of course be pointed out that we do not expect our expression for
$\Delta(T)$ to hold accurately in the 1D region above T_{3D}; but the fact
that we are not able to adjust the three parameters of (4) to give
good fits at temperatures $T \lesssim T_{3D}$ indicate that $\sigma_o(T)$ varies
in a dramatic way. This is emphasized by our fit to ACP(CL) in
Fig.2, where the computed $D(T)$ is obtained as a balance between
obtaining a reasonable narrow peak, without shooting too high at
T_{3D}. For KCP(Br) at a pressure of 32 kbar, we have made two
fits. Firstly $\Delta(T)$ was adjusted to give a good fit outside the
range of structure in the measured $D(T)$. But the calculated
$D(T)$ (full line) is far too broad. Secondly in an attempt to
sharpen $D(T)$, we used (7) and a large value of β. But then
$D(T)$ exceeds the measured one substantially around T_{3D}. We
therefore conclude that the conduction is not diffusive throughout
the whole temperature range, but that there is a more dramatic
change in the conduction mechanism at T_{3D}. This hitherto
unnoticed observation is substantiated by the experimental evidence
that in ACP(Cl), the conductivity changes from frequency
independent above T_{3D} to frequency dependent below T_{3D}.

Because of the rather structureless experimental $D(T)$ of
GCP(Br),[4] the parameters are not very accurately derived; but our
analysis indicates that the experiments were done above T_{3D} and
that owing to the slow decline of $\Delta(T)$, this conductor might be
very 1D like. This is in contrast to the conclusions of Schaffmann
et al. They state that GCP(Br) is far below its Peierls temperature and argue that it is rather 3D like. Clearly, this
discrepancy warrants further investigations, in particular conductivity and diffuse X-ray experiments at lower temperatures, in
order to establish the characteristics of GCP(Br)with respect to its
position in our phase diagram in Fig.2.

The isostructural compounds K(def)CP and Rb(def)CP[5] appear
at low temperatures to be rather similar to GCP(Br). But at high
temperatures the experimental $D(T)$ increases dramatically.
Although less pronounced this feature is also seen in Mg(def)OP
and Co(def)OP. By varying the H_2O vapour pressure around $T=250$ K
in K(def)CP and Rb(def)CP we could establish that this increase
in $D(T)$ (and $\sigma(T)$) correlated with the amount of crystal water.
We tentatively suggest that this feature in the conductivity comes
from the onset of free H_2O rotation.

Mg(def)OP and Co(def)OP are discussed elsewhere at this workshop.[6] Although isostructural, their behaviour in the present context is drastically different. Whereas Co(def)OP gives excellent agreement between theory and experiment, Mg(def)OP is in Fig.3 the exception (proving the rule!) which it seems meaningless to fit. Since X-ray studies[14] show that down to 100 K, Mg(def)OP is in the 1D distorted regime, T_{3D} must be very low. Hence Mg(def)OP and Co(def)OP appears as opposite extremes in Fig.2, the former characterized by a very low interchain coupling η , and the latter having a large η .

CONCLUSION

The analysis presented here supports the view of the Pt-compounds as a unified class of simple quasi one-dimensional conductors. Using a simple model for the band gap $\Delta(T)$ at temperatures $T \lesssim T_{3D}$ we expect reasonable agreement between experiments and theory for the observable parameter $D(T)$. This is useful when placing an actual compound on the phase diagram, relating T_{3D}/T_P to η . We find bandgaps in the range $0.026 \text{ eV} \lesssim D(0) \lesssim 0.127 \text{ eV}$ and transition temperatures $50 \text{ K} \lesssim T_{3D} \lesssim 296 \text{ K}$, indicating that a large portion of Fig.2 is covered by these compounds. Finally our analysis indicates that the conduction mechanism changes at T_{3D} .

REFERENCES

1. H.R. Zeller and A. Beck, J.Phys.Chem.Solids, 35, 77 (1974).
2. M. Thieleman, R. Deltour, D. Jérome, and J.R. Cooper, Solid State Commun. 19 (1976).
3. D.J. Wood and A.E. Underhill, N.Y.Acad.Sci. (in press); and K. Carneiro, A.S. Pedersen, A.E. Underhill, D.J. Wood, G.A. Mackenzie (preprint).
4. M.J. Schaffman, M.B. Salamon, G. Pasquali, A.J. Schultz, and G.D. Stucky, N.Y. Acad. Sci. (in press).
5. K. Carneiro, C.S. Jacobsen, and J.M. Williams (preprint).
6. A.E. Underhill and D.J. Wood, paper presented at this workshop.
7. C.G. Kuper, Proc.R.Soc. A227, 214 (1955).
8. M.J. Rice and S. Stässler, Solid State Commun. 13, 25 (1973).
9. V. Emery, in "Chemistry and Physics of One Dimensional Metals" (ed. H.J. Keller, Plenum, 1977).
10. B. Horovitz, H. Gutfreund, and M. Weger, Phys.Rev. B12, 3174 (1975).

11. R. Comès, M. Lambert, H. Launois, H.R. Zeller, Phys.Rev. B8, 571 (1973).
12. L.B. Coleman, J.A. Cohen, A.F. Garito, A.J. Heeger, Phys.Rev. B7, 2122 (1973).
13. J.W. Lynn, M. Iizumi, G. Shirane, S.A. Werner, and R.B. Saillant, Phys.Rev. B12, 1154 (1975).
14. J.Y. Dubois, R. Comès and A.E. Underhill, Ferroelectrics 16, 147 (1977); and
     J.Y. Dubois, Thesis (Orsay 1975).

A COMPARISON OF THE PROPERTIES OF THE ISOSTRUCTURAL 1D CONDUCTORS

$Co_{0.83}[Pt(C_2O_4)_2].6H_2O$ and $Mg_{0.82}[Pt(C_2O_4)_2].6H_2O$

Allan E. Underhill and David J. Wood

School of Physical and Molecular Sciences

University College of North Wales, Bangor, Wales

The prototype 1D conductor $K_2Pt(CN)_4Br_{0.3}.3H_2O(KCP(Br))$ has been widely studied over the past eight years[1]. Although many theories have been proposed to explain the electrical conduction behaviour of this compound the importance and role of such effects as interchain coupling, partial 3D ordering of the lattice (Peierls transition) and the random potential of the halide ions, are still not understood. It has recently been shown that information about the importance of these effects can be obtained by studying an isomorphous series of compounds in which only one ion in the lattice is varied[2].

We have recently carried out the first detailed study of a pair of isomorphous platino-oxalate complexes $Mg_{0.82}[Pt(C_2O_4)_2].6H_2O$ (Mg-OP) and $Co_{0.83}[Pt(C_2O_4)_2].6H_2O$(Co-OP). This pair of compounds is of particular interest because Co-OP is the first 1D platinum atom chain compound to contain a paramagnetic cation. Detailed X-ray crystal structures have been determined for both compounds and Figure 1 gives a nonperspective view of the structure[3,4]. The $[Pt(C_2O_4)_2]^{-1.66}$ units stack in the $\underline{c}$ direction with a Pt-Pt spacing of $c/2$ equal to 2.841 Å in Co-OP compared with 2.85 Å in Mg-OP. Adjacent $[Pt(C_2O_4)]^{-1.66}$ units are staggered by 55° in Co-OP and 58° in Mg-OP and alternate groups are in an eclipsed configuration. The closest inter-Pt chain separation is 10.94 Å in Co-OP and 10.93 Å in Mg-OP.

The M^{2+} ions are located in between the planes containing the $[Pt(C_2O_4)_2]^{-1.66}$ units and are co-ordinated to six water molecules with a slightly distorted octahedral symmetry. Each M^{2+} site is ~41% occupied in a random array in all three dimensions. Several

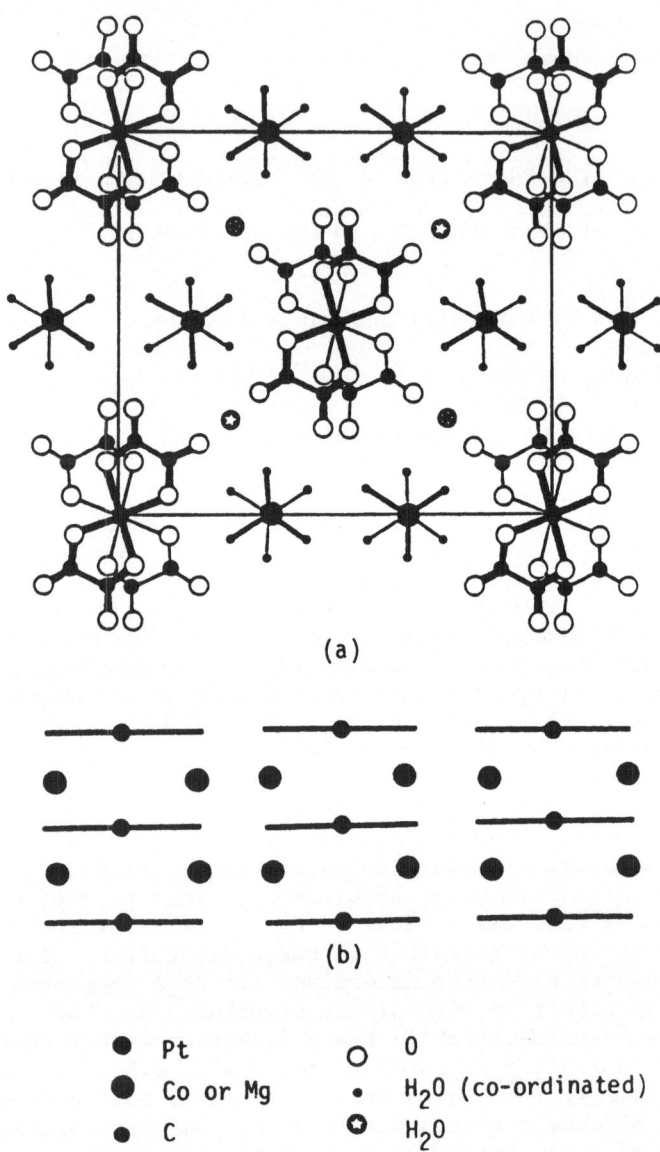

(a)

(b)

● Pt   ○ O

● Co or Mg   · H₂O (co-ordinated)

● C   ✪ H₂O

Figure 1.   Crystal structure of $M_{0.8}[Pt(C_2O_4)_2] \cdot 6H_2O$ (M=Co,Mg)
(a) Projection parallel to, and (b) projection perpendicular to,
Pt chain (the oxalate groups are represented by thick lines in
(b)).

short O–O distances indicate that hydrogen bonding occurs between
the water molecules and the oxalate groups[4].

ELECTRICAL CONDUCTION STUDIES

The crystals were studied by a 4-probe dc technique using both
aquadag (colloidal graphite in water) and Dag 580 (colloidal
graphite in organic solvents) as contact materials between the
crystals and the 0.001" gold wires. All the crystals studied gave
ohmic behaviour under the conditions of the measurements.

Because of the small size of the crystals available for
measurement only the specific conductivity in the platimum atom
chain direction ($\sigma_\parallel$) was determined. At room temperature $\sigma_\parallel$
for Mg–OP was found to lie in the range 0.2–4 Ω^{-1} cm^{-1} and for
Co–OP in the range 2–25 Ω^{-1} cm^{-1}. The absence of significant
electrode effects has been confirmed by a measurement at 35GHz
which gave a value of 38 Ω^{-1} cm^{-1} for $\sigma_\parallel$ for a single crystal of
Co–OP[5]. These values of $\sigma_\parallel$ for both Mg–OP and Co–OP are considerably
lower than that observed for KCP(Br) although the Pt–Pt distance is
shorter in the platino-oxalates. Within the series of partially
oxidised platinocyanides a correlation has been observed between the
intra-chain Pt–Pt separation and the value of $\sigma_\parallel$ at room temperature[6].
This suggests that conducting strands based on the $[Pt(C_2O_4)_2]$ unit
may be inherently less conducting than those based on $[Pt(CN)_4]$ units.
This has been borne out by work on $Rb_{1.5}[Pt(C_2O_4)_2]1.5H_2O$ and
$Cu_{0.82}[Pt(C_2O_4)_2].6H_2O$[7].

Mg–OP

The temperature dependence of $\sigma_\parallel$ has been studied down to 85K
and is shown for a typical crystal of Mg–OP in Figure 2. It has
been found that the temperature dependence of the conductivity
varies from crystal to crystal but any significant deviation from
the behavior shown in Figure 2 appears likely to be due to either
partial dehydration of the crystals or to poor quality crystals.
It can be seen that there is an almost linear ln $\sigma_\parallel$ vs $1/T$
relationship over the whole of the temperature range with, at the
higher temperatures, a suggestion of a change to more temperature
independent behaviour above room temperature. This feature is much
more pronounced with the less perfectly formed or partially
dehydrated crystals. Unfortunately measurements above room
temperature are not possible as the crystals readily lose water on
heating. The activation energy at low temperature is .055 ± .003 ev.

Co–OP

The temperature dependence of $\sigma_\parallel$ for Co–OP has been measured
from 50 K to 330 K and is shown for a typical crystal in Figure 2.

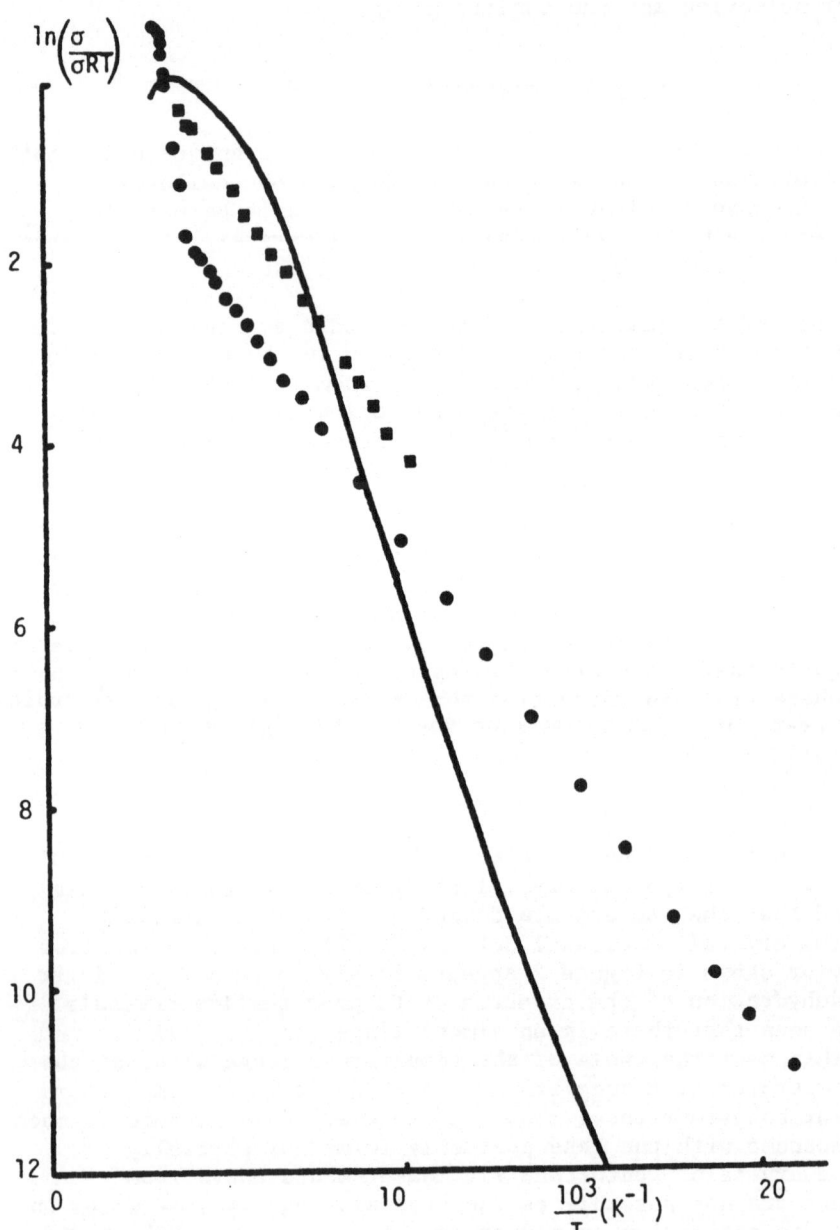

Figure 2. The ratio of the conductivity to the room temperature
conductivity vs inverse temperature for typical crystals of
●, Co-OP; ■ , Mg-OP; —, KCP(Br), (ref. 8).

The temperature dependence of the conductivity again varied slightly
from crystal to crystal with most pronounced variation being seen
around the transition at $\sim$ 250 K. In a number of crystals the sharp
transition depicted in Figure 2 was replaced by a more gradual
transition over a temperature range of up to 70 degrees. Since
this effect was particularly noticeable with crystals subjected
to dehydrating conditions it is not considered to be the intrinsic
behaviour of Co-OP. Above 315 K the conductivity falls, presumably
due to dehydration of the crystals since they never recover their
original room temperature values. Between 315 and 303 K $\sigma_\parallel$ is
almost independent of temperature whilst from 303 to 250 K the
conductivity falls rapidly with decreasing temperature. At about
250 K a fairly sharp transition occurs and then from 250 to 50 K
there is a linear relationship between ln $\sigma_\parallel$ and $^1/T$ with an
activation energy of $\sim$ 0.05 ev.

Figure 2 also shows the temperature dependence of $\sigma_\parallel$ for
KCP(Br)[8]. In KCP(Br) the quasi-metallic behaviour at room
temperature is associated with the presence of purely 1D distorted
platinum atom chains with no evidence of 3D ordering. However,
the change-over to semiconductor behaviour at low temperatures
is accompanied by partial 3D ordering of the lattice[8]. This
transition which is smeared out over a wide temperature range is
centred at about 100 K, and is accompanied by a very rapid fall in
conductivity with decreasing temperature.

The temperature dependence of the conductivity of Co-OP with an
approximately temperature independent behaviour at high temperatures
together with a sharp fall in conductivity to a semi-conductor type
of behaviour at lower temperatures is somewhat analagous to that
observed in KCP(Br). Preliminary X-ray diffuse scattering results
show the presence of spots on the diffuse lines found between the
Bragg spots at room temperature[10]. This suggests some 3D ordering
of the lattice at room temperature in Co-OP and therefore a
considerably higher T_{3D} than in KCP(Br). This is in agreement with
the rapid fall in conductivity associated with T_{3D} being observed
around 280 K in Co-OP and 100 K in KCP(Br).

DISCUSSION

It is obvious from Figure 2 that although Mg-OP and Co-OP are
isomorphous with very similar inter and intra-platinum-platinum
separations, the temperature dependencies of their conductivities
are very different. This is unusual since previous studies have
shown that isomorphous compounds of this type containing similar
cations exhibit very similar temperature dependencies of their
conductivities (e.g. KCP(Cl), RbCP(Cl)[2]; CsCP(FHF), RbCP(FHF)[6]).

The temperature dependence of $\sigma_\parallel$ for Mg-OP shows no evidence
of a phase transition between room temperature and 85 K. X-ray

diffuse scattering has also shown the absence of any 3D ordering of the lattice at room temperature or at 80 K[11]. Thus it appears that T_{3D} for Mg–OP lies below 80 K and therefore at a very much lower temperature than in the isomorphous Co–OP. It is noteworthy that although above 80 K there is no 3D ordering of the platinum atom chains the conductivity continues to rise with temperature up to room temperature.

Since Mg–OP and Co–OP are isostructural with very similar lattice dimensions the large differences observed in their electrical conduction properties must be attributable to differences due to the presence of the paramagnetic Co^{2+} ion (0.74Å)[12] compared with diamagnetic Mg^{2+} ion (0.72 Å)[12]. The M^{2+} ion is situated in an octahedral environment of water molecules and it seems unlikely, therefore, that there will be any magnetic coupling between the isolated Co^{2+} ions or between the Co^{2+} ions and the platinum atoms of the 1D metallic chain in Co–OP. Magnetic studies down to 4 K have confirmed this by showing that there are no magnetic inter-actions even at this low temperature[13]. As in the $M_2[Pt(CN)_4]Cl_{0.3}.3H_2O$ series of compounds interchain coupling via hydrogen bonding may play a vital role. However, a full explanation of these differences in the electrical conduction properties between Co–OP and Mg–OP must await more detailed X-ray studies now being carried out.

ACKNOWLEDGEMENT

We would like to thank the Science Research Council for a maintenance grant (to D.J.W.). A. E. Underhill wishes to acknowledge NATO (Grant No. 1276) which has facilitated the exchange of information with foreign scientists.

REFERENCES

1. H.R. Zeller, Festkorperprobleme, 1973, 13, 31.; 'One Dimensional Conductors' Ed. H.G. Schuster, Springer-Verlag 1975.
2. D.J. Wood and A.E. Underhill, N.Y. Acad. Sci., in press.
3. K. Krogmann, Z. Anorg. Allg. Chem., 1968, 358, 97.
4. A.J. Schultz, A.E. Underhill and J.M. Williams, Inorg. Chem., 1978, 17, 1313.
5. H.J. Pedersen and C.S. Jacobson, personal communication.
6. D.J. Wood, A.E. Underhill, A.J. Schultz and J.M. Williams, unpublished results.
7. D.J. Wood and A.E. Underhill, unpublished results.
8. H.R. Zeller and A. Beck, J. Phys. Chem. Solids, 1974, 35, 77
9. K. Carneiro, G. Shirane, S.A. Werner and S. Kaiser Phys. Rev(B), 1976, 13, 4258
10. K. Carneiro, unpublished results.
11. J.Y. Dubois, R. Comes and A.E. Underhill, Ferroelectrics 1977, 16, 147; J.Y. Dubois, These Zeive cycle Orsay-France 1975.
12. James H. Huheey 'Inorganic Chemistry: Principles of Structure and Reactivity' Harper Intn. Ed. 1972, New York, p.74
13. W.E. Hatfield, personal communication.

RESONANCE RAMAN SPECTROSCOPY AS A PROBE OF THE MIXED-VALENCE STATE

OF LINEAR-CHAIN COMPLEXES

Robin J.H. Clark

Christopher Ingold Laboratories, University College London
20 Gordon Street, London WC1H OAJ, U.K.

INTRODUCTION

A Raman spectrum recorded using exciting radiation whose frequency coincides with that of an electronic transition of a molecule is known as a resonance Raman (r.R.) spectrum. Such a spectrum differs from a normal Raman spectrum in that the intensities of certain bands are enhanced in the resonance case, and overtone and combination tone progressions may be observed. The value of r.R. spectroscopy lies in its close connection with electronic absorption spectroscopy.[1] Those vibrations which are 'active' in an electronic transition, i.e. provide intensity in an absorption band, are also the ones which give rise to the enhanced bands and progressions in the r.R. spectrum. Therefore, since the widths of Raman bands are normally very much less than those of electronic absorption bands, it is possible, by monitoring the changes in the Raman spectrum as the frequency of the exciting radiation is changed, to deduce the modes which are active in the resonant electronic transition. This is especially valuable in the case of mixed-valence compounds for which the absorption bands are typically thousands of wavenumbers in width even at low temperatures.[2] A plot of the

variation of intensity of a Raman band with excitation frequency
is known as an excitation profile. Such profiles not only confirm
that a particular Raman band (and thus the corresponding normal
mode) is connected with the resonant electronic transition but
also can allow separate absorption bands to be resolved in those
cases where two or more are severely overlapped. Excitation
profiles have proved to be of considerable use in previous r.R.
studies of mixed-valence compounds,[2] in that relatively sharp
features have been resolved from very broad absorption bands.

This paper will discuss the application of r.R. spectroscopy,
first to structurally simple inorganic species of axial symmetry
and then to linear-chain halogen-bridged mixed-valence complexes
of platinum. The sort of results obtainable will thus be made
clear, as is the potential of the technique for determining the
nature of the structural changes on the molecule's changing from
the ground to the resonant excited state.

A. $M_2X_8^{n-}$ IONS

Earlier studies of the $Mo_2Cl_8^{4-}$, $Re_2Cl_8^{2-}$, and $Re_2Br_8^{2-}$ ions[3]
have now been completed with the synthesis and study of the $Mo_2Br_8^{4-}$
ion.[4] The $^1A_{2u} \leftarrow {}^1A_{1g}$, $\delta^*(b_{1u}) \leftarrow \delta(b_{2g})$ transition of this ion at
18000 cm^{-1} consists at ca. 80 K of a ten-membered progression in
$\nu(MoMo)$, 320 cm^{-1}. The r.R. spectrum of the ion, obtained by use
of an exciting line whose wavenumber corresponds to that of the
$\delta^* \leftarrow \delta$ transition, is characterised by an enormous enhancement to
the intensity of the $\nu(MoMo)$, $\nu_1(a_{1g})$ band at 335.8 cm^{-1}, together
with the appearance of two long progressions, in each of which it
is the $\nu_1(a_{1g})$ mode which acts as the progression-forming mode.
The progressions observed are $v_1\nu_1$ to $v_1 = 6$ at 300 K and to $v_1 =$
11 at ca. 80 K, and $\nu_2 + v_1\nu_1$ to $v_1 = 3$ at 300 K. The $\nu_2(a_{1g})$ mode,
at 168.8 cm^{-1}, is the symmetric MoBr stretching mode. The observed
maxima for the members of the two progressions for this tetrad of
ions are shown in Table 1.

TABLE 1

Results of r.R. Studies on $M_2X_8^{n-}$ Ions at 300 K

Complex	ν_1/cm^{-1} (MM)	ν_2/cm^{-1} (MX)	Main Progr.	Second Progr.	ω_1/cm^{-1}	x_{11}/cm^{-1}
$[NH_4]_5[Mo_2Cl_8]Cl$	338.0	274.0	$9\nu_1$	$\nu_2+4\nu_1$	339.6	-0.76
$[NH_4]_4[Mo_2Br_8]$	335.8	168.8	$6\nu_1$	$\nu_2+4\nu_1$	336.9	-0.48
$[n-Bu_4N]_2[Re_2Cl_8]$	271.9	356.5	$4\nu_1$	$\nu_2+2\nu_1$	272.6	-0.35
$[n-Bu_4N]_2[Re_2Br_8]$	275.4	209.8	$4\nu_1$	$\nu_2+2\nu_1$	276.2	-0.39

The excitation profiles of the ν_1 bands peak near to the Frank-Condon maximum of the $\delta^* \leftarrow \delta$ transition in each case.

B. $[M_2OX_{10}]^{n-}$ IONS

These ions contain a linear M-O-M bond, and thus have D_{4h} symmetry (as have the $[M_2X_8]^{4-}$ ions discussed in (A). They are thus close precursors to the linear-chain bridged complexes to be discussed in (C).

Studies of the r.R. spectra of the complexes $K_4[Ru_2OCl_{10}]$ and $Cs_3[Re_2OCl_{10}]$ have led[5] in each case to the observation of, and identification of, six progressions at room temperature and eight at ca. 80 K, in all of which it is the $\nu_1(a_{1g})$, symmetric metal-oxygen-metal (MOM) stretching mode which acts as the progression-forming mode (Table 2). The longest progression in each case is the $\nu_1\nu_1$ one, which reaches $\nu_1 = 12$ (Ru) and 14 (Re). The enabling modes are other a_{1g} fundamentals or combination bands, except for certain cases in which the first and third overtones of infrared-active fundamentals so act. The $\nu_9(e_u)$, $\delta(MOM)$, mode is particularly effective in this respect, progressions $\nu_1\nu_1 + 2\nu_9$ and $\nu_1\nu_1 + 4\nu_9$ reaching $\nu_1 = 8$ and 4 respectively (M = Ru), and 11 and 5 respectively (M = Re).

TABLE 2

Progressions observed in the r.R. spectra of $K_4[Ru_2OCl_{10}]$ and $Cs_3[Re_2OCl_{10}]$ at 80 K

Progression	$K_4[Ru_2OCl_{10}]$	$Cs_3[Re_2OCl_{10}]$
	ν_1	ν_1
$\nu_1\nu_1$	12 (ν_1=258.9 cm^{-1})	14 (ν_1=229.8 cm^{-1})
$\nu_1\nu_1 + \nu_2$	9 (ν_2=361.1 cm^{-1})	5 (ν_2=354.7 cm^{-1})
$\nu_1\nu_1 + \nu_3$	5 (ν_3=295.2 cm^{-1})	6 (ν_3=288.2 cm^{-1})
$\nu_1\nu_1 + \nu_4$	2 (ν_4=145.8 cm^{-1})	2 (ν_4=137.1 cm^{-1})
$\nu_1\nu_1 + \nu_2 + \nu_3$	0	4 (ν_2+ν_3=643.6 cm^{-1})
$\nu_1\nu_1 + 2\nu_5$	1 ($2\nu_5$=1779 cm^{-1})	0
$\nu_1\nu_1 + 2\nu_9$	8 ($2\nu_9$=935 cm^{-1})	11 ($2\nu_9$=822.6 cm^{-1})
$\nu_1\nu_1 + 4\nu_9$	5 ($4\nu_9$=1858 cm^{-1})	5 ($4\nu_9$=1635 cm^{-1})
$\nu_1\nu_1 + 728$	3 (728 unassigned)	–
$\nu_1\nu_1 + 552.5$	0	4 (552.5 unassigned)

The spectroscopic constants ω_1, ω_9, x_{12}, x_{13}, x_{14}, x_{19}, and x_{99} have been determined in each case from the data. The symmetry force constant (F_{11}) associated with MO stretching, 5.60 and 6.84 x 10^2 N m^{-1} for Ru and Re respectively, is high and there is extensive mixing of the axial a_{1g} coordinates. The excitation profiles of all four a_{1g} fundamentals of each ion maximise near 20,000 cm^{-1} and, with Raman polarisation data, this leads to the conclusion that the resonant electronic transition is an axially polarised $\pi^* \leftarrow \pi$ transition of the M-O-M π-bond system in each case.

This work has been extended[6] to detailed r.R. studies of ruthenium red and ruthenium brown, viz. $[Ru_3O_2(NH_3)_{14}]^{6+}$ and $[Ru_3O_2(NH_3)_{14}]^{7+}$.

MIXED-VALENCE LINEAR-CHAIN COMPLEXES

Class II type mixed-valence complexes consist of those mixed-valence complexes for which the valences are firmly trapped with distinguishable metal-ion sites, but for which there is sufficient

overlap between orbitals on adjacent metal ions (possibly via the
intermediacy of intervening s and p orbitals of bridging halogen
species) to permit electron transfer between the sites. Such
complexes typically display intense, broad absorption bands in the
visible or near infrared regions which are due to intervalence
charge transfer. Of the many mixed-valence complexes of Class II
known, the halogen-bridged linear-chain complexes of platinum have
so far been studied in greatest detail.

These are the sorts :

$[Pt^{II}L_4][Pt^{IV}L_4X_2]X_4$ e.g. where L=ethylamine or propylamine
      X=Cl, Br or I

$[Pt^{II}L_2X_2][Pt^{IV}L_2X_4]$ e.g. where L=ammonia, X=Cl or Br

$[Pt^{II}(L-L)_2][Pt^{IV}(L-L)_2X_2]Y_4$ e.g. where (L-L) = 1,2-diamino-
      ethane, or 1,2-diaminopropane, X=Cl, Br or I, Y=ClO$_4$

and $[Pt^{II}(L-L)X_2][Pt^{IV}(L-L)X_4]$ e.g. (L-L)=1,2-diaminoethane,
      X=Cl or Br

In these complexes, the two metal-atom sites are structurally
distinguishable, but may be interconverted by a concerted movement
of the axial halogen atoms in phase away from the platinum (IV)
atoms towards the platinum(II) atoms. Such complexes are strongly
dichroic, and possess a number of highly anisotropic properties.
Thus their electrical conductivity is typically about 300 times
greater in the chain direction than in the perpendicular directions.
The complexes are thus considered to be one-dimensional semiconductors.

Excitation of this type of mixed-valence transition within the
contour of the mixed-valence band leads to a resonance Raman spectrum
which is dominated by the band associated with the $\nu_1 (X-Pt^{IV}-X)$
symmetric stretching fundamental, together with its associated overtone progression. This intense progression usually overwhelms the
rest of the Raman-active bands, although in some cases other weak
progressions have also been seen. For such subsidiary progressions,
the enabling mode is another Raman-active mode, while the

progression-forming mode is, as for the main progression, ν_1.

The case of Wolffram's red, $[Pt^{II}(etn)_4][Pt^{IV}(etn)_4Cl_2]Cl_4$-
.$4H_2O$, etn=ethylamine, is typical of those for mixed-valence
complexes of the sort under discussion. Excitation within the
contour of the broad mixed-valence band centred around 21,000 cm^{-1}
(and polarized parallel to the chain axis) leads to the appearance
of a very strong band at 316 cm^{-1} (ν_1) together with its associated
overtone progression (Figure 1). This progression reaches as far
as $9\nu_1$ at room temperature and with 514.5 nm excitation, but as far
as $16\nu_1$ when the sample is held at ca. 80 K.[7] (Figure 1). Similar
results are obtainable for the corresponding bromide.

Results on a large number of linear-chain complexes of
platinum are summarised elsewhere[2,8]. Harmonic wavenumbers and
anharmonicity constants have been determined in all cases. The ν_1
normal coordinate seems to be related to the halogen movements
involved in the proposed hopping process for the conductivity of

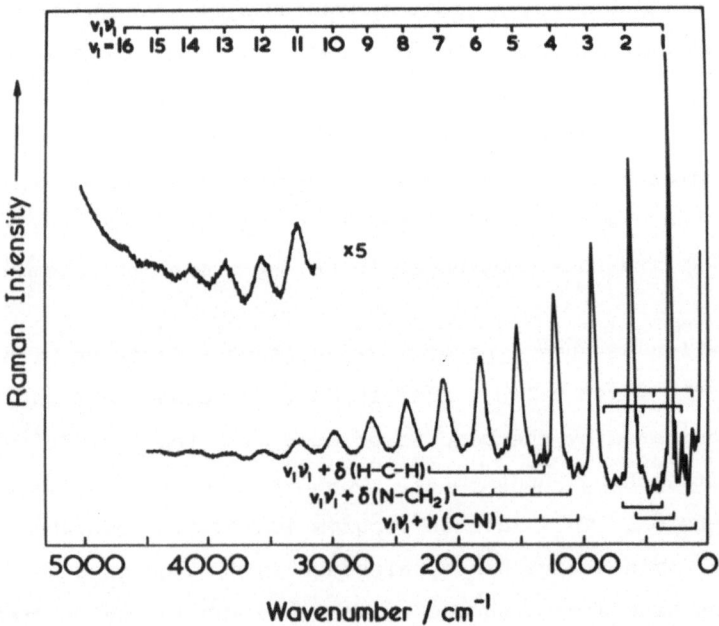

Figure 1.   Resonance Raman spectrum of Wolffram's red at ca. 80 K.

these linear-chain mixed-valence complexes.[9] The chain halogen atoms would need to move, on average, 0.54, 0.38 and 0.22 Å. for chlorides, bromides and iodides, respectively, in order to reach the point midway between the two platinum atoms, i.e. to the situation of a platinum(III) chain. These distance changes are related to the shift in the equilibrium Pt^{IV}-X bond lengths on excitation of the complex from the ground to the resonant excited state in each case.

Mixed-valence linear-chain complexes of palladium display very similar features to those of platinum. Indeed, the great intensity of the ν_1 progression in all these cases means that resonance Raman spectroscopy provides a sensitive and rapid means for detecting the formation of new linear-chain complexes of this sort.

CONCLUSION

The long progression in ν_1 (X-Pt^{IV}-X) observed in the r.R. spectra of the mixed-valence complexes implies that there is a large increase in the metal-halogen bond length on their going from the ground to the resonant excited state, and the appearance of subsidiary progressions implies other smaller and consequential structural changes. R.R. spectroscopy thus provides a sensitive means for probing the nature of the coupling of electronic and vibrational motions, and of structural changes consequent on change of electronic state of the scattering molecule. The technique may confidently be expected to be applied for these and other purposes, not only the mixed-valence complexes discussed herein, and to TTF^+TCNQ^- [10-12] and $(SNBr_{0.4})_x$,[13] but to a variety of other molecular metals.

REFERENCES

1. A.R. Gregory, W.H. Hennecker, W.C. Siebrand, and M. Zgierski, J. Chem. Phys., 1975, 63, 5475.

2. R.J.H. Clark and M.L. Franks, J. Chem. Soc. (Dalton), 1977, 198; R.J.H. Clark, Ann. N.Y. Acad. Sci., 1978, in press.

3. R.J.H. Clark and M.L. Franks, J. Amer. Chem. Soc., 1975, 97, 2691; 1976, 98, 2763.

4. R.J.H. Clark and N. D'Urso, J. Amer. Chem. Soc., 1978, 100,
    3088.

5. R.J.H. Clark, M.L. Franks, and P.C. Turtle, J. Amer. Chem. Soc.,
    1977, 99, 2473; J.R. Campbell and R.J.H. Clark, Mol. Phys.,
    in press.

6. J.R. Campbell, R.J.H. Clark, W.P. Griffith and J. Hall, to be
    published.

7. R.J.H. Clark and P.C. Turtle, Inorg. Chem., in press.

8. R.J.H. Clark, Advances in Infrared and Raman Spectroscopy, vol.
    1, Heyden, London, 1975, p. 143. R.J.H. Clark and B. Stewart,
    Structure and Bonding, in press.

9. L.V. Interrante and K.W. Browall, Inorg. Chem., 1974, 13, 1162.

10. G.R. Anderson and J.P. Devlin, J. Phys. Chem., 1975, 79, 1100.

11. D.L. Jeanmaire and R.P. Van Duyne, J. Am. Chem. Soc., 1976, 98,
    4029, 4034.

12. M.R. Suchanski and R.P. Van Duyne, J. Am. Chem. Soc., 1976, 98,
    250.

13. H. Temkin and G.B. Street, Solid State Comm., 1978, 25, 455.

CONDUCTIVE MOLECULAR CRYSTALS: METALLIC BEHAVIOR IN PARTIALLY

OXIDIZED PORPHYRIN, TETRABENZPORPHYRIN, AND PHTHALOCYANINE

Brian M. Hoffman, Terry E. Phillips, Charles J. Schramm
and Scott K. Wright

Department of Chemistry and Materials Research Center
Northwestern University
Evanston, Illinois 60201

Because of their highly anisotropic and sometimes spectacular
electrical, magnetic and optical characteristics, molecular crystals with strongly one-dimensional interactions are of great
current interest to chemists and physicists.[1] The discovery that
molecular solids can exhibit characteristics previously associated
only with "true" metals presents the opportunity of utilizing the
techniques of synthetic organic and inorganic chemistry to deliberately tailor the properties of the constituent molecules in
order to achieve desired materials properties. We have thus initiated a program of the synthesis and characterization of new
classes of highly conducting molecular crystals composed of planar
transition metal-macrocycle (ML) complexes.[2-7] These are partially
oxidized to a mixed-valency state since this is requisite for good
conductivity in such crystals. We have focused on related classes
of macrocycles which can be viewed as proliferations upon the por-
phin skeleton: porphyrins, tetrabenzporphyrins, and phthalocyanines:

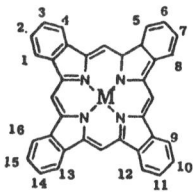

M(tetrabenzporphyrin)   M(porphyrin)   M(phthalocyanine)

M(TBP)   M(P)   M(Pc)

The chemical flexibility provided by these systems gives us a unique ability to purposively and rationally vary the electronic properties of the ML building block, and therefore of the resulting solid, through choice of ligand and peripheral substituents, through a variation in incorporated metal and finally, through a choice of oxidant. We find that halogen oxidation of the parent macrocycle produces lattices composed of one-dimensional arrays of planar donor molecules with fractionally occupied electronic valence shells ("partial oxidation"). With I_2 as oxidant, we have also found that a given ML can often be oxidized to a varying degree, forming MLI_x where x exhibits a range of values.

In this brief report we discuss the properties of partially oxidized nickel complexes: (1) $NiOEPI_{1.2}$ (OEP=1,2,3,4,5,6,7,8-octethyl-P); (2) $NiTBPI_{1.0}$; (3) $NiOMTBPI_{1.05}$, and (4) $NiOMTBPI_{2.9}$ (OMTBP=1,4,5,8,9,12,13,16-octamethyl TBP); (5) $NiPcI_{1.0}$. These oxidized solids exhibit conductivities up to seventeen order of magnitude greater than those of the parent complexes, in several instances with a characteristically metallic temperature dependence and room temperature transport properties as favorable as in any previously known molecular crystal.

RESULTS

Reaction of the ML complexes with I_2 yields powders and sometimes crystals with stoichiometry MLI_x. In order to relate the transport properties to the degree of oxidation and thus to "band filling", the state of the iodine must be determined. Resonance Raman (RR) spectroscopy is particularly useful for this task.[2,8] The RR spectra of (2), (3), and (4) are all virtually identical with the spectrum of (5) in Figure 1. The sharp absorption at $107cm^{-1}$ and the higher order overtone progression are characteristic of the polyiodide species, I_3^-. I^{129} Mossbauer spectroscopy further indicates a complete absence of I^- in NiPc(I)[2] and by analogy we assume that this is also true for the other materials. On the other hand, the RR spectrum of (1) which is characteristic for all the $Ni(OEP)I_x$ series is different (Figure 1). The strong absorption at $167cm^{-1}$ is diagnostic of I_5^-, and the small shoulder at $113cm^{-1}$ indicates that this anion is bent.[8] These spectra show that the proper formulations of the $ML(I)_x$ is $(ML)(I_n^-)_\rho$, where ρ is the actual degree of partial oxidation of ML. Thus, for the materials discussed here,

(1) $= NiOEP(I_5)_{0.24}$,   (2) $= Ni(TBP)(I_3^-)_{0.33}$,

(3) $= Ni(OMTBP)(I_3^-)_{0.35}$,   (4) $= Ni(OMTBP)(I_3^-)_{0.97}$,

(5) $= NiPc(I3)_{0.33}$

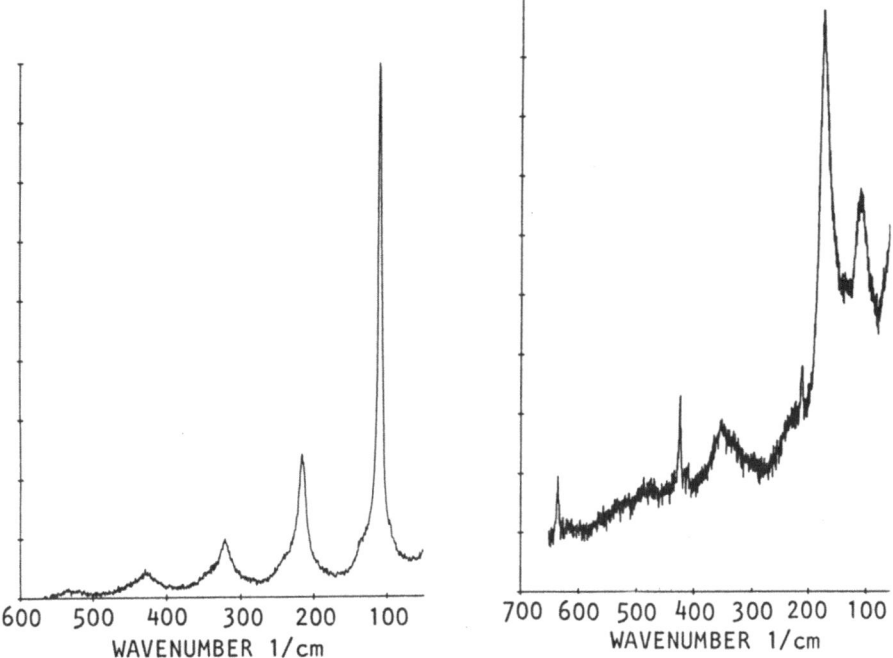

Figure 1:   Representative Resonance Raman spectrum of $ML(I_3^-)$ (left); $ML(I_5^-)$ (right).

Determining the nature of the "hole-state" created by partial oxidation of ML is no less important than determining ρ, the degree of charge transfer, and epr spectroscopy provides the answer to this question.   These $NiLI_x$ systems exhibit epr signals from the partially oxidized ML stacks.   The principle g-values in all cases

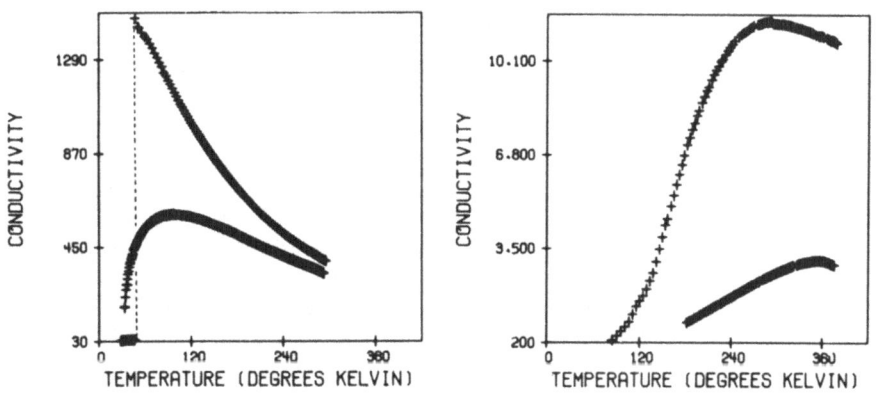

Figure 2: Four-Probe Singe Crystal Conductivity: (2) (left, lower); (3) (right upper); (4) (r, lower); (5) (1, upper).

fall between 2.010 and 2.002. This shows that the unpaired spins of the hole state have the character of a π cation radical. Thus for these complexes, the oxidation is ligand, not metal, centered.

Temperature dependent single crystal conductivities along the chain axis, using a four-probe A.C. phase-locked technique,[9] have been obtained for (2), (3), (4), and (5):

Each one of the four shows a region of metal-like behavior, $d\sigma/dT < 0$, and (2) and (5) are highly conductive and metallic over a wide range of temperature.

The most conductive material, (5), has an ambient temperature, $\sigma_{11} = 550 cm^{-1}$. As T is lowered, the conductivity change is proportional to $T^{-1.9}$ down to $T_c \sim 55°K$; at this temperature microwave conductivity instruments show an abrupt metal-insulator transition.[11] The metallic character of the material for $T > 55°K$ is confirmed by the observation of an essentially temperature-independent spin susceptibility in this regime. Interpreting this data with the simple 1-D tight-binding model gives a transfer integral, $t \sim 0.2 eV$.

The very similar material (2) is also highly conductive, with $\sigma_{11} \sim 350 \Omega cm^{-1}$ at room temperature. However, the temperature dependence of σ is different. Lowering the temperature from ambient, σ increases at $T^{-2.3}$ until it reaches a maximum at $T_m = 95°K$, substantially above the $T_c = 55°K$ for (5). Moreover, σ does not change discontinuously at T_{max}, as it does for (5) at T_c. Below T_{max}, the conductivity decreases in an activated manner, with $\Delta E_{act} = 0.04 eV$.

Curve C depicts the conductivity response of the methylated version of NiTBP(I), (3). The room temperature conductivity is $10 ohm^{-1} cm^{-1}$ at $T_m = 300°K$. Upon further cooling, the conductivity becomes activated with an activation energy 0.045eV. The behavior of the more highly oxidized phase, (4), is qualitatively the same (curve D). The maximum conductivity is decreased to $4 (ohm^{-1} cm^{-1})$ while the T_m is shifted upwards to $340°K$. The activation energy in the lower temperature regime is 0.050eV.

The only conductivity results to date for (1) are from pressed pellets. If single crystals of (1) behave like (5), with the ambient temperature ratio of σ_{11}(crystal)/σ(pellet)$\sim 10^2$, then we expect for (1), $\sigma_{11} \sim 6 \Omega cm^{-1}$, comparable to (3) or (4). The large powder activation energy for (1), $\Delta E \sim 150 eV$, compared to the small value for (5), $\Delta E \sim 0.05 eV$, suggests that for (1) the single crystal T-dependence of σ would also be like that of (3) or (4).

DISCUSSION

These results show the wide range of behaviors available from just a small subset of the partially oxidized ML we are investigating. Compounds (2) and (5) have the same shaped ligand, the same ρ, and crystal structures which are almost indistinguishable:[5,6] stacks of NiL with an ~3.25Å Ni-Ni spacing, each stack separated from its neighbors by chains of I_3^-. Both exhibit a high, metallic-type conductivity, but quite different forms of metal-insulator transition.

Methyl substitution of TBP should have modest effects upon the TBP electronic properties, but in (3) it is found to increase the Ni-Ni spacing in a stack to 3.77Å. This increased spacing at a constant value of ρ presumably lowers the transfer integral, noticeably reducing σ_{11} and also altering the form of σ_{11} vs T. Higher oxidation to (4) further reduces σ without change in the shape of σ_{11} vs T. Finally, (1) employs the smaller, parent macrocycle. It is of substantial interest that the low level of oxidation in (1), and the necessarily great plane-plane spacing forced by the bulky side-chains still produces a conductivity comparable to that of (3).

These observations with a limited subset of available systems clearly demonstrate the utility of this approach for exploring the relationships between molecular composition and solid-state transport properties.

Acknowledgments: This work is supported by the Materials Research Center of Northwestern University and by an NSF grant to BMH. We thank our collaborators, Professors J.A. Ibers, T.J. Marks, T. Poehler, and Dr. R.P. Scaringe.

REFERENCES

1. H.J. Keller, Ed., Chemistry and Physics of One-Dimensional Metals (Plenum, New York, 1977); J.S. Miller and A.J. Epstein, Prog. Inorg. Chem. 20, 1 (1976); A.J. Berlinsky, Contemp. Phys. 17, 331 (1976); J.J. Andre, A. Bieber, F. Gautier, Ann. Phys. 1, 145 (1976).

2. J.L. Peterson, C.J. Schramm, B.M. Hoffman, T.J. Marks, J. Am. Chem. Soc. 99, 286 (1977).

3. T.E. Phillips and B.M. Hoffman, J. Am. Chem. Soc. 99, 7734 (1977).

4. C.J. Schramm, D.R. Stojakovic, B.M. Hoffman, T.J. Marks, Science 200, 47 (1978).

5. C.J. Schramm, R.P. Scaringe, D.R. Stojakovic, B.M. Hoffman, T.J. Marks, J.A. Ibers; in preparation.

6. T.E. Phillips, R.P. Scaringe, B.M. Hoffman, J.A. Ibers; in preparation.

7. S. Wright, D.M. Schollter, C.J. Schramm, T.E. Phillips, B.M. Hoffman; in preparation.

8. R.C. Teitelbaum, S.L. Ruby, T.J. Marks, J. Am. Chem. Soc. 100, 3215 (1978).

9. T.E. Phillips, J.R. Anderson, C.J. Schramm, B.M. Hoffman, Rev. Sci. Inst.; in press.

10. T. Poehler; private communication.

SYNTHESIS AND PROPERTIES OF SEVERAL NEW MOLECULAR CONDUCTORS

David B. Brown and James T. Wrobleski

Department of Chemistry, University of Vermont

Burlington, VT 05405 U.S.A.

The preparation of highly-conducting materials which contain stacks or chains of coordinated metal ions continues to challenge synthetic chemists. As a synthetic aid, Keller (1) has developed guidelines for the preparation of such materials. These guidelines derive from both experimental observations and interpretations of the excitonic theory of superconductivity (2). This theory requires that a spine of delocalized charge carriers interacts <u>via</u> an excitonic mechanism with a proximal array of polarizable molecules (or ions) which will largely determine the exciton frequency. Although Little (3) has proposed several specific model compounds in which the essential features of the model might be realized, to date theoretical expectations are more nearly realized for the "organic metal" conductors than for the columnar transition-metal complexes. The latter materials possess a more-or-less conducting spine which is not strongly coupled (mechanically or electronically) to other molecules or ions in the lattice. Although this "insulation" of the conducting spine may lead to enhanced anisotropy in the electrical conductivity, it may also drastically lower the superconducting transition temperature of the material (by way of a Peierl's distortion or similar mechanisms.)

We are currently investigating the synthesis, electrical conductivity, and magnetic susceptibility of both single-valence and mixed-valence stacked complexes containing bifunctional ligands in order to ascertain the effect of weak <u>interchain</u> and/or donor-acceptor interactions. Our general synthetic outline is shown as two reactions in Fig. 1. In this figure M is a $\underline{d}^8$ transition-metal ion capable of forming square-planar bis complexes with a bidentate ligand containing donor atoms L. Molecule (or atom) A

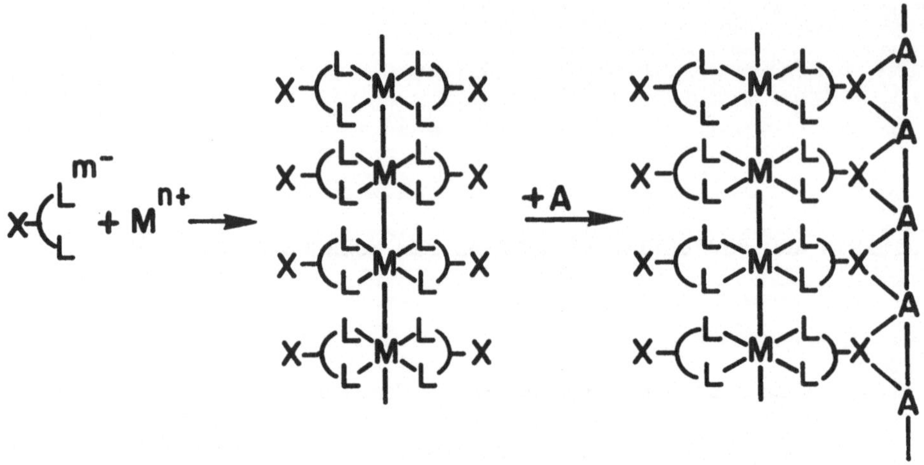

Fig. 1. General synthetic outline for coupled donor–acceptor stacks.

is an electron acceptor capable of partially oxidizing the metal
chain. X is a functional group capable of hydrogen bonding to A.
Acceptor A should be capable of crystallizing in stacks as well.
Although we explicitly treat columnar structures with direct d_{z^2}
overlap, chain structures with ligand bridges are not excluded from
consideration.

Several of the ligands we have used are illustrated in Fig. 2.
Of these ligands, diaminoglyoxime and rubeanic acid (dithiooxamide)
are capable of forming very strong hydrogen bonds with chains of
halide ions and chains of oxygen-containing electron acceptors.
For this reason we have used the "classical" oxidizing agents Br_2
and I_2 as well as the unique oxidants p-quinone (1,4-benzoquinone,
Q) and 1,4-naphthoquinone (NQ) (Fig. 2). Synthesis of the Ni(II),
Pd(II), and Pt(II) single-valence stacked complexes containing
these ligands is normally carried out in aqueous solution although
we often use ethanol-water mixed solvents to facilitate dissolution
of the reactants. In some cases the crystals which are isolated
from the reaction display striking color anisotropy.

We have obtained preliminary room-temperature electrical (dc)
conductivity data (pressure contacts, 4-point probe) by using
pressed pellets of these polycrystalline materials. Some typical
results are given in Table I. This representative set of compounds
contains anionic, neutral, and cationic chains with both nitrogen
and sulfur donor atoms. The conductivities of these single-
valence materials (Table I) are similar to the conductivities of
related stacked d^8 complexes (weak semiconductors). It is interesting to note that although these complexes are all diamagnetic at

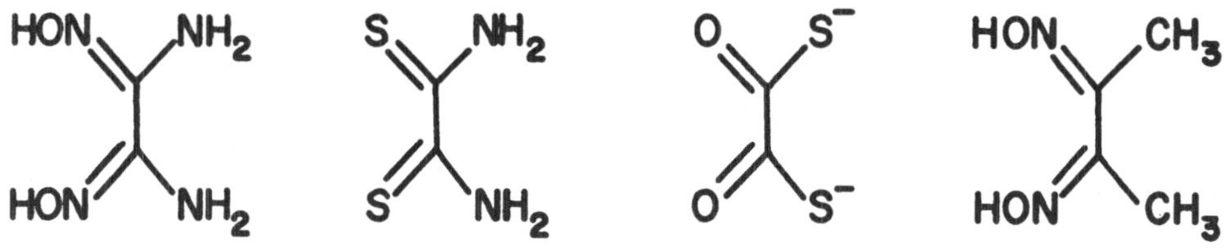

diaminoglyoxime   rubeanic acid,   dithiooxalate,   dimethyl-
    DAGH   RA   DTO   glyoxime,
                                                                    DMGH

p-Quinone,   Naphthoquinone,
    Q   NQ

Fig. 2.   Representative ligands and   electron acceptors (quinones)

Table I.   Single-Valence (Closed Shell) $\underline{d}^8$ Complexes with Columnar
Structures.

Complex	Color	$\sigma, {}^{300K}\Omega^{-1}cm^{-1}$
Ni(DAG)$_2$	red-yellow	3 x 10^{-8}
Pd(DAG)$_2$	brown	1 x 10^{-7}
Pt(DAG)$_2$	red-brown	2 x 10^{-8}
Ni(RA)$_2$Cl$_2$	red-brown	5 x 10^{-10}
Pd(RA)$_2$Cl$_2$	red	9 x 10^{-8}
Pt(RA)$_2$Cl$_2$·xH$_2$0	red-violet	6 x 10^{-8}
K$_2$[Ni(DTO)$_2$]	purple	1 x 10^{-9}
K$_2$[Pd(DTO)$_2$]	red	8 x 10^{-8}
K$_2$[Pt(DTO)$_2$]	violet	6 x 10^{-10}
Ni(DMG)$_2$	red	5 x 10^{-11}
Pd(DMG)$_2$	red	4 x 10^{-10}
Pt(DMG)$_2$	violet	1 x 10^{-9}

300K (to within experimental uncertainty) they do possess a feeble
Curie paramagnetism below ~ 200K. The origin of this paramagnetism
may be due to monomeric impurities in the sample or may have its
origin in other effects which are a reflection of the chain
structure of these materials.

We have used a variety of synthetic techniques for the preparation of mixed-valence analogs of the compounds in Table I.
These techniques are best illustrated by considering specific
examples. For example, $Ni(DAG)_2$ was treated with a stoichiometric
amount of Br_2 in hot CH_3CN. To this solution was added a CH_3CN
solution containing two equivalents of $Ni(DAG)_2$. Upon cooling the
resulting dark-yellow solution we obtain yellow-red crystals which
analyze for $Ni(DAG)_2Br_{0.3}$. This partial oxidation technique has
been previously utilized for the synthesis of KCP(Br) and is
applicable for soluble starting materials.

The iodine oxidation product of $K_2[Ni(DTO)_2]$ was obtained by
a solid-state oxidation of the parent Ni(II) complex with a four-
fold excess of I_2 at 90-100°C in a sealed tube. Although this
technique gives a well-defined product in this case, it suffers
from the use of relatively high temperatures. For example, attempted oxidation of $K_2[Pt(DTO)_2]$ with I_2 in a sealed tube produces
mainly PtS as the platinum-containing product.

We normally carry out oxidations with quinones by reacting a
stoichiometric amount of quinone with the metal complex slurried
in refluxing 1,2,4-trichlorobenzene (TCB). For example, treatment
of 10 mmole of $Ni(DMG)_2$ with 10 mmole of p-quinone results in the
formation of a dark violet product which contains one quinone
unit per $Ni(DMG)_2$, $Ni(DMG)_2Q$. We have also used TCB for I_2 and
Br_2 oxidations of chain compounds.

Several materials obtained by these methods are listed in
Table II. By comparing the pressed-pellet conductivities of these
materials (Table II) with those of the parent $\underline{d}^8$ complexes (Table
I) an impressive increase in the room-temperature conductivity is
evident in proceeding from the starting complex to its oxidation
product. As an illustration, the conductivity of $K_2[Ni(DTO)_2]$
increases as much as eleven orders of magnitude upon iodine
oxidation. We have thus far obtained several samples of this
oxidation product which display metallic temperature dependence
of conductivity in the approximate temperature range 200-300K.
Some typical pressed-pellet, four-probe dc conductivity data for
$K_2[Ni(DTO)_2]I$ are shown in Fig. 3. The illustrated sets of points
correspond to data obtained on two different preparations, A and
B. Whereas samples A and B were prepared under apparently identical
conditions, the former shows metallic conductivity in the temperature range 200-300K while the latter is a semiconductor in the
entire temperature range studied.

Table II. Representative Oxidized Complexes.

Complex	Color	$\sigma, ^{300K}_{\Omega^{-1} cm^{-1}}$
Ni(DAG)$_2$Br$_{0.3}$	yellow-red	4 x 10^{-5}
Ni(DAG)$_2$Q	violet	9 x 10^{-2}
Ni(DAG)$_2$I	red	9 x 10^{-3}
Pd(DAG)$_2$Br$_x$	gold-brown	6 x 10^{-1} [a]
Pt(DAG)$_2$Br$_x$	gold-brown	2 - 200 [a]
Pt(RA)$_2$Cl$_2$Br$_{0.3}$	yellow	8 x 10^{-2}
K$_2$[Ni(DTO)$_2$]I$_{0.3}$	black	0.5 -200 [a]
Ni(DMG)$_2$I	violet	8 x 10^{-6}
Ni(DMG)$_2$Q$_x$	violet	6 x 10^{-3}

[a]Sample dependent value.

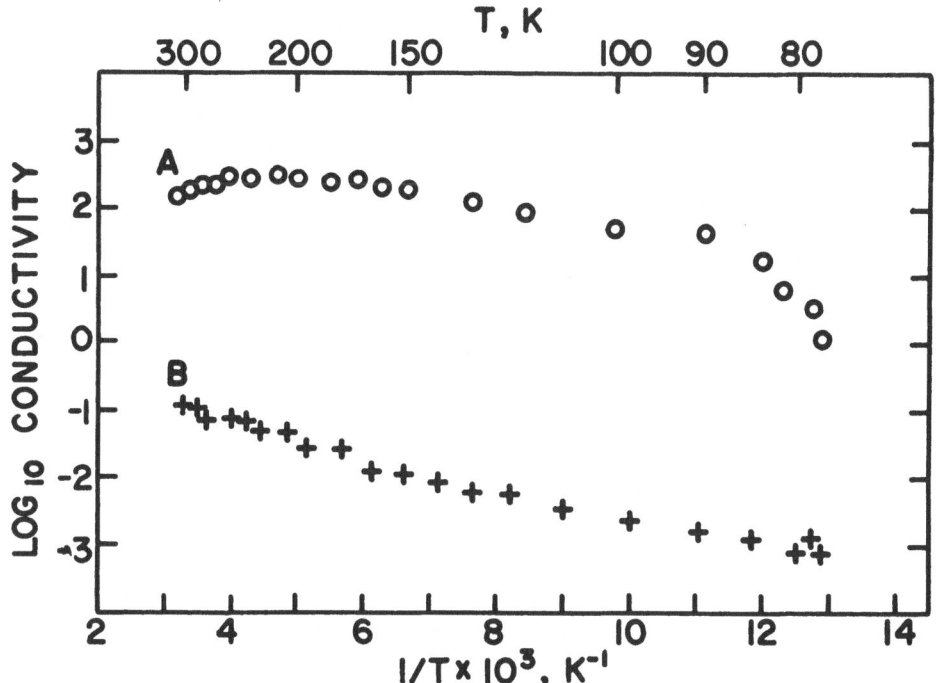

Fig. 3.   Conductivity of two samples of K$_2$[Ni(DTO)$_2$]I

(Sample A represents the most highly-conducting specimen of
K$_2$[Ni(DTO)$_2$]I we have thus far prepared.) We have also measured
the magnetic susceptibility of K$_2$[Ni(DTO)$_2$]I in the temperature
range 20-300K. Both the metallic-conducting sample A and the

semiconducting sample B have room temperature magnetic moments
near 1.73µB. Samples A and B both show Curie-Weiss behavior with
θ = + 26 and + 12K, respectively.

The bromine oxidation products of Pd(DAG)$_2$ and Pt(DAG)$_2$ also
show high room-temperature conductivities (Table II). These
materials are, however, semiconductors with activation energies of
0.15 and 0.06 eV, respectively. We have thus far been unable to
isolate a sample of these oxidation products which displays
metallic conductivity in the temperature range experimentally
accessible to us. These materials lose Br$_2$ at room temperature -
pressure and under vacuum. We have also noticed that passage of
a current in excess of 1mV through these samples leads to enhanced
loss of Br$_2$. Room-temperature magnetic susceptibility data for
these compounds indicate that they are diamagnetic materials.

The p-quinone oxidation products of Ni(DAG)$_2$ and Ni(DMG)$_2$ are
both dark violet in color as compared to the yellow-red color of
the Ni(II) starting materials, and the oxidation products show a
large enhancement in conductivity. Both Ni(DMG)$_2$Q and Ni(DAG)$_2$Q
have room-temperature magnetic moments near 1.73µB indicating one
unpaired electron per Ni. Preliminary ESR spectra suggest that
the electron is at least partly metal based.

Quinone oxidations of metal stacks are particularily interesting when discussed in light of the recent hypothesis of Perlstein
(4) regarding the possible existence of an intermolecular migration
of aromaticity in TCNQ-like salts. A tentative model for quinone
oxidation products of stacked metal complexes is shown in Fig. 4.
In this model the quinone stack is reminiscent of the structural
role of TCNQ stacks in some organic metals. In order to confirm
this proposed stacking arrangement of the quinone molecules it will
be necessary to perform single-crystal X-ray studies on these
materials. It has been our observation that a rather large number
of columnar d^8 complexes may be oxidized by Q and NQ. Such
oxidations may offer new synthetic routes to a different class of
materials with high (possibly metallic) conductivities.

In conclusion we wish to briefly comment on a possible model
of excitonic superconductivity based on mixed-valence stacked
complexes containing bifunctional ligands and "polarizable" acceptor
stacks. We will consider a stack of square-planar complexes bonded to I chains (Fig. 5a) and to p-quinone stacks (Fig. 5b). Whereas the electrical conductivity will still occur along the spine of
metal ions, an excitonic mode will be enhanced by the interaction
of the conduction electrons with the aromatic (charge) migration
along the quinone-hydroquinone or iodine-iodide spine. The formation of electron pairs should be energetically favored by the
nearest-neighbor polarization effects of the charged acceptor spine.

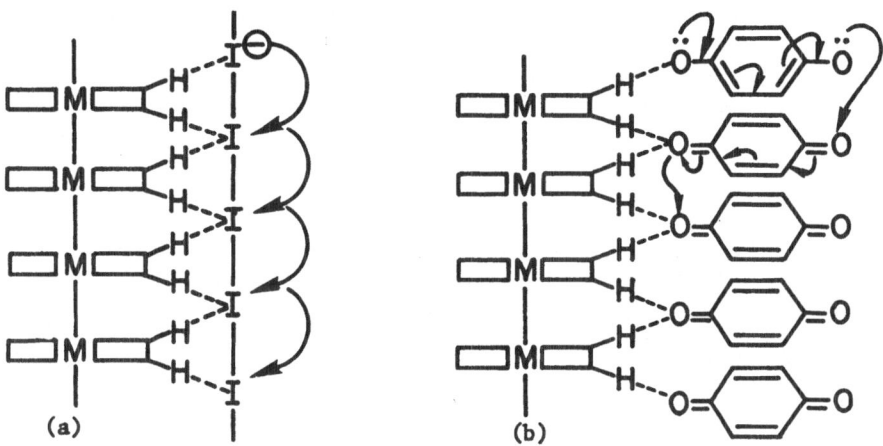

Fig. 4. Possible arrangement of oxidized metal stacks with intervening quinone-hydroquinone columns.

It seems probable that an interactic. between acceptor and donor
stacks will provide an opportunity for excitonic superconductivity.
Unfortunately, very little experimental evidence to date is germane
to this point. The series of oxidized compounds $Ni(DMG)_2I$,
$Ni(DPG)_2I$ (5), $Ni(BQD)_2I_{0.7}$, (6) and $Ni(DAG)_2I$ shows a range of
conductivities (8×10^{-6}, 1×10^{-7}, 2×10^{-5}, 9×10^{-3}, respectively)
which may be explained by the model shown in Fig. 5a. However,
great care must be exercised in attempting to make comparisons as
all the compounds in this series do not contain equivalent Ni-Ni

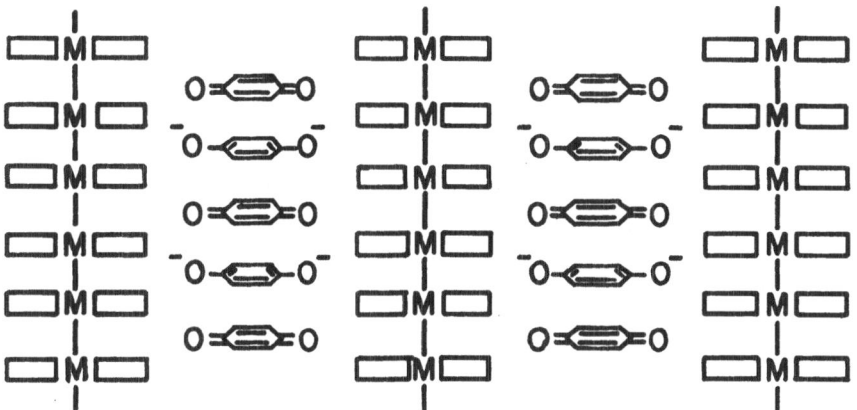

Fig. 5. Model of an excitonic superconductor with acceptor stacks
as (a) iodine and (b) p-quinone.

separations or equivalent I sites.　Future work on more diverse systems (particularily those capable of hydrogen bonding to the acceptor spine) should help to clarify these apparent trends.

REFERENCES

(1)　H. J. Keller, "Low-Dimensional Cooperative Phenomena", H. J. Keller, Ed., Plenum Press, New York, 1974, p. 315.
(2)　W. A Little, Phys. Rev., 134, A1416 (1964).
(3)　H. Gutfreund and W. A. Little, "Chemistry and Physics of One-Dimensional Metals", H. J. Keller, Ed., Plenum Press, New York, 1976, p. 279.
(4)　J. H. Perlstein, Angew. Chem. Int. Ed. Engl., 16, 519 (1977).
(5)　A. E. Underhill, O. M. Watkins, and R. Pethig, Inorg. Nucl. Chem. Letters, 9, 1269 (1973).
(6)　H. Endres, H. J. Keller, M. Mégnamisi-Bélombé, W. Moroni, and D. Nöthe, Inorg. Nucl. Chem. Letters, 10, 467 (1974).

THE STRUCTURE AND MAGNETIC AND ELECTRICAL CONDUCTIVITY PROPERTIES OF THE CHARGE TRANSFER COMPOUND 1,1'-DIMETHYLFERROCENIUM BIS-(TETRACYANOQUINODIMETHANE), $[(CH_3C_5H_4)_2Fe][TCNQ]_2$

Scott R. Wilson,[a] Peter J. Corvan,[a] Reginald P. Seiders,[a]
Derek J. Hodgson,[a] Maurice Brookhart,[a] William E.
Hatfield,[a] Joel S. Miller,[b] Arthur H. Reis, Jr.,[c]
P. K. Rogan,[c] Elizabeth Gebert[c] and Arthur J. Epstein[d]

Kenan Laboratory 045A, University of North Carolina,
Chapel Hill, NC 27514[a]
Rockwell Science Center, Thousand Oaks, CA 91360[b]
Argonne National Laboratory, Argonne, IL 60439[c]
Xerox Webster Research Center, Rochester, NY 14644[d]

Metallocenes react with tetracyanoquinodimethane (TCNQ) to yield charge transfer compounds predominantly of the stoichiometry [metallocene][TCNQ] or [metallocene][TCNQ]$_2$.[1,2] The 1:2 compounds have relatively high electrical conductivities which range from 4 ohm^{-1}cm^{-1} for $[(C_5H_5)_2Fe][TCNQ]_2$ to 0.03 ohm^{-1}cm^{-1} for $[(CH_3C_5H_4)_2Fe]$-$[TCNQ]_2$,[1] but, up to now, difficulties in obtaining high quality single crystals have prevented structural determinations.

The recent report[3] of the use of Sephedax gels for the growth of single crystals led us to use this technique in the case of $[(CH_3C_5H_4)_2Fe][TCNQ]_2$.

Experimental Section:

The title charge-transfer complex was prepared by reaction of 1,1'-dimethylferrocene and TCNQ (molar ratio 1:2, respectively) in hot acetonitrile solution. Dark needles were isolated by filtration of the cooled solution. A portion of this material was dissolved in a minimum volume of hot acetonitrile and the resulting dark green solution was poured into a warm test tube (1.5 x 12 cm). Immediately, Sephadex LH-20 (Pharmacia Fine Chemicals) was added until all the acetonitrile had been absorbed. The test tube was allowed to cool slowly, and within 30 minutes small rectangular prisms of the complex began growing inside the gel. After one day at room temperature, the gel was then broken up with distilled water and individual crystals were removed for study.

For the preparation of dimethylferrocenium hexafluorophosphate, dimethylferrocene (4.0 mmole) was oxidized by stirring in dilute nitric acid for 2 hours at which time dissolution was nearly complete. The addition of ammonium hexafluorophosphate to the filtered solution produced a blue-grey precipitate of the PF_6^- salt. This material was then rapidly recrystallized from anhydrous methanol to give 0.22 g (15% yield) of the PF_6 salt as a green-grey powder, whose formulation was confirmed by elemental analysis.

Magnetic properties were measured using a PAR vibrating sample magnetometer. Temperatures were measured with a calibrated gallium/arsenide diode by observing the voltage on a Dana Model 4700 4.5 place digital voltmeter. Details of the experimental techniques have been given elsewhere.[4]

Crystallographic Section:

Two entirely separate sets of data were collected, one at Argonne National Laboratory and the other at the University of North Carolina; the following discussion pertains to the latter set, since these data were obtained from a crystal synthesized by the method described above. The two data sets are, however, of comparable quality, and the derived structural parameters are equal with standard deviations.

On the basis of precession and Weissenberg photography, the intensely dark blue prismatic crystals were assigned to the triclinic system, the space group being either $P\bar{1}$ or P1; the former choice was confirmed by the successful refinement of the structure. The cell constants, obtained by least-squares methods from the diffractometer settings of twelve reflections with 2θ(Mo) $\geq$ 25°, are a = 7.660(2), b = 7.530(2), c = 14.083(4) Å, α = 83.83(2)°, β = $\bar{1}$10.42(2)°, γ = 94.48(2)°. The observations were made at 22°C with the wavelength assumed as λ(MoKα$_1$) = 0.70926 Å. A density of 1.35(1) g cm^{-3} observed by flotation in chloroform/dichloromethane solutions is in acceptable agreement with the value of 1.370 g cm^{-3} calculated for one formula unit per cell. Hence, in space group $P\bar{1}$ the iron atom is constrained to lie on a crystallographic inversion center.

Diffraction data were collected from a prismatic crystal bounded by the (001), (00$\bar{1}$), (100), ($\bar{1}$00), (0$\bar{1}$0), and (01$\bar{1}$) faces. The approximate crystal dimensions were 0.51 x 0.16 x 0.11 mm in the [100], [001], and [010] directions, respectively. The crystal was mounted approximately parallel to the crystallographic a axis, and intensity data were collected on a Picker four circle automatic diffractometer equipped with a molybdenum tube and a graphite monochromator. Data were collected by the θ/2θ scan technique at a

scan rate of $1.0°min.^{-1}$. To allow for the presence of both $K\alpha_1$ and
$K\alpha_2$ radiations, peaks were scanned from $1.0°$ in 2θ below the calculated $K\alpha_1$ peak position to $1.0°$ in 2θ above the calculated $K\alpha_2$
peak position. Stationary-counter, stationary-crystal backgrounds
were counted for 20s at each end of the scans. A unique data set
having $2\theta(MoK\alpha) \leq 55°$ was gathered, a total of 3747 reflections
being collected. The intensities of three standard reflections,
examined after every 100 reflections, remained essentially constant
throughout the run.

The data were processed by the method of Ibers and coworkers.[5]
After correction for background, the intensities were assigned standard deviations as

$$\sigma(I) = [C + 0.25(ts/tb)^2(B_H + B_L) + (pI)^2]^{\frac{1}{2}}$$

where the symbols have their usual meanings[5] and the value of p was
assigned as 0.04. The values of I and $\sigma(I)$ were corrected for
Lorentz-polarization effects[6] and for absorption.[7] The absorption
coefficient μ for these atoms with MoKα radiation is 5.28 cm^{-1}, and
for the sample chosen the transmission coefficients ranged from
0.92 to 0.96. Of the 3747 intensities recorded, only 1599 were
independent data with $I > 3\sigma(I)$; only these data were used in the
structure analysis.

The structure was solved by direct methods using the highest
256 normalized structure amplitudes (E's) in the program MULTAN.[8]
Because of some pseudo symmetry, the data with h = 2n had to be
scaled separately from those with h = (2n + 1). The resulting E-
map revealed the locations of the TCNQ carbon atoms and the iron
atom (constrained to lie on an inversion center). Isotropic refinement of these positions led to values of the conventional agreement
factors $R_1 = \Sigma||Fo|-|Fc||/\Sigma|Fo|$ and $R_2 = [\Sigma w(|Fo|-|Fc|)^2/\Sigma w(Fo)^2]^{\frac{1}{2}}$
of 0.281 and 0.364, respectively. All least-squares refinements
were on F, the function minimized being $\Sigma w(|Fo|-|Fc|)^2$ with the
weight w being assigned as $4Fo^2/\sigma^2(Fo^2)$. The scattering factors
were taken from the usual sources.[9]

The carbon atoms of the methylcyclopentadienyl moiety were
located in a difference Fourier map. Eventual anisotropic refinement of these atoms led to $R_1 = 0.072$, $R_2 = 0.077$. Addition of the
four hydrogen atoms in the TCNQ moiety, refined isotropically, gave
$R_1 = 0.064$, $R_2 = 0.065$. At this stage of refinement the C-C ring
bond lengths in the methylcyclopentadienyl moiety were rather short,
ranging from 1.33 to 1.39 Å, and the Uij's were very large; these
observations could be attributable to some disorder, which is
extremely common for systems of this type,[10] but attempts to refine
a disordered model were unsuccessful. Addition of the remaining

hydrogen atoms led to R_1 = 0.059; a difference Fourier at this stage was featureless with no peak higher than 0.5 eA^{-3}.

Description of the Structure:

The structure consists of $Fe(CH_3C_5H_4)_2$ and TCNQ units. The iron atom lies on a crystallographic inversion center. The TCNQ moieties also lie on inversion centers, there being two crystallographically independent "half-TCNQ" moieties. The packing of these fragments in the crystal is shown in Figures 1 and 2. As is seen in Figure 2, one TCNQ unit (A) approaches neighboring iron atoms (through its terminal CN groups) much more closely than does the

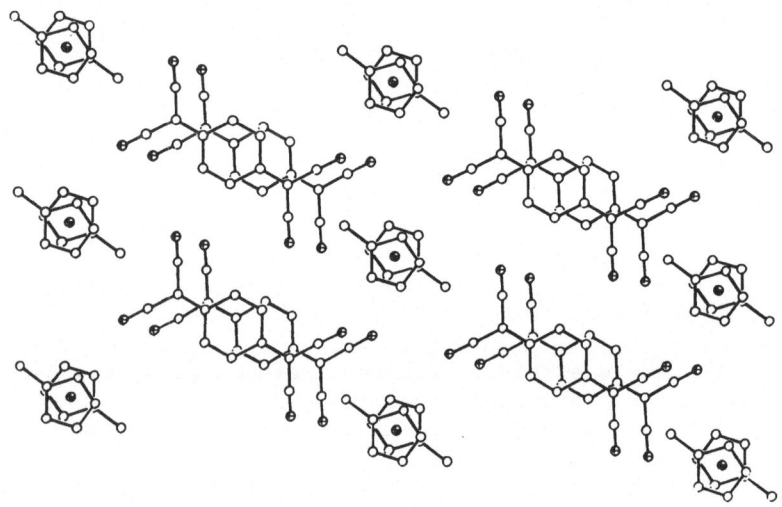

Figure 1.   A view of structure of 1,1'-dimethylferrocenium $(TCNQ)_2$ along the a axis which shows the slipped nature of the TCNQ stacks.

other (B). As is seen in the figures, the structure can be viewed as consisting of chains of $CH_3C_5H_4$ (MeCp) rings separated by paralle chains of TCNQ units; the direction of the chains is the crystallographic a-axis. The MeCp and TCNQ rings are not normal to the chain direction, however; in the MeCp chain the angle between the normal to the planes and the a-axis is 9.1°, while in the TCNQ chain the average value is 14.2°.

In the MeCp chains, there are two distinct separations; the rings involved with any given Fe atom are separated by 3.41 Å

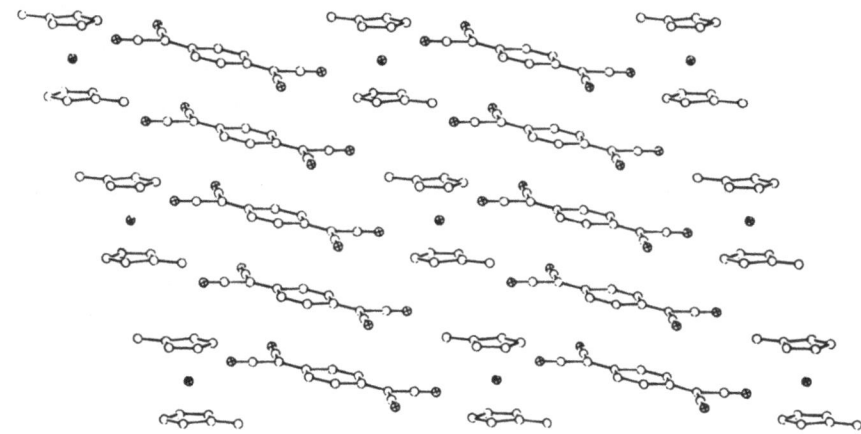

Figure 2. A view of the structure of 1,1'-dimethylferrocenium (TCNQ)$_2$ along the b axis which shows the parallel stacks of ferrocenium ions and TCNQ units.

while neighboring rings are separated by 3.63 Å. The units in the TCNQ chains are equally spaced, the center of each ring lying on an inversion center; the separation between each ring center is simply a/2 as measured along a. Since the rings are not normal to the chain direction, however, the interplanar separation is only 3.23 Å.

The two TCNQ units are crystallographically distinct and are much more precisely determined than the MeCp rings because of the high thermal motion (or disorder) of the latter. The rings each display the expected quinoidal geometry, with two "double" bonds and four "single" bonds. In the A ring these are of lengths 1.343(7) and 1.425(6) and 1.433(6) Å, respectively, while in the B ring the values are 1.334(7), 1.436(6), and 1.438(6) Å, respectively. The four terminal C-N bond lengths are all very similar, with values of 1.139(6), 1.139(6), 1.140(6), and 1.145(6) Å. There appears to be a significant difference, however, between the extracyclic nominally double C-C bonds in the two rings; in ring A this has the length 1.395(6) Å, while in ring B it is 1.378(6) Å. This small but significant difference, which is independently confirmed in the Argonne analysis, has been observed by other workers in related systems[11] and can be attributed to an increase in radical character on ring A relative to that on B. In neutral TCNQ this distance is 1.374 Å, in TCNQ$^-$ it is 1.418 Å and in the system reported as TCNQ$^{\frac{1}{2}-}$ it is 1.388 Å; hence, it is apparent that this C=C distance is

extremely sensitive to electron donation from donors in the crystal.[11] In the present case, the observation of apparently greater
donation to ring A is entirely consistent with the structural feature noted above that this ring approaches the $Fe(MeCp)_2$ donor
moieties much more closely than does ring B. Alternatively, this
crystallographic result of uniformly spaced TCNQ molecules with
alternating intramolecular bond lengths may reflect a Peierls distortion and gap which is stabilized by intramolecular distortion
in this uniform chain band system which is one-quarter filled with
large on-site Coulomb repulsions.[12] The dc conductivity is proportional to exp(-1200/T) in agreement with this model. However,
resonance Raman data taken with 457.9 nm radiation supports the
formulation $TCNQ^{0.5-}$, since the primary Raman active exocyclic C-C
absorption occurs at 1428 cm^{-1}, a value which is consistent with a
negative charge on TCNQ of 0.42 $\pm$ 0.1. [13]

Magnetic Properties:

        The temperature variation of the magnetic susceptibility of
a powdered sample of $[(CH_3C_5H_4)_2Fe][TCNQ]_2$ was determined in the
temperature range 4.2 to 76°K. The data are presented in Figure 3
as a plot of the reciprocal of the molar susceptibility versus
temperature. The data may be fit very precisely by the Curie-Weiss
Law

$$\chi = \frac{C}{T-\theta}$$

where $C = Ng^2\beta^2 S(S+1)/3k$ with g = 2.837 and θ = -0.732°. The best
fit of the Curie-Weiss Law to the data is shown as the solid line
in Figure 3. The average g value determined from the magnetic
susceptibility data agrees nicely with an average g value of 2.79
which may be calculated from the EPR data for 1,1'-dimethylferro-
cenium hexafluorophosphate using the expression

$$g_{ave}^2 = 1/3(g\|^2 + 2g\perp^2)$$

and the observed $g_\|$ = 4.002 and $g_\perp$ = 1.92.[14] For the purpose of
comparison, magnetic susceptibility data for 1,1'-dimethylferro-
cenium hexafluorophosphate are plotted as open squares in Figure 3.
The close agreement of the data for the two compounds leads to
the conclusion that the susceptibility in this temperature region
is dominated by the 1,1'-dimethylferrocenium cation, and more
remarkably, to the conclusion that these cations, although packed
closely in the crystal structure, do not exhibit appreciable intermolecular spin-spin interactions.

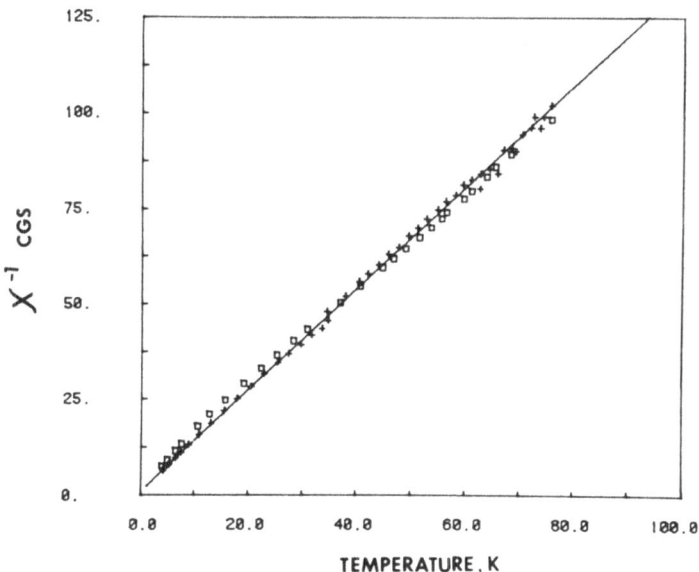

Figure 3. The inverse susceptibility of 1,1'-dimethylferrocenium
(TCNQ)$_2$ (+) and 1,1'-dimethylferrocenium (PF$_6$)$_2$(■) as a
function of temperature. The solid line is the best fit
to the Curie-Weiss law for the TCNQ charge transfer compound.

Acknowledgement

    This work was supported in part by the Office of Naval Research,
the National Science Foundation, and the Department of Energy.

References

1. L. R. Melby, R. J. Harder, W. R. Hertler, W. Mahler, R. E. Ben-
    son, and W. E. Mochel, J. Amer. Chem. Soc., 84, 3374 (1962).

2. J. S. Miller, A. H. Reiss, Jr., and G. A. Candela, Lecture
    Notes in Physics, in press.

3. G. R. Desiraju, D. Y. Curtin, and I. C. Paul, J. Amer. Chem.
    Soc., 99, 6148 (1977).

4. D. B. Losee and W. E. Hatfield, Phys. Rev. B, 10, 212 (1974).

5. P. W. R. Corfield, R. J. Doedens, and J. A. Ibers, Inorg. Chem.,

6, 197 (1967).

6. S. A. Goldfield and K. N. Raymond, Inorg. Chem., 10, 2604 (1971)

7. For a description of the programs used, see D. L. Lewis and
    D. J. Hodgson, Inorg. Chem., 13, 143 (1974).

8. P. Main, M. M. Woolfson, and G. Germain, "MULTAN: A Computer
    Program for the Automatic Solution of Crystal Structures",
    University of York, England.

9. "International Tables for X-ray Crystallography", Vol. IV,
    Kynoch Press, Birmingham, England.

10. F. Herbstein, in "Perspectives in Structural Chemistry", Vol.
    IV, J. D. Dunitz and J. A. Ibers, eds., pp. 166-395.

11. R. P. Shibaeva, L. O. Atovmyan, and V. I. Ponomarev, Zh. Str.
    Khim., 16, 860 (1975) and references therein.

12. E. M. Connell, A. J. Epstein, and M. J. Rice, Proc. Int. Conf.
    on Quasi-One-Dimensional Conductors, Dubrovnik, Yugoslavia,
    Sept. 4-8, 1978. (Springer-Verlay, to be published).

13. T. W. Cape and R. P. Van Dyne, private communication.

14. D. M. Duggan and D. N. Hendrickson, Inorg. Chem., 14, 955 (1975)

PERYLENE-DITHIOLATE COMPLEXES

THE PLATINUM SALT

L. Alcácer *, H. Novais * and F. Pedroso **

* Lab. Física Eng. Nucleares, Sacavém, Portugal

** C. Química Estrutural, Complexo I, I.S.T., Lisboa

In the sequence of previous work on perylene – metal dithiolate complexes [1,2] , we wish to report some preliminary data on the platinum salt.

The material is a molecular complex of perylene, I, and the planar bis-maleonitriledithiolate – platinum chelate, II, with stoichiometry 2:1.

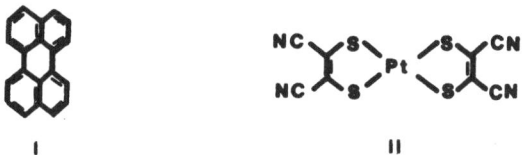

I   II

Similarly to the previous compounds of this series, the complex was prepared by oxidation of perylene with iodine in the presence of $(n-C_4 H_9)_4 N [Pt S_4 C_4 (CN)_4]$, in dichloromethane

$$2C_{20} H_{12} + 1/2 I_2 + (n-C_4 H_9)_4 N [Pt S_4 C_4 (CN)_4] \rightarrow$$
$$\rightarrow (C_{20} H_{12})_2 [Pt S_4 C_4 (CN)_4] + [(n-C_4 H_9)_4 N]I$$

Although we have obtained crystals by diffusion methods similar to those used for TTF TCNQ, the best crystals were obtained by slow cooling of dichloromethane solutions to which a few drops of nitrobenzene were added. This was done, since the crystals are highly insoluble in dichloromethane. By adding a very small amount of nitrobenzene, some solubility is allowed and exchange of molecules between crystals and solution ,

and consequently a better crystalization, are induced.
      Elemental analysis gave the stoichiometry

$$(C_{20} H_{12})_2 [Pt S_4 C_4 (CN)_4] \quad . \text{ Calculated: C, 58.8 ;}$$

H, 2.47 ; N, 5.72 . Found: C, 58.34; H, 2.60 ; N, 5.91
The determination of the crystal structure by X-ray diffraction
methods is under way. It can, however, be expected, from the
magnetic and electrical behaviour of this class of compounds, that
the perylene molecules are stacked in a linear chain sided by the
dithiolate counter ions.

      This platinum salt of the perylene-metal dithiolate
complexes seems to be the most conductive of the series, the other
members being the nickel, the copper, the palladium and the gold
compounds. Values as high as 280 (ohm.cm)$^{-1}$, at room temperature,
have been measured by the four probe method, along the needle axis
of single crystals, using either dc or ac current. Values of the
order of 6 (ohm.cm)-1 have been obtained in compressed pellets.
Attempts to obtain reproducible data on the electrical conductivity
as a function of temperature have failed, so far. Silver paint
contacts seem to generate unusual high noise both using dc currents
of 1 to 45 µA and ac currents of 1 to 10 µA at frequences between
70 and 1 000 Hz with a lock-in amplifier. All results   at   room
temperature point, however, to conductivities of the order of 125
to 280 (ohm.cm)-1 in a great number of crystals, and changing very
little with temperature. In compressed pellets most results agree
with a very weakly activated conductivity.On single crystals,noise
problems are worse and we obtain either a weakly activated
conductivity or even an increasing conductivity as we lower the
temperature. We cannot decide yet on the definite electrical
behaviour of this material. A possible explanation for this unusual
high noise could be the existence of iodine as an impurity in the
crystals, which reacting with the silver paint could give bad
electrical contacts. Contact resistances are indeed unusually high,
of the order of 1000 ohm. The presence of iodine in the crystals ,
in very small amounts, can be expected since iodine enters in the
preparation of this compounds. Other contacts will be attempted as
soon as possible and the detection of traces of iodine is programmed
to be made soon, by activation analysis. The electrochemical method
for the preparation of this material[1] was not attempted yet due to
the usually low yields obtained for the other compounds of the series,
but it will probably give crystals of better quality.

      Magnetic susceptibility measurements were made from 3.5
to 295 K by the Faraday method. In fig 1 a) we plot the values of
the reciprocal of the paramagnetic portion of the susceptibility as
a function of temperature. The data apparently follows a Curie-Weiss

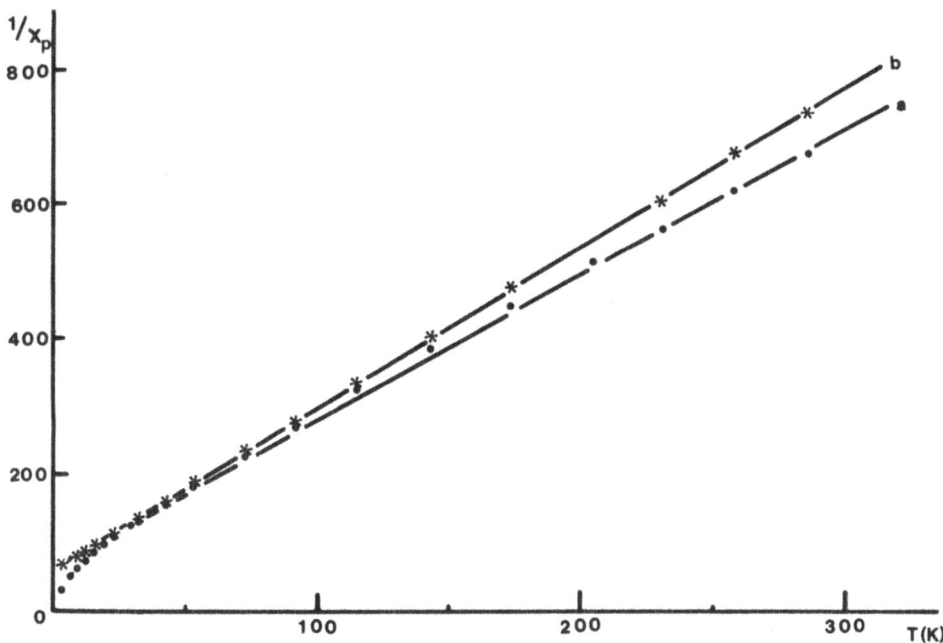

Fig.1. Temperature dependence of the magnetic susceptibility

behaviour with a Weiss constant of the order of 30 K , and with an
efective magnetic moment of 1.87 Bohr magnetons at 295 K .
      In a previous paper[2], the magnetic susceptibility of a
series of the perylene-metal dithiolate compounds were reported.
In the case of the palladium complex, it was shown that the
paramagnetic susceptibility was due to two contributions. One
derived from a perylene stack, almost temperature independent, and
low, and another contribution with Curie-Weiss behaviour, derived
from the palladium-dithiolate anions. In the case of the platinum
salt the susceptibility data can also be explained in terms of two
similar contributions. If, at each temperature, we subtract the
almost constant perylene contribution, we end up with a better
fitting for the Curie-Weiss law of the platinum dithiolate
contribution as can be seen from fig 1 b). Only at temperatures of
the order of 30 K , a Curie tail seems to take over. That could be
atributed to paramagnetic impurities or imperfections.

      An analysis of the infrared spectra[3], particularly the
position of the B_{3u} C-N stretching mode, the B_{3u} C-S stretching and
the B_{3u} and B_{2u} Pt-S stretching modes of the platinum-dithiolate
monoanion, show that the charge on the ion is close to -1, however
slightly shifted towards the values expected for the neutral
species.

      A more detailed study of the electrical and magnetic

properties of this material is under way.

Aknowledgement: We are indebted to Dr. S. Flandrois, for his help on the magnetic susceptibility measurements.

References

1. L. Alcácer and A.H. Maki, J. Phys. Chem., _78_, 215 (1974)
2. L. Alcácer and A.H. Maki, J. Phys. Chem., _80_, 1912 (1976)
3. C.W. Schlapfer and K. Nakamoto, Inorg. Chem., _14_, 1338 (1975)

$Hg_{3-\delta}AsF_6$: INCOMMENSURATE LINEAR CHAINS WITH QUASI-ONE-DIMENSIONAL LATTICE DYNAMICS AND ELECTRONIC PROPERTIES*

A. J. Heeger[†] and A. G. MacDiarmid[‡]

Laboratory for Research on the Structure of Matter

University of Pennsylvania, Philadelphia, PA 19104

ABSTRACT

A brief review of the electronic properties of the incommensurate linear chain metal, $Hg_{3-\delta}AsF_6$, is presented. We discuss the structure, lattice dynamics, temperature dependent stoichiometry, optical properties, nuclear magnetic resonance results, electrical transport, and anisotropic superconductivity.

I. INTRODUCTION

A new class of anisotropic conductors containing metallic chains of mercury atoms has recently been reported.[1] These compounds, with nominal formula Hg_3AsF_6 and Hg_3SbF_6, consist of metallically bonded polymercury cations in a tetragonal lattice consisting of an array of $(AsF_6)^-$ or $(SbF_6)^-$ anions.[2] The mercury chains are situated in non-intersecting channels along the $\vec{a}$

---

* Work at the University of Pennsylvania primarily supported by the National Science Foundation MRL Program under Grant DMR 76-00678.
† Department of Physics
‡ Department of Chemistry

and $\vec{b}$ axes of the tetragonal lattice; there are no chains along $\vec{c}$. The diffuse scattering sheets observed in x-ray[2] and neutron[3] diffraction studies indicate that there is no phase coherence between the polymercury chains and that the Hg-Hg distance within a chain is 2.64 Å. Since the lattice constant is a = b = 7.54 Å, the empirical structural formula (assuming all anion sites are occupied) is $Hg_{2.86}AsF_6$ implying that the mercury atoms in a chain are incommensurate with the unit cell of the tetragonal lattice.

These polymercury cation compounds $Hg_{2.86}AsF_6$ (hereafter $Hg_{3-\delta}AsF_6$) and $Hg_{2.91}SbF_6$ have anisotropic electrical conductivities[4,5] and anisotropic optical properties[6,7] consistent with relatively weak interchain electronic coupling. The $Hg_{3-\delta}AsF_6$ compound remains metallic[4,5,8] to low temperatures with no indication of residual resistance and a linear term in the specific heat[9] which is unusually large compared to that expected from the electronic density of states.[8] Below 4.1 K, $Hg_{3-\delta}AsF_6$ becomes an anisotropic superconductor in low magnetic fields[5] and exhibits an anisotropic Meissner effect.[10]

This brief review emphasizes the recent studies carried out in a collaborative effort at the University of Pennsylvania and with colleagues at Brookhaven National Laboratory and Argonne National Laboratory. The structural studies were carried out by A. J. Schultz and J. M. Williams[3] at Argonne and by J. M. Hastings, J. P. Pouget and G. Shirane[11] at Brookhaven with recent important contributions from I. V. Heilmann, V. Emery and J. Axe. N. D. Miro and P. Nigrey were primarily responsible for the synthesis and crystal growth[12]. The optical reflectance studies[7] were carried out by D. L. Peebles, C. K. Chiang and M. J. Cohen. The NMR studies[8] are the work of E. Ehrenfreund, J. Kaufer, and P. R. Newman. The studies of the electrical transport[5] are the work of C. K. Chiang, R. Spal, A. Denenstein and D. P. Chakraborty. The measurements of the Meissner effect[10] were carried out by R. Spal with important contributions on the ac susceptibility by C. K. Chiang and A. Denenstein. The reader is referred to the original papers for additional discussion of the results and more detailed experimental data.

II. STRUCTURE[2,3,11]

The crystallographic structure of the $Hg_{3-\delta}AsF_6$ compound has been determined at room temperature by Brown et al. [2] and confirmed by Schultz et al. [3] using neutron diffraction. This compound has a body-centered tetragonal space group I4$_1$/amd(D$_{4h}^{19}$) with four formula units of Hg$_{3-\delta}$AsF$_6$ per unit cell. It is convenient to separate the lattice into two parts: the host lattice (AsF$_6$ lattice) and the mercury chains. The host lattice is formed by a body-centered tetragonal array of well separated AsF$_6$ octahedra. The mercury chains lie in non-intersecting channels in the [100] and [010] directions, equivalent by tetragonal symmetry. A sketch of the crystal structure is shown in Figure 1.

In addition to the Bragg reflections, Brown et al. [2] found two sets of equidistant sheets in reciprocal space with very strong diffuse intensity. They associated each direction of diffuse sheet with the corresponding direction of the linear channel in real space. The diffuse intensity arises from the mercury atoms, and its distribution in a sheet implies that there is no phase coherence between the neighboring chains. These diffuse sheets can therefore be viewed as Bragg reflections from a 1 D lattice.

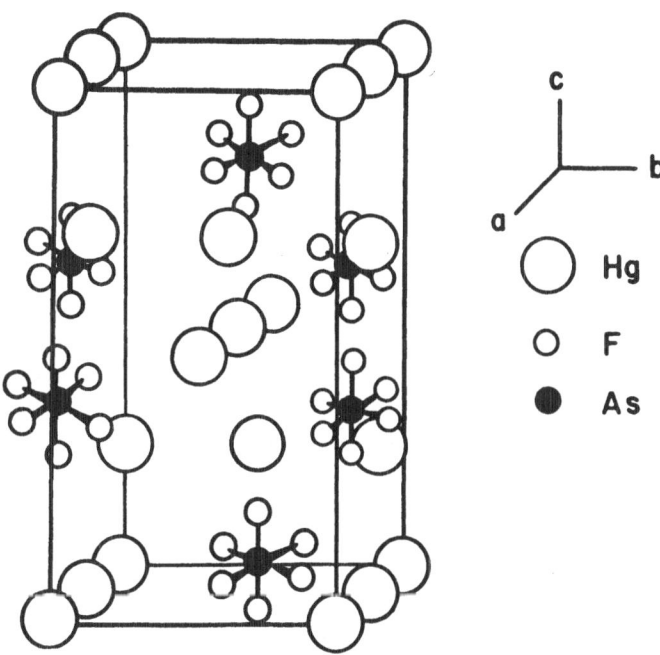

Figure 1: View of the crystal structure of Hg$_{2.86}$AsF$_6$.

Important information concerning the spread of the order of the Hg atoms within individual chains can be obtained by measuring the width of the sheet. [11] Elastic neutron scattering scans along the [k00] direction, for different temperatures ranging from room temperature to 30 K, always yield a resolution limited width of 0.018 a^* ($a^* = (2\pi/a)$ where a = 7.55 Å) for the first sheet. Taking this value as a lower bound for the 1 D order along the chain direction, a coherence length in excess of 400 Å or 150 Hg-Hg spacings is obtained. [11]

In summary, the room temperature structure has the following features:

a) The Hg atoms are well ordered along the chain direction.

b) The intrachain mercury distance a_{Hg} is incommensurate with the host lattice parameter a_L along the chain direction.

c) There is no phase order among neighboring chains.

d) There is periodicity in position of the Hg chains, imposed by the regular network of open channels of the host AsF_6 sublattice.

The existence of the randomly phased incommensurate chains at room temperature implies a new symmetry; translational invariance of the mercury chains with respect to the three-dimensional lattice. As a result, in additional to the usual three dimensional (3 D) lattice phonons, we expect 1 D phonons originating from the diffuse sheets in reciprocal space and associated with the mercury chains.

The phonon dispersion curve at room temperature along the $(\zeta, 0, 0)$ direction has been measured using inelastic neutron scattering techniques. [11] The results are shown in Figure 2. The dispersion curve beginning at the Bragg peak ($\zeta = 2$) is the longitudinal acoustic (L. A.) branch associated with the 3 D lattice. Centered about $\zeta = 3-\delta$ one finds another branch with a very steep slope which is the L. A. mode of the Hg chains. The momentum transfer about the point, $\zeta = 3-\delta$, is designated by X to indicate that the measurements were made over a limited range of values of c^* as shown in the reciprocal lattice diagram included in Figure 2. The observed dispersion curve was independent of the value of c^* as one would expect from a sheet in reciprocal space.

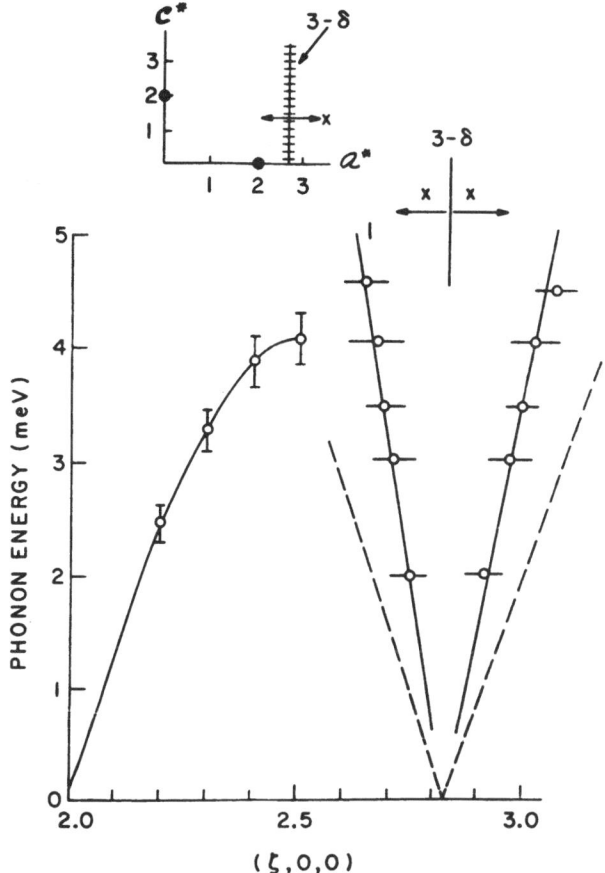

Figure 2: Room temperature phonon dispersion curve. The inset at the top is a diagram of the a*c* plane showing the location of the (3-δ) sheet. The dispersion curve associated with the Hg chains centered about ζ - 2.82 was measured for various values of x, i.e. (ζ, 0, x).

The dashed lines, with the initial slope of the dispersion curve associated with the 3 D lattice have been added to provide a comparison of the lattice and Hg chain stiffness. The sound velocities are obtained from the slopes of the curves: 2.2 x 10^5 cm/sec associated with the 3 D phonons and 4.4 x 10^5 cm/sec associated with the 1 D phonons.

Synthetic metals with anisotropic and/or quasi-one-dimensional (1 D) electronic properties have been intensely studied in recent

years. Examples include the well-known organic metals such as
TTF-TCNQ and related derivative salts, [13] KCP and related
platinum chain compounds, [13] as well as $(SN)_x$ [13] and the $(SNBr_y)_x$
derivatives. [14] However, the polymercury chain compound,
$Hg_{3-\delta}AsF_6$, is a unique example of <u>elastic</u> one-dimensionality with
the 1 D phonon spectrum[11] resulting directly from the incommensurate chain structure.

Analysis of the crystal growth leads to the conclusion that
mass transport of the mercury atoms occurs along the chain
channels at room temperature. Thus the atoms are mobile, and
the presence of interchain interactions can be expected to lead to
phase ordering of the mercury chains. The detailed nature of the
low temperature ordered state will depend on the relative strengths
of the parallel chain and perpendicular chain interactions. In
general, the two kinds of interactions will lead to different order,
and thus compete in the region above the phase transition. Experimentally the perpendicular chain coupling is found to dominate
at low temperatures with parallel chain short range order playing
an important role at intermediate temperatures.

Extensive elastic neutron scattering measurements were
carried out at Brookhaven to characterize the temperature dependence of the Hg chain-chain correlations. The distribution of
intensity in the (hk0) zone has been measured at a variety of temperatures. Figure 3 presents elastic scans along the diffuse line
$(3-\delta, \eta, 0)$. The room temperature scan shows that within experimental errors the distribution of intensity is nearly constant along
the k direction. This uniform distribution of the intensity proves
directly that there is no phase relation between the mercury
positions among neighboring chains. As the temperature is lowered the effect of the chain-chain interactions leads to a modulation
of the uniform intensity implying a progressive phase ordering
between the "rigid" mercury chains. Short range order gradually
builds up, with transverse coherence lengths of a few lattice constants as a result of parallel chain-chain interactions. At 120 K,
the competing orthogonal chain-chain interaction leads to three
dimensional order and a phase ordering phase transition.

This unique phase transition is of broad theoretical interest.
[15, 16] It is more generally characteristic of a coupled chain
problem with phase rigidity within a chain, but weak interchain
phase coupling. An analogous example might be an array of
weakly coupled 1 D superconducting chains with weak Josephson

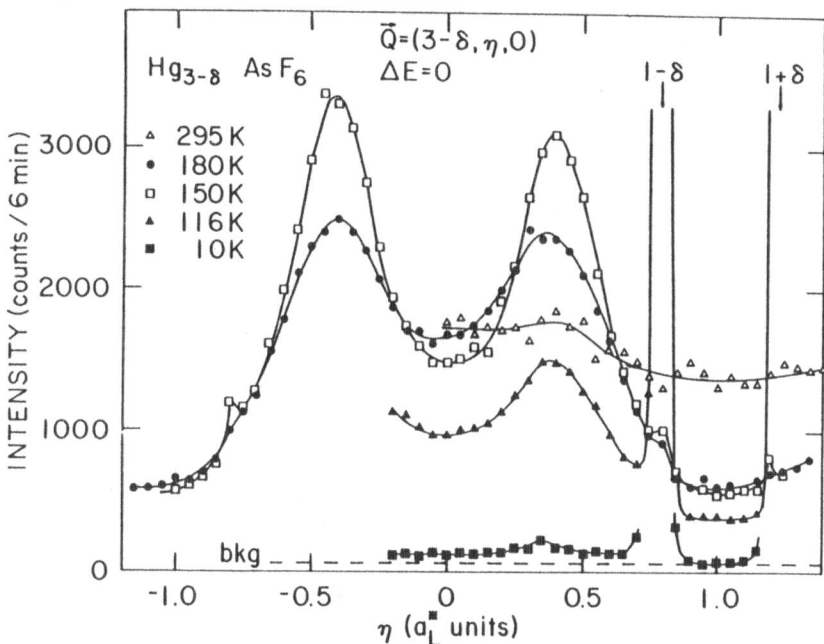

Figure 3: Elastic neutron scans along the diffuse line [3-δ, η, 0]. (see ref. 11)

interchain coupling. At temperatures well below the 1 D mean field temperature there is phase rigidity within a single chain, but random interchain phase. In this fluctuation regime the conductivity would be high along the chains and relatively small transverse to the chains. At lower temperature phase ordering would occur resulting in 3 D superconductivity.

The phase ordering phase transition has been studied in detail using both elastic and inelastic neutron scattering.[17] The elastic scattering results provide information on the growth of order. At temperatures below 120 K where interchain phase coherence (3 D long range order) is established, a gap develops in the dispersion relation corresponding to out-of-phase vibration of neighboring parallel chains with displacements along the chain direction (a phason mode).

The intensity of the maximum of the (3-δ, 0.4, 0) peaks are replotted in Figure 4. It increases until 125 K, then drops abruptly by a factor of two in the five following degrees. Below

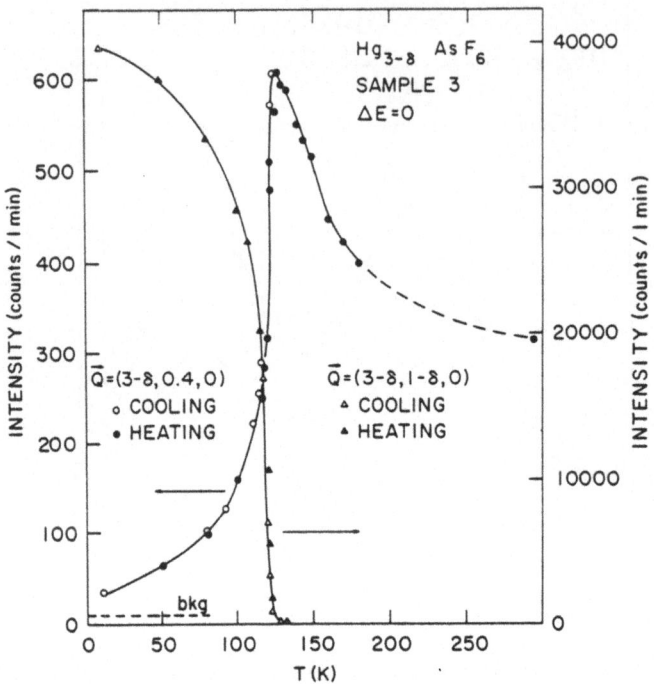

Figure 4: Temperature dependence of the $(3-\delta, 0.4, 0)$ and
$(3-\delta, 1-\delta, 0)$ peak intensity (solid points) and cooling (open points).
The phase transition at 120 K is evident. (see ref. 11).

120 K the decrease continues, but somewhat more slowly. At
10 K the peak is still barely visible, but away from ± 0.4, the
data in Figure 3 show that the diffuse sheet has nearly reached the
background level. The sharper peaks at the $\eta = 1 \pm \delta$ and $\eta =$
$-1 \pm \delta$ positions begin to appear at 150 K (see Figure 3) and
increase strongly in intensity on cooling. The temperature dependence of the intensity of the $(3-\delta, 1-\delta, 0)$ peak is also presented
in Figure 4. A dramatic increase of the intensity of this peak
occurs near 120 K, indicative of a true phase transition. The
intensities at $(3-\delta, 0.4, 0)$ and $(3-\delta, 1-\delta, 0)$ have been plotted in
Figure 4 for comparison to show the correlation in temperature
between the drop in intensity of the first peak and the increase of
the intensity of the second one. There is a transformation from
one type of order to another resulting from competing inter-chain
interactions[15].

As described above, with the onset of interchain phase order, we expect a finite energy gap, Δ, to appear in the dispersion relation[17] as shown in Figure 5a. Figure 5b shows peak positions of inelastic scans at 70 K, performed through the dispersion sheets centered at $(3-\delta)a_L^*$ (see Figure 5a). For energy transfer $\Delta E \gtrsim 0.3$ meV, the peaks of the constant-E scans are resolved and display an approximately linear dispersion with slope $v = 4.4 \pm 0.5 \times 10^5$ cm/sec, equal to the room temperature value. The lowest point in Figure 5b results from a constant-Q scan; a well defined excitation with energy ~ 0.13 meV is observed thus demonstrating the existence of the phason mode gap. The inset to Figure 5b shows the gap parameter Δ as a function of temperature through the 120 K phase transition (points at 40 K, 70 K, 100 K and 150 K). The gap decreases with increasing temperature below T_c and vanishes above T_c, thus displaying the qualitative behavior of an order parameter.

As emphasized by Heilmann et al.[15] the observed temperature dependence of the gap parameter Δ shows that its presence is brought about by the chain-chain interaction, and the observed excitation at the gap below T_c can be considered as a soft mode. However, the observed behavior differs in two ways from the soft-mode response usually related to certain structural phase transitions. Firstly, the soft-mode remains well defined down to energies below 0.1 meV and thus does not become overdamped, in contrast to what is observed in other systems. Secondly, there is no truly elastic "central peak" involved in the soft mode response. These observations suggest that the structure of Hg-ions below T_c is remarkably "perfect" with low-frequency phonon decay and defect mechanisms playing a minor role. The absence of the gap above T_c further indicates that there is virtually no "pinning" of the Hg-chains to host lattice imperfections.

III. TEMPERATURE DEPENDENT STOICHIOMETRY

Since the diffuse sheets can be viewed as Bragg reflections from a 1 D lattice, one can obtain the intrachain repeat Hg-Hg distance, a_{Hg}, from the distance between sheets. The relation between a_{Hg}, the host lattice constant a_L, and the stoichiometry parameter, δ, is given by

$$a_{Hg} = a_L/3-\delta$$

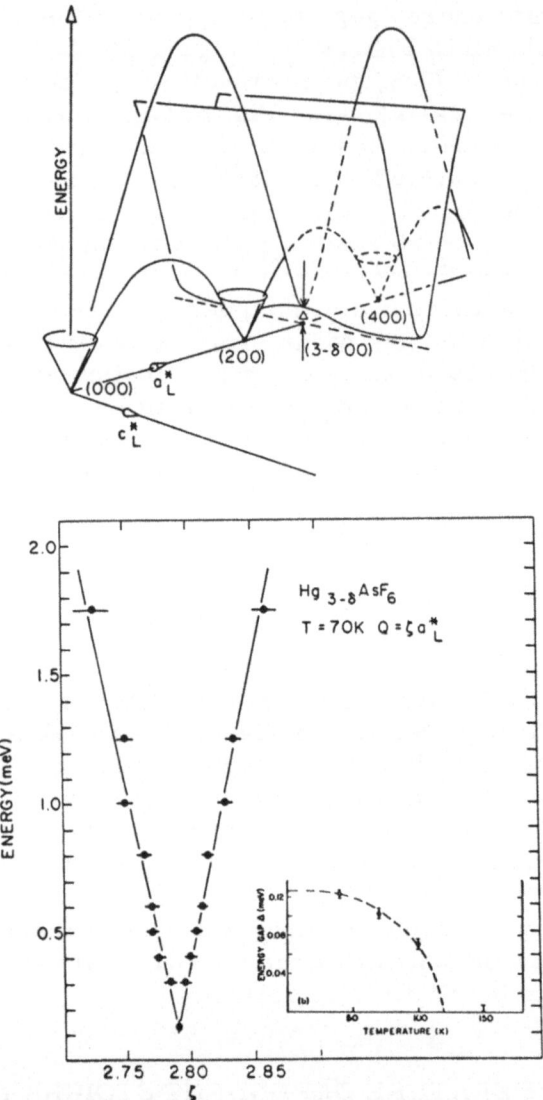

Figure 5: (a) Idealized dispersion relations in the $a_L^* c_L^*$ reciprocal plane. Along the a_L^* direction is shown dispersion curves for acoustic lattice phonons (period $2 \cdot a_L^*$) as well as for phonons in the Hg chains (period $(3-\delta) \cdot a_L^*$). Portions of the dispersion surfaces are likewise indicated. (b) Positions of inelastic neutron scattering peaks at $T = 70$ K showing phonon modes in the Hg chains. The lowest point results from a constant Q scan. The inset shows the gap in the dispersion (phason mode) as a function of temperature.

At room temperature, a_{Hg} = 2.670 ± 0.005 Å, incommensurate with the host lattice parameter as described above, giving the empirical formula Hg$_{3-\delta}$AsF$_6$. At room temperature δ = 0.18. However, the resulting formula, Hg$_{2.82}$AsF$_6$, deduced from the structural measurements is not in good agreement with that obtained from an elemental analysis[12] where the results indicate stoichiometric composition Hg$_3$AsF$_6$.

The resolution of the apparent conflict in the structural and analytical results appears to be in the existence of a defect structure with approximately one AsF$_6$ in twenty missing from the structure.[12] The resulting chemical formula expressed per average unit cell would be Hg$_{2.82}$(AsF$_6$)$_{.94}$. Precision density measurements[12] have confirmed this picture. The measured density is found to be ρ_{meas} = 7.05 ± 0.01 g/cm^3, in close agreement with the calculated value ρ_{calc} = 7.04 g/cm^3 based on the known structure and the above formula. An alternative suggestion of 6% excess Hg disordered in the lattice would lead to a density of 7.49 gm/cm^3. Thus, the mercury chain salt is apparently a stoichiometric compound in an incommensurate structure.

The temperature dependence of the host lattice constant, a_L, and the position of the first diffuse sheet with respect to the host reciprocal lattice constant, 3-δ, are given in Figure 6. The parameter a_L shows a relatively large contraction of 1% between room temperature and 10 K. A similar contraction of the c_L lattice parameter is observed; c_L decreases from 12.395 Å at room temperature to 12.248 Å at 10 K.

The temperature dependence of the Hg stoichiometry, 3-δ entering in the empirical formula Hg$_{3-\delta}$AsF$_6$ is deduced from the scattering measurements. The observed variation in fact reflects completely that of the host lattice parameters a_L. The position of the diffuse sheets are (within experimental error) constant in reciprocal space at all temperatures giving a temperature independent Hg-Hg repeat distance: a_{Hg} = 2.670(5) A. This change in (3-δ), reversible in temperature and reproducible over several cycles, poses an interesting question concerning the behavior of the Hg atoms in the chains on cooling down. Several possibilities must be considered: 1% of the Hg atoms flow out of the channels onto the surface of the sample, 1% of the Hg atoms go into interstitial positions or into the AsF$_6$ vacancy sites described above.

Direct visual observation of single crystals cooled to 77 K in

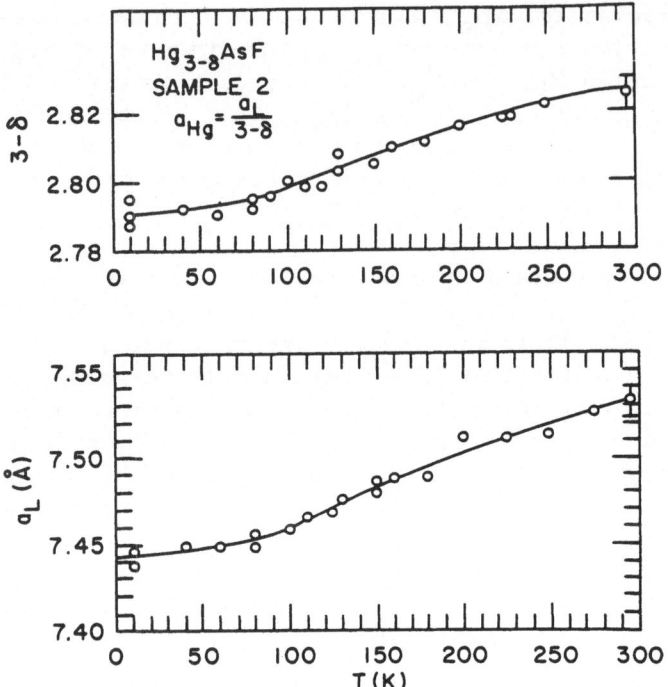

Figure 6: Temperature dependence of the position, 3-δ, of the first sheet and of the lattice parameter, a_L.

a glass bulb showed a few tiny mercury droplets appearing on the surface. However, quantitative measurements[12] with a calibrated microscope indicated that no more than about 10^{-4} volume fraction of Hg comes out in this form.

Datars et al.[18] independently inferred a temperature dependent stoichiometry. Based upon their differential thermal analysis data, they argued that free mercury is extruded onto the surface when the compound is cooled and is reincorporated into the compound during warming. They observe a peak in DTA on warming through the melting point of mercury, but only if the sample is first cooled below 200 K. They estimate between 0.5 and 5 at. % Hg is extruded. Moreover, they quote a change in the coefficient of thermal expansion of a factor of seven (7.6 x 10^{-5} K^{-1} to 1.0 x 10^{-5} K) at 200 K which they attribute to the proposed extrusion.

Examination of Figure 6 shows that the change in stoichiometry is smooth and continuous starting from room temperature.
There is no indication of a major change in thermal expansion
coefficient at 200 K (the lattice constant variation is smooth with
temperature; Figure 6). The maximum change in stoichiometry
is about 1% on cooling to 4.2 K with about 0.5% on cooling to
~ 180 K.

Differential scanning calorimetry (DSC) measurements[19] at
Pennsylvania show features similar to the DTA results[18], but
with important differences. A peak is observed on warming
through the Hg melting temperature. Calibration with pure Hg
indicates 0.6% after first cooling the sample to 180 K; in excellent
agreement with Figure 6. The DSC data are reproducible with
little variation from sample to sample. Although the main peak
is not observed on cooling, a series of small peaks are observed
below the Hg freezing point suggesting freezing of small fractions
of the total 0.6% at varying temperatures.

In an attempt to search for the proposed[18] Hg extruded on the
surface of the sample we have initiated an XPS study[20], in collaboration with W. R. Salaneck, of Hg$_{3-\delta}$AsF$_6$ as a function of temperature. Initial results show no sign of free mercury appearing
on the surface upon cooling either on the $\hat{a}$ - $\hat{c}$ face or the $\hat{a}$ - $\hat{b}$
face. The Hg to As to F ratios are independent of temperature.
Moreover, the Hg 4f core excitation in the compound is shifted by
6 eV from that of free Hg so that trace amounts of free Hg on the
surface would be readily detected even on the scale of a small
fraction of a monolayer. No free mercury signal is observed
either at room temperature or as a function of temperature on
cooling to ~ 100 K. As a check on the measurements, an independent sample was purposely exposed to air to initiate decomposition with the formation of free mercury on the surface. The
free Hg signal then shows up clearly in the XPS data. These
initial results indicate that the proposed[18] idea of mercury
extrusion onto the surface of the crystals is not correct.

The absence of Hg on the surface combined with the continuous variation in $\delta(T)$ shown in Figure 6 suggest that on cooling
the Hg atoms go into interstitial positions or into the AsF$_6$ vacancy
sites. This is consistent with the absence of a major peak in DTA
or DSC on cooling. Evidently, as the sample contracts, Hg atoms
leave the chains in such a way as to keep the Hg-Hg distance (a_{Hg})
constant as a function of temperature. The chain structure

nevertheless remains remarkably perfect with no pinning, as
described above. The individual atoms apparently diffuse and,
upon continued cooling, coalesce into large enough clusters to
exhibit collective freezing. The results imply that only upon
cooling below 200 K are the Hg clusters large enough to exhibit
collective melting and freezing. This would imply extremely
small free mercury clusters with dimensions on the scale of 10 Å-
100 Å distributed through the bulk of the crystals. Additional DSC
and XPS studies are in progress.

IV. OPTICAL PROPERTIES

Polarized reflectance studies from single crystal faces of
anisotropic $Hg_{3-\delta}AsF_8$ were carried out. The results indicate
metallic behavior for electronic excitations along the chains (in
the $\vec{a}$ - $\vec{b}$ plane) with free electron Drude theory giving a satisfactory initial description of the data and imply quasi-1 D electronic
behavior.

The results are summarized in Figures 7a and 7b. Figure
7a shows a fit (dashed curve) to the $\vec{a}$ - $\vec{b}$ plane reflectance (solid
curve) using the Drude dielectric function $[\epsilon(\omega) = \epsilon_\infty - \omega_p^2/(\omega^2 + i\,\omega/\tau)]$
in the Fresnel equation for normal incidence reflectance. To ob-
tain the dashed curve, a least square fit was carried out arbitrarily restricting the input to the range 1.3 eV to 4.2 eV. The free
parameters were ω_p, ϵ_∞ and τ since the absolute reflectance was
determined using a wide angle photovoltaic detector. Trial fits
with a 1% change in reflectance gave only small changes in param-
eters. The results are $\hbar\omega_p$ = 4.8 eV, $\hbar/\tau$ = 0.27 eV and ϵ_∞ = 2.7.

The values for ω_p and τ can be used to obtain a measure of
the optical conductivity through the standard relation $\sigma_{opt} =$
$(4\pi)^{-1}\,\omega_p^2\tau$. Using the values for ω_p and τ leads to $\sigma_{opt} \doteq 1.2 \times 10^4$
$(\Omega\text{-cm})^{-1}$ to be compared with the room temperature dc and con-
tactless ac (100 kHz) values of 10^4 $(\Omega\text{-cm})^{-1}$. The excellent agreement provides some justification of the simple Drude approach.

The anisotropic reflectance data are summarized in Figure
7b. Analysis of the results requires a treatment of normal incidence reflectance from a plane which is not coincident with the
principal optical axes of the solid. For a uniaxial (tetragonal)
crystal, the dielectric tensor is defined

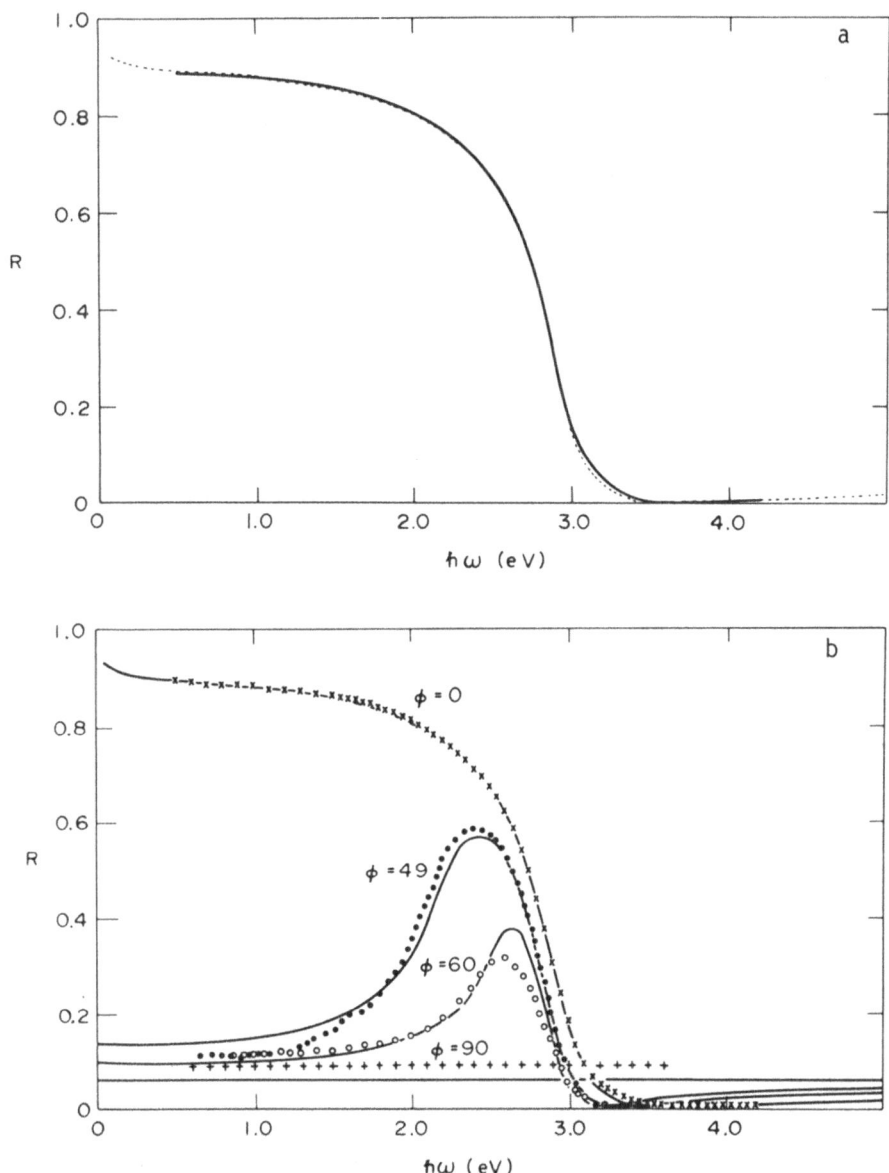

Figure 7: (a) Drude fit (dashed line) to the $\vec{a}$ - $\vec{b}$ face reflectance (solid line). (b) Normal incidence reflectance of the anisotropic faces.

$$\underset{\sim}{\epsilon} = \begin{pmatrix} \epsilon_{ab} & 0 & 0 \\ 0 & \epsilon_{ab} & 0 \\ 0 & 0 & \epsilon_{c} \end{pmatrix}$$

where $\epsilon_{ab} = \epsilon_c - \omega_p^2/(\omega^2 + i\omega/\tau)$ corresponds to the dielectric function along the chains and $\epsilon_c = \epsilon_\infty$ corresponds to the assumed constant perpendicular value. The anisotropic reflectance can be calculated directly from the dielectric tensor in terms of ϵ_{ab} and ϵ_c. The theoretical results for $R_{(\omega, \omega)}$ are shown in Figure 7b using the parameters listed above for ω_p, ϵ_∞, and τ. Figure 7b presents a direct comparison of the theoretical curves to the data with <u>no</u> <u>additional</u> <u>parameters</u>. The overall good qualitative and quantitative agreement provides additional confirmation of the validity of the simple one-dimensional Drude treatment of this anisotropic metal.

V.   NUCLEAR MAGNETIC RESONANCE[8]

Nuclear magnetic resonance (NMR) is a measure of the local susceptibilities at the sites of the nuclei and may be used as a probe of the respective amplitudes of the conduction-electron wave function at the various sites. The Knight shift and the spin-lattice relaxation are related to the Fermi surface parameters of the metal and can yield information about the static and dynamic properties of the conduction electrons.

Within experimental accuracy, the [199]Hg Knight shift is independent of temperature and magnetic field; K = +1.9%. The [199]Hg Knight shift is anisotropic with $K_{ax} = \frac{1}{3}(K_{\parallel} - K_{\perp}) = -0.07\%$. This is evidence that the conduction electron wavefunction is not spherically symmetric around the nucleus but rather has a size-able axial component. The negative sign of K_{ax} indicates that the conduction electron density is smaller in the chain direction and larger perpendicular to it consistent with substantial p_π character to the conduction electron wavefunction. Because of the incommensurate structure containing 6% anion vacancies, there is a distribution of Hg sites along the chain, which results in a distribution of Knight shifts of width $\sim 0.1\%$ about the average, $K_o = 1.9\%$.

The nuclear spin-lattice relaxation rate, $1/T_1$, is proportional to the temperature both for the ^{199}Hg line and for the ^{19}F line. However $[T_1(^{19}F)T] = 340$ sec K whereas $[T_1(^{199}Hg)T] = 7.5 \times 10^{-3}$ sec K; a ratio of approximately 5×10^4.

The overall features are characteristic of metals. Furthermore since K and (T_1 T) measure the local fields at the nucleus, comparison of the ^{199}Hg and ^{19}F relaxation rates would give direct information about the spatial distribution of the conduction-electron wave functions. Since $T_1^{-1} \propto \gamma \langle |\Psi_{(0)}|^2 \rangle$ where γ is the nuclear gyromagnetic ratio and $\langle |\Psi_{(0)}|^2 \rangle$ is the mean square (at the nucleus) of the conduction-electron wave function at the Fermi energy. Comparing the two rates at 4.2 K, we obtain $\langle \Psi_F^2(0) \rangle$ / $\langle \Psi_{Hg}^2(0) \rangle \approx 10^{-3}$, indicating that the conduction electrons reside primarily on the mercury chains. Thus one may expect the metallic properties to be dominated by electronic band structure originating from the mercury electrons.

The Knight shift in metals may be written as

$$K \equiv \Delta H/H_{res} = \mu_B^{-1} H_{hfs} \chi,$$

where χ is the susceptibility per atom and H_{hfs} is the hyperfine field which for s electrons is given by $H_{hfs} = (8\pi/3)\mu_B \langle \Psi^2(0) \rangle$ and $\Psi^2(0)$ is normalized in an atomic volume. The temperature independence of $K(^{199}Hg)$ thus shows that the susceptibility is temperature independent in the range 1.5 to 300 K and in external magnetic fields ranging from 5.6 to 13 kOe in agreement with direct measurements of the susceptibility using a Faraday balance.[9] The hyperfine field for the 6s electrons for the free mercury atom is estimated as $H_{hfs}^{at.} = 25.8 \times 10^6$ Oe. In the solid, the 6s wave functions have a greater spatial extent than in the stom and thus a somewhat smaller hyperfine field (say, 20×10^6 Oe) is expected. Assuming that all of $K(^{199}Hg)$ arises from the mercury 6s electrons, we obtain for the magnetic susceptibility of the conduction electrons: $\chi = 1.5 \times 10^{-5}$ cm^3 per mole of Hg$_{2.86}$AsF$_6$ units. Assuming no enhancement of the Pauli susceptibility $\chi = 2\mu_B^2 N(0)$, we obtain for the density of states at the Fermi energy, $N(0) \cong 8 \times 10^{-2}$ per eV per Hg atom for single spin direction.

The NMR results thus indicate that the conduction electrons are mainly confined to the mercury chains and their wave functions

do not spread much outside these chains onto the AsF_6^- anions.
A two-dimensional network of interpenetrating weakly interacting
linear chains of conduction electrons is implied. A simplified
band structure for this network has been modeled[8] with the conclusion that even for relatively strong interchain parameters the
Fermi surface remains essentially "one-dimensional". This 1-d
band structure can account for the short T_2 observed for ^{199}Hg.
Calculations[8] were made of the second and fourth moments of the
^{199}Hg line due to the dipolar and RKY interactions and it was
found that because of the quasi-one-dimensionality the indirect
RKY interaction falls off like x^{-1} and therefore dominates the
^{199}Hg linewidth. The calculations indicate that the one-dimensional
indirect interaction can account for the relatively short $T_2 \simeq 50$
μsec measured by the spin-echo method.

VI. ELECTRICAL PROPERTIES: MAGNETIC FIELD INDUCED RESIDUAL RESISTIVITY

Transport studies of this linear chain compound obtained by
four-probe dc techniques showed that the $\hat{a}$-$\hat{b}$ plane resistivity
varied approximately as $\rho_{ab} \sim T^n$ where $n \simeq 3/2$ with $\rho_{ab}(300\ K) \simeq$
10^{-4} Ω-cm; ρ_{ab} showed no sign of residual resistivity even at
temperatures down to 1.2 K.[5] The anisotropy, $\rho_c/\rho_{ab} \sim 10^2$,[4,5]
and is essentially independent of temperature. Recent results of
resistivity and magneto-resistivity were obtained using contactless ac mutual inductance techniques. The data imply a magnetic
field dependent residual resistivity which appears to go to zero as
the field is reduced to zero.

Figure 8 shows the temperature dependence in zero field.
The straight line (slope = 3/2) on the log-log plot is representative
of the dc data above 20 K. The contactless ac data agree in magnitude and temperature dependence with the earlier dc results.
There is a change in slope below 30 K where a stronger T-
dependence is observed. The absolute value of the conductivity
at 4.2 K is approximately 40 (μΩ-cm)$^{-1}$.

The magnetic field dependence of the resistance at 4.2 K is
plotted as $\Delta\rho/\rho$ vs H in Figure 9. The inset shows the higher field
data at the same temperature. An important feature of the data
is the non-quadratic form in low fields; this is a universal aspect
of the data in the temperature range investigated, 2 K < T < 25 K.
Note that at the lowest temperatures $\Delta\rho$ does not saturate; there

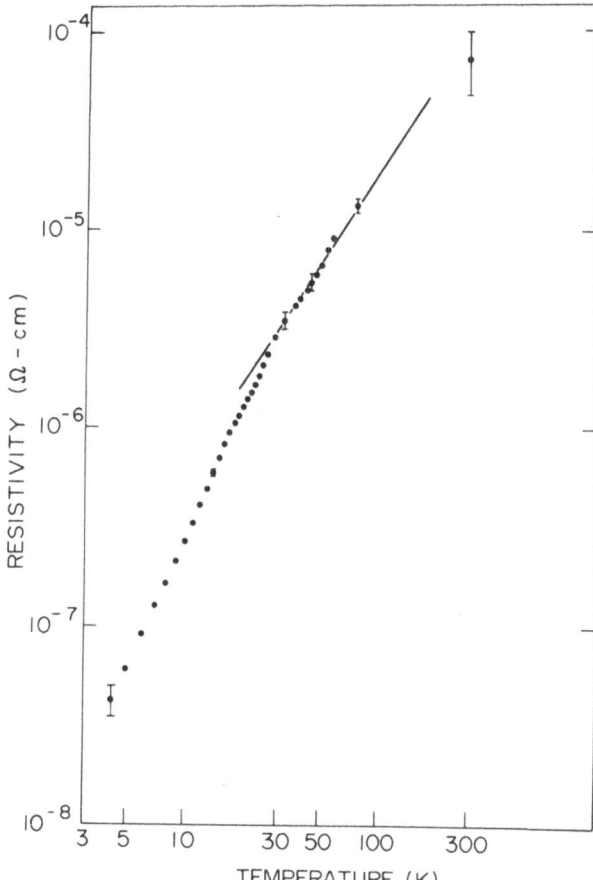

Figure 8: Temperature dependence of ρ_{ab} in zero magnetic field. The straight line (slope = 3/2) is representative of the dc data above 30 K. Below 25 K, the contactless ac data were obtained with a continuous temperature sweep; points are plotted at 1 K intervals.

is a 'knee' separating the initial concave downward region from a higher field regime where $\Delta\rho$ is roughly linear in H. This latter feature is lost on going to higher temperatures, e.g., at 7 K the 'knee' is above 40 kG.

Data obtained by allowing the temperature to drift up slowly in constant fields, 0 and 20 kG, are shown in the inset to Figure 10. The curves look parallel suggesting the empirical relation:

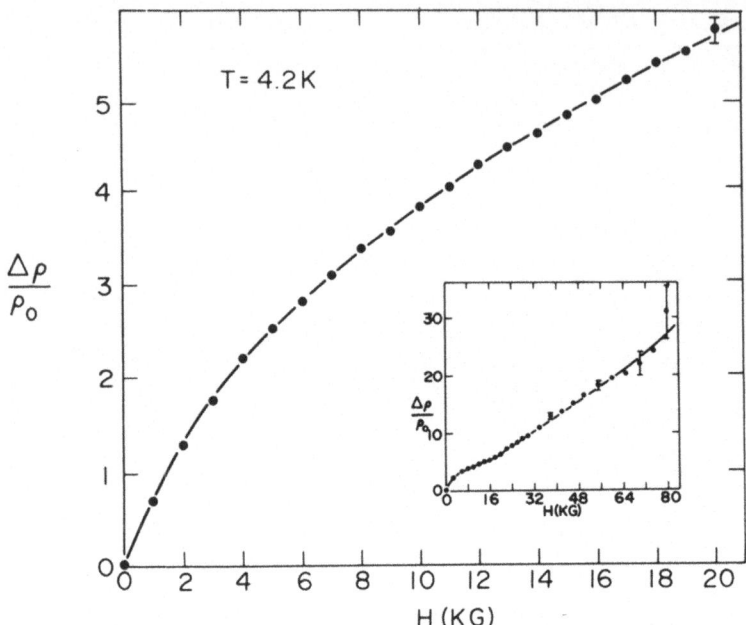

Figure 9: Magnetic field dependence of the resistivity at 4.2 K plotted as $\Delta\rho(H)/\rho_{ab}$ vs H.   The inset shows higher field data at the same temperature.

$$\rho(H, T) \; = \; \rho_0(H) + \rho_1(T) \tag{1}$$

where $\rho_0(H)$ $[\rho_1(T)]$ is a function of $H[T]$ only, going to zero as $H[T]$ goes to zero. If eq. 1 is valid, one should be able to collapse the data for different magnetic fields onto a single universal plot by simply subtracting $\Delta\rho(4.2, H) \equiv \rho(4.2, 0)$ from $\rho(T, H)$. Figure 10 shows such a plot of data taken at several fields; the agreement with the above expression is seen to be excellent. Note that eq. 1 and Figure 10 are inconsistent with conventional magnetoresistance arising from the Lorentz force, where $\Delta\rho(H)$ would be a strong function of $\omega_c \tau$ where ω_c is the cyclotron frequency and τ is the scattering time. Over the temperature interval of Figure 10, $\rho_1(T)$ changes by more than a factor of thirty whereas $\rho_0(H)$ is constant.

These unusual transport properties must be examined in the context of the established 1-d aspects of this material as described in previous sections. The quasi-(1d) nature of the

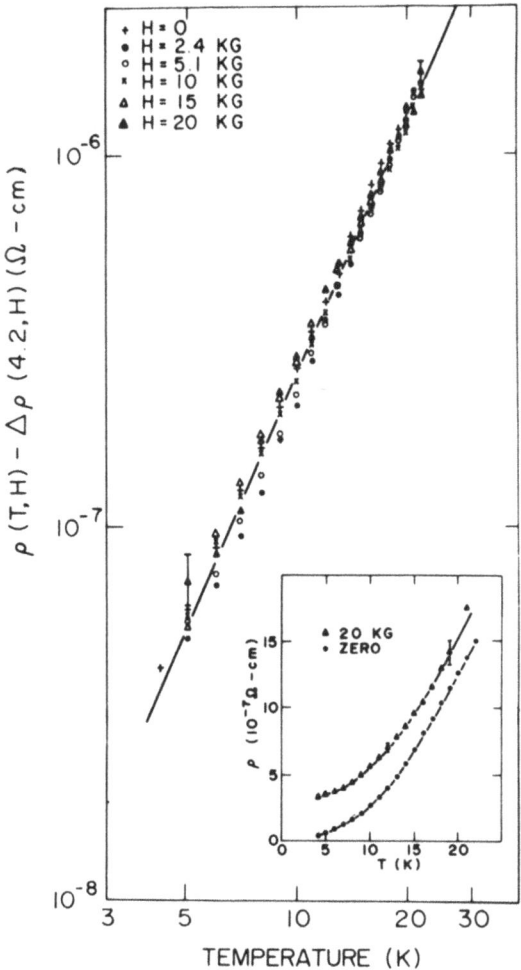

Figure 10: Universal plot of ρ_1(T) [with ρ_1(H) subtracted, see text]. The collapse of all data onto a single cruve indicates that ρ_0(H) can be represented as a field dependent residual resistivity. The inset shows parallel curves of ρ_{ab}(H, T) vs T at fixed fields.

electronic properties is to some extent evident from the incommensurate linear chain structure. The 1 d aspects are directly implied by initial band structure calculations, and they are seen in some experimental detail through studies of the anisotropic optical reflectance. The weak interchain coupling implied by the long coherence length randomly phased chains and the associated

1 d phonons at room temperature provide additional evidence of unusually anisotropic behavior.

The resistivity data summarized in Figure 10 imply a magnetic field dependent residual resistivity which appears to go to zero as the external field is reduced to zero. The apparent absence of residual resistivity in zero field in this linear chain compound is remarkable in view of the well-known sensitivity of quasi-1 d systems to impurities and defects. This is especially true in the context of the chemical, structural and density measurements which imply the presence of several per cent charged anion vacancies.[12]

Theoretical studies of 1 d models[21-24] have provided insight into the manner in which residual resistivity can be avoided through collective effects. Writing g_1 and g_2 for the electron-electron coupling constants corresponding to backward and forward scattering respectively, one has the results that $\rho \rightarrow 0$ as $T \rightarrow 0$ if either (a) $g_1 - 2g_2 > 0$ or (b) $g_1, g_2 < 0$. The actual situation may be considerably more complicated on account of the competing interchain couplings (parallel and perpendicular) which presumably are playing a role in the anisotropic superconductivity observed below 4.1 K.[5, 10] The former case (a) corresponds to the regime of the triplet pairing superconductive ground state.[23] In the fluctuation regime ($T > 0$) calculations have shown that the residual resistivity goes to zero with a power law dependence on (T/T_F) where T_F is the Fermi temperature.[21, 23] The latter case (b) corresponds to a singlet pairing superconductive ground state with corresponding pairing fluctuations at $T > 0$. In this regime, it has been shown that backward scattering by impurities goes to zero as $T \rightarrow 0$.[22, 24]

The effect of a magnetic field on the resistivity provides important information relevant to the possible application of the results of these models to the $\hat{a}$ - $\hat{b}$ plane transport. Kharadze et al.[25] have investigated the effect of an applied field for $g_1 > 0$, $g_1 - 2g_2 > 0$. They find that as a result of spin polarization, the 1 d divergent screening is removed and the residual resistivity is continuously restored by the field so long as $g\mu_B H/k_B T >> 1$. At higher temperatures the field is ineffective due to thermal smearing of the divergent response. These theoretical results are in qualitative agreement with eq. 1. However, the observation (Figure 10) that $\rho_o(H)$ is independent of T at temperatures greater than 20 K ($g\mu_B H/kT < 10^{-2}$ in 1 kG) is difficult to understand

within the context of this theory. Moreover, the apparent anisotropy observed in the dc Montgomery measurements[5] is
inconsistent with a spin polarization effect.

In the context of pairing fluctuations,[26-29] magnetic field
dependence might arise from pair-breaking by the magnetic field.
[29] The large effects observed in low fields are qualitatively consistent with this picture. However, it is difficult to understand
the central result that the effect of the magnetic field is additive
in resistivity (not conductivity).

VII. ANISOTROPIC SUPERCONDUCTIVITY[5, 10]

At low temperature, the c-axis resistivity drops abruptly
over a temperature interval less than 0.2 K; actual dc measurements (two contacts on the top face and two contacts on the bottom)
have demonstrated a decrease in ρ_c by at least three orders of
magnitude to a value below our present measurement capability.
This unusual transition with apparent superconductivity along $\bar{c}$
while ρ_{ab} remains normal and continuous was verified[5] using the
contactless ac mutual inductance technique.

Magnetization measurements on single crystals and powders
of Hg$_{3-\delta}$AsF$_6$ show flux expulsion when samples are cooled in a
small magnetic field, indicating superconductivity below 4.1 K.
The observation of anisotropic flux expulsion with magnitude
dependent upon the orientation of the external field with respect
to the crystalline axes confirms the anisotropic superconductivity
discovered in conductivity measurements.

The Meissner effect was observed on single crystal samples
and on powders. In each case, the sample was cooled in a constant external field, and the induced magnetic moment was measured with a SQUID magnetometer. The magnetization results
differ from the usual Meissner effect in three respects. First,
the flux expulsion is anisotropic; i.e., the magnitude depends
strongly on the orientation of the applied field with respect to the
crystalline axes. Second, the temperature dependence of the flux
expulsion is not a step function, but rather a continuously
increasing function for $T < T_c$. Third, $(-4\pi M/H)$ is field dependent even at fields less than 10^{-4} of the thermodynamic bulk
critical field.

The data for $(-4\pi M/H)$ are shown in Figure 11 for an applied magnetic field of $H_0 = 0.015$ gauss. When the external field is applied in the crystallographic $\vec{a}$ - $\vec{b}$ plane, a large temperature dependent diamagnetism is observed indicative of 20% flux expulsion at 1.6 K in 0.015 gauss. Even at $T/T_c < 0.5$, the magnitude is increasing toward larger values. For $\vec{H} \| \vec{c}$, the onset of flux expulsion occurs at precisely the same temperature. However, as shown in Figure 11, the measured magnetic moment is more than an order of magnitude (approximately a factor of 15) smaller with $\vec{H} \| \vec{c}$ than with $\vec{H} \| \vec{a}$.

A smaller anisotropy is observed in the $\vec{a}$ - $\vec{b}$ plane. Measurements with $H = 0.015$ G applied along a (110) direction, $\vec{H} \| (\vec{a} + \vec{b})$, gave results approximately 20% larger than for $\vec{H} \| \vec{a}$. Demagnetization corrections estimated from the sample shapes and measured magnetization are less than about 5%.

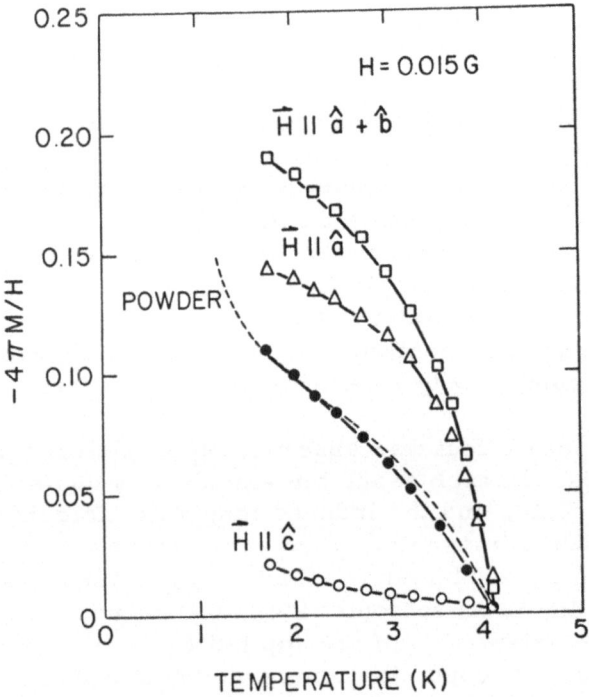

Figure 11: Temperature dependence of $(-4\pi M/H)$ at $H = 0.015$ G for single crystals and a powder sample. The dashed curve represents the ac susceptibility ($\nu = 500$ Hz) data for the powder sample.

In addition to single crystal studies measurements were carried out on fine and coarse powdered samples. In all cases the data from powdered samples are consistent with a simple powder average, $M(\text{powder}) \simeq \frac{1}{3}M(\vec{H} \| \vec{c}) + \frac{2}{3}M(\vec{H} \| \vec{a})$, to within an accuracy limited by the small anisotropy observed in the $\vec{a}$-$\vec{b}$ plane. The powder data for H = 0.015 G are shown in Figure 11.

A comparison of the dc flux expulsion and the ac susceptibility is also shown in Figure 11. Because of the high $\vec{a}$-$\vec{b}$ plane conductivity, low frequency ac measurements on single crystals inevitably lead to eddy current effects; whereas the powder data avoid such difficulties. The ac and dc results are in agreement. The ac data extend the temperature range down to 1.2 K and show a clear indication of an upturn toward complete flux expulsion at the lowest temperatures.

The experimental results for ($-4\pi M/H$) exhibit an unusual magnetic field dependence. The data for $\vec{H} \| (\vec{a} + \vec{b})$ are shown in Figure 12 for applied magnetic fields of H = 0.015 G, 0.12 G, 1.0 G, and 100 G. Figure 12 demonstrates that the magnitude of the incomplete Meissner effect is a function of field even at $H/H_c < 4 \times 10^{-5}$ where H_c = 380 G is the critical field obtained from resistance measurements at 1.4 K. The uniform increase in ($-4\pi M/H$) with decreasing field makes extrapolation to zero field impossible, but complete flux expulsion is not ruled out by the data.

The strong field dependence shown in Figure 12 is reminiscent of flux penetration in a type II superconductor above H_{c_1}. However, the observation from resistance measurements of critical fields ($H_c \simeq 380$ gauss at 1.4 K) consistent in magnitude with $T_c \simeq 4$ K appears to rule out such an interpretation. Moreover, the results shown in Figure 12 would imply $H_{c_1} < 0.015$ gauss, or $H_c/H_{c_1} = \sqrt{2}\,\varkappa > 2 \times 10^4$ where H_c is the thermodynamic critical field and $\varkappa$ is the Ginsburg-Landau parameter. Assuming a typical value for the penetration depth, $\lambda \simeq 500$ Å, such a low value for H_{c_1} would require ($H_c/H_{c_1} \simeq \lambda/\xi$) a coherence length $\xi < 10^{-2}$ A. Finally, simple type II effects will not explain the large anisotropy, the shape of the magnetization curves, or the continuity of σ_{ab} through the phase transition.

Meissner effect measurements[30] have been extended to lower temperatures (T > 0.28 K) through use of a helium-3 refrigerator. The ac susceptibility data for a powder sample are shown in

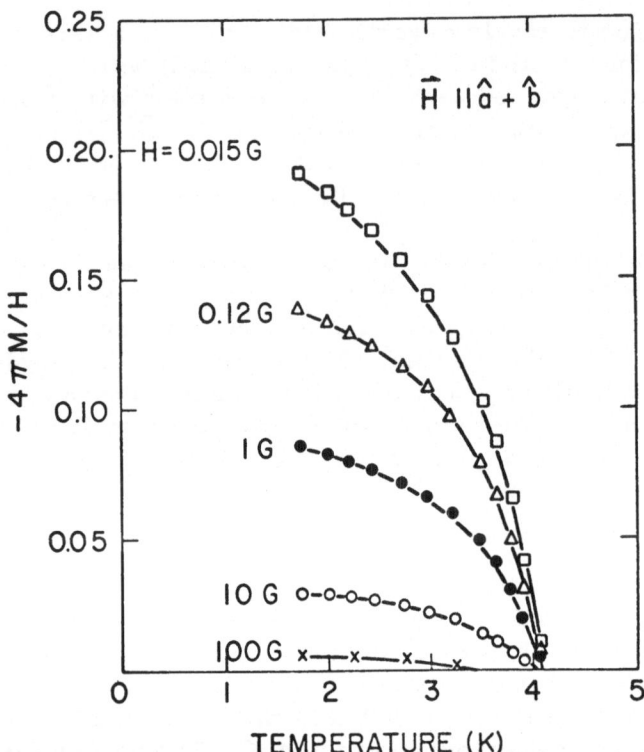

Figure 12: Temperature dependence of (-4πM/H) at different magnetic fields for a single crystal sample; $\vec{H} \parallel \hat{a} + \hat{b}$).

Figure 13.   The diamagnetic susceptibility (mutual inductance technique) approaches -(1/4π) below 0. 4 K indicative of complete flux exclusion; i. e. perfect diamagnetism.   These results demonstrate complete three-dimensional superconductivity in $Hg_{3-\delta}AsF_6$ at low temperatures.

Magnetization curves were obtained from single crystal samples by integrating the response to a slowly swept quasi-dc field as shown in Figures 14, 15, and 16.   At T = 0. 28 K flux penetration occurs near H_c = 8 gauss.   However, the penetration is incomplete (Figures 14 and 15), somewhat anisotropic (Figure 16) with considerable anisotropy in magnetization above H_c.   At successively higher temperatures (Figures 14 and 15), H_c decreases in magnitude until above 1 K the behavior is monotonic characteristic of the data in the anisotropic incomplete

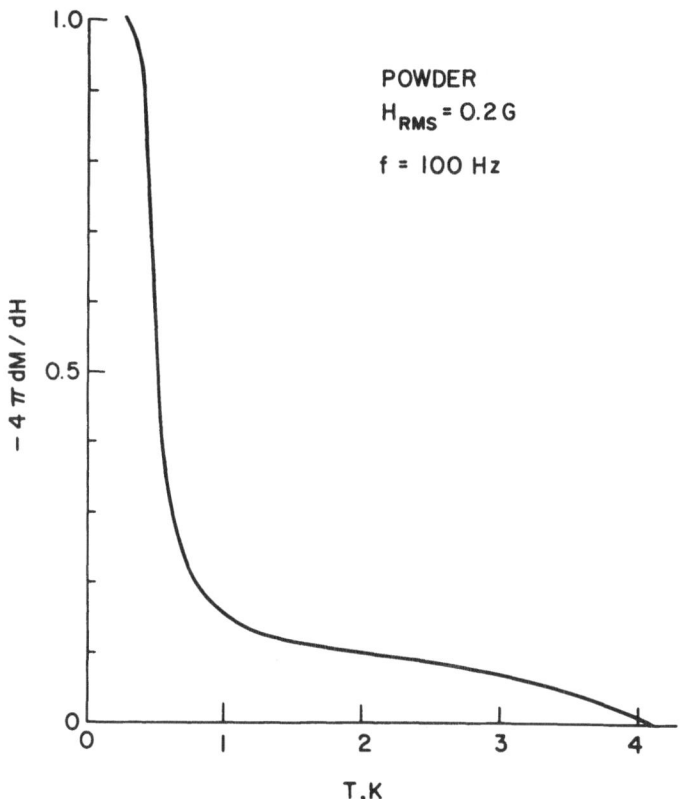

Figure 13: AC susceptibility for a powder sample. The diamagnetic susceptibility approaches (-1/4π) below 0.4 K.

Meissner effect regime at higher temperatures (1 K < T < 4.1 K;
see Figures 11 and 12).

   The low temperature behavior appears to be indicative of a
phase transition near 0.4 K. The critical fields, obtained from
Figures 14 and 15 are plotted as a function of temperature in
Figure 17. The results for H ∥ â are linear, extrapolating to zero
at 0.43 K. Thus, from the Meissner effect data appear to be two
phase transitions; T_{c_1} = 4.1 K being the onset of the anisotropic
superconducting regime and T_{c_2} = 0.43 K indicating the transition
to complete flux exclusion and three-dimensional superconductivity.

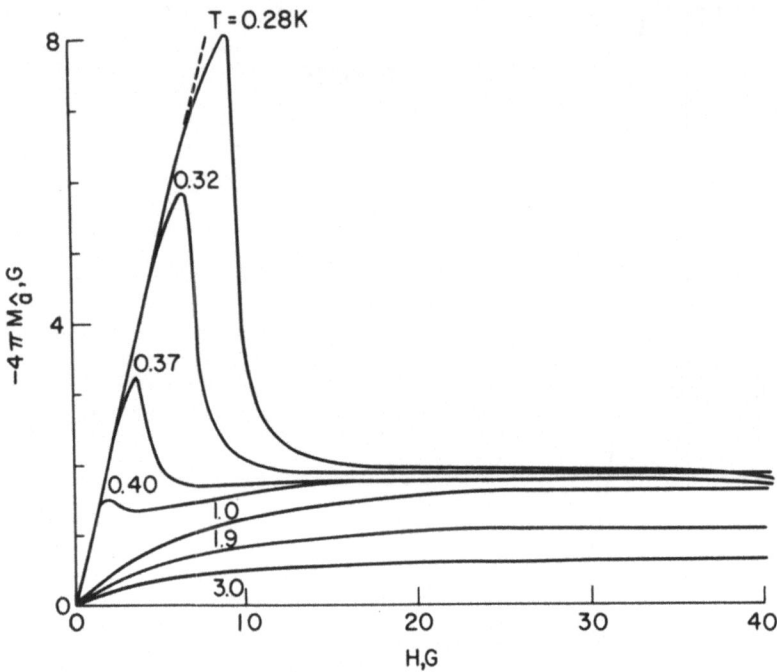

Figure 14: Magnetization vs. field at various temperatures for $\hat{H} \parallel \hat{a}$.

VIII.   THE ROLE OF FREE MERCURY IN THE SUPERCONDUCTIVITY

The observation of complete flux exclusion below 0. 4 K demonstrates unambiguously that the $Hg_{3-\delta}AsF_6$ compound is superconducting at low temperatures.   However, the situation in the anisotropic regime, 0. 4 < T < 4. 1 K, is less clear.   Since resistivity measurements cannot distinguish bulk superconductivity from a filamentary effect the origin of the c-axis superconductivity must be questioned, especially in the context of the temperature dependent stoichiometry (see Sec. III) since T_c = 4. 1 K is close to that of free Hg (T_c = 4. 15 K).

Indeed, Datars et al. [18] have argued that the anisotropic superconductivity results from extruded mercury.   As shown in Section III this argument is evidently oversimplified and incorrect. The temperature dependent stoichiometry does not lead to a surface layer of free mercury, but rather to approximately 1%

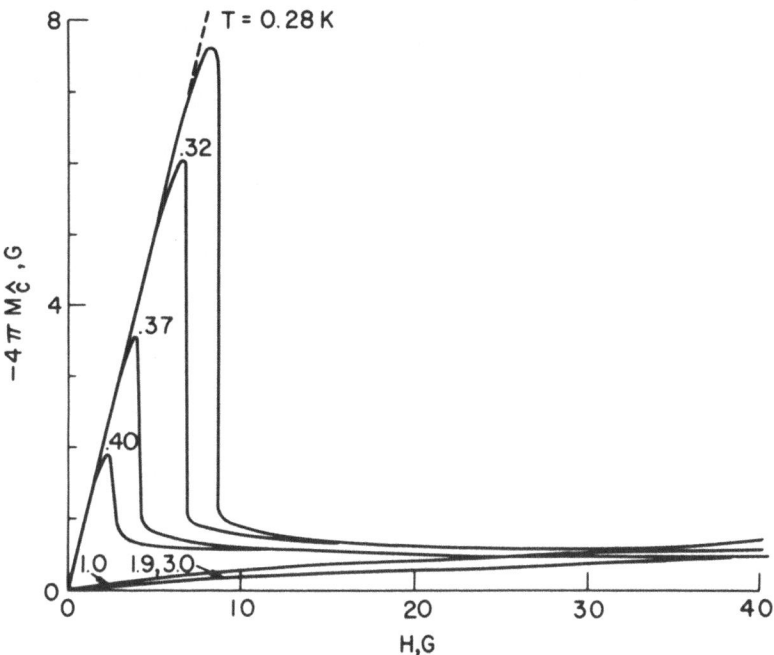

Figure 15: Magnetization vs. field at various temperatures for H ∥ ĉ.

free Hg apparently in small clusters distributed throughout the bulk of the crystal. There is at present no direct measurement of the inferred cluster size. However, details of the DTA and DSC data when compared to the continuous reversible change in stoichiometry (Figure 6) imply a cluster size much less than the penetration depth for Hg. Such clusters would therefore not be expected to show a Meissner effect. This is consistent with our attempts to detect free Hg through the Meissner effect at T_c(Hg) = 4.15 K. Careful examination with the ac technique using a sample sealed in a glass container showed no effect setting an upper limit of 10^{-4} volume fraction of free Hg in particle sizes greater than the penetration depth.

The magnitude of flux expulsion in the anisotropic regime is far greater than the 1% free Hg in the samples. Moreover, the experimental results are not consistent with a small volume of interconnected free Hg filaments. The latter would be expected to trap flux rather than expel it. The agreement between the static flux expulsion and ac susceptibility indicated in Figure 11

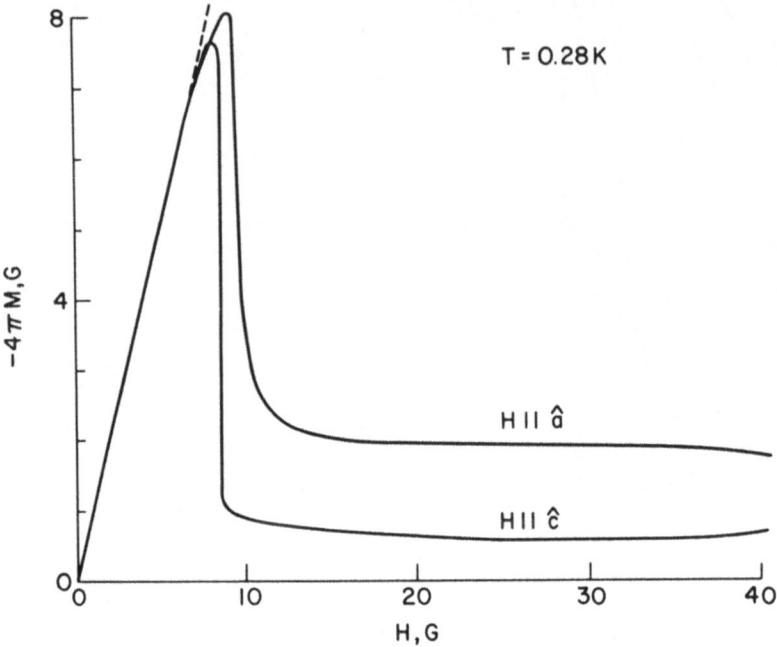

Figure 16: Comparison of magnetization curves for $\hat{H} \parallel \hat{a}$ and $\hat{H} \parallel \hat{c}$ at 0.28 K.

shows that there is no evidence of multiply connected superconducting regions. Moreover, the large anisotropy and unusual
temperature and field dependence appear to be inconsistent with
filamentary effects. An array of superconducting Hg filaments
weakly coupled by intervening metal might lead to bulk superconductivity through the proximity effect. However such a system
would be expected to show isotropic flux expulsion particularly at
low fields. Anisotropy can arise from internal field corrections
due to the demagnetization field. However, the existence of a
surface sheet (due to extrusion) is not consistent with the XPS
data. Detailed analysis of the field and temperature dependences
shown in Figures 11 and 12 indicate that the results are inconsistent with filaments with arbitrary shape.

Our conclusion regarding the anisotropic superconducting
regime is that the data are inconsistent with free Hg as the source
of superconductivity.

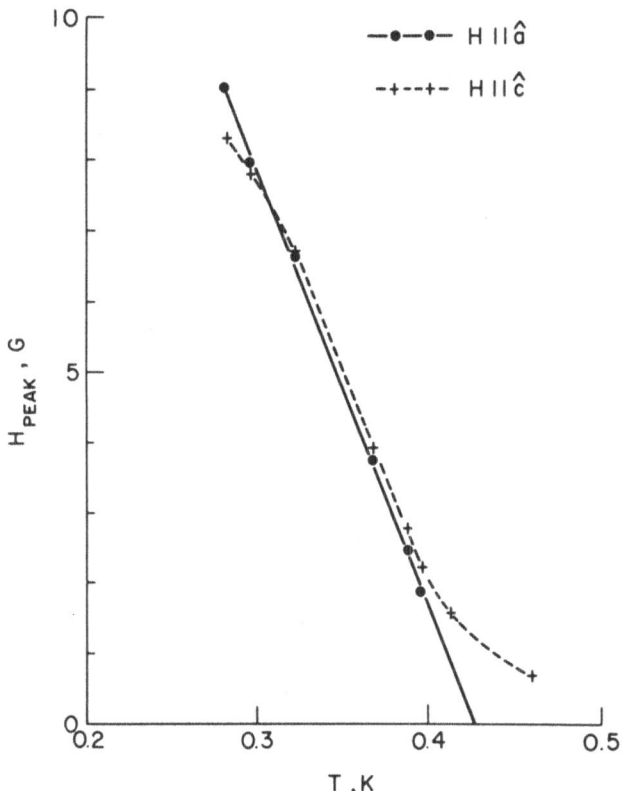

Figure 17: Peak (critical) magnetic field, from Figures 14 and 15, vs. temperature.

Although the superconductivity does not arise from free Hg in the samples, a more subtle and interesting possibility is that the observed phenomena result from proximity effect driven by free Hg and extending into the bulk of the Hg$_{3-\delta}$AsF$_6$ compound. In this situation, the transition at 4.1 K would be that of Hg (depressed by contact with the compound) and that at 0.43 K would be the intrinsic T$_c$ of the compound. However, once again, the anisotropy is difficult to explain in these terms, and the magnitude of flux exclusion (for example at 1.5 K) is far greater than observed in proximity effect samples[31] under comparable conditions. Continued evaluation in terms of possible proximity effect contributions is underway.

The inferred micro-clusters may play a role in providing increased coupling between 1-d metallic chains in the compound. This suggestion is particularly interesting in the context of the field induced residual resistivity described in Section VI. The latter implies the importance of 1 d fluctuation effects over a broad temperature range.

IX. SUMMARY AND CONCLUSIONS

The interesting and unusual electronic properties of $Hg_{3-\delta}AsF_6$ result directly from the remarkable incommensurate polymercury chain structure. The quasi-one-dimensional metallic behavior of the orthogonal sets of chains, their interaction along the $\vec{c}$-axis and the related coupling to the one-dimensional chain phonons are fundamental features to be considered in any attempts toward microscopic theoretical understanding of the transport properties, magnetic field induced residual resistivity and anisotropic superconductivity.

In this brief review we focused on $Hg_{3-\delta}AsF_6$ where a wide range of data on the structural and electronic properties are now available. Unusual features of special fundamental interest include the following:

1) Incommensurate linear chain structure

2) Unique "phase only" phase transition near 120 K due to weak competing interchain interactions.

3) One-dimensional phonons at room temperature developing transverse phason modes below 120 K.

4) Optical properties of quasi-one-dimensional metallic chains.

5) Anisotropic metallic transport properties remaining metallic to low temperatures.

6) Magnetic field induced residual resistivity which goes to zero as the field is reduced to zero.

7) Anisotropic superconductivity (0. 4 K $<$ T $<$ 4. 1 K) including Meissner effect in the presence of finite chain axis conductivity.

8) Complete isotropic flux exclusion below 0. 4 K suggesting a second superconducting phase transition.

These results imply the importance of one-dimensional fluctuation phenomena over an extended temperature range in this upper experimental system.

Extension of the experimental results to related compounds and to alloy systems with known concentrations of impurity atoms is planned in order to further study the phenomena discovered in $Hg_{3-\delta}AsF_6$ and to provide information on the mechanisms involved.

REFERENCES

1. B. D. Cutforth, Ph. D. Thesis, McMaster University, Hamilton, Ontario (1975); B. D. Cutforth, W. R. Datars, R. J. Gillespie and A. van Schyndel, Advances in Chemistry Series 150 (1976)

2. I. D. Brown, B. D. Cutforth, C. G. Davies, R. J. Gillespie, P. R. Ireland, and J. E. Vekris, Can. J. Chem. 52, 791 (1974)

3. A. J. Schultz, J. M. Williams, N. D. Miro, A. G. MacDiarmid and A. J. Heeger, Inorganic Chemistry 17, 646 (1978)

4. B. D. Cutforth, W. R. Datars, A. van Schyndel, and R. J. Gillespie, Solid State Commun. 21, 377 (1976)

5. C. K. Chiang, R. Spal, A. Denenstein, A. J. Heeger, N. D. Miro and A. G. MacDiarmid, Solid State Commun. 22, 293 (1977)

6. E. S. Koteles, W. R. Datars, B. D. Cutforth, and R. J. Gillespie, Solid State Commun. 20, 1129 (1976)

7. D. L. Peebles, C. K. Chiang, M. J. Cohen, A. J. Heeger,
    N. D. Miro, and A. G. MacDiarmid, Phys. Rev. B 15, 4607
    (1977)

8. E. Ehrenfreund, P. R. Newman, A. J. Heeger, N. D. Miro
    and A. G. MacDiarmid, Phys. Rev. B 16, 1781 (1977); E.
    Ehrenfreund, J. Kaufer, A. J. Heeger, N. D. Miro and A. G.
    MacDiarmid, Phys. Rev. (in press)

9. T. Wei, A. F. Garito, C. K. Chiang and N. D. Miro, Phys.
    Rev. B 16, 3373 (1977)

10. R. Spal, C. K. Chiang, A. Denenstein, A. J. Heeger, N. D.
    Miro and A. G. MacDiarmid, Phys. Rev. Lett. 39, 650 (1977)

11. J. M. Hastings, J. P. Pouget, G. Shirane, A. J. Heeger,
    N. D. Miro and A. G. MacDiarmid, Phys. Rev. Lett. 39,
    1484 (1977); J. P. Pouget, G. Shirane, J. M. Hastings, A. J.
    Heeger, N. D. Miro and A. G. MacDiarmid, Phys. Rev. (in
    press)

12. N. D. Miro, A. G. MacDiarmid, A. J. Heeger, A. F. Garito,
    C. K. Chiang, A. J. Schultz and J. M. Williams, J. Inorg.
    and Nucl. Chem. (in press); N. D. Miro, A. G. MacDiarmid,
    A. J. Heeger, A. F. Garito, C. K. Chiang and A. R. McGhie,
    5th International Conference on Crystal Growth, Boston,
    Massachusetts, July, 1977

13. Low Dimensional Cooperative Phenomena, Edited by H. J.
    Keller, Plenum Press, New York (1975); Lecture Notes in
    Physics 34; One-Dimensional Conductors (Springer, New York,
    1975); A. J. Berlinsky, Contemp. Phys. 17, 331 (1976); J. J.
    Andre, A. Bieber, F. Gautier, Ann. Phys. 1, 145 (1976);
    Chemistry and Physics of One-Dimensional Metals, Edited by
    H. J. Keller, Plenum Press, New York (1977)

14. M. Akhtar, J. Kleppinger, A. G. MacDiarmid, J. Milliken,
    M. J. Moran, C. K. Chiang, M. J. Cohen, A. J. Heeger and
    D. L. Peebles, Chem. Commun. 473 (1977); Solid State
    Commun. (in press); G. B. Street, W. D. Gill, R. H. Geiss,
    R. L. Greene and J. J. Mayerle, Chem. Commun. 407 (1977);
    W. D. Gill, W. Bludau, R. H. Geiss, P. M. Grant, R. L.
    Greene, J. J. Mayerle and G. B. Street, Phys. Rev. Lett.
    38, 1305 (1977)

15. V. J. Emery and J. D. Axe, Phys. Rev. Lett. (in press)

16. G. Theodorea and T. M. Rice, (to be published)

17. I. V. Heilmann, J. M. Hastings, G. Shirane, A. J. Heeger and A. G. MacDiarmid, Phys. Rev. Lett. (to be published)

18. W. R. Datars, A. van Schyndel, J. S. Lass, D. Chartier and R. J. Gillespie, Phys. Rev. Letters **40**, 1184 (1978)

19. A. R. McGhie, private communication

20. W. R. Salaneck, R. Spal, A. J. Heeger and A. G. MacDiarmid, (to be published)

21. N. Menyhard and J. Solyom, J. Low Temp. Phys. **12**, 529 (1973)

22. A. Luther and I. Peschel, Phys. Rev. B **9**, 2911 (1974)

23. H. Fukuyama, T. M. Rice, C. M. Varma and B. F. Halperin, Phys. Rev. B **10**, 3775 (1974); P. A. Lee, T. M. Rice and R. Klemm, Phys. Rev. B (in press)

24. D. Mattis, Phys. Rev. Lett. **32**, 714 (1974)

25. G. A. Kharadze, G. E. Gurgenishvilli, and A. A. Nerseyan (preprint)

26. R. A. Craven, G. A. Thomas, and R. D. Parks, Phys. Rev. B **7**, 157 (1973)

27. J. S. Langer and V. Ambergaokar, Phys. Rev. **164**, 498 (1967)

28. D. E. McCumber and B. I. Halperin, Phys. Rev. B **1**, 1054 (1970)

29. Introduction to Superconductivity, Michael Tinkham (McGraw-Hill, New York 1975); Chapter 8

30. R. Spal, A. Denenstein, C. K. Chiang, A. J. Heeger, and A. G. MacDiarmid, Bull. Amer. Phys. Soc. **23**, 305 (1978); and to be published

31. D. P. Seraphim, F. M. D'Heurle and W. R. Heller, "Free Energy of Composite Wires in the Superconducting State", Rev. Modern Phys., Jan. 1964, p. 323-327

ONE-DIMENSIONAL FLUCTUATIONS AND THE CHAIN-ORDERING TRANSFORMATION

IN $Hg_{3-\delta}AsF_6$ *

V. J. Emery and G. Shirane

Department of Physics, Brookhaven National Laboratory

Upton, NY 11973

I. INTRODUCTION

The mercury ions in $Hg_{3-\delta}AsF_6$ form arrays of interpenetrating infinite chains with a mean separation between ions which is incommensurate with the AsF_6 sublattice. This structure leads to a number of quite remarkable properties. According to our present understanding, based mainly upon a recent series of neutron scattering experiments[1] and related theoretical work[2], the chains are essentially one-dimensional liquids at room temperature, and are free to flow down extremely narrow channels ($\sim$7.5 Å in diameter) formed by the AsF_6 sublattice. At about 120 K, the interchain coupling induces a cooperative freezing transition, which takes place in a quite unusual manner. The theory[2] predicts that the first stage consists in the formation of an <u>ion</u> charge-density wave of wave vector d*. As the order develops, further charge-density waves with wave vectors nd* are generated harmonically, and their amplitudes grow until the whole assembly becomes an ordered lattice at low temperatures. A complete description of the evolution of long-range order in the chains involves a number of distinct physical processes which will be explored in more detail in subsequent sections. At high temperatures, the chains are essentially independent and they offer the possibility of studying the behavior of one-dimensional liquids. (Sec.2). Between room temperature and 120 K, the growth of

* This paper consists of a review of recent experimental and theoretical work carried out at Brookhaven and represents the work of a number of people. The research was supported by the Division of Basic Energy Sciences, Department of Energy, under Contract No. EY-76-C-02-0016.

transverse order is a consequence of coupling between both parallel
and perpendicular chains. The result is a close competition between two different kinds of long-range order, leading to a transition which is almost tricritical (Sec.3). In the ordered phase,
a growing charge-density wave in the mercury ions generates strong
harmonics which eventually turn into an ordered lattice. This
process is accurately described by a mean field which acts on the
individual chains to produce a Sine-Gordon Hamiltonian. The thermodynamic properties and the energy spectrum of the excitations
(solitons and soliton-antisoliton bound states) may be calculated
and compared with experiment (Sec. 4)

2. THE INDEPENDENT CHAINS

X-ray[3] and neutron diffraction[4] studies have shown that the
room-temperature crystal structure of $Hg_{3-\delta}AsF_6$ consists of a tetragonal arrangement ($I4_1/amd$) of AsF_6 octahedra which form open
channels running parallel to both basal-plane edges. These channels
are filled with tightly packed mercury ions to form a set of orthogonal nonintersecting linear chains shown schematically in Fig.1(a).
A basal-plane projection in which, for simplicity, only one pair
of parallel chains ($Z = 0,1/2$) is considered is shown in Fig. 1(b).
The observed intrachain mercury-mercury distance, d of 2.665±0.005A
is such that only $3-\delta$ mercury ions can be accommodated within the
unit cell formed by the AsF_6 octahedra; i.e., $(3-\delta)d = a_L$, where
a_L is the lattice parameter. This incommensurability, which is
indicated schematically in Fig. 1(b), leads to the empirical molecular formular $Hg_{3-\delta}AsF_6$.

The scattering intensity in reciprocal space from a single
pair of parallel incommensurate infinite chains ($Z = 0, 1/2$) with
random interchain phasing is shown in Fig. 1(c). The scattering
is uniformly distributed in reciprocal-lattice planes [$k = n(3-\delta)$;
$n = 1,2,...$] normal to the chains in direct space. However, as a
result of the periodicity in position of the Hg chains imposed by
the regular network of open channels of the host AsF_6 sublattice,
the intensity in the zeroth-order plane (n=0) coalesces to Bragg
spots. The spots can therefore be thought of as arising from the
interaction of individual chains of the mercury sublattice with the
host sublattice.

Figure 2 shows that the diffuse sheets are the origins of
inelastic scattering surfaces with dispersion $\omega = \pm c|q|$, where
$Q = (2\pi n/d) + q$ is the component of momentum along the chain
direction. The measured value of c is 4.4×10^5 cm/sec, twice as
large as the sound velocity for three-dimensional phonons.

These experiments allow us to conclude that it is sufficient

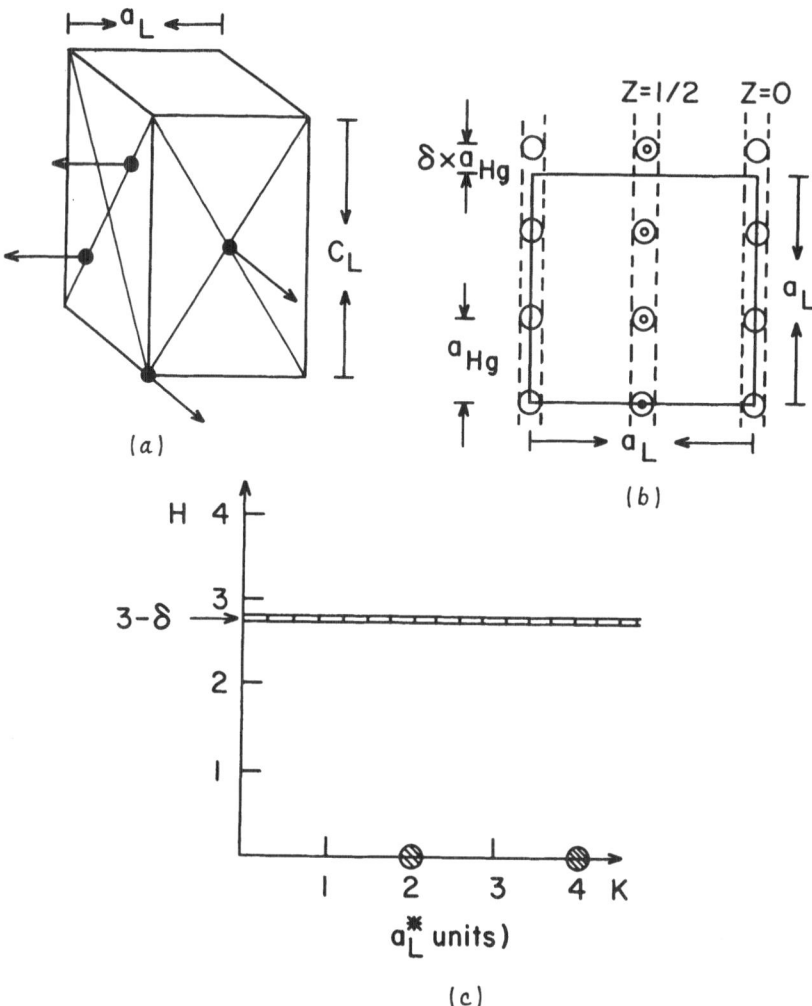

Fig. 1.   (a)   Schematic representation of the Hg$_{3-\delta}$AsF$_6$ unit cell showing the location of Hg chains (arrows).   (b)   Basal-plane projection of one pair of parallel Hg chains (Z = 0, 1/2).   The dashed lines outlining the chains are intended to indicate the randomness in the relative position of Hg ions between the chains, corresponding to the room-temperature structure.   (c)   Reciprocal space corresponding to projection shown in (b).

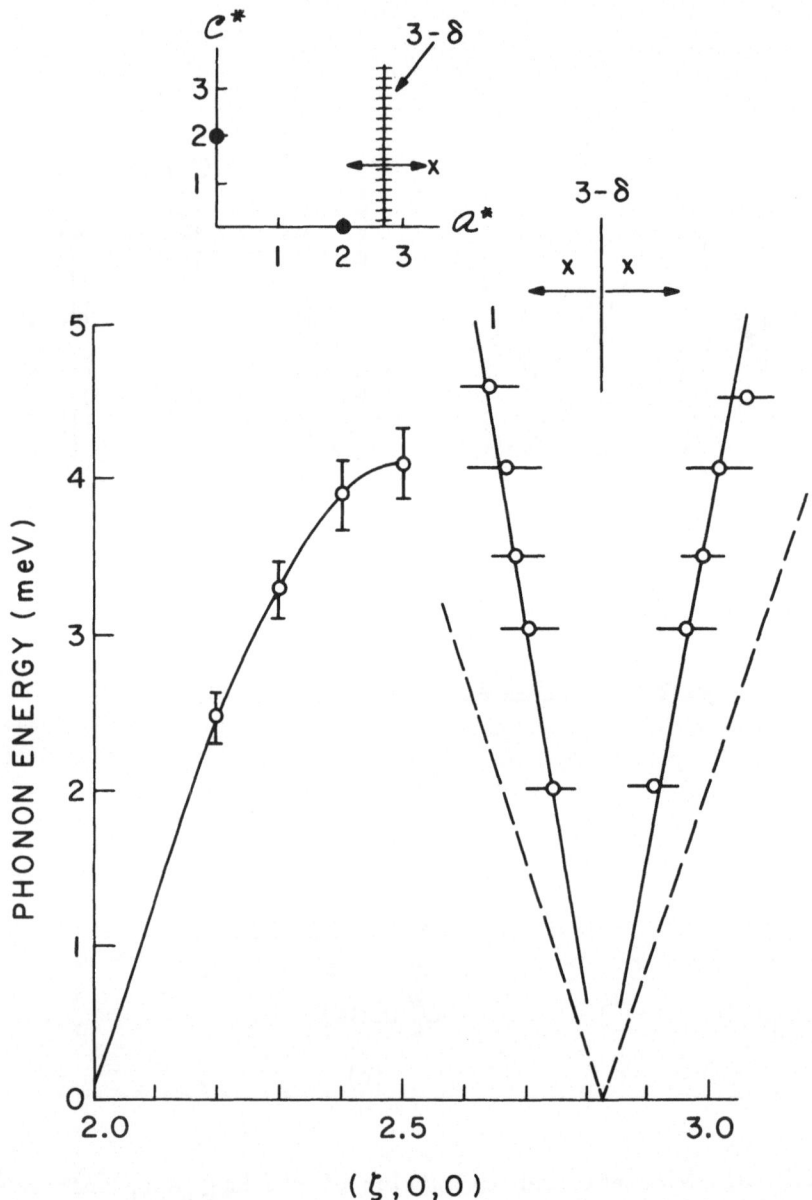

Fig. 2. Room-temperature phonon dispersion curve. The dispersion curve associated with the Hg chains centered about $\zeta=2.82$ was measured for various values of x, i.e. $(\zeta,0,x)$. The dashed lines, drawn with the initial slope of the lattice dispersion curve have been added to facilitate comparison between the slopes. After Hastings *et al.* (Ref.1).

to use a harmonic Hamiltonian for the individual chains:

$$H_0 = \frac{1}{2} \sum_\alpha \left\{ \frac{\pi_\alpha^2}{m} + \frac{mc^2}{d^2}(x_{\alpha+1}-x_\alpha-d)^2 \right\} \tag{1}$$

where m is the mass of a mercury ion and (π_α, x_α) are the momentum and position of the α'th particle. The point is that at temperature T, the root mean square deviations in the relative separations of mercury ions are given by $<(x_{\alpha+\ell}- x_\alpha-\ell d)^2>= \sqrt{\ell}\,\sigma^2$ where $(\sigma/d)^2 = k_B T/mc^2$ and is equal to 6.4×10^{-4} at room temperature. As $\ell \to \infty$, this expression diverges so there is no long-range order, but, for $\ell = 1$, it is very small so the harmonic approximation is justified. Furthermore, in evaluating the properties of the system, $\hbar$ occurs in the combination $\theta \equiv \pi\hbar/2mcd$ which is so small(4.2×10^{-4}) that a classical approximation is valid except, possibly, close to T = 0.

Once this is understood, it is not difficult to evaluate[2] the Fourier transform of the pair distribution function:

$$S^0(Q) = \frac{\sinh(Q^2\sigma^2/2)}{\cosh(Q^2\sigma^2/2)- \cos Qd} \tag{2}$$

which has a sequence of peaks centered at $Q = Q_n \equiv 2\pi n/d$ and almost Lorentzian, with width κ_n given by $\kappa_n d = 2\pi^2 n^2 (\sigma/d)^2$. The shape of this function is characteristic of a liquid. In the neighborhood of the peaks the dynamical structure factor, which determines the neutron scattering from a single chain, is given by[2]

$$S^0(Q,\omega)= \frac{4\kappa_n^2/\pi dc}{\left[(q+\omega/c)^2 + \kappa_n^2\right]\left[(q-\omega/c)^2 + \kappa_n^2\right]} \tag{3}$$

This is an exact result for the Hamiltonian given in Eq. (1). It is not possible to introduce the usual phonon expansions because the Debye-Waller factor is infinite for a one-dimensional system. Eq.(3) shows that the diffuse sheets occur at the intersections of the phonon branches where $\omega = 0$, and that there is no truly elastic scattering.

The study of these one-dimensional liquids is an interesting subject in itself and has only just begun. It should be possible to carry out experiments to measure the widths of the higher sheets and to verify that κ_n is proportional to $n^2 T$. It would also be of interest to check the detailed form of the dynamical structure factor given in Eq. (3), for it is quite different from the usual expression obtained by phonon expansions.

If the system remained classical, the Lorentzian peaks would become δ-function Bragg peaks at T = 0, and the individual chains would be solid. However, long before this happens, the interaction between chains takes over, and the freezing transition occurs in a quite different way, which will now be explored.

3. THE DEVELOPMENT OF CORRELATIONS BETWEEN THE CHAINS

As the temperature is lowered, the intensity on the first sheet acquires a modulation, which is illustrated in Fig. 3 for the reciprocal lattice line (h, 3-δ, 0) and several representative temperatures. The modulation implies that the different chains are becoming correlated and, since the associated intensity increases as the temperature decreases, there appears to be an impending transition to an ordered phase. However, according to Fig. 4, the intensity reaches a maximum at 125 K and, below $T = T_c \equiv 120$ K, well-defined Bragg peaks appear (Fig. 3), but their location (1±δ,3-δ,0) bears no obvious relation to the precursor behavior for $T > T_c$. The intensity of the Bragg peaks, given in Fig. 4 and inset in Fig. 3, rises rapidly but continuously, indicating a second order transition. It is evident that there are two competing kinds of order, but the main puzzle is the apparent lack of critical scattering which should presage the state which finally is established at low temperatures. Any theory must account for this behavior as well as the origin and shape of the peaks shown in Fig. 3 and the nature of the ordered phase.

In order to account for these effects, it is necessary to introduce a coupling between the chains, assumed to be local, so that it may be written in the form

$$H' = \frac{1}{2} \sum v_{ij} (\underset{\sim}{p}, \underset{\sim}{p}') \rho_i(\underset{\sim}{p}) \rho_j (-\underset{\sim}{p}') \tag{4}$$

where

$$\rho_i(\underset{\sim}{p}) \equiv \sum_{\alpha,n} e^{i\underset{\sim}{p} \cdot \underset{\sim}{r}_{i\alpha n}} \tag{5}$$

is the Fourier transform of the density operator for the i'th array of chains. (i = 1,2 specifies the a-axis or b-axis array). The position of the α'th mercury ion on the m'th chain of the a-axis array is $\underset{\sim}{r}_{1\alpha m} = (x_{m\alpha}, \bar{y}_m, \bar{z}_m)$ and the corresponding positions in the b-axis array are $\underset{\sim}{r}_{2\beta n} = (\bar{x}_n, y_{n\beta}, \bar{z}_n + c_{T/4})$. The coordinates with a bar denote the discrete lattice points at which the chains are located, and the corresponding wave-vector components are restricted to the first Brillouin zone. The third wave-vector component is unrestricted since it is conjugate to the continuously varying position of a mercury ion within a chain. The interactions

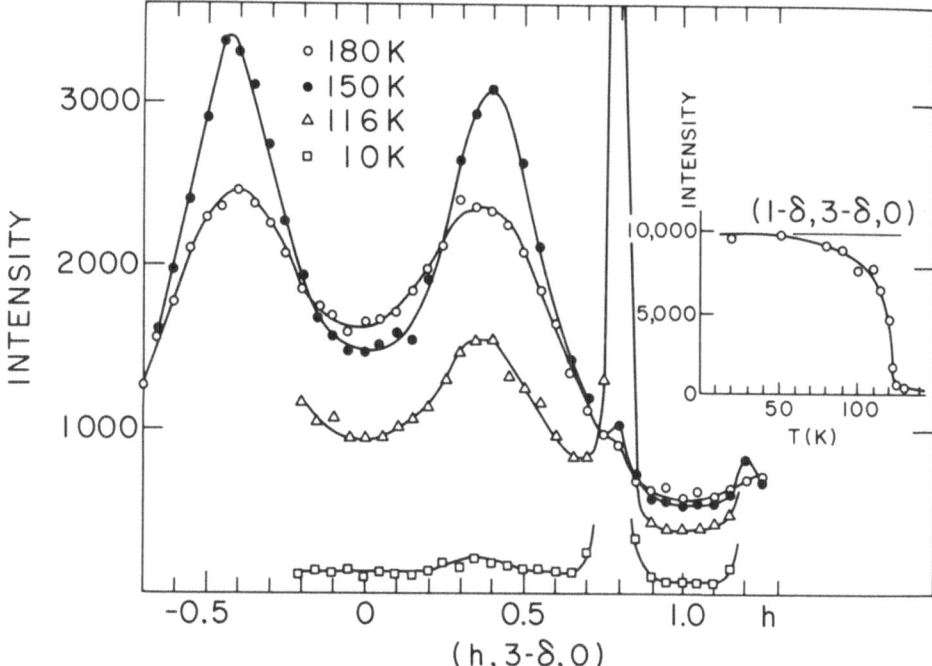

Fig. 3. Elastic scattering along the line (h, 3-δ,0) at some representative temperatures showing both short-range and long-range order. The inset shows the temperature dependence of the long-range-order peak (1-δ,3-δ,0). After Hastings *et al.*(Ref.1).

are of short range, and it is necessary to keep first and second neighbors within an array (since they are approximately equidistant) and near neighbors from different arrays. The full interaction will not be written down explicitly but it will prove useful to state the coupling within the b-axis array:

$$v_{22}(\underset{\sim}{p},\underset{\sim}{p}') = [2v_2(k)\cos2\pi h + 4v_1(k)\cos\pi h \cos\ell\pi]\delta_{\underset{\sim}{p}\underset{\sim}{p}'} \qquad (6)$$

where (h,k,ℓ) are the components of $\underset{\sim}{p}$ in the tetragonal basis.

Given the Hamiltonian of Eqs. (1) and (4), the many-chain problem may be solved by expansion in the small-parameter κ$_1$d. In lowest order the harmonic motion along the chains is treated exactly but time-dependent mean-field theory (or, equivalently, the random phase approximation) is used for the interchain coupling.[5]

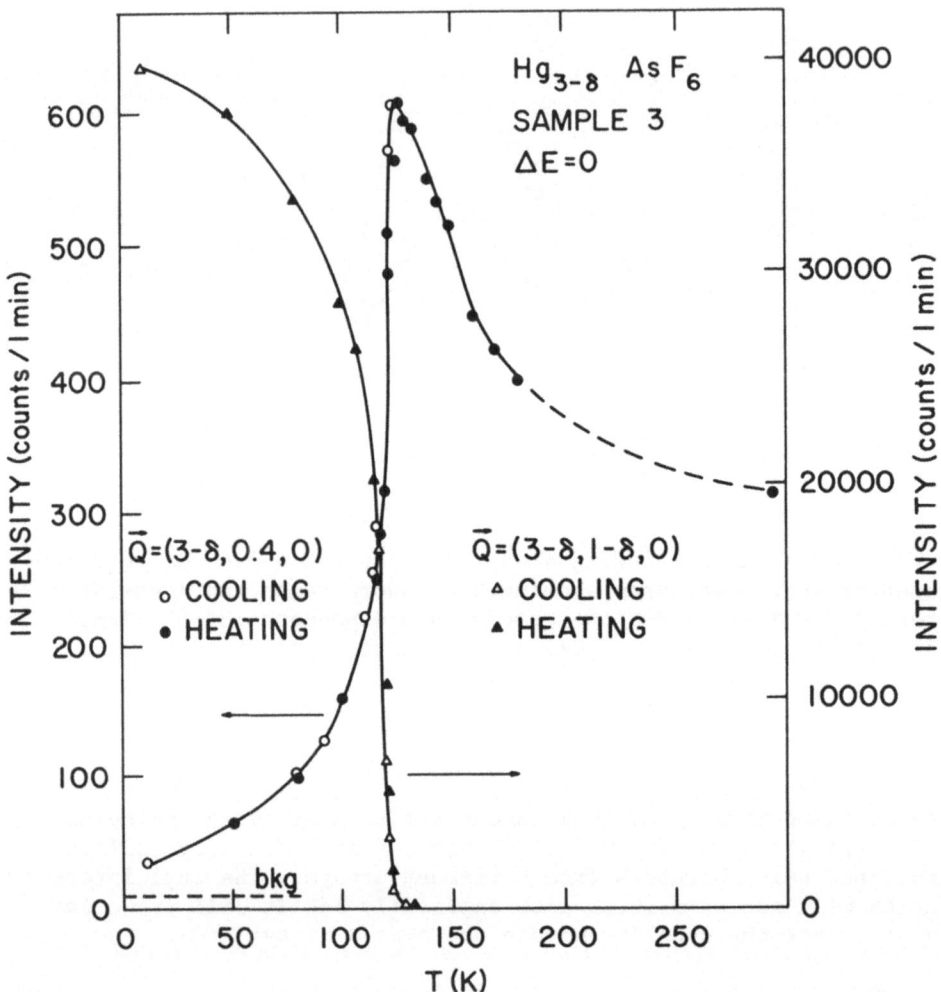

Fig. 4. Temperature dependence of the $(3-\delta,0.4,0)$ and $(3-\delta,1-\delta,0)$ peak intensity on heating (solid points) and cooling (open points). After Pouget et $al.$(Ref. 1).

However, near to T_c, it will prove necessary to include fluctuations of order $\kappa_1 d$, since they are almost divergent.

The results may be expressed in terms of the dynamical susceptibility matrix $\hat{\chi}(\omega)$ which has elements $-i\theta(t)<[\rho_i(p,t),\rho_i(-p')]>$, describing the response of the system to an external probe (such as a neutron). For the disordered phase, the solution has the well-known RPA form

$$\hat{\chi}(\omega) = \hat{\chi}^0(\omega)[1 + \hat{v}\hat{\chi}^0(\omega)]^{-1} \tag{7}$$

where $\hat{v}$ has matrix elements $v_{ij}(p,p')$ and $\hat{\chi}^0(\omega)$ is the susceptibility of the noninteracting chains with diagonal elements given by $\mathrm{Im}\chi^0_{ii}(p_i,\omega) = \pi(1 - \exp(-\beta\omega))S^0(p_i,\omega)$. The right-hand side of Eq. (7) has peaks when the external probe excites collective or single-chain modes, and its evaluation requires the inversion of a large matrix, since $\hat{v}$ is non-diagonal. However, it is possible to use perturbation theory to obtain two kinds of mode, which are essentially independent, away from T_c.

A.   Independent Arrays

The ·potential $\hat{v}$ is sufficiently weak that the denominator in Eq. (7) does not differ significantly from the unit matrix unless the elements of $\hat{\chi}^0(\omega)$ are large. It follows that peaks related to the first sheet are most significant, which is why the expansion parameter is $\kappa_1 d$ and not some other $\kappa_n d$. Also, unless h = k', the coupling between the arrays is very weak, and the separate susceptibilities are given by

$$\chi_{ii}(p,p',\omega) = \frac{\chi^0(p_i,\omega)\delta_{p,p'}}{1 + v_{ii}(p,p)\chi^0(p_i,\omega)} \tag{8}$$

where p_1 and p_2 are respectively the x- and y- components of the wave vector. For elastic experiments, as illustrated by Fig. 3, it is necessary to set $\omega = 0$, and, since for a classical system $kT\chi_{ii}(p,p',0)$ is just the pair distribution function $S_{ii}(p,p')$, Eq. (8) may be written

$$S_{ii}(p,p') = \frac{S^0(p_i)\delta_{p,p'}}{1 + v_{ii}(p,p')S^0(p_i)/k_B T} \tag{9}$$

At high temperatures, the denominator is close to unity and the scattering is essentially that of the separate chains. There is no dependence on the transverse components of p and there are uniform sheets of scattering, as described earlier. Since the peaks

of $S^0(p_i)$ have height proportional to $\kappa_n^{-1} \sim (n^2 T)^{-1}$, the denominator
of Eq. (9) will begin to depart from unity as T decreases, and this
gives rise to the observed modulation, brought about by the variation of $v_i(p,p')$ with the transverse wave vector. Eqs. (6) and (9)
predict a very characteristic shape for the modulation, since the
denominator in Eq. (9) is a linear combination of $\cos\pi h$ and $\cos 2\pi h$.
The curves in Fig. 3 may be fitted[2] with two parameters $v_1((3-\delta)b_T^*)$
and $v_2((3-\delta)b_T^*)$ (since χ^0 peaks strongly) to obtain a reasonably
good representation of the data at all temperatures. Any differences can be attributed to resolution and absorption effects, and
the general features are reproduced so well that there is no doubt
that the modulation comes from parallel-chain couplings.[1,2]

B. Coupled Modes

When the two basal-plane components of the wave vector are
equal ($h = k'$), the collective oscillations of the two arrays are
degenerate and even a weak coupling between them may not be neglected, since it gives off-diagonal matrix elements which break the
degeneracy. Then for $\omega = 0$, the solution becomes

$$\sum_{i,j} \chi_{ij}(p,p',0) = \frac{2S^0(ha_T^*)}{1 + (v_{22} + v_\perp)S^0(ha_T^*)/k_B T} \qquad (10)$$

Here $v_\perp$ is the coupling between perpendicular chains and has the
form $u(h)\cos\ell\pi/2$. Peaks of the right-hand side of Eq. (10) correspond to joint oscillations of the a-axis and b-axis arrays and they
also will give rise to a modulation of the diffuse sheets. How-
ever, the width will be $2\kappa_1$ in all directions in the (a_T^*, b_T^*) plane,
and, since this is smaller than the experimental resolution, the
peaks will contribute a relatively weak integrated intensity to the
scattering, giving the appearance of Bragg peaks persisting above
T_c, which actually have been observed.[1] However, their true nature
may be determined by looking for a finite extension of the scattering in a direction perpendicular to the basal plane, for which
there is no long coherence length.

From this analysis, it is clear that there are indeed two
distinct kinds of collective mode, and that one will be much more
evident than the other in a neutron scattering experiment. They
are competing in that they would lead to different phase relations
between the chains if they condensed into ordered states. If the
modes remained independent, there would be a phase transition at
the highest temperature for which a denominator of Eq. (8) or
Eq. (10) vanished. The details of the interchain coupling determine which mode becomes unstable, and since the observed Bragg
peaks have indices $(1+\delta, 3-\delta)$ in the basal plane, it is clear that
the actually established order involves the coupled arrays

(Eq. (10)) although the fit of v_{22} to the experiments above T_c indicates that the independent a-axis and b-axis arrays would have
ordered just a few degrees below the actual transition temperature.[2]
This suggests that it may be possible to change the nature of the
ordered state by varying the interchain coupling, for example
through application of pressure. Moreover, near to T_c, the interaction between the two kinds of mode may not be neglected and this
will modify Eqs. (8) and (10). The analysis, which will be presented in detail in a future publication[6], is typical of a phase
transition involving coupled order parameters. It shows that the
transition is almost tricritical and might possibly become so on
application of pressure.

The fit to the experiments for $T > T_c$ involves three parameters-
v_1 and v_2 in Eq. (6) together with $\bar{v_1}$ in Eq. (10), all evaluated on
the first sheets. The value of v_1 is consistent with Coulomb interactions[2], but this cannot be true of v_1 and v_2 since v_1 is attractive and both are much too large in magnitude.[2] It is likely that
the AsF_6 sublattice plays a rôle in determining the magnitude of
v_1 and v_2, and, indeed, all of the foregoing analysis has assumed
that the trace over the AsF_6 variables has been carried out, giving
an effective Hamiltonian for the mercury chains.

4. THE ORDERED PHASE

Since the divergence in Eq. (10) occurs on the first sheet,
the primary order parameter describes a static mercury-ion charge-
density wave (CDW) with wave vector Q_1. Interchain coupling does
not become significant for the higher sheets until well below T_c,
and other CDW's with wave vectors Q_ν are generated as harmonics
whose amplitude grows as the order develops. For an individual
chain the mercury-ion density is given by

$$\rho(x) = 1 + 2 \sum_{\nu=1}^{\infty} A_\nu \cos(2\pi\nu x/d + \delta) \tag{11}$$

where Q_ν has been written explicitly as $2\pi\nu/d$ and the A_ν are temperature-dependent amplitudes. At low temperatures, all the A_ν
saturate to unity and $\rho(x)$ assumes the ordered lattice form
$\sum_{n=-\infty}^{\infty} \delta(x-nd+\delta')$ for which Eq. (11) gives the Fourier transform when
$A_\nu = 1$, all ν.

The evolution of the ordered phase may be worked out by evaluating the static mean field acting on a single chain. This is obtained by replacing one of the density operators in Eq. (4) with
its expectation value, in the usual way. On defining

$x_\alpha = x_0 + \alpha d + u_\alpha$, the single-chain Hamiltonian becomes

$$H = H_0 - V \sum_\alpha \cos(Q_1 u_\alpha + \phi) \tag{12}$$

where H_0 is given by Eq. (1) and V is to be determined self-consistently.

The phase ϕ is determined by choosing x_0 to make $\phi = 0$ so that (classically) $u_\alpha = 0$ when T = 0. The experimental results can be accounted for by assuming that there are two types of domains in which the Hg ions in successive chains at Z = 0 are shifted by a fraction $\pm\delta$ while maintaining the body-centering conditions for the chains at Z = 1/2. The resulting two monoclinic unit cells of Hg ions would give rise to Bragg peaks with indices h$\pm$kδ,k(3-δ),0, respectively, with h+k even in accord with observation.

Assuming the same type of phase-ordering scheme and domain structure for the pair of Hg chains at Z = 1/4 and 3/4, one can account for the observed peaks with indices h(3-δ),k$\pm$hδ,0. Furthermore, one can determine the relative phase of the two pairs of orthogonal mercury chains by noting that there are peaks with indices h(3-δ), k(3-δ),0 to which both pairs contribute. Since the peak at (3-δ), (3-δ), 0 is absent while the one at 2(3-δ), 2(3-δ), 0 is present and four times as intense as, e.g., the one at 2(3-δ), 2(3-δ), 0, one concludes that the (3-δ, 3-δ, 0) planes from the two pairs of chains are out of phase.

Equation (12) gives a Sine-Gordon Hamiltonian for a lattice. In the neighborhood of the first diffuse sheet, the length scale is $\kappa_1^{-1}(T_c) \simeq 200d$, and it is a good approximation to replace the differences $(x_{\alpha+1} - x_\alpha - d)/d$ in Eq. (1) by derivatives $\partial u_\alpha/\partial x$ to obtain the continuum Sine-Gordon Hamiltonian for which the classical partition function and quantum-mechanical energy spectrum are known. This is sufficient to work out the properties of the ordered phase.

The Sine-Gordon equation is Lorentz-invariant and the excitations have the form $c(M^2 c^2 + \hbar^2 q^2)^{1/2}$ where Mc^2 is the "rest energy" or energy gap. They are of two kinds – solitons with "mass" $M = M_s \equiv (8\hbar V/\pi\theta c^3)^{1/2}$ and a spectrum of soliton-antisoliton bound pairs with "mass" M_ν where

$$M_\nu = 2M_s \sin \pi\nu\theta/2 \tag{13}$$

for $\nu = 1, 2,.<\theta^{-1}$ (neglecting the small renormalization of θ). For ν larger than θ^{-1}, the pairs break up into a free soliton and antisoliton. Since θ is small, $M_\nu \approx \pi\theta\nu M_s$ for the lowest levels and they are not too different from the harmonic phonons obtained by expanding Eq. (12) to second order in u_α. Higher order terms

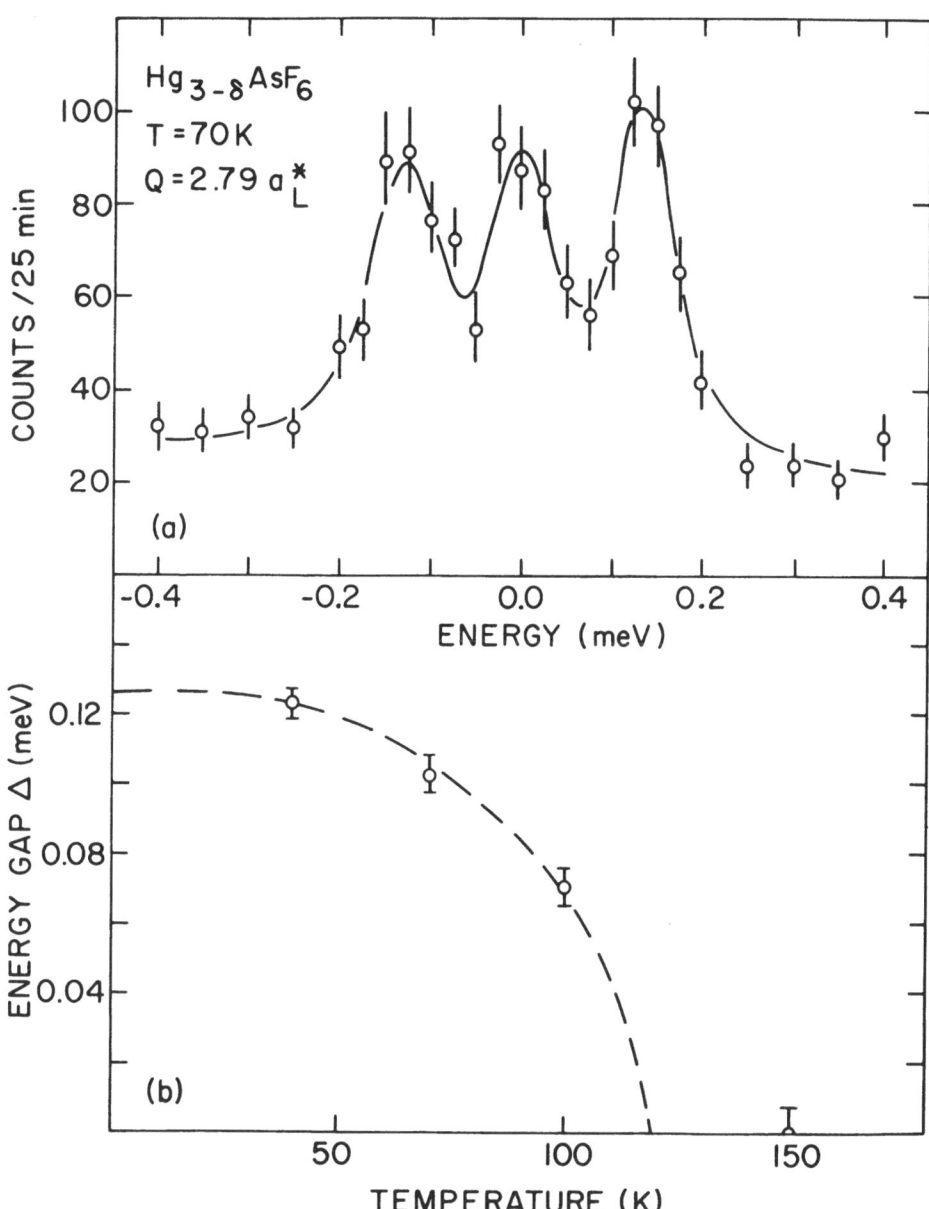

Fig. 5. Constant-Q scan at 70 K for Q = 2.79 a_L^*, showing the existence of an energy gap in the Hg chain dispersion relation (a).
Temperature dependence of the energy gap Δ found from computer fits
to constant-Q scans (b). Solid and dashed curves are guides to the
eye. After Heilmann *et al.* (Ref. 9).

give rise to interaction between the phonons, and Eq. (13) may equally be regarded as a spectrum of bound phonon states.[7]

These excitations constitute defects in the ordered ground state, and it should be possible to observe at least some of them in a neutron scattering experiment. In order to work out their energies, it is necessary to know the magnitude of the self-consistent field V, and this may be obtained from the partition function for the Sine-Gordon Hamiltonian. Once again, the smallness of θ implies that the classical limit may be used, so that it is possible to express[8] the partition function and hence V in terms of the lowest eigenvalue of Nathiew's equation. In this way[2], it is found that for $t \equiv (1 - T/T_c) \ll 1$, the solution gives $M_s c^2/k_B = 16 T_c \sqrt{t}$ whereas, at low temperatures, $M_s c^2/k_B = 2^{5/2} T_c \simeq 700$ K and $M_\nu c^2/\nu k_B \simeq 0.93$ K. To simplify the argument, the coupling between the two-order parameters ahs been omitted, since it requires a more detailed discussion of the fint to the existing experiments. Inclusion of this effect would increase the value of M_s.

Figure 5 shows the result of a measurement[9] of the energy of the lowest of the excited states, described above. It can be seen that the magnitude and temperature dependence are consistent with the theoretical predictions. The work described in Ref. 9 is the first of an extensive program of experiments designed to check our current understanding of this most interesting material. Further results and also a more expanded version of the theory will be presented in future publications.

ACKNOWLEDGMENTS

The work described here stemmed from the collaboration of a number of people. We have profited from discussions with R. P. Comès, J. M. Hastings, A. J. Heeger, I. U. Heilmann, and J. P. Pouget. In particular, we should like to acknowledge collaboration and discussion with J. D. Axe, who provided a good deal of physical insight and is responsible for many of the ideas and calculations which were presented.

REFERENCES

1.   J. M. Hastings, J. P. Pouget, G. Shirane, A. J. Heeger, N. D.
      Miro, and A. G. MacDiarmid, Phys. Rev. Lett. **39**, 1484(1977);
      J. P. Pouget, G. Shirane, J. M. Hastings, A. J. Heeger, N. D.
      Miro, and A. G. MacDiarmid, Phys. Rev. B (in press).
2.   V. J. Emery and J. D. Axe, Phys. Rev. Lett. **40**, 1507 (1978).
3.   I. D. Brown, B. D. Cutforth, C. G. Davies, R. J. Gillespie,
      P. R. Ireland, and J. E. Vekris, Can. J. Chem. **52**, 791(1974);

B. D. Cutforth, Ph.D. thesis, McMaster University, Hamilton, Ontario, 1975 (unpublished); B. D. Cutforth, W. R. Datars, R. J. Gillespie, and A. van Schyndel, in <u>Organic Compounds with Unusual Properties</u>, Advances in Chemistry Series No. 150, edited by R. Bruce King (American Chemical Society, Washington D. C., 1976).

4. A. J. Schultz, J. M. Williams, N. D. Miro, A. G. MacDiarmid, and A. J. Heeger, Inorg. Chem. <u>17</u>, 646 (1978).

5. D. J. Scalapino, Y. Imry, and P. Pincus, Phys. Rev. B <u>11</u>, 2042 (1975).

6. V. J. Emery and J. D. Axe, to be published.

7. R. F. Dashen, B. Hasslacher, and A. Neveu, Phys. Rev. D <u>11</u>, 3424 (1975); A. Luther, Phys. Rev. B <u>14</u>, 2153 (1976).

8. S. F. Edwards and A. Lenard, J. Math. Phys. <u>3</u>, 778 (1962); N. Gupta and B. Sutherland, Phys. Rev. A <u>14</u>, 790 (1976).

9. I. U. Heilmann, J. M. Hastings, G. Shirane, A. J. Heeger, and A. G. MacDiarmid, to be published.

POTENTIAL APPLICATIONS OF MOLECULAR METALS

Susumu Yoshimura

Matsushita Research Institute Tokyo, Inc.

Ikuta, Tama-ku, Kawasaki 214, Japan

INTRODUCTION

Design and synthesis of novel materials with extremely high electrical conductivity has been the main concern for those involved in this new subfield of solid state chemistry of molecular solids (1). Even though most of the investigators have been concerned with academic sciences, they would never hesitate to exploit possible technological applications. In fact, after Professor Inokuchi conceived the concept of organic semiconductors in 1954 (2), realization of new devices such as organic transistors was expected to result from the extensive research around 1960 (3). Recent explosive activities have also been much stimulated by the interest in a high-temperature superconductor (4).

Although these years have not been marked by very active practical applications of molecular metals, several unique devices have been conceived and developed, and continued efforts have been devoted to make these new materials technologically recognized. Especially, iodine charge-transfer complexes, which are examples of the typical organic semiconductors explored by Professor Akamatsu's school in the past (5), have resurged as cathode materials in a lithium-iodine cell, and are having significant economic impact in the area of cardiac pacemakers (6). Intercalation reactions in some layered compounds have also shown great promise for use in high energy batteries (7). These and other examples give us good reason to believe that there will be decent activity in technological applications in the next few years. Such activity generally requires some background. First, materials engineering sciences will have to flourish. In this respect, solid state chemistry of molecular solids is receiving increased attention, and a growing number of experiments on the

471

understanding and control of physical properties with chemical
methods are being accumulated (8,9). Secondly, the thermal stability
of these weakly bound crystals needs to be characterized and improved
(8), because it has partially limited the reliability and/or performance of novel devices. Although this appears to be a matter of
technology, applications of new materials often require results of
fundamental endeavors. The third and the most important requirement
is that unique properties of newly prepared materials or their by-
products be suitably linked with social needs. There are a number
of areas that are open to new materials; such as energy saving and
conversion, medical electronics, sensing, control, and display, for
example.

In the following I shall try to outline the present status of
technological applications of molecular, metallic or semiconductive
solids. In particular, as contrasted to a recent review on the
same subject (10) which was concerned with topics and examples of
applications, this presentation will focus on various physical and
chemical phenomena having the possibility of practical use; those
being ELECTRONIC PHENOMENA, PHASE TRANSITIONS and NONSTOICHIOMETRY.

ELECTRONIC PHENOMENA

A. Polymeric Metals

One of the most interesting questions concerning the potential
application of molecular solids is whether they can be substituted
for real metals, since lighter substances in weight that have good
mechanical strength and high electrical conductivity have long been
wanted. Although some of the mechanical properties of metals have
been replaced by polymeric materials, their electrical, magnetic
and optical attributes have only partly been assumed by molecular
solids. Recent trends of research on highly conducting polymeric
materials such as polysulfur nitride, $(SN)_x$, and polyacetylene,
$(CH)_x$, are in line with the above mentioned expectation. Shirakawa
et al. (11a) and Chiang et al. (11b,c) have found that polycrystal-
line films of $(CH)_x$ can be rendered metallic (as high as $10^3 \ \Omega^{-1} \ cm^{-1}$
when doped with halogens, arsenic pentafluoride or Na, with the
metal-insulator (MI) transition occurring at dopant concentrations
near 1%. Of equal interest is the discovery of the metallic conductivity of polymer-carbon black compounds. Jachym et al. (12) have
synthesized polyester resin-acetylene black compounds from the
liquid polymer applied with carbon black at temperatures between 90
and 150°C. The metallic conductivity $(dR/dT > 0)$ was obtained at a
sudden MI transition with a very insignificant amount of carbon
black (about 0.35 parts by weight per 100 parts of polymer resin).
The presence of positive carriers has been indicated as a conduction mechanism, and a free motion of carriers along polymer macromolecules connected by means of the carbon black particles has

been postulated. Since such polymerized carbon blacks or carbon-
polymer composites have been widely used as heating elements substituting for those based on nickel-chromium alloys (13) or as
flexible or elastomeric conductors, both their chemistry and conduction mechanisms (14) should be investigated. Recently, Perlstein
et al. (15) have grown fine filamentary structures of

$$(TCNQ)^{\overline{\cdot}} \qquad (\sigma_{RT}= 10 \sim 30 \ \Omega^{-1} \ cm^{-1})$$

within an insulating polymer matrix. In spite of a very low loading
of the TCNQ salt (5% by weight), the polymer film remains highly
conductive (1 Ω^{-1} cm^{-1}) and the conductivity is virtually flat near
300°K. This composite material is of special interest because of
its capability to provide unique polymeric metals or semiconductors
of completely one-dimensional nature. For such purposes, there has
been relatively less effort in the study of the interactions between
polymer and organic fillers.

B. Thin Metallic Films

The advantage of molecular solids may be the ease of deposit-
ing thin films of them on substrates; the techniques include solution deposition with binders (15,16), vacuum deposition at lower
temperatures (17), and photochemical (18) and electrochemical (19)
depositions. Engler et al. (20) have prepared nearly transparent,
antistatic coatings of TTF-TCNQ and TTFBr on copier tonar sensor
windows. Halogen complexes of TTF and its analogues have been
synthesized with various processes (21), among which the photochemical process has been applied to a method for printing highly
conductive images (22). The method is characterized by the photo-
conversion of TTF dissolved in a solution or liquid of halocarbons
to TTFX$_n$ according to the following scheme:

$$(TTF)(CCl_4) \xrightarrow{h\nu} (TTF\cdot+)(CCl_4\cdot-) \longrightarrow (TTF)Cl_{0.77} + CCl_3. \quad \text{This}$$

invention provides a technique for the simple and rapid formation
of metallic images with high resolution (better than 20 lines/mm)
and will develop a unique field of imaging or printing technology
(23).

The very large optical anisotropy of epitaxially grown films
of low-dimensional metals may be combined with their high conductivity, to lead to a polarization-sensitive pyroelectric detector,
as has been suggested by Flygare et al. (24). By the way, the spectral selectivity in the metallic reflection finds important uses in
solar energy convension. Some metallic oxides have been employed
as heat mirrors of solar energy absorbers and as electrodes of
solar cells, and this need has opened up a new trend of materials
engineering of inorganic oxides (25). In order for molecular

metals to find some applications in this area, it would be desirable
to demonstrate the possibility of synthesizing transparent films
of metallic polymers, and there needs to be great progress made in
the improvement of their environmental ruggedness.

C. Electrode Materials

The metallic nature of molecular solids permits them to be
utilized as electrode materials. $(SN)_x$ has recently been found by
Nowak et al. (26) to have good electrode characteristics in liquid
electrolytes. The electrode is featured by its strong surface interactions with ions and by its amenability to modification for the
production of surface electrocatalysts (27). Semiconducting salts
of TCNQ have already been characterized as electrode materials (28),
and Sharp (29) has prepared solid-state ion-selective electrodes
with AgTCNQ, PbTCNQ, KTCNQ or CdTCNQ for the detection of cations.

The counterelectrode of solid electrolytic capacitors, manga-
nese dioxide, may be replaced by organic metals, such as quinolinium-
$(TCNQ)_2$, $Q(TCNQ)_2$, N-methylphenazinium(TCNQ), or N-methylacridinium-
$(TCNQ)_2$ (16). A schematic diagram of the capacitor is shown in Fig.
1 along with that of a lithium-iodine cell based on organic conductors (vide infra). The use of metallic conductors in this device
has lead to a considerable improvement in a dissipation factor or
impedance at low temperatures and at high frequencies as demonstrated
in Fig. 2 (30). Among other values added, it has been found that the
organic conductors behave like solid electrolytes formed in situ
which anodically oxidizes aluminum in the presence of water (31).
This new type of capacitor may advantageously be used in hybrid circuits, and very active patent application suggests further development in this field.

As mentioned in the INTRODUCTION, a solid-state lithium-iodine
cell (32) constructed with a poly(vinylpyridine)-I_2 cathode (33)
(See Fig. 1b) has already been put into commercial use. It has
offered great advances in the performance of cardiac pacemakers (6)
as a long-life power source of high energy density (about 120 WH/kg),
and is quite compatible with liquid crystal display devices (33b,
34). The iodine-complex cathode serves as an iodine source for the
cell reaction: $2Li + I_2 \rightarrow 2LiI$; where the reaction product functions
as a solid electrolyte. Many other iodine charge transfer complexes
have been investigated since the pioneering work by Gutmànn, Hermann
and Rembaum (35), and monomeric complexes, such as benzidine-,
phenothiazine-, and N-methylphenothiazine-I_2 (36), have also been
characterized as being excellent cathodes. The electrical conductivity of the cathode layer should be high on account of the internal
resistance of the cell, and the metallic nature of conductivity is
advantageous in those cases where the cell is supposed to be used at
lower temperatures. Some of the molecular metals such as TTF-halogens (37), TTT_2-I_3 (38), phthalocyanines-I_n (39), $(SNX_n)_x$ (40), or

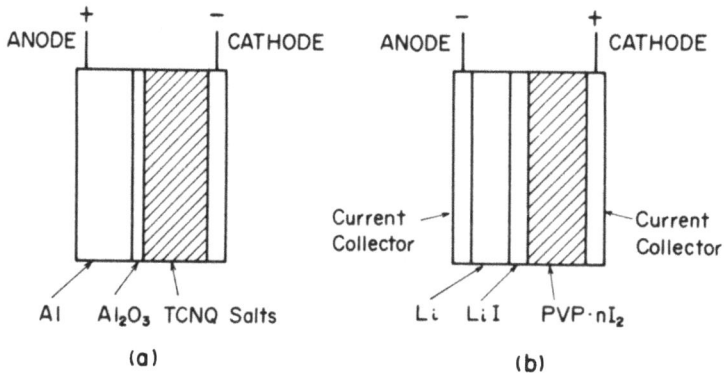

Fig. 1 Schematic diagrams of a solid electrolytic aluminum capacitor (a) and a lithium-iodine cell (b).

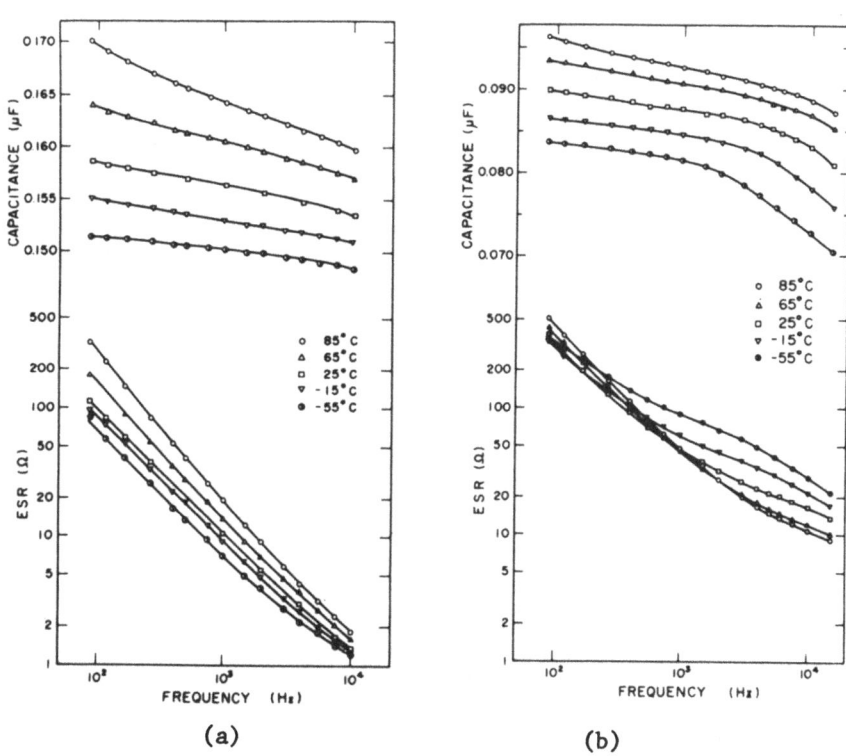

Fig. 2 Capacitance and equivalent series resistance, ESR, as a function of frequency for solid electrolytic capacitors having $Q(TCNQ)_2$ (a) and MnO_2 (b) as a counterelectrode. (After Ref. 30)

$(CHX_n)_x$ (11) may be shown to be still better candidates.

D.   Thermoelectric Phenomena

A thermoelectric material should have as high an electrical conductivity as possible since the figure of merit, Z, for the thermoelectric conversion is expressed by $\sigma S^2/\kappa$, where S is the Seebeck coefficient and κ is the thermal conductivity. Conventional semimetallic materials such as $(Bi, Sb)_2 Te_3$, $Bi_2(Te, Se)_3$, have limited values of Z (less than 3×10^{-3} K^{-1}) and added shortcomings of heavy weight and high cost. Since all the quantities appearing in Z are a function of free carriers and 'theoretically' the Z of conventional semimetals has its maximum in a specific region of carrier density (41), the development of new materials with a greater Z from semimetals is unlikely. It has been shown, however, that organic molecular conductors are featured by an extremely low κ (of the order of 10^{-3} W/cm°K) (42) as compared with that of typical thermoelectric materials (1.5×10^{-2} W/cm°K for $(Bi,Sb)_2 Te_3$, for example), resulting in a smaller Wiedemann-Franz coefficient. Petrov et al. (42a) have explained that the low κ is due to the low lattice thermal conductivity which results from weak intermolecular (translational) vibrations. Thus in the molecular metals, σ can be enhanced with no essential change in κ (43). Miles, Wilson and Cohen (44) have reported in the patent literature that TTF-TCNQ has a Z value of $(0.2 \sim 1.6) \times 10^{-4}$°$K^{-1}$ setting $\sigma = (1 \sim 5) \times 10^2$ Ω^{-1} cm^{-1}, S = $- (20 \sim 26)\mu V/$°K and $\kappa = 2 \times 10^{-3}$W/cm°K, and revealed a possibility of Peltier cooling using a $TTF-TCNQ/Bi_2Te_3$ junction. In order that the thermoelectric refrigerator might be competitive with a vapor-compression system, materials having a greater Z (hopefully greater than 10^{-2}°K^{-1}) need to be developed. Deficiencies of molecular metals for such purposes are 1) the relatively low Seebeck coefficient ($20 \sim 60$ V/°K) (45), and 2) the lack of well-defined p-n junctions. The latter problem may be solved rather easily as hole-conducting metals have recently been found (11c,12,46) and various attempts have been made to prepare molecular junctions. As an example, Fig. 3 shows a current-voltage curve for a p-n junction composed of $(CHNa_n)_x/(CH(AsF_5)_n)_x$ which has been demonstrated most recently by Chiang et al. (11c). They have suggested, by the way, that the p-n junction may be utilized as a solar battery on account of its suitable band gap (1.6eV) for the solar spectrum. Consequently, search for new thermoelectric materials is of both academic and practical interest, because it will not only stimulate materials engineering sciences which deal with the most important physical quantities of solids, but also the research may produce light, flexible and low-cost refrigerators with a high coefficient of performance, which may find a variety of technological applications.

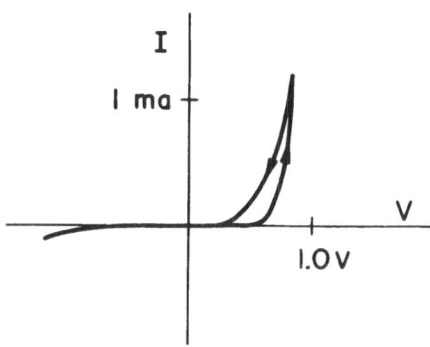

Fig. 3. I-V curve for a doped polyacetylene p-n junction. (Ref. 11c).

E.   Nonlinear Transport Phenomena

The one-dimensional metals TTF-TCNQ, TSeF-TCNQ, TTFI$_2$, and
(TTF)$_7$I$_5$ show strongly nonohmic behavior with applied fields of
a few V/cm in the semiconductor phase (47). Cohen and Heeger (48)
have explained the behavior in terms of one-dimensional nature of
the electronic state, claiming that the nonlinear transport is
caused by field-induced excitation of pinned charge-density waves
(CDW). Ong et al. have also found anomalous transport properties
in the linear-chain metal NbSe$_3$ (49). NbSe$_3$ exhibits not only
nonohmic conductivity below 145°K, but also the resistivity anomalies at 145 and 59°K may be suppressed by an applied field as low
as 0.1V/cm (breakdown effect). While the nonohmic behavior has
been accounted for by the formation of CDW state with an energy gap
(BCS gap) of the order of several 10mV, the striking breakdown
effect remains to be elucidated. The nonlinear phenomena may permit the materials to be used as variable resistor (varistor) elements (50) based on bulk conductivity in contrast to conventional
ceramic varistors which are based on a grain boundary effect. The
interaction between CDW and electromagnetic fields is of particular
interest because here there are promising systems for applications,
such as superconductor or surface accoustic wave devices, logical
switches, memory devices, delay lines, or amplifiers. Of course,
much development work must be done to furnish technological significance to the novel elementary excitation.

It is known that some organic thin films undergo reversible
bistable switching between insulating and highly conducting states
with applied voltages. Several different mechanisms for the
memory switching have been proposed; classical electrical breakdown,

diffusion of electrode constituents with formation of filaments of
a metallic nature, phase changes caused by Joule heating, and field-
assisted reduction of space charges (51). Even though no evidence
for an Ovshinsky type mechanism, (the phase change from amorphous
state to a semicrystalline one) has been reported, work in this
area to yield new switching devices needs to be continued further.

PHASE TRANSITIONS

Molecular metals exhibit various phase transitions and extensive work has been done on the interpretation and control of MI
transitions. MI transitions which are accompanied by a sudden
change in conductivity may be employed as temperature switches provided that the transition temperature T_c lies in a suitable range
for use. Two typical phase transitions which have been utilized
for a number of purposes are known; one as a critical temperature
resistor (CTR) (52) and the other as a positive temperature coefficient (PTC) resistor (53) (See Fig. 4). The former device utilizes
the MI transition of vanadium oxides found by Morin in 1959 (54),
and facilitates temperature regulation at ca. 60°C. The PTC characteristics, which have been exhibited typically by barium titanates
(53), have recently attracted much attention as very simple, self-
regulating heating elements. It should be noted in this connection
that there is a growing market for a flat heating element using
polymer-carbon black composites. These display remarkable PTC
phenomena which results from carbon particles which are in direct
contact at the phase transition temperature of the polymer (55).

Molecular solids which undergo phase transitions accompanied
by a sharp conductivity change are summarized in Table 1; substances
with either CTR or PTC characteristics are included. One of the
interesting features of molecular solids is that their physical
properties, in particular the phase transition temperature, can be
modified by the introduction of disorders or by alloying. The
effect of disorders on T_c is not simple in the case of TTF-TCNQ,
as T_c is shifted to lower temperature with methyl-substituted TCNQ
(56), but to higher temperatures with a mixture of diselenium analogs
of TTF (57). With organic alloys, on the other hand, T_c tends to
vary continuously with the content of either component as demonstrated by $(MePh_3P)_{1-x}(MePh_3As)_x(TCNQ)_2$ (58) and $(TTF)_{1-x}(TSeF)_x$-
(TCNQ) (8b). Attempts have been made in our laboratory to control
the MI transition of N-n-propylpyridinium $(TCNQ)_n$, $(NPPy)(TCNQ)_n$,
because this material is potentially useful as a thermistor element
having CTR characteristics suitable for protection of overheating
(See Fig. 5) (59). With alloying or doping with related materials
such as N-alkylsubstituted pyridinium and thiazolium salts of TCNQ,
two groups of new alloys have been prepared. Figures 6 and 7 show
temperature dependence of resistivity, ρ-T, for these two sets of
alloys. The first group includes materials with parallel shift of

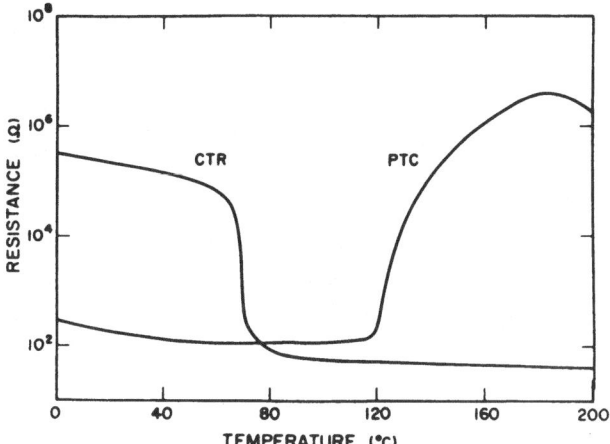

Fig. 4 Temperature dependence of resistance for two typical
thermally sensitive switches; CTR based on vanadium oxides
(Ref. 52) and PTC resistor based on barium titanates
(Ref. 53).

Table 1 Data on phase transitions with a sudden change in the
electrical conductivity.

Compound	Tc (°K)	ΔT(°K)	Ref.
TSeF–TCNQ	28		9
TTF–TCNQ	38, 53	1	9
TTF–TNAP	185		a
TTF–$I_{0.71}$	286	50~100	37c
$(TTF)_{12}(SCN)_7$	169		64
$(TTF)_{12}(SeCN)_7$	173		64
$(MePh_3P)(TCNQ)_2$	316	5	58
N–n–butylthiazolium$(TCNQ)_n$	371*	20	10
N–n–propylpyridinium$(TCNQ)_n$	381	11	10, 59
N–n–propylthiazolium$(TCNQ)_n$	405	65	10
N–n–butylpyridinium$(TCNQ)_n$	407*	23	10, 59

* PTC characteristics, ΔT = temperature hysteresis.
(a) P.A. Berger, D.J. Dahm, G.R. Johnson, M.G. Miles, and
J.D. Wilson, Phys. Rev. B12, 4085 (1975).

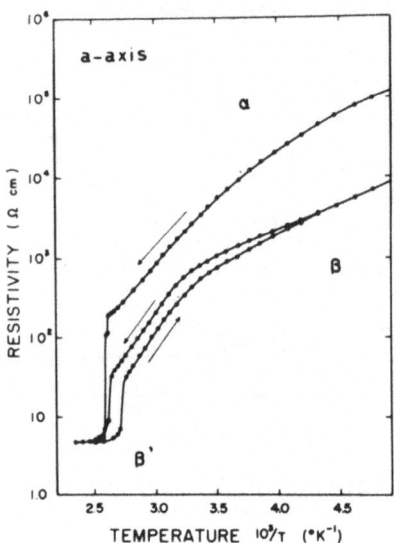

Fig. 5   Temperature dependence of the electrical resistivity of
         (NPPy)(TCNQ)$_n$ single crystal.

T_c between 60 and 108°C, while the other has a great temperature
hysteresis in the ρ-T curve.   These and other related organic solids
can easily be put into a thermistor element, molded into a variety
of forms and manufactured at low cost.   Although devices utilizing
these materials would have several limitations in, say, lifetime,
environmental requirements or measuring range, they will probably
find their own area of use.   Since the degree of conductivity change
at T_c should be as large as possible, molecular design to increase
both the resistivity of the insulating phase and the conductivity
of the metallic phase is needed, comparable to a successful improvement of CTR characteristics of VO_2 (60).

   Temperature dependent electrical conductivity is known to lead
to striking nonohmic current-voltage characteristics.   Switching
and negative resistance of CTR thermistors have found various circuit applications where conventional semiconductor devices are
unusable (61).   Presentation of this kind of phenomena by molecular
metals is not impossible; Kahlert (47a) has observed a switching
current in TTF-TCNQ single crystals with applied fields between 340
and 660 V/cm.   Although the conventional thermistor has a very high
sensitivity to temperature, the thermistor characteristics,
expressed as $R = R_o exp(B/kT)$ where B is a constant, are not always
convenient for general temperature-sensing circuits.   There have
been needs for other temperature dependences of resistance, such as
$R \propto T$ with a large coefficient, log $R \propto T$, or for high, set values
for the constant B.   Molecular metals may have some possibilities

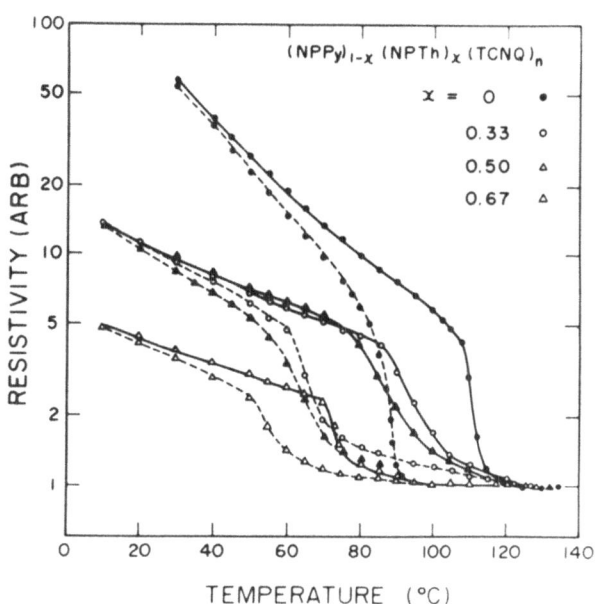

Fig. 6 $\rho - T$ curves for $(NPPy)(TCNQ)_n$ doped with $(NPTh)(TCNQ)_n$.

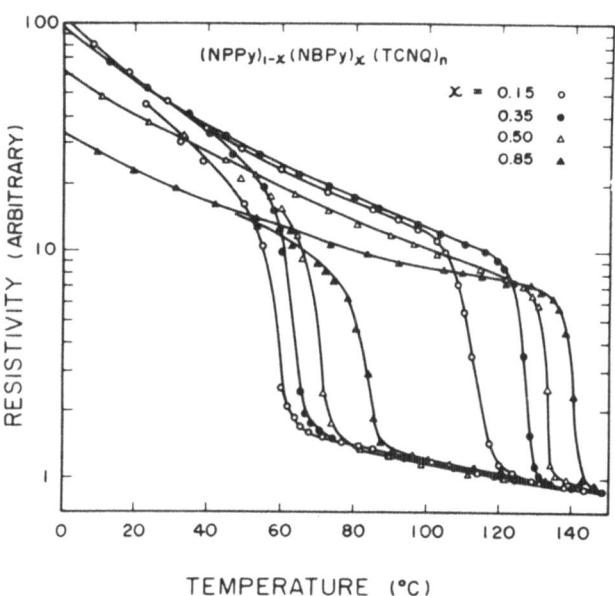

Fig. 7 $\rho - T$ curves for $(NPPy)(TCNQ)_n$ doped with $(NBPy)(TCNQ)_n$.

in presenting unique ρ-T characteristics.

The striking hysteresis behavior shown by the $(NPPy)(TCNQ)_n$ alloys may be another exploitable aspect of phase transitions in molecular solids as has been observed in $TTFI_n$ (37b,c) and K-chloranil (62). Figure 8 shows the DSC traces for $(NPPy)_{0.8}$-$(NBPy)_{0.2}(TCNQ)_n$ powders, which clearly shows the existence of first order phase transitions expressed as

$$\beta \xrightarrow[\underset{57.2°C}{\longleftarrow}]{111.6°C} \beta' \xleftarrow{154.8°C} \alpha (\text{metastable}).$$

We would suggest that the unique feature of the ρ-T curves (Fig. 7) fulfils some demands of safety in temperature controlling systems. We have also been interested in utilizing the switching phenomenon coupled with memory or storage effect to new functional devices. In particular, since the $\beta' \longrightarrow \beta$ transition occurred isothermally, time is an important variable in the transition kinetics, and a very sluggish change in resistivity is important in long-term memory and/or delay devices.

NONSTOICHIOMETRY

A. Appearance of Nonstoichiometry

A number of nonstoichiometric molecular complexes have been reported. Phenomenologically, the appearance of nonstoichiometry is due to the fact that the molecular complexes are composed of

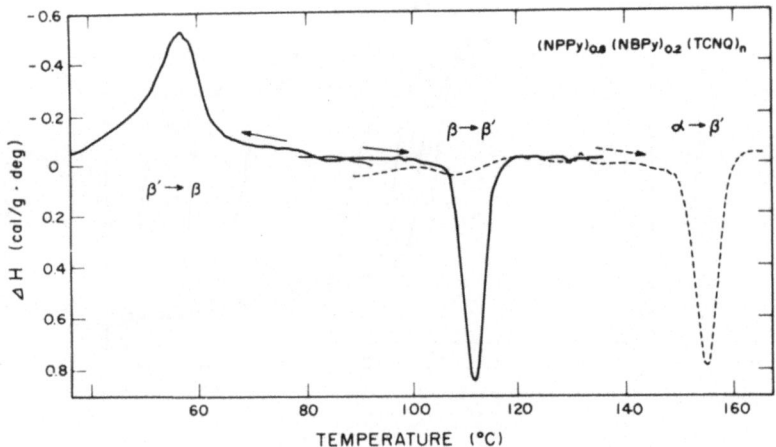

Fig. 8. Differential scanning calorimetry traces for $(NPPy)_{0.8}(NBPy)_{0.2}(TCNQ)_n$ powders.

mixed-valency compounds with bulky molecular structures, and to disordered crystal structures or the coexistence of multiple phases (63). Partial charge transfer between donors and acceptors has been the major origin of new nonstoichiometric compounds like TTF-halogens (37), TTF-SCN (64) or TTF-halogens (38). We have found from studies on the crystallization processes of substituted quinolinium- and pyridinium-TCNQ salts (59, 65) that the deviation of the cation-to-TCNQ ratio from 1:1, 2:3 or 1:2 stoichiometries is large for those salts that have polymorphs, and that most of the nonstoichiometric salts crystallize in metastable modifications. Although great differences in electrical properties between various nonstoichiometries in the metastable state have been noted, the nonstoichiometry is not likely to have an essential influence on those in the stable state. For example, $(NPPy)(TCNQ)_n$ crystallizes in two metastable forms with n = 1.9 and 2.5, but on heating both crystals are irreversibly transformed into stable phases, which have nearly the same ρ-T characteristics (59). The same situation arises in the case of N-methylquinolinium $(TCNQ)_n$, $(NMQ)(TCNQ)_n$ (65b). Control of stoichiometry and research on its relation to physical properties may be useful for elucidating the role of both disorder and valency in conduction mechanisms. A higher or lower valency state may be promoted by the introduction of fixed valent ions a technique which serves as another approach to the control of physical properties of molecular solids. Examples of recent research in this area include $(SN)_x$-halogens (46), $(CH)_x$-halogens and -Na (11), (trimethyl ammonium)$(TCNQ)I_{1/3}$ (66), and related three component compounds (67).

B. Inclusion Compounds

Molecular solids which have highly anisotropic crystal structures are open to occlusion of foreign atoms or molecules, thereby leading to new nonstoichiometric molecular solids, that is, to inclusion or intercalation compounds (63). The inclusion compounds and/or reactions potentially have promise for various applications which are based on remarkable differences in physical and chemical properties between the host and the compound.

Layered compounds of metal chalcogenides undergo topotactic intercalation reactions with lithium, and these have offered great promise for a new class of secondary batteries having high energy densities (7). Wittingham et al. (68) have reported on a cell, Li│ n-butyl lithium in propylene carbonate │ TiS_2, the discharge of which is based on a reaction $xLi + TiS_2 \rightarrow Li_x TiS_2$, proceeded by the intercalation of lithium into the lattice of TiS_2. The energy density was found to be 480 WH/kg, which is comparable to that of a Na/S high temperature system and much higher than that of the lithium-iodine cell (see above). The feasibility of using this high-rate, ambient-temperature secondary battery for electric

vehicle propulsion has been suggested (68). Electrochemical characterization of a number of related compounds have been carried out
and at the same time new types of intercalation compounds based
on FeOCl (69), MI_2 (M=Pb, Bi) (70), or MCl_4 (M=Cu, Mn, Cd) (71)
have been explored recently. These studies contribute to the further
development of lithium batteries.

Materials which can form inclusion compounds are essentially
absorbers which have some relevance for the storage of gases and
vapors. Also the changes in the free energy on going through the
inclusion reaction can be utilized for heat storage or heat pump
systems. Especially, in the area of absorption refrigerators, a
new combination of refrigerant and absorber needs to be developed
which could be employed to replace ammonia-water or LiBr-water combinations (72). In the course of fundamental studies, many workers
may have recognized both the inclusion of foreign molecules in
molecular solids and anomalous physical properties. Molecules of
recrystallization solvent such as acetonitrile are included into a
number of TCNQ complexes, e.g., benzidine-TCNQ (73), some substituted benzimidazolium $(TCNQ)_2$ (74), $NMQ(TCNQ)_x$ (65, 74b, 75) or
$[Pd(CNMe)_4](TCNQ)_4$ (76), and water is also included in inorganic
square planar compounds(1) and $[1,2-Di(N-ethyl-4-pyridinium)-$
ethylene$](TCNQ)_4$ (77). It has been found in most cases that the
guest molecules are located in the hollow spaces surrounded by the
1D chain of the host lattice, forming a structure like a channel
inclusion compound. Reversible inclusion with notable changes in
the electrical and thermal properties have been found in benzidine-
TCNQ (73), $NMQ(TCNQ)_x$ (65) and KCP (78). We have demonstrated with
thermal analysis on $(NMQ)(TCNQ)_x$ that acetonitrile is bound to the
host lattice with a force ($\Delta H = 19.1$ kcal/mol) higher than that in
the liquid state (7.13 kcal/mol) (65a). This result suggests that
a low-temperature heat storage system may be constructed using the
inclusion compounds of molecular solids. Inclusion reactions are
rather specific with respect to the included substances. This property may be combined with the remarkable conductivity change, and
provide a new method for selective detection of gases and vapors.
In this sense, the molecular design is fascinating and promising.

CONCLUSIONS

We have concerned with some physical and chemical phenomena of
highly conducting molecular solids which have been of considerable
interest in both the study of fundamental features and the development of devices. To sum up the potential utility, the author would
identify the following areas as being especially significant for
future advances in this field:
1) Search for new polymeric metals to form mechanically and
thermally stable films;
2) Development of innovative techniques to prepare thin films;

3) Electrochemical characterization of molecular metals and development of their own field of use;

4) Design of molecular (semi)metals having higher Seebeck coefficient suitable for the thermoelectric conversion and preparation of the p-n junctions;

5) Studies on the interactions of CDW with electromagnetic fields to assess their applicability to microelectronics;

6) Solid-state-chemical modification of phase transitions, so that the variations in physical quantities can be rendered usable; and

7) Application of the inclusion reactions with ions, gases and vapors, and preparation of new multicomponent complexes to bring forth unique properties.

As a final remark, it seems that the research trends in this field are going to assume more practical color, and it would not be too optimistic to believe that the power and versatility of organic and solid-state chemistry will bring about the next great stimulus to the academic sciences of molecular metals.

ACKNOWLEDGEMENTS

The author wishes to thank many stimulating and helpful discussions with his colleague, M. Murakami. He also wishes to acknowledge valuable information and preprints transferred to him by N. P. Ong, F. Wudl, R. B. Somoano, A. J. Epstein, M. G. Miles, E. M. Engler, M. Fütöss-Wéhner, B. Jachym, H. Kahlert, J. H. Perlstein, and A. J. Heeger.

REFERENCES

1. J. S. Miller and A. J. Epstein, Prog. Inorg. Chem. 20, 1 (1976); J. S. Miller, Annal N.Y.A.S., 313, 25 (1978).
2. H. Inokuchi, Bull. Chem. Soc. Jpn. 27, 22 (1954).
3. F. Gutmann and L. E. Lyons, Organic Semiconductors (John Wiley and Sons Inc., N.Y., 1967) p.632.
4. L. B. Coleman, M. J. Cohen, D. J. Sandman, F. G. Yamagishi, A. F. Garito, and A. J. Heeger, Solid State Commun. 12, 1125 (1973).
5. H. Akamatsu, H. Inokuchi, and Y. Matsunaga, Nature 173, 168 (1954).
6. A. M. Hermann and E. Luksha, J. Cardiovasc. Pulm. Tech. 6, 15 (1978).
7. D. W. Murphy and F. A. Trumbore, J. Cryst. Growth 39, 185 (1977).
8. J. H. Perlstein, Angew. Chem. 16, 519 (1977).
9. E. M. Engler, V. V. Patel, J. R. Andersen, Y. Tomkiewiez, R. A. Craver, B. A. Scott, and S. Etemad, to be published.
10. S. Yoshimura and M. Murakami, Annal N.Y.A.S., 313, 269 (1978).
11. (a) H. Shirakawa, E. J. Louis, A. G. MacDiarmid, C. K. Chiang,

and A. J. Heeger, J.C.S. Chem. Comm., 579 (1977); (b)
C. K. Chiang, C. R. Fincher, Jr., Y. W. Park, A. J. Heeger,
H. Shirakawa, E. J. Louis, S. C. Gau, and A. G. MacDiarmid,
Phys. Rev. Lett. 39, 1098 (1977); (c) C. K. Chiang, S. C. Gau,
C. R. Fincher, Jr., Y. W. Park, A. G. MacDiarmid, and
A. J. Heeger, Appl. Phys. Lett., 33, 18 (1978).

12. B. Jachym, H. Sodolski, T. Słupkowski, and R. Zieliński, phys.
stat. sol. (a) 24, K159 (1974); ibid 34, 657 (1976).

13. K. Ohkita, Enbi to Polymer 17, 21 (1977).

14. P. Sheng, E. K. Sichel, and J. I. Gittleman, Phys. Rev. Lett.
40, 1197 (1978).

15. J. H. Perlstein, J. A. VanAllan, L. C. Isett, and G. A. Reynolds,
Annal. N.Y.A.S., 313, 61 (1978).

16. S. Yoshimura, Y. Itoh, M. Yasuda, M. Murakami, S. Takahashi, and
K. Hasegawa, IEEE Trans. Parts Hybrids Packg. PHP-11, 315 (1975).

17. (a) S. Yoshimura, M. Murakami, Y. Itoh, and K. Hasegawa, Chem.
Lett., 835 (1972); (b) P. Chaudhari, B. A. Scott, R. B. Laibowitz,
Y. Tomkiewicz, and J. B. Torrance, Appl. Phys. Lett. 24, 439
(1974); (c) A. A. Bright, M. J. Cohen, A. F. Garito, and
A. J. Heeger, Appl. Phys. Lett. 26, 612 (1975); (d)
E. E. Simonyi, J. F. Graczyk, and J. B. Torrance, IBM J. Res.
Develop. 22, 315 (1978).

18. B. A. Scott, F. B. Kaufman, and E. M. Engler, J. Amer. Chem. Soc.
98, 4342 (1976).

19. (a) F. B. Kaufman, E. M. Engler, D. C. Green, and J. Q. Chambers,
J. Amer. Chem. Soc. 98, 1596 (1976); (b) J. S. Miller and
D. Cobb, Science 194, 189 (1976).

20. E. M. Engler, V. Y. Merritt, B. A. Scott, and R. H. Skovlin, IBM
Tech. Discl. Bull. 18, 4177 (1976).

21. B. A. Scott, S. J. LaPlaca, J. B. Torrance, B. D. Silverman, and
B. Weber, Annal N.Y.A.S., 313, 369 (1978).

22. E. M. Engler, F. B. Kaufman, and B. A. Scott, U.S. Patent
4,036,648 (1977).

23. J. C. McGroddy and B. A. Scott, U.S. Patent 4,052,272 (1977).

24. W. H. Flygare, G. D. Stucky, and H. Ehrenilich, Report on Mat.
Res. Council Conf. One-and Two-Dimensional Conductors. La Jolla,
Ca., July 10 - 11, 1975.

25. J. B. Goodenough, J. Solid State Chem. 12, 148 (1975).

26. R. J. Nowak, H. B. Mark, Jr., A. G. MacDiarmid, and D. Weber,
J.C.S. Chem. Comm., 9 (1977).

27. A. N. Voulgaropoulos, R. J. Nowak, W. Kutner, and H. B. Mark, Jr.,
J.C.S. Chem. Comm., 244 (1978).

28. P. Weidenthaler and E. Pelinka, Collect. Czech, Chem. Commun.
34, 1482 (1969).

29. M. Sharp and G. Johnsson, Anal. Chim. Acta. 54, 13 (1971);
M. Sharp, ibid. 85, 17 (1976).

30. Y. Itoh and S. Yoshimura, J. Electrochem. Soc. 124, 1128 (1977).

31. S. Yoshimura and M. Murakami, Bull. Chem. Soc. Jpn. 50, 3153
(1977).

32. W. Greatbatch, J. H. Lee, W. Mathias, M. Eldridge, J. M. Moser,
and A. A. Schneider, IEEE Trans. Bio-Med. Eng. EME-18, 317 (1971).

33. (a) J. R. Moser, U.S. Patent 3,660,163 (1972); (b)
    A. A. Schneider, W. Greatbatch, and R. Mead, Proc. Int. Power
    Sources Conf., 651 (1974).
34. T. Wada, Japan. Kokai 49-56132 (1974).
35. F. Gutmann, A. M. Hermann, and A. Rembaum, J. Electrochem. Soc.
    114, 323 (1967); ibid. 115, 359 (1968); Japan. Patent 47-30769
    (1972).
36. B. Scrosati and M. Torroni, Electrochimica Acta 18, 225 (1973);
    M. Pampallona, A. Ricci, B. Scrosati, and C. A. Vincent,
    J. Appl. Electrochem. 6, 269 (1976).
37. (a) S. J. LaPlaca, P. W. Corfield, R. Thomas, and B. A. Scott,
    Solid State Commun. 17, 635 (1975); (b) R. J. Warmack, and
    T. A. Callcott, Phys. Rev. B12, 3336 (1975); (c) R. B. Somoano,
    A. Gupta, V. Hadek, T. Datta, M. Jones, R. Deck, and
    A. M. Hermann, J. Chem. Phys. 63, 4970 (1975).
38. (a) C. C. Isett and E. A. Perez-Albuerne, Solid State Commun.
    21, 433 (1977); (b) R. B. Somoano, S.P.S. Yen, V. Hedek,
    S. K. Khanna, M. Movotry, T. Datta, A. M. Hermann, and
    J. A. Woolam, Phys. Rev. B17, 2853 (1978).
39. J. L. Petersen, C. S. Schramm, D. R. Stojakovic, B. M. Hoffman,
    and T. J. Marks, J. Amer. Chem. Soc. 99, 286 (1977).
40. M. Akhtar, J. Kleppinger, A. G. MacDiarmid, J. Milliken,
    M. J. Moran, C. K. Chiang, M. J. Cohen, A. J. Heeger, and
    D. L. Peebles, J.C.S. Chem. Comm., 473 (1977).
41. R. R. Heikes and R. W. Ure, Jr., Thermoelectricity: Science
    and Engineering (Interscience Publishers, N.Y., 1967).
42. (a) A. V. Petrov, Yu. G. Nurullaev, R. M. Vlasova, V. V. Kuzina,
    and A. I. Sherle, Soviet Phys.-Solid State 14, 1594 (1972);
    (b) M. B. Salamon, J. W. Bray, G. DePasquali, and R. A. Craven,
    Phys. Rev. B11, 619 (1975).
43. B. Jachym, A. Szumilo, H. Sodolski, and R. Zielinski, Acta Phys.
    Pol. A50, 125 (1976).
44. M. G. Miles, J. D. Wilson, and M. H. Cohen, U.S. Patent
    3,779,814 (1973); M. H. Cohen, M. G. Miles, and J. D. Wilson,
    U.S. Patent 4,026,905.
45. J. F. Kwak, G. Beni, and P. M. Chaikin, Phys. Rev. B13, 641
    (1976); G. R. Johnson, M. G. Miles, and J. D. Wilson, Mol. Cryst.
    Liq. Cryst. 33, 67 (1976); P. M. Chaikin, J. F. Kwak,
    R. L. Greene, S. Etemad, and E. M. Engler, Solid State Commun.
    19, 1201 (1976).
46. G. B. Street, S. Etemad, and R. D. Smith, Annal N.Y.A.S. 313, 737
    (1978).
47. (a) H. Kahlert, Solid State Commun. 17, 1161 (1975); (b)
    M. J. Cohen, P. R. Newman, and A. J. Heeger, Phys. Rev. Lett. 37,
    1500 (1976).
48. M. J. Cohen and A. J. Heeger, Phys. Rev. B16, 688 (1977).
49. P. Monçeau, N. P. Ong, A. M. Portis, A. Meerschaut, and
    J. Rouxel, Phys. Rev. Lett. 37, 602 (1976); N. P. Ong and
    P. Monceau, Phys. Rev. B16, 3443 (1977); N. P. Ong, to be
    published.
50. C. F. Wahlig, U.S. Patent 3,255,329 (1966); S. Ikeno,

H. Mikawa, and M. Yokoyama, Japan. Kokai 52-118,596 (1977).

51. M. Füstöss-Wegner, Thin Solid Films 36, 89 (1976); Proc. 6th
    Czech. Conf. Elect. Vac. Phys., Bratislava, Czech., Aug. 23 - 26
    (1976)· p301.
52. H. Futaki, Japan. J. Appl. Phys. 4, 28 (1965).
53. E. Andrich, Phy. Tech. Rev. 30, 170 (1969).
54. F. J. Morin, Phys. Rev. Lett. 3, 34 (1959).
55. I. Tsubata and N. Takahashi, 10th Biennial Conf. Carbon, CA148
    (1971).
56. A. N. Bloch, J. P. Ferraris, D. O. Cowan, and T. O. Poehler,
    Solid State Commun. 13, 753 (1973).
57. S. Etemad, T. Penney, E. M. Engler, B. A. Scott, and
    P. E. Seiden, Phys. Rev. Lett. 34, 741 (1975).
58. Y. Iida, J. Chem. Phys. 59, 1607 (1973).
59. M. Murakami and S. Yoshimura, Chem. Lett., 929 (1977).
60. H. Sasaki and A. Watanabe, J. Phys. Soc. Jpn. 19, 1748 (1964).
61. B. Fisher, J. Phys. C: Solid State Phys. 8, 2072 (1975).
62. J. J. Andre and G. Weill, Chem. Phys. Lett. 9, 27 (1971).
63. H. L. Mandelcorn ed. Non-stoichiometric Compounds (Academic
    Press, N.Y., 1964).
64. R. S. Somoano, A. Gupta, V. Hadek, M. Novotry, J. Jones, T. Datta,
    R. Deck, and A. M. Hermann, Phys. Rev. B15, 595 (1977).
65. (a) M. Murakami and S. Yoshimura, Bull. Chem. Soc. Jpn. 48, 157
    (1975); (b) J. Phys. Soc. Jpn. 38, 488 (1975).
66. (a) A. Cougrand, S. Flandrois, P. Delhaes, P. Dupuis,
    D. Chasseau, J. Galtier, and J. L. Miane, Mol. Cryst. Liq. Cryst.
    32, 165 (1976); (b) M. A. Abkowitz, A. J. Epstein,
    C. H. Griffiths, J. S. Miller, and M. L. Slade, J. Amer. Chem.
    Soc. 99, 5304 (1977).
67. P. Dupuis, S. Flandrois, P. Delhaes, and C. Coulon, J.C.S. Chem.
    Comm., 337 (1978).
68. M. S. Whittingham, Science 192, 1126 (1976).
69. F. Kanamaru, M. Shimada, M. Koizumi, M. Takano, and T. Takada,
    J. Solid State Chem. 7, 297 (1973).
70. V. M. Koshkin, A. P. Mil'ner, V. V. Kukol', Yu. R. Zabrodskii,
    Yu. N. Dinitriev, and F. I. Brintsev, Soviet Phys.-Solid State
    18, 354 (1976).
71. L. H. de Jongh and A. R. Miedema, Adv. Phys. 23, 1 (1974).
72. E. Farber, Workshop Proc. Sol. Cooling for Building, Los Angeles,
    Ca., Feb. 6 - 8 (1974) P. 109.
73. M. Ohmasa, M. Kinoshita, M. Sano, and H. Akamatsu, Bull. Chem.
    Soc. Jpn. 41, 1998 (1968).
74. (a) D. Chasseau, J. Gaultier, C. Hauw, and J. Jaud, C. R. Acad,
    Sci., Ser. C 276, 661 (1973); C. R. Acad. Sci., Ser. D 267, 751
    (1973); (b) S. Flandrois, P. Libert, and P. Dupuis, Phys. Stat.
    Sol (a) 28, 411 (1975).
75. G. Mihály, K. Holczer, G. Grüner, and L. D. Kunstelj, Solid
    State Commun. 19, 1091 (1976); Y. Matsunaga and T. Tanaka, Bull.
    Chem. Soc. Jpn. 49, 2713 (1976).
76. S. Z. Goldberg, R. Eisenberg, J. S. Miller, and A. J. Epstein,

J. Amer. Chem. Soc. 98, 5173 (1976).
77. G. J. Ashwell, phys. stat. sol. (b) 85, K7 (1978).
78. S. Drosdziok and M. Engbrodt, Solid State Commun. 17, 1339 (1975).

STUDY GROUP REPORTS

REEVALUATION AND EXTENSIONS OF EXISTING SYSTEMS

K. Bechgaard (Chairman), A. Bloch, P. Delhaes, L.
Interrante (Recording Secretary), D. Jerome, S. Kagoshima, D. MacDonald, J.S. Miller, and M. Weger

I. Introduction

In the intense development of molecular and polymeric conductors during the past five years, traditionally segregated areas of
synthetic chemistry and solid-state physics have been effectively
joined. Among the results are a rapidly expanding series of qualitatively new materials, substantial refinements in sample preparation and experimentation, new theoretical avenues, and important
changes in the interdisciplinary structure of materials research
itself.

When dramatic advances are made in so short a time, it is
practically inevitable that progress be uneven. In this report we
propose to critically review that progress, to identify areas of
current strength and weakness, and to recommend steps toward solidification of our understanding and systematic development of new
materials.

The appeal of these substances needs no elaboration here.
They are famous for their inherent electrical anisotropy, often so
extreme that they display in classic detail the special physics of
electrons in lower dimensions. Perhaps more important, they extend
quasi-metallic electrical and magnetic behavior to parts of the
periodic table where they had never before been observed. We
believe that the synthetic flexibility of these elements holds the
promise of a class of materials whose electrical and magnetic properties are unique, useful, and subject to systematic chemical control.

Development of such control, of course, will require the kind
of detailed understanding that can only come from continuous,

balanced interplay among synthetic chemistry, materials preparation, physical experiment, and theory. In this respect it is probably true that the field is still far from mature. Except in a
few important areas, the physics of these materials has not yet
developed a full identity of its own, but has usually progressed in
spurts of ad hoc attention to specific new materials as they have
appeared. Theory is rarely systematic and carries little predictive force from compound to compound. Chemical synthesis, for its
part, has been inspired largely by the promise of new physics
rather than new chemistry, and it may be in part for this reason
that established synthetic chemists have not entered the field in
numbers comparable to their counterparts in solid-state physics.
Meanwhile, the physical chemists in the area, who have a central
natural role to play in bridging the gaps of language and concept
which separate the physical and synthetic extremes, number only a
handful.

The committee is inclined to regard these shortcomings as the
temporary growing pains of a newly emergent field. Our optimism
is based in part on the inherent interdisciplinary character of the
area, and the kind of close, long-term cooperation that it has
generated among disciplines and among research institutions throughout the world. In both universities and industry, the field has
fostered the rapid development of unified interdisciplinary
research groups consisting of chemists, physicists, and engineers
who are learning to speak one another's language and fertilize one
another's thinking. The committee regards this phenomenon as a
genuine turning point in the development of the science, and
expects that its ultimate significance will extend far beyond the
importance of any particular class of materials.

II. Scope of the Report

Because the new physics and chemistry developed in the feild
bears implications for other systems, the boundaries of the discipline are necessarily diffuse. Hence the committee, limited by
time, space and its own expertise, has chosen arbitrarily to confine its attention to four prototypical groups of materials: metal
chain compounds, quasi one dimensional organic conductors, conducting polymers and the chemistry of intercalated compounds. Although
chemically disparate, the four groups share much common physics,
and are generally perceived as seminal to the development of the
field.

Our report is divided into five parts. Part III reviews the
chemical situation for each group of compounds, and Part IV considers the physics of the class as a whole. In Part V we summarize
our conclusions and suggestions.

III. Structural and Chemical Considerations

The materials covered in this review vary widely in structure
and chemical composition, however, there are some important features

which extend across the area as a whole and provide a common basis for discussion. The most obvious is the anisotropic character of the basic structures which lead to anisotropic electronic interactions of a quasi-one dimensional (1-D) or two dimensional (2-D) nature. This tendency toward anisotropic interactions follows directly from orbital overlap considerations. In those cases where molecules or ions form the conducting array the most effective overlap of partially occupied orbitals is achieved in the colinear stacking of planar molecules where the energetically accessible band forming orbitals are spatially extended out from the molecular plane. These can be either delocalized π-orbitals or localized d_z^2 orbitals on metal atoms in the unit. This stacking and the directional character of the interacting orbitals then leads to a strong directionality of the electronic conduction bands. Larger atoms usually lead to wider bands.

In the covalent polymer and graphite intercalate systems, in addition to the out-of-plane π-orbitals which constitute the conduction band system there is a strong covalent σ-bonded framework which determines the underlying dimensionality. The opportunity for significant interchain or interlayer orbital interactions also exists here and apparently manifests itself in cases like $(SN)_x$ and the Li graphite intercalate, perhaps due to the availability of π-orbitals of larger spatial extent or a host atom with the requisite bonding capacity.

Another major factor in determining effective orbital overlap is the relative configuration of the interacting molecules. In the case of planar molecular with extended π-orbital systems the most effective overlap will be in the completely eclipsed configuration; however, this is usually modified by the need to avoid excessive non-bonding repulsive or electrostatic interactions which are also maximized in this configuration. The result is usually a non-eclipsed compromise with less effective but still satisfactory overlap. Apparent exceptions are many of the TTF, TSeF, TTT and TSeT halides, where a completely eclipsed stacking configuration occurs, perhaps due to an unusually favorable π-orbital overlap situation in these cases. In this same context it is clear that any features of the molecular units which tend to interfere with optimum overlap (both in terms of the intermolecular separation and relative configuration) should be avoided. Thus bulky substituents or groups which project out of the molecular planes are, in general, undesirable in systems designed to achieve high electrical conductivity. These considerations of intrachain orbital overlap can, and often are, overridden in many systems by broader structural and crystal packing considerations in which cases the orbital interactions may be simply dictated by overall crystal energy requirements.

Another common feature of the highly conductive molecular systems is a uniformity in the basic molecular framework. Structural or electronic situations which would prevent uniform stacking should be avoided.

Finally to achieve metal-like conductivity behavior it is usually necessary to avoid commensurate electronic situations, i.e. one electron per site (or even 1 per 2 sites), thus, forming a partially occupied conduction band. As a result, the highly conducting substances are either non stoichiometric ($K_2Pt(CN)_4Br_{0.3} \cdot 3H_2O$, $TTFBr_{0.76}$, etc) or have nonintegral electron transfer (TTF·TCNQ).

A. Metal Chain Compounds

1. Quasi-metallic Conductor

Highly conducting 1-D inorganic substances are primarily based on Pt, Ir, and Hg chains. The prototype materials contain either tetracyanoplatinate (e.g., $K_2Pt(CN)_4Br_{0.3} \cdot 3H_2O(KCP)$, bis(oxalato)-platinate, halocarbonyliridate square planar units or covalent chains of Hg atoms (e.g. $Hg_{2.82}AsF_6$). The planar complexes exhibit d_{z^2} orbital overlaps and possess metal-metal separations less than 3A. The larger spatial extent of the $5d_{z^2}$ orbital and the shorter interplanar spacings gives rise to larger bandwidths and higher anisotropies for these inorganic materials with respect to the 1-D organic conductors. These highly conducting inorganic materials exhibit characteristically nonstoichiometric molecular formulas indicating a nonintegral oxidation state for the chain forming metal ion.

The synthesis and characterization of the tetracyanoplatinates is fairly well established and the initial problems associated with water loss have been basically overcome. Considerable progress has been made in exploring the full range of chemical variables (cation, anion and water content) which are available in this system. The corresponding structural changes have been evaluated and progress has been made in the characterization of the effects on properties. A continuing effort here, particularly in the area of property evaluation, should be encouraged. It is particularly notable in light of these investigations that the degree of variation in the basic parameters of this system, i.e., the degree of partial oxidation and the basic features of the $Pt(CN)_4^{n-}$ chain structure, are surprisingly small. The reasons for this, as well as for the apparent preference for a particular degree of partial oxidation by this system are not at all well understood at present.

From a chemical and structural point of view the bis(oxalato)-platinate (OP) and iridiumhalocarbonyl are even less well understood. The preparation of high quality single crystals remains a key problem, however, some progress has been made here particularly in the OP case and furthermore physical measurements on this substance can be anticipated in the near future. The iridium halocarbonyls are, in contrast, poorly characterized both from a chemical and physical property standpoint due mainly to the lack of suitable single crystal samples.

   2. Semiconductors

   Poorly conducting linear chain systems possess an even number of electrons per repeat unit and consequently the substances are narrow to wide band gap semiconductors. Two distinct structural forms exist, namely linear chain substances where all the metals are in the same oxidation state and exhibit moderately short M-M spacings (3.2 - 3.5 Å) and ligand bridged materials, where the metals may or may not be in equivalent oxidation states, and where metal-ligand(s)-metal chains, are observed. These substances exhibit relatively long intrachain metal-metal separations.

   The former substances, e.g., $SrPt(CN)_4 \cdot xH_2O$, $Ir(CO)_2(acac)$ (acac = acetylacetonate), $Ni(HDMG)_2$ (DMG = dimethylglyoximate), are comprised of equivalent planar molecules stacked on top of each other with 3.2 -3.5 Å interplanar separations. In the case of the $Pt(NH_3)_4(PtCl_4)$ complex the structure contains alternating cations and anions.

   The ligand bridged complexes e.g., $CuCl_2(pyridine)_2$, $(H_3C)_4$-$NMnCl_3$, $Pd(NH_3)_2Cl_3$, and $AuBr_2 S(CH_2C_6H_5)_2$ contain examples of both the mixed valence and single valence type. These are, in general, poorly conductive materials which have been of interest mainly from the viewpoint of their spectral and low-dimensional magnetic characteristics.

B. Organic Conductors

   Highly conducting 1-D organic materials are based on planar molecules with delocalized π-systems derived from carbon p_z orbitals, e.g. 7,7,8,8-tetracyano-p-quinodimethane (TCNQ), which forms a chemically stable anion radical on reduction. $TCNQ^-$ anions tend to form uniform segregated stacks which may exhibit high conductivity in 1-D. Typical interplanar separations are 3.1 - 3.3 Å and are significantly larger than in the highly conductive 1-D metal chain compounds. Consequently they exhibit significantly smaller bandwidths. High conductivity is observed in two main groups of materials; namely single and double stack substances. The former is exemplified by $Cat^+(TCNQ)_2^-$, $TTFBr_{0.79}$, $Ni(phthalocyanine)I$, $N(CH_3)_3H(I)(TCNQ)$. Double conducting stack systems are exemplified by $(TTF)(TCNQ)$ and its analogues. Since the discovery of these materials in which semiconducting or metallic behavior is observed, a large number of derivatives have been prepared in order to evaluate the effect of steric and electronic properties on the electrical and dynamical structural properties.

   Notable successes are that the substitution of S with Se in several donor molecules and the preparation of a large number of TCNQ derivatives. The main chemical problems are the synthesis of

reactive molecules and purification to a level appropriate for the
study of the intrinsic electrical, optical, and magnetic properties.

The synthetic methods developed to prepare TTF and TCNQ molecules (and their analogs) are at a rather high level of sophistica-
tion and many desired structural types can be prepared. However,
purification techniques, and especially the identification of
organic impurities at a ppm level are still not satisfactory. Likewise crystallization techniques are not well developed except in a
few systems. Consequently, although many 1-D organic systems have
been characterized in some detail the question remains whether
these are intrinsic properties of certain members of this class of
substances.

C. Conducting Polymers

There are two types of conducting covalent polymers of
interest here, namely poly(sulphurnitride), $(SN)_x$, and the doped
polyacetylenes, $(CHX_n)_x$. In addition to the delocalized π-system,
these substances, in contrast to the 1-D metal chain and organic
compounds, have intrachain covalent σ-bonding. They are basically
insoluble in organic media. Poly(sulphurnitride) is prepared by
the solid state polymerization of S_2N_2 and the fibrous structure of
the polymer is the direct consequence of the method of preparation.
Attempts to extend the system to $(SeN)_x$ have been unsuccessful as
have attempts to prepare $(SN)_x$ copolymers. $(SN)_x$ is a reasonably
well characterized material, although the crystals usually have
many defects. All of the SN distances are effectively equivalent
and the substance is best described by a semi-metallic electronic
structure with significant interchain interaction. Thus $(SN)_x$ is
metallic over the entire temperature range studied and is superconducting below $0.3°K$. Addition of halogens, e.g., Br_2, forming
$(SNBr_{0.4})_x$ removes electrons from the conduction band increasing
the conductivity. This derivative is also superconductive below
$\sim0.3°K$.

The cis- and trans-forms of polyacetylene are prepared by
solution polymerization of acetylene with a Ziegler-Natta catalyst.
The polymer has a very open "spagetti-type" fibrous structure containing many air-spaces and the material is not well characterized.
Although highly reflective toward visible light, both isomers are
poorly conductive when pure but doping with many acceptor species
(e.g., AsF_5, I_2, Na) leads to very highly conducting, metal-like
solids even at low temperature.

Relative to the other molecular metals polyacetylene is poorly
characterized and preparative differences may lead to significant
differences in the resultant physical properties. Doping of the
polymers is fairly straightforward; however, again the detailed
nature of the substance produced under various conditions may be
different and thus reflect different physical properties. The

reactivity of $(SN)_x$, $(CH)_x$ and their halogen adducts toward oxygen and water (etc.) has not been studied in detail, however, preliminary data suggests reaction does occur resulting in alteration of the physical properties.

In contrast to $(CH)_x$, polydiacetylenes, prepared by the solid state polymerization of monomer crystals, are extremely well characterized both chemically and structurally, but efforts to dope the crystals to produce highly conducting materials have failed. They are, however, materials of considerable interest for their optical and mechanical properties.

D. Intercalated Compounds

Several highly conducting 2-D (layered) substances permit the introduction of a molecular (or atomic) species between the layers (intercalation). The best characterized of these are graphite and transition metal dichalcogenides; e.g., TaS_2. Upon reaction with the intercalant, e.g., AsF_5, Br_2, $AlCl_3$, K, etc. for graphite, and amines, metallocenes, and alkali metals for the metal dicalcogenides, molecular or ionic species are introduced between the layers. In the case of graphite this produces an increase in conductivity presumably by virtue of a higher carrier density in the carbon atom layers due to the addition or removal of electrons by the host species. Detailed structural information is available for the 2-D starting materials, however, disorder has thrwarted the detailed structural investigation of the intercalated species. Although the structures are not known in detail, the physical characterization and reproducible preparation of the various intercalation stages is well under control. For graphite large samples well ordered perpendicular to the graphite planar as HOPG (highly orientated pyrolytic graphite).

IV. Physics

In this section, we limit ourselves to linear chain metals, of the KCP type, organic two-stack (TTF-TCNQ), etc.), one stack (TTT_2I_3, etc.), and polymer (polyacetylene, and $(SN)_x$). The doped polyacetylenes are relatively new materials and on the basis of the information available so far, their physics seem to be similar to that of classical heavily-doped semiconductors.

A. Structural Properties

The crystallographic structure of most of these materials has been determined by classical methods. The thermal and elastic properties are somewhat anisotropic, but basically three-dimensional. The lattice is strongly anharmonic, typical of molecular crystals. The cohesive energy involves a balance between ionic, metallic, and polarization (van der Waals) forces; the covalent bonds inside the molecules may also be affected by the charge transfer.

In the organics, electrical conduction requires incomplete-

shell π-electrons. This can be achieved by stacking planar molecules on top of each other, so as to obtain appreciable overlap,
thus resulting in a one-dimensional array. A similar situation
occurs in the KCP like materials where overlap of the d_z^2-orbitals
is the important electronic feature. The restricted dimensionality
gives rise to electronic instabilities and sensitivity to disorder.
When interatomic distances are small, characteristic of covalent or
metallic bonds, the electronic bandwidth is large. On the other
hand large, close to van-der-Waals, distances give rise to narrow
(1/2 eV) bands. Standard molecular-orbital and band calculations
give the correct bandwidth, but are not yet able to account for the
degree of charge transfer.

Conductivity measurements indicate the presence of the metallic
state for structures with regular segregated stacks; they become
semiconductors at low temperatures, depending sometimes on disorder
overcoming interchain coupling. The conductivity transition is
sometimes associated with a structural instability, which can be
seen by neutron and diffuse x-ray scattering. Above the structural
instability (Peierls distortion), diffuse $2k_F$ sheets are seen, due
to the Kohn anomaly, and the phonon frequency can be measured by
neutron inelastic scattering. The value of $2k_F$ is a precise measure
of the charge transfer. Just above the transition, short-range
transverse order appears, and long-range transverse order may appear
at the transition. Additional diffuse x-ray scattering at $4k_F$ is
seen in TTF-TCNQ, HMTTF-TCNQ (but not the Se analogues) up to room
temperature, and several interpretations have been proposed for
them.

B. Electronic Properties

1. Thermodynamic Properties

Specific heat anomalies indicate the occurrence of the phase
transitions. The electronic contribution to the specific heat is
usually very small compared with the lattice part. Magnetic susceptibility measurements confirm the presence of the phase transitions; however, the spin-susceptibility of TTF-TCNQ is not yet
properly understood. Excess diamagnetism has been observed at low
temperatures in HMTSF-TCNQ.

The magnetic analogue of the Peierls transition, namely, the
spin-Peierls transition, has been observed in magnetic systems by
specific heat and susceptibility measurements and recently by X-ray
and neutron scattering studies. Spin susceptibility measurements
in insulators, extended to very low temperatures, have shown the
existence of random exchange coupling in disordered materials.

2. Static and Low-Frequency Transport Properties.

The electrical conductivity is unusually high for organic
materials. In molecular charge transfer complexes, conductivities

up to about 3×10^4 $(\Omega \text{ cm})^{-1}$ have been observed in high-quality crystals; a value of 10^5 $(\Omega \text{ cm})^{-1}$ at helium temperature has been reported, but not yet confirmed. In $(SN)_x$, the conductivity rises to 10^6 $(\Omega \text{ cm})^{-1}$, and superconductivity is observed below about 0.3 K. The temperature, pressure, and frequency dependences of the conductivity in the metallic state are strange.

Below the Peierls transition; the conductivity is usually activated. However, HMTSF-TCNQ, HMTSF-TNAP, TTT_2I_3 remain conducting down to the lowest observed temperatures (10 mK). It is not yet clear whether this conductivity is extrinsic or intrinsic. The anisotropies range from about 10-20 (HMTSF-TCNQ, $(SNBr_{0.4})_x$) up to 10^5 (KCP).

The thermoelectric power is usually measured, and confirms the metallic behavior, which is sometimes electron-like and sometimes hole-like. In some materials of intermediate conductivity a large, nearly temperature independent S, suggestive of electron-electron correlations, has been observed.

The thermal conductivity of TTF-TCNQ confirms the Wiedemann-Frantz law. Hall effect and magnetoresistance measurements (in TTF-TCNQ, HMTSF-TCNQ) show a continuous transition from the quasi-one-dimensional metallic electronic behavior to an anisotropic semi-metallic state, as predicted by band structure calculations for finite interchain coupling. Non-linear I-V characteristics have been observed recently. These have been interpreted either as collective or conventional hot electron effects.

C. NMR and EPR

The NMR relaxation time T_1 is pretty well understood, in terms of the peculiar properties of one-dimensional systems. The EPR g-shift, and NMR ^{13}C Knight shift, have been used for the decomposition of the spin susceptibilities of the two chain families. The EPR linewidth is correlated with the dimensionality. Its temperature dependence follows approximately the transverse conductivity in two-chain compounds, but decreases strongly at low temperatures in one-chain compounds such as TTT_2I_3.

D. High Frequency and Optical Properties

The conductivity of TTF-TCNQ possesses a minimum in the far-infrared; the dielectric constant in the microwave region is very large (2000-6000) and positive. This behavior is poorly understood. A maximum in the conductivity in the near infrared is a general feature of one-dimensional electronic systems. Several explanations for it have been proposed (such as a charge-transfer band). At higher frequencies, a Drude-like behavior is observed, which makes it possible to determine the plasma frequency ωp quite accurately, and the collision time τ_{opt} as well. τ_{opt} is only weakly temperature dependent, and probably dominated by high frequency phonon emission processes. Intramolecular vibrations are also responsible for interference effects in the IR spectrum. In the visible range, the absorption spectrum is characteristic of the molecular species.

Electron energy-loss spectra have been measured for TTF-TCNQ and $(SN)_x$, to determine the plasmon dispersion, which provides information about the optical anisotropy of the system. The spectra provide an upper limit for the electronic bandwidth (0.55 eV for TTF-TCNQ).

E. Interpretation and Models

The band-structure of one-dimensional metals is strongly affected by static and dynamic disorder. The band-structure is usually one dimensional and metallic. When the scattering rate $\tau_{||}^{-1}$ along the chain becomes smaller than the interchain coupling t , a (usually) semi-metallic, three-dimensional anisotropic band structure is predicted.

A key question which has not yet been resolved, is whether the dominant interactions are electron-electron Coulomb repulsions, or electron-phonon interactions. The electron-electron intramolecular repulsion, U, which plays a key role in the Hubbard model, has been suggested to be large, comparable, or small compared with the electronic bandwidth. As for the phonons, rigid molecular translations, rotations, or intra-molecular vibrations, have been invoked as the ones mostly affecting the electronic properties. Half-filled bands apparently give rise to magnetic insulators, in accordance with the Hubbard Hamiltonian.

The effective interactions (whether Coulomb, or electron-phonon) may be momentum (q) dependent, and the relative values at q = 0 and q = $2k_F$ play an important role in certain theoretical models ("q-ology"). TFF-TCNQ (as a prototype) has several phase transitions, and a complicated phase diagram. Several theories have been proposed to account for it; some by Landau theory, and others by attributing a dominant role to fluctuations. All of these theories consider the evolution of three-dimensional order in chains weakly coupled via their mutual coulomb interaction.

Disorder plays a special role in restricted dimensionality systems. Strong disorder localizes the electronic states, limits the conductivity, smears out the phase transitions, and gives rise to characteristic magnetic and thermal effects. The role of the weaker disorder, usually encountered in these materials, remains to be determined. Radiation-damage experiments may be of particular use in this regard.

F. Conclusion

The experimental situation has by now reached a very sophisticated stage, where the access to almost all desired information is limited only by the availability of good samples. The field has

generated, or refined, important new experimental techniques, such
as diffuse X-ray scattering and microwave conductivity measurement.
It has also helped to develop important new theoretical avenues.
However, the overall fit of any given theory, with all the structural and electronic experimental data, is not yet satisfactory.
Also, the systematic behavior of a large class of materials has to
be accounted for.

V. Conclusions

This study group's overall impression is that the area of
molecular conductors is a robust, dynamic field in rapid expansion.
We are especially encouraged that so much progress has already been
made at a time when the wealth of synthetic possibilities has been
barely tapped, and the preparation of well characterized samples of
suitable size and quality is just coming into its own.

Much more, of course, needs to be done. The problem of crystal
growth and microscopic characterization still requires attention,
particularly in light of the dramatic sample variations documented
in compounds such as NMP-TCNQ. Mechanistic information about
nucleation and crystal growth is also important. Measurements of
defect and impurity levels and the effects of annealing have been
much less common than they should, particularly in light of the
sensitivity of lower dimensional systems to disorder.

Progress has also been hampered by a lack of fundamental understanding in several areas. It is an embarassing fact that we still
have practically no quantitative knowledge of the cohesive forces
responsible for the existence of these materials. It is therefore
hardly surprising that we do not know why it is so very difficult
to produce significant variations upon a compound like KCP, for
example, while the synthesis of organic conductors remains so very
flexible.

Other questions of equal importance also remain to be answered.
How significant are coulomb correlations, especially in the organic
conductors? What are the roles of structural and dynamic disorder?
How does dimensionality affect transport, fluctuations, and phase
transitions? What are the origins of the striking electronic differences between chemically and structurally similar systems, as in
the extreme case of highly conducting HMTSF-TCNQ and insulating
HMTSF-TCNQ(F_4)?

Particularly among the organic conductors, it is difficult to
think of a single significant set of experiments whose detailed
interpretation now commands a theoretical consensus. Some of this
difficulty will undoubtedly pass with the further refinement of both
theory and experiment. Nevertheless, the study group emphasizes
that while intensive studies of isolated materials are certainly
essential, they can usually add but limited perspective to the
field as a whole. Unified understanding of these materials will
depend critically upon systematic investigations of chemical and

physical trends within each class.   Encouraging studies have already
appeared of the effects of controlled variations of dimensionality,
disorder, bandwidths and band filling, and have elucidated some
very preliminary relationships between molecular and solid-state
properties.   Nevertheless, much of this insight has come after the
fact, and has yet to prove its a priori predictive value.   Further
systematic studies of this sort are of control importance.

By such means, and with the increasingly formidable array of
integrated interdisciplinary resources at hand, this study group
sees no fundamental impediment to eventual realization of the full
potential of these materials.   This evaluation is certainly rein-
forced by the impressive series of new materials, results, and
potential areas of application reported at the present meeting.

NEW SYSTEMS COMPRISED OF NON-METALLIC ELEMENTS

R.H. Baughman, R.L. Greene (Secretary), A.G. MacDiarmid (Chairman), M. Schmidt, Z. Soos, G.B. Street and F.L. Vogel.

I. Introduction

The motivation for producing new conductors is in part a desire to achieve unusual, or even unprecedented, materials properties of technological importance. Metallic conductors are based on open-shell atoms, while nonmetallic compounds typically exhibit covalent or polar bonding based on closed shells. Since open-shell electronic bands are a prerequisite for any conductor, molecular or polymeric conductors will almost certainly be unusual chemically. The stabilization of interacting open-shell non-metals in a solid, so as to prevent the exclusive formation of polar, covalent bonds, is an enormously challenging task. On the other hand, recent developments with doped polythiazyl, $(SN)_x$, and polyacetylene, $(CH)_x$, suggest that the judicious incorporation of electron donating and withdrawing groups may yet provide a fairly general scheme for achieving incompletely filled bands, as required for conductors.

The stabilization of partly-filled bands in nonmetallic conductors is primarily a chemical and structural problem. The synthesis, characterization, optimization and exploitation of conducting systems provides many opportunities for both experimental and theoretical studies. When nonmetallic open-shell systems are available, traditional chemical ideas about crystal packing, about isoelectronic substitutions, or about analogous reactions have generated essentially all of the currently known "families" of molecular metals. These have been derived from a few conductors that were originally prepared for other reasons. Finding new conductors among nonmetallic elements or compounds will consequently demand great ingenuity, with only very modest theoretical suggestions. On the other hand, the theoretical appreciation of the role of dimension-

ality in phase transitions and in magnetic phenomena has greatly
increased interest in model compounds, whose experimental study
has further enriched theory.

The bulk properties of most known conducting polymers are quite
different from their intrinsic crystal properties. These bulk properties typically reflect the state of crystallite aggregation as much a.
the crystallite properties. For example, a variety of indirect
measurements indicate metallic conductivity for $(SN)_x$ and for doped
$(CH)_x$ in directions in which d.c. conductivity measurements indicate a non-metallic temperature dependence. This difference
between bulk and intrinsic properties, which includes everything
from mechanical strength and environmental stability to superconductivity, is a consequence of gross structural inhomogeneity in
these materials. As a result, basic uncertainties exist in our
knowledge of their fundamental materials properties, which limit
our ability to understand their polymer physics and to optimize
their properties for possible applications.

This section of the report concentrates on three presently
known classes of metallic compounds derived from non-metals, namely
those based on $(SN)_x$, $(CH)_x$, and graphite.

II.   Critical Assessment of Existing Knowledge in the Field

The following paragraphs are brief summaries. The reader is
referred to the review papers for more detailed information.

A. $(SN)_x$ and Derivatives

Polythiazyl, $(SN)_x$, is the first example of a covalent polymer containing no metal atoms which has been shown to be a conduct-
ing ($\sigma_{RT} \approx 5 \times 10^3$ ohm^{-1}cm^{-1}) and a superconducting polymer
($T_c \approx 0.3°K$). These unique electronic properties have opened up
new vistas for the scientific investigation of polymers. Currently
available $(SN)_x$ consists of crystalline microfibrils which are
fully oriented in a common chain direction, but poorly coupled
together electronically. Gross twinning occurs and about 20% of
the chains in each microfibril are in defect positions. However,
band structure calculations based on the ideal $(SN)_x$ structure have
led to a semi-quantitative understanding of its properties; in
particular $(SN)_x$ has been shown to be an anisotropic semi-metal,
stabilized against the Peierls distortion by interchain interactions.

Presently, the only known conducting derivatives of $(SN)_x$ are
obtained by partial oxidation with halogens and interhalogens. Of
these compounds $(SNBr_{0.4})_x$ has been the most thoroughly investigated. This material exhibits an order of magnitude higher conductivity than $(SN)_x$ and other interesting electronic properties.
However, details of the crystal structure and the role of bromine
are not yet entirely clear. Halogen derivatives of $(SN)_x$ have been
obtained also by halogenation of S_4N_4 which eliminates the intermediate synthesis of $(SN)_x$.

B. $(CH)_x$ and Derivatives

Polyacetylene, $(CH)_x$, is the simplest possible conjugated organic polymer and is therefore of special fundamental interest. It may be prepared in cis- or trans-isomeric forms as silvery poly-crystalline, flexible films by the polymerization of acetylene, C_2H_2, using a Ziegler catalyst. Both isomeric forms are semi-conductors; however, through chemical doping, their electrical con-ductivity may be controllably varied over an extraordinary 13 orders of magnitude with properties ranging from semiconducting ($\sigma < 10^{-10}$ ohm^{-1} cm^{-1}) to metallic ($\sigma > 1 \times 10^3$ ohm^{-1} cm^{-1}). The metallic state is preserved down to at least 40 mK. The semiconductor to metal transition occurs at a few mole percent dopant concentration. Both electron acceptors (e.g. I_2, Br_2, AsF_5, $AgClO_4$, etc.) and electron donors (e.g. Li, Na, K, etc.) can be used to yield p-type or n-type material respectively. The $(CH)_x$ films can be partially chain-aligned by mechanical stretching and may also be doped. Such films exhibit anisotropic electrical and optical properties with an increase in conductivity in the direction of alignment of up to one order of magnitude. Preliminary band structure calculations suggest that $(CH)_x$ is an anisotropic semiconductor with a small band gap ($\sim$ 1eV) but with wide conduction and valence bands (5-10eV in the chain direction). Presently available $(CH)_x$ is morphologi-cally complex and is not well understood.

C.   Graphite Intercalation Compounds

Graphite and its intercalates exhibit interesting anisotropic two dimensional properties in contrast to the more one-dimensional properties of $(SN)_x$ and $(CH)_x$. The electronic properties of intrinsic graphite crystals are reasonably well established with one main exception: the specific effects of crystal defects on the scattering mechanism. Both donor and acceptor intercalation com-pounds of graphite exhibiting a wide variety of interesting elec-tronic properties have been made. We mention just three general pro-perties of these compounds which are of particular interest.

(1)   Acceptor (e.g. HNO_3, SbF_5, AsF_5, Br_2) compounds have higher in-plane electrical conductivity ($\sigma_{RT} \approx 10^6$ ohm^{-1} cm^{-1}) than the donor (e.g. K, Cs, Rb) compounds ($\sigma_{RT} \approx 10^5$ ohm^{-1} cm^{-1}).

(2)   The maximum in-plane conductivity occurs at a stage higher than one in acceptor compounds.

(3)   There is an increase in anisotropy (ratio of in-plane conductivity to conductivity normal to the planes) which accompanies increased in-plane conductivity. For example, with AsF_5 doping, an exceedingly large anisotropy of $\sim 10^6$ is observed.

In general, the donor (group I) compounds are better under-stood than the acceptor compounds but the acceptor compounds appear to have the potentially more interesting properties, e.g. high electrical conductivity and anisotropy.

D. Other Materials

Here we list a few other materials composed of non-metallic elements which have been demonstrated to exhibit interesting properties that are worthy of future study; however, they presently fall outside the scope of the subject of molecular metals and therefore they will not be discussed. (1) Amorphous silicon, $(SiH_y)_x$; (2) Polydiacetylenes; (3) Boron nitride intercalation compounds; (4) Polyvinylcarbazole-type semiconductors (and photoconductors); (5) Porphyrin and phthalocyanine complexes. There are obviously other materials which could have been included in this list.

III. Scientific and Technological Potential

A. $(SN)_x$ and Derivatives

$(SN)_x$ is a unique material and has served as a model compound for understanding the metallic and superconducting properties of polymers. $(SN)_x$ and its derivatives have scientific and possible technological potential. It does not appear likely that high superconducting transition temperatures will be achieved with polymers unless novel methods for increasing the electronic density of states are discovered. $(SN)_x$ films have been used in the formation of Schottky barrier photovoltaic devices. Other potential technological applications are under investigation.

B. $(CH)_x$ and Derivatives

The scientific potential of doped $(CH)_x$ derives from the fact that this new class of materials has by far the highest conductivity of any known organic polymer. An understanding of the conduction and doping mechanism and the effect of alignment on the electrical and optical properties presents fundamental scientific challenges. p-n junctions having typical rectifier diode characteristics have been prepared by pressing together films of p-$(CH)_x$ and n-$(CH)_x$. A photovoltaic effect has been observed with an n-Si/p-$(CH)_x$ heterojunction. Furthermore, the ease of fabricating large sheets of p-n junction $(CH)_x$ films suggests possible use for terrestial solar cells, etc.

C. Graphite Intercalation Compounds

On the scientific level these compounds have significant potential for improving our understanding of anisotropic two dimensional systems in which electrical conductivity and anisotropy can be controlled over wide ranges. Stage 3 graphite-SbF_5 has a conductivity comparable to the best elemental metal at room temperature. As far as technological potential is concerned, three areas can be mentioned:

(1) Stage 3 graphite-SbF_5 can be swaged into a practical wire form having a room temperature conductivity of 6.5×10^5 ohm^{-1} cm^{-1}. In addition, the conductivity decreases less rapidly with increasing temperature than for the best elemental metals. The stage 3

compound has a density less than 3 grams/cm^3 so that considerable weight saving is possible compared to copper.

(2) The electrical conductivity of graphite fibers of the kind used to reinforce epoxy composites can also be increased by intercalation, while maintaining their high strength and elastic modulus. Conductivities of 1×10^5 ohm^{-1} cm^{-1} have been obtained and values as high as that of copper are forecast. These high conductivity fibers have been incorporated into epoxy composites which then demonstrate an increase of conductivity of about a factor of ten. The electrical conduction of the composite suggests increased applications that require high strength, light weight and conductivity such as aircraft structural members, large antennae and others.

(3) On intercalation with acceptors, the reflectivity of graphite on the c face – the outside surface of graphite fibers – is increased to substantially 100% in the infra-red (specifically at the 10.6 μ laser wavelength).

(4) Stereospecific halogenation of aromatic molecules and stereospecific catalytic polymerization of intercalated species are being evaluated. Also, reversible redox reactions of intercalated layer structures have found commercial application in battery technology.

IV. Useful Research Directions

A. (SN)$_x$ and Derivatives

There are several directions for significant research on this polymeric system: the chemistry of halogen derivatives and its relationship to the electronic properties needs further study; demonstration of donor intercalation; separation of the (SN)$_x$ chains to change the dimensionality; attempted synthesis of isoelectronic analogues; a better understanding of the "S_4N_4-halogen route" to conducting polymers, which may provide the key to obtaining a much wider class of materials based on the (SN)$_x$ polymer.

B. (CH)$_x$ and Derivatives

There are a number of important directions in which research should proceed. Essentially only one set of polymerization conditions for the synthesis of (CH)$_x$ have been investigated to date. Undoubtedly (CH)$_x$ films with considerably different electronic, optical, mechanical, etc. properties will be prepared using different conditions. Only a very few different types of dopants, which give widely different conductivities, have been investigated so far. The chemical properties of the doped and undoped (CH)$_x$ need to be studied, particularly those related to stability to heat, air and light. It is extremely important to devise methods of synthesizing completely aligned (CH)$_x$ in order to ascertain its intrinsic conductivity parallel to the (CH)$_x$ chains. Much more theoretical work needs to be done in order to gain an understanding of the conduction process.

An almost unlimited number of chemical derivatives of $(CH)_x$ are potentially available in which the hydrogen atoms of $(CH)_x$ have been replaced in whole or in part by organic, inorganic or organo-metallic groups. It therefore appears that it may be possible to synthesize a very large number of semiconducting and metallic poly-mers based on $(CH)_x$ whose electronic, optical, mechanical, etc. properties may be controlled and pre-determined by chemical fine tuning of the system.

C. Graphite Intercalation Compounds

Several significant scientific problems for these materials are briefly outlined below.

(1) Identification of the chemical species and the state of ionization in acceptor compounds.

(2) Determination of role of defects and defect-intercalant interactions in transport scattering mechanisms.

(3) Examination of other systems a) with intercalants of high electron affinities, b) with donor systems other than Group I, and c) intercalation into boron nitride, $(BN)_x$, an isoelectronic system, for comparison with the conductivity of graphite intercalates;

D. Structural Characterization

Future research on (A), (B), (C) should fully characterize the relationship between defect structure and electronic properties; no comprehensive studies of this sort exist for $(SN)_x$, $(CH)_x$ and graphite intercalation compounds. This probably reflects the scarcity of research facilities which are equipped to accomplish all of the following: (1) prepare new materials under carefully controlled experimental conditions (2) fully characterize defect structure and (3) evaluate electronic properties. In this light, collaboration between researchers in different facilities and coun-tries should be encouraged. Major efforts should be devoted to the development of methods for producing conductive materials as high-perfection single crystals suitable for electronic and structural measurements.

V. Conclusions

It appears that the new field of polymeric conductors and the closely related area of graphite intercalation compounds are both advancing at an extraordinarily rapid rate. However, the synthesis of new chemical systems is absolutely fundamental for the continued growth of the field. The potential for increasing the scientific knowledge in these fields by the synthesis of new compounds, by better characterization of presently-known materials, and by the-oretical analysis of the new phenomena is extremely high. It pre-sents a challenge and opportunity for collaborative interaction between chemists, physicists and materials scientists. These materials have important technological potential in electronic devices. Present knowledge suggests that we are seeing only the tip of the iceberg and that the next few years will bring forth important new materials with unusual properties.

NEW ORGANIC MATERIALS

D. Bloor, R. Comes (Chairman), V.J. Emery, E. Engler,
S. Flandrois, A. Garito (Recording Secretary), J.
Gaultier, Y. Tomkiewicz, J. Torrance, G. Wegner, and
L. Weiler

I. Introduction

The field of quasi-one dimensional (1D) organic metals remains
an active area of fundamental research, spanning experimental and
theoretical solid state physics and organic synthetic chemistry
that has led to the development of important new techniques and concepts in each of these areas.

The design and synthesis of a rich variety of organic conducting solids has enabled experimental observation of a significant
number of unusual solid state phenomena. These novel results have
generated considerable developments in the theory of 1D many-body
interactions. General understanding of various kinds of long-range
ordered states such as change density waves (CDW), spin density
waves (SDW), singlet and triplet superconductivity (SS and TS) has
advanced rapidly. Ideas and concepts such as incommensurate lattices; pinning; intramolecular electron-electron correlations;
electron-intramolecular phonon coupling, and strong disorder due to
phonon localization have originated from these 1D studies. Correspondingly, major discoveries have been achieved in organic synthetic reaction pathways particularly in organo-sulfur and -selenium
chemistry.

It is this kind of continued development associated with organic
metals that defines the high potential of the field. Accordingly,
the committee has reviewed and analyzed the present status and has
indicated the directions of research-experimental, theoretical and
synthetic - which seem likely to enhance our fundamental understanding of the unusual phenomena that have emerged in this field.
From this review, a series of future directions have been summarized.

II.   Current Developments

A.   Structural Properties

It is well known that organic metals crystallize into a 3D structure consisting of segregated stacks of donor cation and acceptor anion molecules, which usually are tilted relative to the crystallographic stacking axis to form a herring-bone pattern.

The most extensively investigated systems are TTF-TCNQ and TSeF-TCNQ, in which the $2K_F$ giant Kohn anomaly of 1D metals could clearly be observed. However, the features are much richer than for a monatomic chain, because the electron transfer integral can be modulated in first order by either longitudinal or transverse displacements of the tilted molecules, and both have been observed. It also is conceivable that rotations (or librations) of molecules could play a role; however, preliminary analyses of the experiments indicate that they are unimportant.

For several compounds, it has been shown that a metal-insulator transition is associated with a 3D ordering of the $2K_F$ charge density waves, and, in a few examples, the associated incommensurate super-lattices have been characterized in detail. In TTF-TCNQ, it is shown that a sequence of phase transitions in the range 38K < T < 54K was associated with the successive ordering of CDWs on TTF and TCNQ chains observed in the $2K_F$ scattering at higher temperatures.

A new effect was observed in a few compounds, but mainly in TTF-TCNQ. One-dimensional scattering with wave vector $4K_F$ was observed in the metallic phase that subsequently condenses at lower temperatures. It appears this observation constitutes strong evidence for strong repulsive interactions between electrons on at least one of the molecular stacks.

B.   Electronic Properties

It is natural to assess the electronic properties of organic metals by summarizing recent measurements of the electrical conduc-tivity, since a primary motivation of the field is to obtain a general understanding of charge transport in solids, and fluctua-tion phenomena in 1D systems. Of particular importance are the possible effects of CDW or SDW formation and of disorder on the properties of the conduction electrons.

Continued high-pressure studies of the dc electrical conduc-tivity and the optical plasma excitations in the metallic state (54K < T < 300K) of TTF-TCNQ have demonstrated that, when the ther-mal expansion (2.5%) is taken into account, the conductivity along the stacking axis increases less rapidly with decreased temperature than was previously thought and is almost linear in temperature. Moreover, the 1D-3D crossover transitions (38K < T < 54K) to the insulating incommensurate phases, observed at atmospheric pressure apparently are reduced to a single transition above 15kbar. These results are having a marked impact on the theoretical understanding of all the electronic properties of quasi-1D systems, and it is expected that similar results may be encountered in the structural properties and the behavior of CDW states in TTF-TCNQ.

Further experimental and theoretical studies of the frequency dependence of the conductivity in the stacking axis direction have shown that the far infrared minimum in TTF-TCNQ is less pronounced than previously thought, and this result and the observed fine structure have been discussed in terms of Fano resonances and Holstein scattering from intramolecular modes. The notion of electron-intramolecular phonon coupling has advanced considerably, both in experiment and theory.

The unique property of organic conductors in exhibiting both band-like and molecular properties has allowed us to distinguish the magnetic excitations occurring on individual donor cation and acceptor anion chains by using g-value analyses and Knight-shift measurements. These studies demonstrated the existence of enhanced susceptibilities in the metallic state of organic conductors arising from electron-electron correlations.

Thus the emerging experimental picture of organic conductors is exceedingly complex, with sufficient evidence being obtained to demonstrate CDW formation, important electron-electron correlations, electron-intramolecular phonon coupling, and the possibility of strong disorder due to phonon localization in TTF-TCNQ and its derivatives.

C. Theory

The experimental studies of the organic metals have provided a real impetus to the development of a corresponding theory of 1D many-body systems. Some exact results for the energy spectra of certain fermion models and spin chains were known already, and ideas on the disorder localization of noninteracting electrons were well established. But the availability of experimental results has stimulated significant extensions of the existing theory and the development of new mathematical approaches, particularly those related to quantum field theories. There is now a good general understanding of the asymptotic forms of the correlation functions that are related to the various kinds of long-ranged order such as SS, TS, CDW and SDW which may develop at low temperatures, thereby enhancing or inhibiting electrical conductivity and modifying magnetic and other properties. When these developments are coupled to the improved understanding of interchain coupling and its role in the crossover from 1D to 3D behavior, it is possible to provide considerable theoretical guidance for the design of new materials.

At the same time, progress in carrying out microscopic calculations of the physical properties of real materials has been more limited. A soluble model inevitably omits much of the complexity of a system in which electrons, lattice modes, and molecular

vibrations interact amongst each other in a significant way.   Progress has been made by incorporating the new ideas into phenomenological models which are much more useful in interpreting both
static and dynamic experiments, particularly when the general
requirements of structural symmetry are taken into account.   This
general procedure is widely used in theory of strongly interacting
systems and is well exemplified by the study of TTF-TCNQ.   Whereas
the temperature dependences of the electrical conductivity and
the magnetic susceptibility are not well understood, the theoretical
analysis of X-ray and neutron scattering experiments has further
stimulated the study of interpenetrating, incommensurate periodic
structures which is presently an expanding area of research in
solid state physics.

D.   Molecular and Polymer Systems

     Recent studies focusing on the synthesis of new organic π-
donors for use in organic metals have now developed a systematic
set of elegant synthetic pathways.   This has brought a deeper understanding to synthetic chemistry, particularly in the cases of organo-
sulfur and -selenium reactions.   The number of new such π-donors
which have been prepared presently is well over 50, and even the
first example of a conducting organotellurium π-donor salt has been
realized.   Moreover, organoselenium donors have provided a variety
of conducting salts such as HMTSeF-TCNQ and TSeT-I that remain
semi-metallic down to low temperatures.

     In the course of these studies, chemical and structural
criteria for organic metals that have proved of significant utility
include the following conditions:

-   the existence of unpaired electrons within the molecular units.
-   molecular units exhibiting intramolecular electron correlation
    to allow charge density to reside on diametrically distant parts
    of the molecule.
-   highly polarizable molecular units to diminish costly electron-
    electron repulsive interactions.
-   molecular donor units whose first ionization potential (electron
    affinity in the case of acceptors) does not assume large values
    on an established relative scale so as to avoid the Madelung
    energy contribution dominating the charge transfer process, thus
    favoring fractional charge transfer.
-   a uniform crystal structure consisting of linear, parallel
    stacked columns of flat planar molecules with significant inter-
    molecular overlap leading to band widths such that the associated
    electronic band structure would be metallic.

     Further singular developments include the discovery that large
π-systems, such as phthalocyanines, form metal-like charge transfer
salts which may provide further examples where on-site Coulomb
repulsions are considerably reduced.   Neutral π-radicals, such as
1,9-dithiophenylene, and new planar molecules that act as both

donor and acceptor on single sites have been prepared as attempts to realize single chain conductors.

The design of molecular metals requires that molecules of different shape, polarity, and/or charge be brought together in a well-defined geometry and that they are fixed in a 3D lattice with longe-range order. To date traditional crystallization procedures have been mainly used to obtain these structures required for high conductivity. However, the occurrence of charge transfer salts with the requisite separated donor and acceptor stacks necessary for high conductivity remains a fortuitous rather than an antici-pated, predictable event. There are, however, a number of other techniques available which are now being considered in order to achieve the desired architecture of molecules in real lattices.

There are among others organic solid state reactions in which precursor molecules are arranged in space and then transformed to the desired structure without disturbing the supermolecular struc-ture. More specifically the chemistry of chemical inclusion com-pounds, intercalated, and layered organic compounds are being considered with emphasis on detecting and/or designing new host/guest systems. Another promising technique is the organization of molecules by buildup of multilayers using film forming techniques.

The synthesis of polyconjugated polymers and their structurally complex adducts with oxidizing or reducing molecular species is of considerable interest. However, a severe problem is posed by the usual quite complex morphology and texture of real polymer structures which does not allow one to directly relate observed physical properties to the molecular structure independently of the nature of the supermolecular structure e.g., the degree of crystal-linity; crystal size; size distribution; and conformation of the chains in the liquid crystal and amorphous phases. Therefore, emphasis should be given to the synthesis of model oligomers and to the study of their physical behavior as well as methods which will lead to polymers with well-defined super molecular structures, e.g., pressure crystallization. There is little hope that fixing the usual charge transfer moities, such as TTF or TCNQ, to polymer chains with the aim of getting metal-like conductivity will be successful in the light of the difficulties encountered in mixing of polymers of different structure, in compatibility problems and polymer crystallization. Moreover, it is becoming apparent that further understanding may emerge from conformational calculations of molecular interactions versus steric hindrance. In this way, guidelines are expected to emerge as to how to link adjacent sub-units of a molecular metal by covalent carbon-carbon bonds.

III. Future Directions

A. Experimental Physics

- continued measurements of the solid state properties of existing and new organic and polymeric conductors.

- high pressure studies of the electronic and structural properties to obtain the phase diagram of TTF-TCNQ and other organic metals.
- topographic studies of defects in organic metals.
- detailed inelastic neutron studies of TTF-TCNQ.
- determination of the low temperature modulated structure of TTF-TCNQ including satellites.

B. Molecular and Polymer Systems

- continued design and synthesis of new organic and polymeric conductors.
- synthetic efforts to probe the importance of interchain coupling; intramolecular and off-site coulomb correlations; and intramolecular phonon and libron modes for transport properties and phase transitions.
- systematic studies of the conditions for fractional charge transfer.
- development of neutral single chain conductors.
- a controlled and systematic study of the role of disorder particularly that produced by defects and impurities.
- produce simpler systems showing incommensurability effects.
- thorough characterization of polymers such as $(CH)_x$ for the levels of structural order and the interaction of complexing agents with individual structures.
- polymer conformational calculations to guide appending donors and acceptors on polymer chains.

C. Theoretical Physics

- study models with both coulomb and electron-phonon coupling while evaluating correlation functions and allowing for possible disorder.
- understand the conditions for partial charge transfer.
- understand the behavior of electrical conductivity and magnetic susceptibility in existing systems.
- understand the factors producing particular kinds of molecular stacking.

IV. Conclusions

    In summary, the study of organic metals has experienced continued active development and increased depth of understanding. The future strength of the field will be clearly determined by continued experimental measurements of existing and newly synthesized molecular systems; refinements in the design of new organic and polymeric conductors, particularly toward elucidating existing solid state phenomena; and increased progress in the theoretical understanding of 1D many-body interactions.

NEW ORGANOMETALLIC SYSTEMS

L. Alcacer, R.J.H. Clark, W.E. Hatfield (Chairman),
K. Klabunde, J.S. Miller, G.D. Stucky (Recording
Secretary), G.A. Toombs, F. Wudl and S. Yoshimura

Scope

The two unique properties of organometallic systems which
make them of particular interest for use in the synthesis of molecular metals and related solid state materials are their ability
(1) to support a wide variety of oxidation states and to undergo
reversible one-electron transfer reactions, and (2) to exist as
chemically independent stable free radicals. The materials discussed in this report are limited to systems in which there are
either direct metal-carbon bonds (excluding metal cyanide complexes)
or in which a transition metal complex acts as either a donor or
acceptor with respect to an organic group i.e. as a charge transfer
complex. Main group metal systems such as sodium doped polyacetylene or graphite are excluded from the discussion as they are
included in the reports of other study groups.

Assessment of Existing Knowledge of the Field

Transition metal organometallic chemistry is a relatively
young field. By far the biggest expansion of our knowledge in this
area has occurred during the past ten to fifteen years. It is
fair to say that progress in the development of solid organometallic
materials has lagged behind that of either organic or inorganic
substances because of a lack of a systematic understanding of the
synthetic aspects of organometallic chemistry.

The greatest application of transition metal organometallic
complexes in the development of molecular metals has been their
use as electron reservoirs, either as acceptor or as donor species.
Some examples in this area include the early studies of ferricenium
and cobalticenium TCNQ complexes by Melby et al. in 1962. An important variation on this approach was the synthesis in the early

517

1970's of "mixed valence" complexes such as (bisfulvalenediiron)$^{+}$-
(TCNQ)$_2$ which has a compressed pellet conductivity of 10-40 ohm^{-1}
cm^{-1}.^{2} An example of the use of transition metal complexes to support unusual oxidation states is illustrated by the work of Miller
and coworkers who have reported the isolation of the previously
postulated $[Nb_3Cl_6(C_6Me_6)_3]^{2+}$. This was the first case in which
the 2+ ion had been isolated in the solid state as a consequence
of the unusual properties of (TCNQ)$_2^{2-}$. The structure of the 2+
monomer complex in the solid state is that of a bent chain configuration as indicated below, with the triangular cluster, [Nb$_3$Cl$_6$-
(C$_6$Me$_6$)$_3$]$^{2+}$ alternating with a TCNQ$^-$ dimer. Although this material
is not metallic, it is reminiscent of the organic charge-transfer

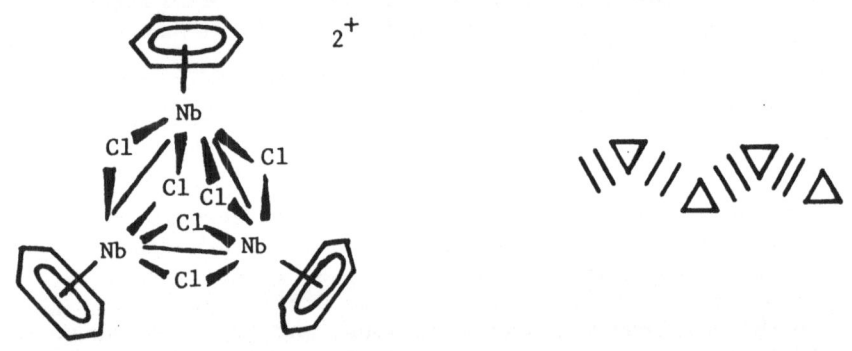

TCNQ complexes with donor-acceptor stacking which were discovered
early in the 1960's.

The majority of attempts to synthesize molecular metals from
organometallic materials have not been successful. Nevertheless,
these studies have led to the preparation of particularly interesting weakly coupled systems which show unusual magnetic properties.
For example, a 1:1 complex formed between decamethylferrocene and
TCNQ has been found to behave as a metamagnet at 2.5 K. The
material switches from an antiferromagnet to a ferromagnet at
approximately 1500 Gauss.

Another example of a system which exhibits previously unre-
ported properties is provided by the complex formed between TTF and
Cu[S$_2$C$_2$(CF$_3$)$_2$]$_2$. This material was prepared by Interrante in an
attempt to synthesize a transition metal charge transfer complex
with TTF which was expected to behave as a molecular metal. Although
he was unsuccessful in this attempt, subsequent low temperature
magnetic and structural studies revealed that it was the first
example of a system which exhibited a spin Peierls instability.
Further experimental and theoretical work as described in the poster session has led to an enhanced understanding of the relationship

between weak interactions and structural instabilities. Similar
types of solids have been reported to be reasonably good molecular
metals, e.g. (perylene)$_2$Pt$[S_2C_2(CN)_2]_2$ has a conductivity of approximately 200 ohm^{-1} cm^{-1}. The conductivity increases with decreasing
temperature with a maximum at 120 K.

The above examples demonstrate the versatility of organometallic molecules in that they have a profound influence on the electronic and magnetic properties of the solids which they form with
either organic donors or acceptors. Organometallic materials also
have the ability to promote unusual solid state chemical processes.
For example, $[Pt(bipy)_2]^{2+}(TCNQ)_2^{2-}$, where bipy =

has been recently shown by Keller et al. to undergo reversible carbon-carbon bond formation between $\overline{TCNQ}$ molecules. As the temperature is raised two radical anions are formed, and as the temperature is lowered a covalent bond is reformed so that the material
is readily converted from a diamagnetic insulator to a paramagnetic
substance.

It should also be mentioned that numerous attempts have been
made to attach organometallic groups to polymer backbones and to
generate non-integral oxidation states in the resulting polymeric
species. The conductivities of these investigated substances are
in the semiconducting region primarily because there are relatively
weak interactions between the adjacent metal atom sites in the
polymers. Pittman has, for example, attached ferrocene in different
oxidation states to organic polymers.

Scientific Potential

An attractive feature of organometallic materials is the
extensive chemistry available for organic molecules augmented by
chemistry associated with the changes in the oxidation states of
the metal atoms. It should also be noted that one important driving force for the generation of new metal-carbon systems, particularly those involving metal clusters, is the recent increasing
research activity and interest in catalysts. It seems very likely
that this research will also lead to new organometallic materials
with interesting transport properties. Recent advances in synthetic
procedures, such as those applied in the syntheses of large metal
clusters by Chini,and in metal-vapor deposition techniques, are
particularly important in this respect. The majority of the synthetic work in organometallic transition metal chemistry has been
directed towards the development and understanding of synthetic
processes, both catalytic and stoichiometric. It is important
that synthetic organometallic chemists be aware of the problems
and needs in the area of solid state chemistry and physics so that
a greater effort can be made to develop materials whose study would
result in a better understanding of solid state phenomena and which
would have new and useful physical characteristics. We strongly
recommend that a group of predominantly synthetic organometallic

chemists along with a smaller number of physicists and physical chemists be called together to promote the development of this very important area of research. Such a program could appropriately be held under the auspices of the NATO ASI program.

For the most part, the limitations of utilizing organometallic materials are similar to those for organic and inorganic substances but there are the added advantages of increased flexibility in synthetic design and electronic properties. The following are two additional limitations which appear to be readily surmountable: (1) Dissociation of metal-carbon sigma bonds which are relatively thermodynamically stable leads to decomposition with formation of highly reactive organic radical species. The problem can be resolved by using fluorocarbons, delocalized organic groups which form stable anions, or π acceptor ligands such as CO and CNR   (2) Electronic and structural properties of small organometallic molecular systems are not sufficiently well understood to enable one to design or predict the efficiency of electron transfer in macroscopic systems built up from the smaller molecular systems. Extensive studies of magnetic interactions in weakly coupled organometallic systems have laid an important foundation which together with future studies on strongly coupled interactions should enable preparative chemists working closely with physicists to prepare materials with specific magnetic, electrical, and optical properties. As noted below, additional research needs to be carried out which is directed towards this problem.

Technological Potential

We anticipate that the technological potential of organometallic materials is similar to that of organic and inorganic molecular metals. Consequently, the text of Yoshimura's lecture and reports of other study groups who have dealt with this important subject in detail should be consulted. However, we call attention to the following: An application that we expect to arise indirectly from the study of molecular metals derived from organometallic systems could make use of the unusual magnetic properties described above. Extended research on systems related to the very simple metamagnetic material $[(C_5Me_5)_2Fe]$ TCNQ may lead to the development of magnetic switching devices. In addition, the study of electron transfer in these compounds should be useful in the design of electrochromic substances because of the variety of available oxidation states to which both the metal atoms and ligands have access. The recent report of metallocenes intercalated into $TaSe_2$ also suggest potential application in solid state electrodes.

Research Directions

A broader base of experimental data on weak and strong inter-actions between radical organometallic species is badly needed before real progress in this area can be made. We wish particularly to emphasize that systematic studies of the structural and chemical

features necessary to vary the electronic properties of semiconducting systems would be useful. An example is the series $[(C_5H_xR_y)_2M]^{n+}$,
M = Fe, Co or Ni (x + y = 5) R can be varied almost at will so that
both electronic and structural effects can be studied. An example
of one such system with interesting magnetic properties $[(C_5Me_5)_2Fe]$-
(TCNQ) has been given above. Another type of system of considerable
interest includes the mixed-valence complexes such as those formed
with bisfulvalene diiron. Mössbauer spectroscopy reveals only one

1a: n = m = 2, i = 0

1b: n = 2, m = 3; i = 1; X = BF_4,
    picrate, $(TCNQ)_2$

1c: n = m = 3; i = 2, X = BF_4

1

type of iron atom, and the conductivity of the compacted powder 1b
with X = $(TCNQ)_2$ is of the order of 20 $ohm^{-1}cm^{-1}$. The mechanism
of the conductivity is unknown, i.e. whether it is via TCNQ chains,
or through the mixed valence iron units or a combination of the two.
Studies of partially oxidized extended chains based on units such
as the diiron complex above could certainly be expected to contribute significantly to our knowledge of cooperative transport
behavior in low dimensional materials in general and organometallic
species in particular. While we recognize that there are considerable difficulties in the synthesis and manipulation of these compounds, we expect that efforts focused on these problems would
lead to solutions.

Another promising research area is the preparation of materials
with direct metal-metal interactions. For example, an interesting
and answerable question is the following: given large metal clusters such as $M_x(CO)_y$ = A, where x ≤ 20, are known, what would be
the physical and chemical properties of polymeric species such as
(-A-A-A-), where the linkage consists of one or more covalent metal-
metal bonds?

Also, since it was demonstrated by Teo, et al., and Engler,
et al., that molecules which form organic conductors can be converted to ligand-metal bonded polymers (shown below), a large
variety of polymers containing a combination of metal-metal and
metal-ligand bonded solids could be prepared by reaction of $M_x(CO)_y$
with a specifically chosen ligand; viz,

In the same vein, the compound below would be expected to form extended metal-metal bonded chains provided R is small and the

n = 0, +1, +2, +3, +4

molecule is partially oxidized. In support of this, Teo and Wudl have shown that partial oxidation of the triphenylphosphine analog leads to stable intermediate oxidation-state species.

New techniques have been developed which allow the vapors of metals to be used as reagents for carrying out syntheses of organo-metallic species, many of which are related to the molecular metals field. These techniques allow the interaction of metal atoms such as Fe, Co, Ni, Pd, or Pt as well as those of other elements, with vapors of organic or inorganic compounds at low temperatures, or with solutions of nonvolatile compounds such as polymers. These new developments should serve to stimulate many new ideas in the area of organometallic molecular metals and metal dopants. Some specific ideas are enumerated below:

(1) The interactions of metal atoms with weakly complexing organic ligands followed by warming and partial M-M bond reformations has allowed the production of new organometallic clusters. An example is a new Pd_4 cluster:

$$Pd + CF_3C \equiv CCF_4 \xrightarrow{\quad CO \quad} Pd_4(CO)_4(CF_3C \equiv CCF_3)_3$$

Larger metal clusters where the nucleation is such that the metal aggregates become "particles" of 10-30 Å diameter have also been prepared. These clusters are stabilized by the presence of organic ligand fragments. In general the metal vapor technique holds great promise for the production of new metal clusters either by trapping growing clusters, or by inserting metal atoms into already formed smaller clusters:

$$M(atom) + M'L_n{}_m \longrightarrow M'ML_n{}_m$$

(2) The metal atom method has been used for the generation of new "coordinatively unsaturated" organometallics. Thus, the insertion of a metal atom such as Pd or Ni into a carbon halogen bond to form an R-M-X species, where M is two ligands short of its normal tetracoordinate state, has been carried out. Molecules such as C_6F_5NiBr, C_6F_5PdBr, $(C_6F_5)_2Ni$, $(C_6F_5)_2Co$, CF_3PdI, and C_6F_5PtBr have been prepared. These are polymeric materials (structures unknown) until an additional ligand is added. The unsaturated organometallic compounds are very reactive, and even a ligand such as benzene is coordinated:

$$[(C_6F_5)_2Ni]_n \xrightarrow{\quad benzene \quad}$$

It is possible that these coordinatively unsaturated molecules would form metal-metal bonded chains under the proper redox conditions and with the correct choice of R and X.

Even TCNQ itself has been deposited with metal atoms yielding films of M-TCNQ. Further studies with TCNQ and other electron acceptors are desirable, with an initial emphasis needed on structural studies.

(3) Several reactive small organometallics have recently been prepared that decompose clearly upon warming of their solutions to deposit metal particles or films. Could these be used as new reagents for intercalation by magnetic metals of graphite or other important layered molecular metals? Successful experiments have already been carried out where Ni has been deposited in the pores of Al_2O_3.

(4) Recent metal atom studies have shown that Ni atoms serve to polymerize styrene with the incorporation of the Ni atom. This may be a method of doping polyacetylenes and/or diacetylenes and at the same time polymerizing them:

$$M \text{ (atom)} + RCH_2-C{\equiv}C-C{\equiv}C-CH_2R \longrightarrow M \text{ doped polymer}$$

(M atoms serve as polymerization catalyst and dopant). Alternatively, polymer solutions could be treated with metal atoms using new rotating metal-atom reactors containing the partially dissolved polymer. In this way metal atoms may be incorporated in the polymer lattice, especially if the polymer swells in the solvent and if the solvent serves to carry the metal atom into the polymer (weak solvent-M complex).

(5) Vapor techniques for the preparation of active small molecules that polymerize to form the host are also possible. For example, the codeposition of C vapor and M vapor is now possible to form a $C_x M_y$ film. It is not presently clear whether the product would be an intercalated graphite or a new metal-carbon species. Other small molecules could be studied similarly. All that is required is that a reactive small molecular fragment be generated and codeposited with M atoms on a cold wall. Some possible studies involving reactive molecular fragments would be CS + Ni, SN + Ni, CF_2 + Ni, SO + Fe, PN + Co. These studies are now experimentally possible and some interesting new conducting films probably could be generated on a synthetic scale.

The final example given here is the use of organometallic compounds as synthetic precursors for metallic films. This technique has been described in the literature and is the basis for at least one commercial process. Passing a volatile organometallic compound such as $(C_6H_6)_2Cr$ over a surface heated to relatively mild temperature ($\sim$300°C) results in the formation of a thin metal film. The thickness of the film is readily controlled even to dimensions as small as 1000 Å. Schrauzer has applied a similar technique to organometallic compounds of niobium and tantalum to produce thin film alloys which have critical superconducting temperatures that are reported to be higher than those of alloys with the same chemical composition prepared by conventional techniques.

Summary

The application of organometallic compounds to solid-state physics and chemistry is only beginning to be examined in detail. The relatively few studies that have been done have been with organometallic molecules acting as electron donors or acceptors. Solid state materials with unique electric, magnetic and optical properties of significant potential have been produced. In particular organometallic molecules can be used to build systems which,

due to their relative simplicity, can bring a better understanding
of ths solid state. We believe that support should be given to
this area of research and the the meeting proposed should
be held in the near future.

NEW HIGHLY CONDUCTING COORDINATION COMPOUNDS

D.B. Brown, K. Carneiro, P. Day (Recording Secretary),
B. Hoffman, H.J. Keller (Chairman), W.A. Little,
A.E. Underhill and J.M. Williams

I. Introduction

The past 10-15 years have seen an extensive search in many
laboratories for coordination complexes which exhibit high conductivity in the solid state. Indeed the chemical systems which have
given us properties illuminating most clearly the basic physics of
the one dimensional state have been of this type. However, until
recently, physical measurements on highly conducting metal coordination compounds have concentrated on the single inorganic prototype one-dimensional (1-D) conductor $K_2Pt(CN)_4 \cdot Br_{0.3} \cdot 3H_2O$, KCP,
simply because of a scarcity of other well characterized examples.
This unusual situation was, in retrospect, of great benefit. The
strikingly unusual properties of KCP and related compounds have
been an intellectual attraction, particularly for the physicists
and the challenging problems relating to fluctuations, disorder,
competing instabilities and broken symmetry have attracted many of
the most distinguished solid state scientists to the field. Their
contributions, although relatively abstract, have played a key role
in building our present understanding of these materials. In
addition the field has served to develop a variety of interdisciplinary collaborations, from which each discipline is now benefit-
ting. In particular, because the field has proved challenging to
synthetic inorganic chemists, it has been possible to screen the
conducting behavior of a sufficiently large number of complexes to
allow one to establish some guidelines which may now be summarized.

II. Basic Considerations

A fundamental requirement for high conductivity in any solid
material is the attainment of a high intermolecular transfer
integral, t, so we begin by pointing out some ways in which that

may be established in coordination complexes. In this type of
compounds one normally has a metal ion (M) in a positive oxidation
state (+1 to +4), typical ligands (L) being mono atomic (halide or
oxide ions), diatomic (CN^-), or conjugated polyatomic from organic
isocyanides through bifunctional aromatics such as pyrazine and
bipyrimidine up to the very large planar ring systems such as porphyrin and phthalocyanines. Contact between the metal centers in
the solid state may come about either by direct overlap or via a
bridging group. To maximize the former, one needs planar complexes
which can be stacked on top of one another or bridging groups which
subtend a sufficiently small MLM-angle. Planar complexes are associated with quite specific d-electron configurations, principally low
spin d^8, unless rigid ligands are employed to force the metal ion
into a planar coordination. Direct MM interaction is inherently
desirable because the presence of a bridging group at once attenuates
the MM transfer integral by the square of the ML overlap integral.
Where polyatomic bridging ligands like pyrazine are concerned, intramolecular $p\pi$-$p\pi$ overlap within the conjugated aromatic system is
sufficiently large that electron migration from one end of the
bridging molecule to the other is not hindered. In this context
lower oxidation states generating greater ML covalency via larger
orbital extension might be a desirable goal.

A second requirement for high mobility is to lower both the on-
site and inter-site Coulomb repulsion. In the context of metal complexes this is better achieved in partially oxidized "non integral
oxidation state" (NIOS) systems than in ones where the electron:
atomic ratio is integral (IOS). The NIOS crystals based on coordination compounds which have been studied to date can be broken down
into two classes. On the one hand, there are systems, exemplified
by the tetracyanoplatinite salts, where the transport properties can
almost completely be understood in terms of carriers confined to a
conducting spine of metal atoms. The coordinated ligands (eg CN^-),
although essential for defining the properties of the parent IOS
molecule or ion, play little or no direct part in the charge transport
process. On the other hand there are systems, such as the metallo-
porphyrins and their analogues, in which the carriers are essentially
ligand based; the metal ion acts in a decisive manner to modify the
ligand properties, but in this case the transport process does not
appear to directly involve the metal.

III. Integral Oxidation State Compounds

IOS compounds in which a molecule, or ion, bridges adjacent
metal centers have rarely been found to exhibit any appreciable conductivity, although substantial magnetic exchange interactions
between partly occupied localized metal orbitals are frequently
observed. When the metal atoms are in direct contact too, high conductivity remains the exception when the oxidation state (electron/
atom ratio) is integral. Examples are the well known metal atom
chain compounds of Ni^{II}, Pd^{II} and Pt^{II} such as the dimethylglyoximates

and tetracyanides which are excellent insulators at ambient pressure though there are indications that they can be rendered conducting if a large (100 kbar) applied pressure is used to bring the metal atoms closer together. On the other hand such compounds do show very unusual optical properties indicative of a substantial MM interaction in their excited states. A further example would be $Ir(CO)_3Cl$, now thought to be an IOS compound, though quite a good conductor at ambient pressure.

Turning to 1D compounds in which the electron/atom ratio varies from one metal ion site to the next, but in which the metal ion sites are connected through ligands, trapped valency and localized electron behavior remains the rule. Compounds like $Pt(NH_3)_2Cl_3$ remain insulators at atmospheric pressure though they, too, can be rendered conducting under pressure. We point out, however, that in some discrete dimeric mixed valence metal complexes with bridging groups like pyrazine and N_2 the electrons may be delocalized between the metal atoms in a manner which, if extended to an infinite chain, would lead to formation of an energy band with a finite density of states at the Fermi surface. Further synthetic efforts in this direction would be highly desirable.

IV. Non Integral Oxidation State Complexes, Direct MM Overlap

From the discussion so far it is clear that high conductivity is most likely to be found in NIOS compounds. Three primary variables must be considered in synthesizing materials with directly interacting metal centers: the nature of the IOS precursor, the identity of the oxidant, and the synthetic technique for combining the two. Because of the geometrical constraints, most new highly conducting materials are likely to derive from the oxidation of square-planar d^8 complexes.

Three basic synthetic techniques have been used to generate NIOS materials. The first, involving crystallization from solutions containing both oxidized and reduced forms of the metal complex, has thus far been successful in only a few cases, most notably KCP itself. The second, involving electrochemical oxidation of solutions of the IOS precursor, has been remarkably successful in generating pure single crystals of a variety of KCP analogs, but has not been utilized in many other systems. The third approach, solid state oxidation of appropriately stacked precursors has rarely produced high quality crystals, but has the advantage of providing a range of redox potentials for the oxidant and a variety of geometrical and chemical features which may be incorporated into the resulting structure. At this point, there are not sufficient experimental data available to determine the effects of these variations, and much more experimental effort is required.

Further, the nature of the ligands, which potentially provides the greatest synthetic variability, must be examined in the future

in much more detail. We believe that the following specific features of ligand variation require investigation. First, there may
be a relationship between conductivity and such donor atom properties as polarizability. For instance we note the significant
increase in conductivity upon changing from an oxygen donor (as in
the platinum oxalates) to a carbon donor (as in KCP and its analogs).
Second, the extent of conjugation or delocalization between atoms
of the ligand remote from the donor atom must be examined more
thoroughly. Third, the synthesis of materials which allow significant interchain coupling, either directly through substituents on
the perimeter of the square-planar moieties, or indirectly through
interaction with suitable oxidants, requires systematic investigation. A caveat should be added that ligand variations may be
restricted by a possible change in the conduction mechanism, from
metal-based to ligand-based.

At this point it should be pointed out that in all the theoretical treatments of conductivity and superconductivity of low dimensional metals the on-site Coulomb potential U plays an important
role. High conductivity can be expected when the ratio U/t is
small. It is believed that U can be reduced by surrounding the
conductive metal spine by a polarizable molecular environment.
This requires bulky ligands, which normally would contain a conjugated carbon substructure. With compounds containing such ligands
the MM spacing can be no smaller than ~ 3.25 Å. An important question, then, is whether significant metallic conduction can occur at
such large MM spacing. Theory predicts that it can but this has
yet to be confirmed by experiment. If good conductive behavior can
occur with such bulky ligands it would provide the incentive for
the study of a rich variety of new compounds with a large range of
physical properties. These general considerations are illustrated
by the recent developments in the chemistry and physics of the KCP
analogues.

As outlined in some of the papers presented at this Institute
there has been a great advance in the range of well characterized
materials, in particular the preparation of the first anhydrous
tetracyanoplatinites $Rb_2Pt(CN)_4FHF_{0.4}$ and $Cs_2Pt(CN)_4Cl_{0.31}$ and compounds exhibiting a much wider range of partial oxidation than seemed
possible a few years ago. This very much increased range of compounds has shown the importance of:

(i) the Pt-Pt intrachain separation on the value of the intrachain
conductivity at room temperature,
(ii) hydrogen bonding between the chains on the three dimensional
ordering temperature, and
(iii) the size of the hydrated or anyhdrous cation on the intrachain Pt-Pt separation. Thus there now exists the potential for
the directed synthesis of new compounds in the KCP series in which
short intrachain Pt-Pt separations are combined with enhanced interchain interaction.

The anhydrous nature of some of these new compounds has removed many of the experimental difficulties associated with the study of 'hydrated' KCP. It should be stressed that what appears to be fairly small chemical changes in these series of compounds do produce very significant changes in the physical properties, and the very closely controlled chemical changes within an isostructural series of compounds has proved particularly valuable in understanding the effects of structural changes on the electrical conductivity.

Turning to other NIOS compounds of the platinum group, work on the oxalato platinites should be aimed at the characterization of compounds with a wide range of cations, and at attempts to prepare anhydrous compounds. In the case of the partially oxidized iridium carbonyl compounds the main restriction at present is the lack of single crystals for detailed physical study. Indeed in all this area of work the growth of good quality crystals is a necessary prerequisite.

Although most of the highly conducting coordination compounds which have been thoroughly characterized are of the KCP type, several new types of highly conducting complex have been reported at this meeting. It is probable that many more such complexes can be prepared if proper combinations of the experimental parameters, described above, are selected. It is clear that a vast amount of synthetic effort is required in order to fulfill the potential of this area.

V.   Nonintegral Oxidation State Complexes: Direct LL Overlap

Truly conductive NIOS systems with ligand-based transport properties have only recently been characterized. Iodine oxidation of ML complexes, where L are the highly conjugated porphyrins (P), tetrabenzporphyrin (TBP) and phthalocyanines (Pc) yields highly conductive NIOS crystals which experiment shows to have ligand-centered carriers. Preliminary studies indicate that the transport properties depend on the ligand electronic structures (MTBP $\underline{vs}$ MPc), on the intrachain separation (MTBP $\underline{vs}$ M-(octamethyl TBP)) and on the ligand size (MTBP $\underline{vs}$ MP). In addition, not only d^8 metals, but other, including d^0 (eg, Mg) ions give conductive systems. Finally in one case crystals with two widely different degrees of partial oxidation have been obtained, and there exists the possibilities of investigating the conductivity for a range of stoichiometries.

Comparing the results to date for the metal- and ligand-based conducting coordination compounds highlights one of the most exciting prospects provided by this field. The systematic variation of metal and ligand should permit an exploration of the nature of the dependence of crystal properties on these and other molecular features. Moreover, we might hope to find systems which span the spectrum between the extremes of metal- and ligand-dominated transport. $K_2Ni(DTO)_2I$ (DTO = dithiooxalate) may be such a system.

VI. The Impact of Highly Conducting Coordination Complexes on
     Physics

The occurrence of materials with totally unexpected physical
properties has stimulated intense studies, from which the physics
society at large has profitted greatly. The impact of highly conducting coordination compounds has been felt in several areas such
as one-dimensional phenomena, electron-electron coupling, pinning
effects and superconductivity. Results obtained from such studies
are now providing guidelines for future preparative work.

One of the greatest advances in solid state physics obtained
over the last few decades is our understanding of the universal
behavior of phase transitions. Critical fluctuations, i.e. precursor effects occurring close to a phase transformation, appear
to be governed by only a few parameters, such as the type of interaction and the symmetry of the chemical structure. However, as
the dimensionality is lowered critical fluctuations become increasingly important, and hence the extreme one-dimensional character
of the conduction in these compounds has offered unusually
stringent tests of the theory of critical phenomena. Also, specific to the physics of one dimensional systems is the occurrence of
persistent solitary waves as opposed to ordinary harmonic wave
propagation, and studies, mentioned at this Institute, are providing a opportunity to confront theory and experiment.

The coupling between electrons in condensed matter has been
at issue for many years, and it is now known that the KCP type
crystals exemplify a variety of cases of electron-phonon coupling
in the conducting chain, and dominant Coulomb repulsion between
chains. This leads to a metal-semiconductor phase transition with
the appearance of an electronic charge density wave at low temperature.

Pinning effects also have recently attracted considerable
theoretical attention. In one dimensional systems, pinning may
arise either from disorder or from so called commensurability
effects. Since compounds exhibiting both commensurate and incommensurate charge density waves in disordered materials have now
been prepared, the similarity between experimental results on the
two classes indicates that disorder effects dominate. Furthermore,
recent studies in ordered systems show that dramatic changes occur
in the electronic behavior when disorder is removed. The occurrence
of $Cs[Pt(CN)_4](FHF)_x$, with variable x, will in the future be a key
to the understanding of these effects.

One of the great challenges to the experimentalist has been
the theoretical predictions that such compounds might show superconductivity at high temperatures. There are in our view two ways
in which superconductivity might occur in KCP-type materials. One
would be to adjust the electronic interchain coupling such that a
semimetallic, rather than semiconducting phase would prevail to

temperature at which the superconductivity could occur. The other way of producing superconductivity would be to change the ligand so as to obtain an electron-exciton coupling, but this would require an increase in the intrachain Pt-Pt distance to accommodate the bulky ligands required for such exciton coupling. At present all known compounds with such large MM distance are insulators. However, experience gained from comparison of oxalato- and cyano-platinites shows that the critical distance for direct overlap metallic conduction depends upon the ligand, and therefore it cannot be ruled out that the ligands, necessary for the support of electron exciton coupling would preclude the metallic state.

VII. Conclusion

In conclusion we wish to remark on what we believe to be a major obstacle to further progress in this field. Despite the generalizations we have been able to make above, synthesis of coordination compounds with interesting collective properties remains basically a serendipitous activity. Ample expertise for preparation of new compounds exists within the chemical community, but is not being fully utilized in this field, since historically, preparative coordination chemists have concentrated their efforts on compounds of interest for their reactivity or local electronic properties, rather than on designing molecules which interact with one another in a crystal to produce novel collective states. We believe that efforts should be made to attract a wider cross-section of preparative inorganic chemists to this new science of low dimensional collective phenomena.

METAL-RICH SYSTEMS

N. Bartlett (Chairman), P.A. Cox, J.E. Fischer, W.B.
Fox, A.J. Heeger (Recording Secretary), K.J. Klabunde,
and G. Shirane

I. General Remarks

This study group represented only a few of the areas which
could be classified under this heading within the framework of
"Molecular Metals". Because of this we have limited our appraisal
to layer compounds (with emphasis on graphite systems and relatives)
and metal chain systems analogous to $Hg_{3-\delta}AsF_6$. Other areas may be
equally important.

Much of the present vitality of this area of research derives
from the symbiotic relationship between those involved in the preparation of materials and those who carry out physical studies.
New and perhaps unanticipated physical properties indicate new
directions for the synthesizers to pursue. Clearly, materials
should be characterized as fully as possible and this is usually
best done by passing the material on to appropriate experts with
special facilities. As our knowledge and minute description of
materials is extended, we require more sophisticated facilities and
broader expertise.

The numerous physical studies commonly applied to these materials show that the definition of a material is becoming an increasingly subtle matter. It is essential to describe as precisely as
possible the chemical composition, detailed structure and morphology.

A. Linear Chain Mercury Compounds

The interesting and unusual structural and electronic properties of the mercury chain compound $Hg_{3-\delta}AsF_6$ result directly from
the remarkable incommensurate polymercury chain structure. This
incommensurability leads to two interpenetrating and independent
lattices: the rigid mercury chains within the tetragonal AsF_6

lattice.   This is one member of what appears to be a growing class of incommensurate and independent lattice compounds.   Such compounds represent an entirely new concept in materials science.

In the prototype mercury chain system, the atomic Hg-Hg distance (2.67 Å) is not commensurate with the tetragonal lattice (a = 7.54 Å) defined by the AsF_6^- anions.   This is brought about by sufficiently strong metal-metal bonding within the chain that the Hg atoms determine their own periodicity.   Other examples may be $TTFX_\alpha$ where X = Br, I, SCN and α is non-integral because of the incommensurability.

Unusual features of special fundamental interest include the following:
(1)   Optical and electrical properties of metallic chains.
(2)   One-dimensional elastic properties including additional 1d phonon branches arising from the incommensurate chains.

This remarkable interpenetrating chain system may be unique and one of a kind.   However, the physical and chemical phenomena involved serve to challenge the synthetic chemist to expand this class of materials.

It is possible that 'free' chains, unencumbered by ligands, such as occur in Hg_3AsF_6, may not be found outside of mercury systems, but mixed systems involving related metals should be examined.

It may be possible to constrain other metal-atom types to form linear chains by confining them to tunnels in suitably nonbonding host lattices.   If necessary, provision could be made for electron addition or withdrawal from such a chain by making an appropriate choice of host lattice.

It may also be possible by ligand attachment, to cause certain metal atoms to 'mascarade' as mercury atoms such that they would then be disposed to form chains of the Hg_3AsF_6 type or even mixed chains with mercury itself.   This brings one close to the systems represented by KCP and organometallic systems.   Metal atoms such as Co, Ni, Pd, Pt, etc; have been shown to react with carbon-halogen bonds to generate "coordinatively unsaturated" R-M-X and R_2M species.

e.g.

These materials have open coordination sites and in the absence of additional ligands form polymers (the structure of which are presently unknown). These and related materials may be useful for the generation of new systems containing chains or sheets. Almost any added ligands can generate new square planar species R-M-R which could then form one-dimensional chains like KCP. The fact that RMX, R_2M as well as new R-M-R species, can now be produced, opens up this potentially large area.

Such systems are expected to be of importance for a variety of physical reasons. They should show one-dimensional lattice dynamics and one dimensional electronic properties, both areas are anticipated to be of special scientific interest in the next few years. Associated with these one dimensional electronic systems one may anticipate the usual $2k_F$ Fermi surface instabilities observed in other more conventional chain compounds. The simultaneous features of incommensurate (independent) lattices with $2k_F$ driven periodic lattice distortions is of particular interest to the future development of understanding in this area. Depending on the electronic properties (i.e. ionic state of metal atoms etc.) such independent lattice systems might, furthermore, lead to ionic transport (i.e. solid state electrolytes) with relatively high mobility.

B. Graphite Intercalation Compounds

As can be appreciated from the extensive literature as well as the papers presented at this Institute, a number of metal-rich graphite layer compounds are known. The metals are presently limited to the more electropositive metals; the alkali and alkaline earth intercalates are well established. The two-dimensionality of the graphite intercalation compounds in addition to their high electrical conductivity gives them special technological as well as scientific interest.

These graphite-metal intercalates follow the pattern established for other graphite intercalates in that guests (in this case metal atoms) reside between the graphite layers, and the intercalate layers separate to maximum extent governed by stoichiometry. Thus there exist, first stage compounds, second stage compounds, etc. The weight of physical evidence indicates that the metal-atom guest species give electrons to the host lattice.

Although this qualitative picture is generally accepted, the quantitative picture is less clear. Some infer that the electron transfer from the electropositive guest atoms to the graphite sheets is essentially complete, with the alkali species becoming monopositive cations. Others prefer to believe in a rather smaller electron transfer. The reasons for the quantitative evaluation difficulties are manifold.

The preservation of order is a major experimental difficulty, even when one starts with single-crystal-quality material. The prime causes of this difficulty must be associated with the gross

mutual displacement of the graphite sheets, as the guest species
intercalate into the graphite structure. Although mutual repulsive interactions may well preserve a high degree of order within
an intercalate layer, it is evident that in many compounds the contents of one layer are not in register with those in another.
This means that reliable structural studies on graphite-metal compounds are rare (and this is true for graphite intercalation compounds generally).

Obviously one cannot provide a theoretical model of significant
utility if one is neither confident of the nature of the intercalated species nor of the disposition of these species with respect
to the atoms of the host network.

The first challenge to be met is a synthetic one. Approaches
must be worked out to provide the highly ordered material essential
for the highest quality structural work. With a firm structural
base, the other physical studies (which should, as far as possible,
be carried out on the same sample) can be more surely interpreted.
Electical conductivity, heat capacity, X-ray-absorption edge and
a variety of other studies can then contribute to the definition
of the charge on the host and guest.

The payoff for such effort is enhanced insight into what is
possible, and what is not possible, in synthesis and physical properties.

Such is our limited knowledge at this time that we cannot be
confident that metals, other than those that have been intercalated
can be introduced into graphite. Purposeful, efficient tailoring
of materials, to satisfy technological as well as scientific needs,
requires that we have a good theoretical basis of understanding.

It is probable that synthetic chemists will achieve new metal
intercalations by proceeding in an empirical manner using new synthetic methods and reagents. Labile monatomic metallic compounds,
which can eliminate their ligands as relatively inert species, provide an attractive approach. One can imagine that transition metal
atoms would provide charge transfer of one electron to the carbon
host network and simultaneously, through other valence electrons,
cause guest-guest bond formation with chain, sheet, or cluster
formation.

The possibility of the polymerization of guest species in
graphite is an especially enticing idea, since it offers the possibility of generating new incommensurate lattice systems. Moreover if chains and sheets (rather than clusters) of metal atoms
were generated, the resultant systems, both guest and host, could
be electrically conducting.

It might be that the alkali metal and alkaline-earth-metal
graphite materials are already capable of providing incommensurate lattice systems, thus far not detected because of

imperfections and stacking disorder. Indeed it could be that these classical systems themselves would reveal markedly changed physical behavior as high order is achieved.

The recent preparation of first stage salts of layer boron nitride, which are analogous to known graphite salts, raises the question of the possibility of introducing metal atoms into boron nitride. Since BN is an insulator, such compounds would likely be semi-metals or metals. Such synthetic attempts are worthwhile at least for the more electropositive metals, but cleavage of the BN network is much more facile than in graphite. Hence such synthetic endeavors must be carried through with subtlety and ingenuity. One might also foresee that hydrids of boron nitride and graphite may be preparable (with B and N replacing C atoms in graphite sheets and vice versa). These materials should have intercalating capabilities somewhat like those of carbon and layer BN. Those hybrids rich in carbon will surely take up at least the electropositive metals.

Some of what has been said about the need for better structural characterization and the need for physical studies on such charac-terized material, can be carried over to the layer chalcogenides, but we are not in a position to give an extensive appraisal of these systems.

POTENTIAL TECHNOLOGY DIRECTIONS OF MOLECULAR METALS

E.M. Engler (Chairman), W.B. Fox, L.V. Interrante, J.S.
Miller, F. Wudl, S. Yoshimura, A. Heeger and R.H. Baughman

In the past decade, anisotropic molecular conductors have
been found which possess unusual electrical, optical, magnetic,
and, in some cases, mechanical properties. Exploitation of these
properties for specific devices is inevitable, and a number of diverse applications of molecular metals have been reported over the
past few years. Specifically, these materials have found use as
components in batteries, electrolytic capacitors, thermistors,
electrochromic displays, optical printing techniques, electrophoto-
graphy, as electrodes, antistatic coatings, etc. A more critical
and detailed review of the state of the art in this area is given
in a paper by Dr. Yoshimura at this NATO-ARI. The scope of this
study group is to evaluate the status of applications, to indicate
problem areas, and to identify some fruitful directions for future
work. But, it must be kept in mind that this is a young field
where potential applications are still in a very early stage of
development. Consequently, a certain amount of speculation and
generalization is unavoidable.

While new applications to conventional problems continue to
emerge, their ultimate potential may lie in the rather <u>unique</u> combination of several novel and useful properties possessed by these
materials. Some of these properties other than high conductivity
are:
(1) anisotropic optical and electrical properties
(2) low weight (density)
(3) ease of fabrication via vacuum deposition and solution casting
    techniques.
(4) sensitivity of conductivity to external influences (e.g. elec-
    tric fields, temperature, gases, doping)

When the conductivity is considered in connection with some of
these above-mentioned properties, interesting applications can be
envisioned. For example, TTF donors were found to undergo photo-
oxidation in halocarbon solvents to give conducting TTF halides.
Light could therefore be used to "turn on" high conductivity in a
film of insulating TTF-halocarbon. This effect has been applied
as a method of optically printing highly conducting, multicolored
patterns. Thus, information can be stored optically (input), and
this written information either optically or electrically sensed
(output). In other applications, the sensitivity of the conductivity to external influences has suggested uses as sensors or
switches. The use of metal-insulator phase transitions in some
molecular metals as temperature-controlling devices, and the sensitivity of conductivity in anhydrous KCP salts to water vapor as
humidity sensors are just two early examples of this combination
of useful properties.

Eventual commercial applications of molecular metals will
depend to a large degree on breakthroughs in materials fabrication
and processing techniques. The evaluation of long-term stability,
toxicity, mechanical properties, etc., and the ability to modify
and improve properties will be important in determining the feasibility of specific applications. In this regard, a promising development has been the discovery that free-standing, large-area,
films of polyacetylene can be readily prepared. These films can be
either acceptor or donor doped to yield both p- and n-type materials
which exhibit a wide range of conductivities, spanning insulating
to metallic behavior.

Attempts to combine the useful mechanical properties of polymers (e.g. film-forming, adhesion, flexibility, etc.) with the
electrical properties of these crystalline molecular metals is at
least one area which deserves more attention. One approach to this
problem has been to incorporate small amounts of a conducting
organic charge-transfer salt (as low as 5% weight) within an insulating polymer matrix. By stretching the polymer film, conductivities
as high as 1/ohm-cm have been observed. In another approach,
insulating π-donor polymers have been prepared which maintained
many of the desired polymer properties, such as solubility and good
film-forming properties. After films have been cast, the material
is converted to a moderate conductor by exposure to halogen.

In terms of research directions on the applications of molecular metals, the following seem of special interest:

(1) the study of junction characteristics (p-n junctions, Schottky
     barriers) between synthetic molecular metals and traditional
     metals and semiconductors.
(2) the use of these materials in energy storage (batteries) and
     generation (solar cells).
(3) development of mechanically and thermally stable,highly con-
     ducting polymers and/or composites.

(4)   study of the modification of electrical properties by external
      influences such as gases, temperature (phase transitions) and
      applied electric fields.
(5)   the study of non-linear transport phenomena in these materials
      and their potential use in microelectronics.
(6)   development of improved materials fabrication and processing
      techniques for molecular conductors, such as in the prepara-
      tion of large area thin films.

Continued basic research on these molecular conductors is very
important in gaining a better appreciation of the real potential of
this field.   In particular, a better understanding of how molecular
properties influence solid state properties will be helpful in the
design of molecular solids with specific properties.

SUBJECT INDEX